ULTRAFAST FLIGHT

OF

POLYPHASE SPACE VEHICLES

RICHARD A. WEISS

K & W Research Company
Tucson, Arizona

K & W PUBLICATIONS
Tucson, Arizona

2005

TO BETH

Published by K & W Publications
A subsidiary of K & W Research Company
Tucson, Arizona

Library of Congress Catalog Card Number: 2005928830

ISBN 0-9624789-7-0

Printed in the United States of America

PREFACE

As part of the centennial celebration of the 1905 publication of Einstein's theory of special relativity I decided some time ago to write a book on the possibility of ultrafast flight of space vehicles within the limits of special relativity and Lorentz invariance. To do this it is only required to extend the validity of special relativity and the Lorentz transformation to complex number space and time coordinates which arise in matter and vacuum in the presence of ambient phase fields. The result is my eighth book ULTRAFAST FLIGHT OF POLYPHASE SPACE VEHICLES in which I suggest that, with proper engineering and tuning of the ordinary fields and the ambient phase fields within a polyphase space vehicle, this type of manmade vehicle can undergo ultrafast flight while natural objects cannot. In its early stages I thought of this book as a practical design manual for ultrafast polyphase space vehicles. However, now it reminds me of the PRODROMO… that was published in 1670 by the Jesuit priest Francesco Lana Terzi (1631-1687) whose picture appears on the front cover of this book and one of whose designs of a flying boat supported by four evacuated hollow copper spheres appears on the back cover. It is only a beginning. So, with Lana, let your imagination fly up to the speed of light in vacuo and beyond.

RICHARD A. WEISS

Tucson, Arizona
April 2005

CONTENTS

INTRODUCTION

The present state of understanding of our solar system and of stellar and galactic systems has been based on the development of space vehicles carrying instruments above and beyond the Earth's atmosphere, and this has only been possible due to the developments in the fields of rocket science and engineering. However, a complete understanding of the nature of stars and galaxies will only come when space vehicles are sent to the vicinity of these objects. But with present day technology the great distance of these objects prevents us of from going there except in our science fiction imaginations. Consider the great spiral galaxy M101 NGC 5457 in Ursa Major that lies at a distance of about 23×10^6 light years.[1,2] Traveling at 10^{-3} of the speed of light in vacuo would get you there in about 23 billion years. It would be more convenient to get there in 23 minutes. True space exploration will only be possible if space vehicles can travel billions of times faster than the light speed in the vacuum.

At present space travel faster than the speed of light in vacuo is thought to be impossible primarily due to the advent of the special theory of relativity by Albert Einstein who discarded the idea of a universal Newtonian time by combining space and time into a mathematical continuum in which the space and time coordinates of an event as seen by two observers assume apparent values that depend on the relative motion of the observers.[3-5] In other words, depending on their relative motion each observer of an event would measure different values of the space and time coordinates for the event and both measurements would be correct. Einstein based his theory on two postulates, the constancy of the light speed in vacuo (which agreed with the Michelson-Morley experiment) and the invariance of all the laws of physics as seen by observers on different inertial systems.[3-7] The special theory of relativity has passed many experimental tests and is used as standard practice in molecular, atomic and nuclear physics.[8] Questions about the validity of special relativity in high-energy physics have been raised along with special relativity's possible contradiction with the deeper concepts of space and time in quantum mechanics such as the violation of causality.[9-11] Rather than making an attempt to tear down special relativity, this book prefers to examine the effects of generalizing the validity of Lorentz and Einstein covariance to active coordinates that are complex numbers in an internal space and result from the presence of a cosmic or a manmade ambient phase field (APF).

An APF is a field that requires all physical quantities including space and time coordinates, but with the exception of the fundamental constants of nature such as $\hbar$, c, G ···, to be complex numbers in an internal space.[12-17] It has been suggested that a manmade APF applied to atomic nuclei leads to the possibility complex number nucleon numbers and complex number dimensions of space and time and ultimately to the concepts of polyphase fission and the polyphase fission map.[13-17] And it has been suggested that a cosmic APF may pervade all space and time, and Lorentz transformations are generalized to the complex number space and time coordinates that are associated with an APF.[12] It is assumed that special relativity expressed in terms of complex number space and time coordinates is valid in the presence of an APF. Finally, it has been suggested that the presence of several manmade APF in a rocket can be used to enhance the liftoff of a heavy-lift rocket, and this type of rocket has been called a polyphase rocket.[16] As with the polyphase rocket, each component of a polyphase space vehicle is associated with an APF which induces internal phase angles to the space and time coordinates within the space vehicle, and this book shows that these complex coordinates are quantities that enter the complex number Lorentz transformations and the transformations of the space and time coordinate factors that allow ultrafast flight of a polyphase space vehicle.

The space and time coordinates in matter or vacuum that is located in an APF have been represented as complex numbers in an internal space.[12-17] A simple theory of the origin and nature of these complex number coordinates has been developed in Paper A.[16] Consider the constant real number spatial parameters L_α^c for the complex number space coordinates $\bar{\alpha}$ in an ambient field in bulk matter, where L_α^c can be taken to be multiples of the Planck length. The real number constant spatial parameters L_α^c are divided into $\bar{n}_\alpha$ parts of length $\Delta\bar{L}_\alpha$ where the complex number spatial segmentation number $\bar{n}_\alpha$ is represented as[16]

$$\bar{n}_\alpha = n_\alpha \exp(j\theta_{n\alpha}) \tag{1}$$

so that

$$\Delta\bar{L}_\alpha = \Delta L_\alpha \exp(j\theta_{\Delta L\alpha}) = L_\alpha^c / \bar{n}_\alpha = L_\alpha^c / n_\alpha \exp(-j\theta_{n\alpha}) \tag{2}$$

and

$$\Delta L_\alpha = L_\alpha^c / n_\alpha \qquad \theta_{\Delta L\alpha} = -\theta_{n\alpha} \tag{3}$$

where $\alpha = x,y,z;\ r,\varphi,z$ or ρ,φ,ψ. Then the complex number space coordinate $\bar{\alpha}$ can be written as

$$\bar{\alpha} = \alpha \exp(j\theta_\alpha) = \bar{\eta}_\alpha \Delta\bar{L}_\alpha = \bar{\eta}_\alpha L_\alpha^c / \bar{n}_\alpha = \eta_\alpha L_\alpha^c / n_\alpha \exp[j(\theta_{\eta\alpha} - \theta_{n\alpha})] \tag{4}$$

where the complex number spatial counting number is represented as

$$\bar{\eta}_\alpha = \eta_\alpha \exp(j\theta_{\eta\alpha}) \tag{5}$$

so that

$$\alpha = \eta_\alpha L_\alpha^c / n_\alpha = \eta_\alpha \Delta L_\alpha \qquad \theta_\alpha = \theta_{\eta\alpha} - \theta_{n\alpha} \tag{6}$$

where η_α and n_α are not integers in the presence of an APF but are integers under zero ambient field conditions.

The time coordinate factor, which is analogous to that of the space coordinate factor, is developed in Paper A as follows.[16] Consider a real number constant L_t^c that is used to measure intervals of the complex number time coordinate $\bar{t}$, where L_t^c can be taken to be multiples of the Planck time. In the presence of an ambient field in bulk matter the real number constant temporal parameter L_t^c is divided into $\bar{n}_t$ parts of length $\Delta\bar{L}_t$ where the complex number time segmentation number is represented as[16]

$$\bar{n}_t = n_t \exp(j\theta_{nt}) \tag{7}$$

and where

$$\Delta\overline{L}_t = \Delta L_t \exp(j\theta_{\Delta Lt}) = L_t^c / \overline{\imath}_t = L_t^c / \imath_t \exp(-j\theta_{\imath t}) \tag{8}$$

and therefore

$$\Delta L_t = L_t^c / \imath_t \qquad \theta_{\Delta Lt} = -\theta_{\imath t} \tag{9}$$

The complex number time coordinate is then written as

$$\overline{t} = t\exp(j\theta_t) = \overline{\eta}_t \Delta\overline{L}_t = \overline{\eta}_t L_t^c / \overline{\imath}_t = \eta_t L_t^c / \imath_t \exp[j(\theta_{\eta t} - \theta_{\imath t})] \tag{10}$$

where the complex number time counting number is written as

$$\overline{\eta}_t = \eta_t \exp(j\theta_{\eta t}) \tag{11}$$

so that

$$t = \eta_t L_t^c / \imath_t = \eta_t \Delta L_t \qquad \theta_t = \theta_{\eta t} - \theta_{\imath t} \tag{12}$$

where η_t and $\imath_t$ are not integers when an APF is present in matter, and the complex number nature of the time coordinate is $\overline{t}$ due only to the complex number time counting number $\overline{\eta}_t$ and the complex number segmentation number $\overline{\imath}_t$ because the constant L_t^c is a real number in equation (10). The real number constants L_α^c and L_t^c that appear in equations (4) and (10) can be taken to be multiples (generally very large multiples) of the Planck length and Planck time respectively because an APF is a field that induces internal phase angles in all physical quantities except the fundamental constants of nature (h, c and G). Equations (6) and (12) allow the possibility of $\theta_\alpha \leq 0$ and $\theta_t \leq 0$ which has been suggested is the case for the Earth's gravity field.[12-17]

The space and time coordinates in vacuum or matter that is located in an APF are represented as complex numbers of the form[12-17]

$$\overline{\alpha} = \alpha\exp(j\theta_\alpha) \qquad \overline{t} = t\exp(j\theta_t) \tag{13}$$

where the coordinate magnitudes are α = x,y,z; r,φ,z or ρ,φ,ψ for Cartesian coordinates, cylindrical polar coordinates or spherical polar coordinates respectively. For instance equation (13) can describe the space and time coordinates within an atomic nucleus in an ambient γ ray field. The space and time coordinate magnitudes and internal phase angles for regions of space and time that are located within matter or vacuum and in an APF are given in Paper A by[16]

$$\alpha = \alpha_o C_{\sigma\alpha} \qquad \theta_\alpha = \theta_{\eta\alpha} - \theta_{\imath\alpha} \tag{14}$$

$$t = t_o C_{\sigma t} \qquad \theta_t = \theta_{\eta t} - \theta_{\imath t} \tag{15}$$

where α_o and t_o = values of the space and time coordinates in matter or vacuum for zero values of the APFs, σ = 1,2,3 and where the space and time coordinate factors $C_{\sigma\alpha}$ and $C_{\sigma t}$ are given by[16]

$$C_{1\alpha} = \cos\theta_{\eta\alpha} \sec\theta_{\pi\alpha} \qquad C_{1t} = \cos\theta_{\eta t} \sec\theta_{\pi t} \qquad [hd] \qquad (16)$$

$$C_{2\alpha} = \cos\theta_{\eta\alpha} \cos\theta_{\pi\alpha} \qquad C_{2t} = \cos\theta_{\eta t} \cos\theta_{\pi t} \qquad [ld] \qquad (17)$$

$$C_{3\alpha} = \sec\theta_{\eta\alpha} \cos\theta_{\pi\alpha} \qquad C_{3t} = \sec\theta_{\eta t} \cos\theta_{\pi t} \qquad [ld] \qquad (18)$$

where [hd] = high-density matter and [ld] = low-density matter and where $\theta_{\eta\alpha}$, $\theta_{\pi\alpha}$ and $\theta_{\eta t}$, $\theta_{\pi t}$ are the internal phase angles of the complex number space and time counting and segmentation numbers. For bound matter in an ambient phase field[12-17]

$$\theta_{\pi\alpha} \geq \theta_{\eta\alpha} \qquad \theta_{\pi t} \geq \theta_{\eta t} \qquad (19)$$

For matter, the values of $C_{\sigma\alpha}$ and $C_{\sigma t}$ can also be simply expressed in terms of the internal phase angles θ_{N} and θ_{n} of the complex particle number and the complex particle number density, while for the vacuum $C_{\sigma\alpha}$ and $C_{\sigma t}$ can also be expressed in terms of the angles θ_{Np} and θ_{np} of the complex number and number density of the virtual particle pairs.[16,17]

As described in Paper B, a polyphase rocket consists of a chemical rocket that has several APFs of different intensities and phases applied to the various component systems.[16] It is shown that the application of ambient energy fields leads to enhancement of the rockets thrust and an increase in the liftoff parameter of the polyphase rocket. For a conventional rocket, which corresponds to zero APFs, and for the adapted condition of maximum thrust, the liftoff parameter is given by[16]

$$s^{o}_{L} = (m^{o}_{R} g^{o}_{E})^{-1} \rho^{o}_{e} (v^{o}_{e})^{2} A^{o}_{a} = (m^{o}_{b} Q^{o}_{R} g^{o}_{E})^{-1} \rho^{o}_{e} (v^{o}_{e})^{2} A^{o}_{a} \qquad (20)$$

which is a simple parameter which can be used for practical calculations because rocket liftoff occurs when $s^{o}_{L} = 1$, and where m^{o}_{R} = conventional total instantaneous mass of a rocket, m^{o}_{b} = mass of the solid part of the rocket body and the liquid or solid fuel, $Q^{o}_{R} = m^{o}_{R} / m^{o}_{b}$ = conventional mass factor, g^{o}_{E} = conventional acceleration of gravity at the Earth's surface, ρ^{o}_{e} = conventional mass density of the exhaust gases in the exhaust tube, v^{o}_{e} = conventional velocity of the gases in the exhaust tube and A^{o}_{a} = conventional area of the aperture of the rocket exhaust tube. For a polyphase rocket the complex number liftoff parameter for the adapted condition is given by[16]

$$\overline{s}_{L} = s_{L} \exp(j\theta_{sL}) = (\overline{m}_{R} \overline{g}_{E})^{-1} \overline{\rho}_{e} \overline{v}^{2}_{e} \overline{A}_{a} = (\overline{Q}_{R} \overline{m}_{b} \overline{g}_{E})^{-1} \overline{\rho}_{e} \overline{v}^{2}_{e} \overline{A}_{a} \qquad (21)$$

where the complex number mass factor is given by

$$\overline{Q}_{R} = Q_{R} \exp(j\theta^{R}_{Q}) = \overline{m}_{R} / \overline{m}_{b} \qquad Q_{R} = m_{R} / m_{b} \qquad \theta^{R}_{Q} = \theta^{R}_{m} - \theta^{b}_{m} \qquad (22)$$

where $\overline{m}_{R}$ = complex number instantaneous mass of the polyphase rocket, $\overline{m}_{b}$ = complex number mass of the rocket body including the fuel, θ^{R}_{m} = internal phase angle of the total mass of a polyphase rocket, θ^{b}_{m} = internal phase angle of the mass of the rocket body, $\overline{g}_{E}$ = complex number acceleration of gravity at the Earth's surface, $\overline{\rho}_{e}$ = complex number mass density of the exhaust gases in the exhaust tube, $\overline{v}_{e}$ = complex number exhaust gas velocity and where $\overline{A}_{a}$ = complex number area of the exhaust area. It has been shown in Paper B that the magnitude of the complex number liftoff parameter for a polyphase rocket can be written as the following general result[16]

$$s_L = s_L^o S_R \sec\theta_n^b \sec\theta_n^e \qquad s_L \geq s_L^o \tag{23}$$

where s_L^o = liftoff parameter for a conventional rocket as given by equation (20) and where S_R is a complicated factor that satisfies $S_R \approx 1$, and therefore liftoff of a polyphase rocket is possible even when the conventional liftoff parameter $s_L^o \leq 1$ provided that the internal phase angles θ_n^b and θ_n^e of the particle number densities of the rocket body and exhaust gases respectively are sufficiently large as to make $s_L \geq 1$.

Special relativity has a long history.[3-8] As described in Paper C, the primary postulates of the special theory of relativity is that of Einstein invariance of the laws of physics and the constancy of the speed of light, which, for the propagation of a spherical light wave states that the quantity $c^2t_o^2 - x_o^2$ is an invariant of the coordinate frame of reference chosen, and therefore[3-8]

$$c^2t_o^2 - x_o^2 = c^2(t_o')^2 - (x_o')^2 = 0 = \text{invariant} \tag{24}$$

which is an invariant that is independent of the motion of the observer, where one observer is located at the origin of the stationary (unprimed) coordinate system and the second observer is located at the origin of the moving (primed) coordinate system which is moving relative to the stationary coordinate system with velocity v_o. The subscript "o" refers to zero APF condition under which special relativity and Einstein invariance were historically developed. Equation (24) gives rise to the Lorentz transformation of coordinates that relates the space and time coordinates of a moving body as observed from the moving and stationary coordinate systems and which is written as[3-8]

$$x_o' = \gamma_o(x_o - v_o t_o) \qquad x_o = \gamma_o(x_o' + v_o t_o') \tag{25}$$

$$t_o' = \gamma_o(t_o - v_o c^{-2} x_o) \qquad t_o = \gamma_o(t_o' + v_o c^{-2} x_o') \tag{26}$$

which leads to the following expression for the apparent mass of a moving body as see from the rest frame of reference

$$m = \gamma_o m_r \tag{27}$$

where m_r = rest mass (proper mass) and where the Lorentz-Einstein factor γ_o is given by

$$\gamma_o = (1-\beta_o^2)^{-1/2} = \begin{cases} \gamma_{oL} & 0 \leq \beta_o < 1 \\ \infty & \beta_o = 1 \\ -i\gamma_{oG} & 1 < \beta_o \leq \infty \end{cases} \tag{28}$$

where $\beta_o = v_o / c$ and where

$$\gamma_{oL} = (1-\beta_o^2)^{-1/2} \qquad 0 \leq \beta_o < 1 \qquad 1 \leq \gamma_{oL} < \infty \tag{29}$$

$$\gamma_{oG} = (\beta_o^2 - 1)^{-1/2} \qquad 1 < \beta_o \leq \infty \qquad 0 \leq \gamma_{oG} < \infty \tag{30}$$

Equation (28) shows that special relativity for the case of zero APF requires that the speed of a body be less than the light speed, that is $\beta_o < 1$ because, as shown in equation (141), the Lorentz-Einstein factor γ_o has a singularity for $\beta_o = 1$ and the case $\beta_o > 1$ leads to an imaginary value of the Lorentz-Einstein factor γ_o and this in turn requires that the space and time coordinates in equations (25) and (26) and the apparent mass in equation (27) are imaginary numbers which does not make sense. Thus for matter, only $\beta_o < 1$ and $\gamma_o = \gamma_{oL}$ are possible for special relativity under zero APF conditions.

About fifteen years ago it was suggested that space and time coordinates of spacetime may have, under some conditions such as the presence of a cosmic APF, a broken symmetry in the sense that these coordinates are complex numbers in an internal space as shown in Paper C.[12] Einstein covariance and the constancy of the speed of light are assumed to be valid in this complex number spacetime, and in particular was applied to light propagation in the vacuum with an APF as follows[12]

$$c^2\bar{t}^2 - \bar{x}^2 = c^2(\bar{t}')^2 - (\bar{x}')^2 = 0 = \text{invariant} \tag{31}$$

Equation (31) gives rise to the complex number Lorentz transformations[12]

$$\bar{x}' = \bar{\gamma}(\bar{x} - \bar{v}\bar{t}) \qquad \bar{x} = \bar{\gamma}(\bar{x}' + \bar{v}\bar{t}') \tag{32}$$

$$\bar{t}' = \bar{\gamma}(\bar{t} - \bar{v}\bar{x}c^{-2}) \qquad \bar{t} = \bar{\gamma}(\bar{t}' + \bar{v}\bar{x}'c^{-2}) \tag{33}$$

where

$$\bar{\gamma} = (1 - \bar{\beta}^2)^{-1/2} \qquad \bar{\beta} = \bar{v}/c \tag{34}$$

where in c = real number light speed in the vacuum which is a constant of nature and is assumed to have the same real value when observed from the stationary or moving inertial systems even in the presence of an APF which induces internal phase angles in the space and time coordinates and in the kinematical quantities such as the relative velocity $\bar{v}$ which now a complex number in internal space. In the presence of an APF the space and time coordinates that appear in the Lorentz transformation equations (32) and (33) are represented as[12]

$$\bar{x} = x\exp(j\theta_x) \qquad \bar{t} = t\exp(j\theta_t) \qquad \bar{x}' = x'\exp(j\theta'_x) \qquad \bar{t}' = t'\exp(\theta'_t) \tag{35}$$

while the kinematic quantities $\bar{v}, \bar{\beta}$ and $\bar{\gamma}$ are written as

$$\bar{v} = v\exp(j\theta_v) \qquad \bar{\beta} = \beta\exp(j\theta_\beta) \qquad \bar{\gamma} = \gamma\exp(j\theta_\gamma) \tag{36}$$

Then equations (34) and (36) give the following magnitudes and internal phase angles[12]

$$\beta = v/c \qquad \theta_\beta = \theta_v \tag{37}$$

$$\gamma = [1 - 2\beta^2\cos(2\theta_\beta) + \beta^4]^{-1/4} \qquad \tan(2\theta_\gamma) = \beta^2\sin(2\theta_\beta)[1 - \beta^2\cos(2\theta_\beta)]^{-1} \tag{38}$$

An examination of equation (38) shows that the magnitude γ has no singular points and therefore special relativity in a spacetime with broken internal symmetries that are induced by the presence of an APF places no restrictions on the velocity of a body, and a body can go from subluminal flight to superluminal flight without any mathematical disasters.[12] Instead of a singularity at $\beta = 1$ the function γ has the value[12]

$$\gamma_1 = [2\sin\theta_\beta]^{-1/2} \qquad \beta = 1 \tag{39}$$

The maximum value of γ is given by[12]

$$\gamma_{max} = [\sin(2\theta_\beta)]^{-1/2} \qquad \beta_{max} = [\cos(2\theta_\beta)]^{1/2} \tag{40}$$

and occurs at the point $\beta = \beta_{max}$. The complex number relativistic apparent mass corresponding to the complex number boost $\bar{\gamma}$ is given by

$$\bar{m} = m\exp(j\theta_m) = \bar{\gamma}m_r \qquad m = \gamma m_r \qquad \theta_m = \theta_\gamma \tag{41}$$

However in the present theory that is promoted in this book, γ_o is not discarded as a useless artifact and still plays an important role in the theory of ultrafast flight. In fact, it will be shown in Paper D of this book that the Lorentz-Einstein factor γ_o, with its singularity at $v_o = c$, as described by equation (28), requires that natural objects cannot obtain subluminal flight up to and including $v = c$ and superluminal flight with $v > c$. In particular it will be shown that the Lorentz-Einstein factor $\gamma_o = \gamma_{oL}$ and given in equation (28) forbid ultrafast subluminal flight with $v \le c$ in natural objects and that only polyphase space vehicles that contain artificial adjustable (internal space) APFs that are tuned to a specific form to cancel the singularity effects of γ_{oL} can undergo ultrafast subluminal flight with $v \le c$. Paper D also shows that for superluminal flight with $v \ge c$, a polyphase space vehicle must contain a combination of properly tuned artificial and adjustable conventional (external space) fields and properly tuned artificial and adjustable (internal space) APFs of specific forms that cancel the effects of $\gamma_o = -i\gamma_{oG}$ which appears in equation (28) and which includes the double effects of the – i term and the singularity of the γ_{oG} term.

ULTRAFAST SUBLUMINAL FLIGHT OF POLYPHASE SPACE VEHICLES

Paper D of this book gives the tuning conditions on the engineered artificial APFs that are generated in a polyphase space vehicle and that are required for ultrafast subluminal flight with $v \le c$ and also shows that natural objects cannot undergo ultrafast subluminal flight even if they contain a natural APF, and this is done by developing the equations for the apparent space and time coordinate factors for polyphase space vehicles and the conditions for ultrafast subluminal flight with $v \le c$. For subluminal flight the zero APF Einstein-Lorentz factor γ_o is given by $\gamma_o = \gamma_{oL}$ where γ_{oL} is given by equation (29). For the case when a cosmic APF is assumed to exist throughout spacetime, the Einstein-Lorentz factor γ_{oL} transforms into γ for $v \le c$ which is given by equation (38) and which does not have a singularity at $v = c$. Paper D shows that the fact that γ is a well-behaved function of relative velocity is not sufficient for ultrafast subluminal motion $v \le c$ to be possible, and in fact it will be shown that even in the presence of a cosmic field only manmade polyphase space vehicles can travel at speeds up to and including the light speed. In the presence of a cosmic APF, only a polyphase space vehicle with a properly engineered and tuned APFs in the various materials of a polyphase space vehicle can travel up to and including the light speed by canceling the effects of γ_{oL}. In Paper D it is shown that these conclusions arise from the calculation of the apparent space and time coordinate factors of the space vehicle which are obtained for the case of a constant APF wave within the space vehicle as follows

$$\bar{x}_L^{sv} = x_L^{sv} \exp(j\theta_{xL}^{sv}) \qquad \bar{t}_L^{sv} = t_L^{sv} \exp(j\theta_{tL}^{sv}) \tag{42}$$

$$x_L^{sv} = x_o^{sv} C_{\sigma x}^{sv} \qquad t_L^{sv} = t_o^{sv} C_{\sigma t}^{sv} \tag{43}$$

$$C_{\sigma x} = \gamma\gamma_{oL}^{-1}(v/v_o)C_{\sigma t}^{sv} = \gamma\gamma_{oL}^{-1}\xi_{xL}^{sv}C_{\sigma x}^{sv} \tag{44}$$

$$C_{\sigma t} = \gamma\gamma_{oL}^{-1}(s_{xo}^{sv}/s_{xL}^{sv})C_{\sigma x}^{sv} = \gamma\gamma_{oL}^{-1}C_{\sigma t}^{sv} \tag{45}$$

where s_{xL}^{sv} and ξ_{xL}^{sv} are given by

$$s_{xL}^{sv} = dx_L^{sv}/dt_L^{sv} = (dx_o^{sv}/dt_o^{sv})(C_{\sigma x}^{sv}/C_{\sigma t}^{sv}) = s_{xo}^{sv}(C_{\sigma x}^{sv}/C_{\sigma t}^{sv}) \tag{46}$$

$$\xi_{xL}^{sv} = (v/v_o)(s_{xo}^{sv}/s_{xL}^{sv}) = \text{real number} \tag{47}$$

where j refers to an internal space, where x_L^{sv} = coordinate variable in the x-direction within the space vehicle traveling at $v_o \le c$, t_L^{sv} = time coordinate variable within the polyphase space vehicle traveling at $v_o \le c$, where v and v_o are the relative velocities of the space vehicle with and without a cosmic APF present, and where s_{xo}^{sv} = characteristic average phase velocity of a mechanical, acoustic or electromagnetic wave propagating in the x direction in the polyphase space vehicle relative to a frame of reference co-moving with and attached to the rear of the polyphase space vehicle under zero APF conditions and s_{xL}^{sv} = real number average velocity of a mechanical, acoustic or electromagnetic wave in the polyphase space vehicle relative to a frame of reference attached to the rear of the space vehicle traveling at $v_o \le c$ and in which there is a local artificial APF and where also for ultrafast subluminal flight the velocity ratio function ξ_{xL}^{sv} is given by equation (47). Equations (44) and (45) contain both γ and γ_{oL}, and therefore the Einstein-Lorentz factor γ_{oL} with its singularity is still there to haunt the space vehicle design engineer, so that this singularity needs to be canceled by tuned APFs.

In Paper D the space vehicle is moving in a weak uniaxial cosmic APF in the x-direction in a small volume (on a cosmic scale) of low-density vacuum so that the apparent space and time coordinate factors $C_{\sigma x}$ and $C_{\sigma t}$ of the distant space vehicle depend on the vacuum in the space between the fixed observer and the space vehicle and are related to the angles $\theta_{\text{Ɲ}p}$ and θ_{np}. The space vehicle itself is assumed to be a small (on a cosmic scale) high-density object containing a manmade uniaxial APF in the x-direction, so that the space and time coordinate factors for the space vehicle $C_{\sigma x}^{sv}$ and $C_{\sigma t}^{sv}$ are specified in terms of $\theta_{\text{Ɲ}}$ and θ_n. Within these assumptions equation (44) becomes

$$\cos\theta_{\text{Ɲ}p}\cos\theta_{np} = \gamma\gamma_{oL}^{-1}\xi_{xL}^{sv}\cos\theta_{\text{Ɲ}}^{sv}\sec\theta_n^{sv} \qquad [ln_{po}, SN_p], [hn_o^{sv}, SN] \tag{48}$$

where ξ_{xL}^{sv} is given by equation (47). Equivalently, these equations can be written in terms of the internal phase angles of the dimensions of spacetime as

$$\cos\theta_{Dst}\cos\theta_{nDst} = \gamma\gamma_{oL}^{-1}\xi_{xL}^{sv}\cos\theta_{\text{Ɲ}}^{sv}\sec\theta_n^{sv} \qquad [ln_{Dst}^{o}, SN_{Dst}], [hn_o^{sv}, SN] \tag{49}$$

Solutions to each of these equations are not in general possible if a space vehicle acted like a natural body with fixed values of the angles $\theta_{\aleph}^{sv}$ and θ_{n}^{sv}, however for a polyphase space vehicle with an internal APF generator the magnitude of the uniaxial APF can adjusted to the situation that the angles $\theta_{\aleph}^{sv}$ and θ_{n}^{sv} have the values for a uniaxial artificial APF determined by

$$\sec\theta_{\aleph}^{sv}\cos\theta_{n}^{sv} = \gamma\gamma_{oL}^{-1}(s_{xo}^{sv}/s_{xL}^{sv}) \qquad s_{xL}^{sv} \geq s_{xo}^{sv} \tag{50}$$

where s_{xL}^{sv} is the characteristic velocity of mechanical, acoustic or electromagnetic waves in the polyphase space vehicle and is given by equation (46) and where $s_{xL}^{sv} \geq s_{xo}^{sv}$ for the $[hn_{o}^{sv}, SN]$ polyphase space vehicle, and in which case the remaining vacuum equations obtained from equations (48) through (50) are

$$\cos\theta_{\aleph p}\cos\theta_{np} = v/v_{o} \tag{51}$$

or

$$\cos\theta_{Dst}\cos\theta_{nDst} = v/v_{o} \tag{52}$$

which are the equations for the phase angles of the cosmic APF in the $[ln_{po}, SN_{p}]$ or $[ln_{Dst}^{o}, SN_{Dst}]$ vacuum with $v \leq v_{o}$. Equations (50) through (52) give the conditions for ultrafast subluminal flight. The angles $\theta_{\aleph}^{sv}$ and θ_{n}^{sv} are always related by the equation[12-17]

$$\theta_{n}^{sv} = \theta_{\aleph}^{sv} - \theta_{V}^{sv} = \theta_{\aleph}^{sv} + |\theta_{V}^{sv}| \tag{53}$$

where θ_{V}^{sv} = internal phase angle of a volume element within the space vehicle, so that these angles are not independent. Combining equations (50) and (53) gives

$$\sec(\theta_{n}^{sv} + \theta_{V}^{sv})\cos\theta_{n}^{sv} = y_{L} \qquad s_{xL}^{sv} \geq s_{xo}^{sv} \tag{54}$$

where

$$y_{L} = \gamma\gamma_{oL}^{-1}(s_{xo}^{sv}/s_{xL}^{sv}) \tag{55}$$

Assuming a constant parameter value for $\theta_{V}^{sv} \leq 0$ allows solutions to the internal phase angle equation (53) and to the tuning equation (54) to be written as[16,17]

$$\tan\theta_{n}^{sv} = -\csc\theta_{V}^{sv}(y_{L}^{-1} - \cos\theta_{V}^{sv}) = \csc|\theta_{V}^{sv}|(y_{L}^{-1} - \cos|\theta_{V}^{sv}|) \tag{56}$$

$$\theta_{\aleph}^{sv} = \theta_{n}^{sv} + \theta_{V}^{sv} = \theta_{n}^{sv} - |\theta_{V}^{sv}| \tag{57}$$

Nevertheless, as a crude approximation it can be assumed that $\theta_{\aleph}^{sv}$ and θ_{n}^{sv} are independent in equation (50), and this gives the following approximate conditions for ultrafast subluminal flight

$$\sec\theta_{n}^{sv} = \gamma_{oL} \qquad \sec\theta_{\aleph}^{sv} = \gamma(s_{xo}^{sv}/s_{xL}^{sv}) \qquad s_{xL}^{sv} \geq s_{xo}^{sv} \tag{58}$$

and then θ_{V}^{sv} is calculated using equation (53). All of these equations are only valid if the amplitudes and frequencies of the APFs in the space vehicle and also the angles $\theta_{\aleph}^{sv}$ and θ_{n}^{sv} can be engineered so as to satisfy equation (50). Equations (50) through (58) show that the engineered APF within a $[hn_{o}^{sv},SN]$ polyphase space vehicle cancels the effects of the singularity of γ_{oL} at $v_{o}=c$ by requiring

$$\theta_{n}^{sv}=\theta_{\aleph}^{sv}+\left|\theta_{V}^{sv}\right|=\pi/2 \qquad \theta_{\aleph}^{sv}=\pi/4 \qquad \theta_{V}^{sv}=-\pi/4 \tag{59}$$

at this point, and therefore polyphase space vehicles can undergo ultrafast subluminal flight. The condition $\theta_{n}^{sv}=\pi/2$ and $\theta_{\aleph}^{sv}=\pi/4$ for a $[hn_{o}^{sv},SN]$ polyphase space vehicle at the point $v_{o}=c$ corresponds to an APF of a specific energy density and frequency.

SUPERLUMINAL FLIGHT OF POLYPHASE SPACE VEHICLES.

By developing the equations for the apparent space and time coordinate factors and the engineered conditions required by polyphase space vehicles for superluminal flight with $v \geq c$, Paper D of this book gives the mathematical conditions on the engineered artificial APF that are required for superluminal flight of a polyphase space vehicle with $v \geq c$ and shows that natural objects cannot undergo superluminal flight even if they contain a natural APF.

For superluminal flight the wave fields within the polyphase space vehicle must be in part an ordinary wave in external space and an APF wave in internal space and must be tuned so as to induce in the materials of the polyphase space vehicle a set of complex number space and time coordinates of a general form that can be engineered for the case $v_{o}\geq c$ so that the conventional, external space, parts of the wave fields are tuned to cancel the $-i$ portion of the $-i\gamma_{oG}$ term that appears in the conventional special relativity equation (28), while the internal space APF parts of the wave field are tuned to cancel the singularity of the γ_{oG} portion of the $-i\gamma_{oG}$ term, as explained in Paper D. The general forms of the coordinates required for this are

$$\bar{x}_{G}^{sv}=x_{G}^{sv}\exp(j\theta_{xG}^{sv})\exp(i\bar{\varphi}_{x}^{sv}) \qquad \bar{t}_{G}^{sv}=t_{G}^{sv}\exp(j\theta_{tG}^{sv})\exp(i\bar{\varphi}_{t}^{sv}) \tag{60}$$

where j refers to an internal space and i refers to the ordinary external space, where the index G refers to a space vehicle traveling at a speed greater than the light speed in vacuo and finally where, for constant phase angles, the velocity of the wave is

$$\bar{s}_{xG}^{sv}=s_{xG}^{sv}\exp(j\theta_{sxG}^{sv})=d\bar{x}_{G}^{sv}/d\bar{t}_{G}^{sv} \tag{61}$$

$$\bar{s}_{xG}^{sv}=s_{mxG}^{sv}\exp(j\theta_{sxG}^{sv})\exp(i\bar{\varphi}_{sxG}^{sv})=d\bar{x}_{G}^{sv}/d\bar{t}_{G}^{sv} \tag{62}$$

$$s_{xG}^{sv}=dx_{G}^{sv}/dt_{G}^{sv}=s_{xo}^{sv}(C_{\sigma x}^{sv}/C_{\sigma t}^{sv}) \tag{63}$$

where s_{mxG}^{sv} is the magnitude of the complex number velocity s_{xG}^{sv} which is a complex number in external space for superluminal flight, so that

$$s_{mxG}^{sv}=\left|s_{xG}^{sv}\right|=s_{xo}^{sv}(C_{\sigma x}^{sv}/\left|C_{\sigma t}^{sv}\right|) \tag{64}$$

where the complex number time coordinate factor $C^{sv}_{\sigma t}$ is a complex number in external space for superluminal flight and which is written as

$$C^{sv}_{\sigma t} = |C^{sv}_{\sigma t}| \exp(i\varphi^{sv}_{C\sigma t}) \tag{65}$$

and the tuned value for superluminal flight is given by

$$C^{sv}_{\sigma t} = -i|C^{sv}_{\sigma t}| = |C^{sv}_{\sigma t}|\exp(-i\pi/2) \qquad \varphi^{sv}_{C\sigma t} = -\pi/2 \tag{66}$$

and where

$$\theta^{sv}_{sxG} = \theta^{sv}_{xG} - \theta^{sv}_{tG} \qquad \overline{\varphi}^{sv}_{sxG} = \overline{\varphi}^{sv}_{x} - \overline{\varphi}^{sv}_{t} - \varphi^{sv}_{C\sigma t} \tag{67}$$

In Paper D it is shown that the first condition for superluminal flight is that $\overline{\varphi}^{sv}_{sxG}$ must be a real number in internal space and given by

$$\overline{\varphi}^{sv}_{sxG} = \varphi^{sv}_{sxG} = \pi/2 \tag{68}$$

so that the conventional external field part of the wave field induces only external space phase angles and does not change the internal space phase angles created by the APF portions of the wave fields which will be subsequently tuned again, so that the external space part of the wave field phase velocity is tuned to the condition

$$s^{sv}_{xG} = is^{sv}_{mxG} \qquad |s^{sv}_{xG}| = s^{sv}_{mxG} \qquad \varphi^{sv}_{sxG} = \pi/2 \tag{69}$$

and for this constant intensity APF and for the tuned condition of equation (69), the time coordinate factor is given by

$$C^{sv}_{\sigma t} = (s^{sv}_{xo}/s^{sv}_{xG})C^{sv}_{\sigma x} = [s^{sv}_{xo}/(is^{sv}_{mxG})]C^{sv}_{\sigma x} \qquad |C^{sv}_{\sigma t}| = (s^{sv}_{xo}/s^{sv}_{mxG})C^{sv}_{\sigma x} \qquad \varphi^{sv}_{C\sigma t} = -\pi/2 \tag{70}$$

and it is these expressions that are used in equations (71) and (72) to remove the imaginary index i from the equations, and that at least mathematically allows the superluminal flight of a polyphase space vehicle.

Superluminal flight under zero cosmic APF conditions is impossible and the Einstein-Lorentz factor γ_o for this case is given by $\gamma_o = -i\gamma_{oG}$ where γ_{oG} is given by equation (30). The Einstein-Lorentz factor γ_{oG} has a singularity at $v_o = c$. The imaginary value of γ_o indicates that superluminal flight for a zero cosmic APF is impossible. For the case when a cosmic APF is assumed to exist throughout spacetime, the factor γ, which is given by equation (38), does not have a singularity at $v = c$ and is analytic in the range $v \geq c$. But the fact that γ is a well-behaved function of relative velocity is not sufficient to show that superluminal motion is possible because the factor γ_{oG} as well as γ appears in equations (71) and (72), and in fact it will be demonstrated that even in the presence of a cosmic field natural bodies cannot move faster than light in vacuo. However, it will be shown in Paper D that in the presence of a cosmic APF a polyphase space vehicle, with a properly engineered mechanical, acoustic or electromagnetic APF component of the wave field within the vehicle, can travel faster than the light speed in vacuo. In order to see this one must look at the equations that give the apparent space and time coordinate factors of the space vehicle as

$$C_{\sigma x} = i\gamma\gamma_{oG}^{-1}(v/v_o)C_{\sigma t}^{sv} = i\gamma\gamma_{oG}^{-1}\xi_{xG}^{sv}C_{\sigma x}^{sv} = \gamma\gamma_{oG}^{-1}\xi_{mxG}^{sv}C_{\sigma x}^{sv} \tag{71}$$

$$C_{\sigma t} = i\gamma\gamma_{oG}^{-1}(s_{xo}^{sv}/s_{xG}^{sv})C_{\sigma x}^{sv} = i\gamma\gamma_{oG}^{-1}C_{\sigma t}^{sv} = \gamma\gamma_{oG}^{-1}\left|C_{\sigma t}^{sv}\right| \tag{72}$$

where the tuned imaginary number characteristic velocity s_{xG}^{sv} of a mechanical, acoustic or electromagnetic wave in a polyphase space vehicle and the imaginary number velocity ratio function ξ_{xG}^{sv} are given by

$$s_{xG}^{sv} = dx_G^{sv}/dt_G^{sv} = is_{xo}^{sv}(C_{\sigma x}^{sv}/\left|C_{\sigma t}^{sv}\right|) = is_{mxG}^{sv} = \text{ imaginary number} \tag{73}$$

$$\xi_{xG}^{sv} = (v/v_o)(s_{xo}^{sv}/s_{xG}^{sv}) = (v/v_o)[s_{xo}^{sv}/(is_{mxG}^{sv})] = -i\xi_{mxG}^{sv} = \text{imaginary number} \tag{74}$$

and where

$$s_{mxG}^{sv} = s_{xo}^{sv}(C_{\sigma x}^{sv}/\left|C_{\sigma t}^{sv}\right|) \tag{75}$$

$$\xi_{mxG}^{sv} = (v/v_o)[s_{xo}^{sv}/(s_{mxG}^{sv})] \tag{76}$$

where x_G^{sv} = characteristic length of space vehicle traveling at a speed $v_o \geq c$ in the x-direction and t_G^{sv} = characteristic time within the polyphase space vehicle traveling at a speed $v_o \geq c$, where the tuned imaginary number characteristic velocity of a mechanical, acoustic or electromagnetic wave in a polyphase space vehicle s_{xG}^{sv} and the imaginary number velocity ratio function ξ_{xG}^{sv} are given by equations (73) and (74) respectively, where s_{mxG}^{sv} = magnitude of the average velocity of a mechanical, acoustic or electromagnetic wave in a polyphase space vehicle which is traveling at a relative speed $v_o \geq c$ and where ξ_{mxG}^{sv} = magnitude of the complex number velocity ratio function. Equations (71) and (72) contain both γ and γ_{oG}, and therefore the Einstein-Lorentz factor γ_{oG} with its singularity at $v_o = c$ is still there to haunt the space vehicle design engineer, however there is now no longer any imaginary numbers in external space to worry about because they been canceled out by the conventional external space component of the wave fields within the polyphase space vehicle.

Paper D assumes that the vacuum in which the space vehicle is moving contains a weak uniaxial cosmic APF in the x-direction and that the volume of low-density vacuum traversed is small (on a cosmic scale) so that the apparent space and time coordinate factors $C_{\sigma x}$ and $C_{\sigma t}$ of the distant space vehicle can be represented by the internal phase angles $\theta_{\text{Ɲ}p}$ and θ_{np} of the number and number density of virtual particle pairs or the internal phase angles θ_{Dst} and θ_{nDst} of the number and number density of the dimensions of the vacuum in the space between the fixed observer on Earth and the space vehicle. The polyphase space vehicle itself is assumed to be a small (on a cosmic scale) high-density object so that the space and time coordinate factors of the polyphase space vehicle $C_{\sigma x}^{sv}$ and $C_{\sigma t}^{sv}$ can be described by the angles $\theta_{\text{Ɲ}}^{sv}$ and θ_n^{sv}. The polyphase space vehicle contains a manmade uniaxial APF in the x-direction within the space vehicle. Within these assumptions, and after using equation (74), equation (71) can be written as

$$\cos\theta_{\text{Ɲ}p}\cos\theta_{np} = \gamma\gamma_{oG}^{-1}\xi_{mxG}^{sv}\cos\theta_{\text{Ɲ}}^{sv}\sec\theta_n^{sv} \qquad [ln_{po}, SN_p],\ [hn_o^{sv}, SN] \tag{77}$$

where ξ^{sv}_{mxG} is given by equation (76). Equivalently, these equations can be written in terms of the internal phase angles of the dimensions of spacetime as

$$\cos\theta_{Dst}\cos\theta_{nDst} = \gamma\gamma^{-1}_{oG}\xi^{sv}_{mxG}\cos\theta^{sv}_{N}\sec\theta^{sv}_{n} \qquad [\ln^{o}_{Dst},SN_{Dst}],\ [hn^{sv}_{o},SN] \qquad (78)$$

where ξ^{sv}_{mxG} is given by equation (76). Solutions to each of these equations are not in general possible if a space vehicle acted like a natural body with fixed or arbitrarily varying values of the angles θ^{sv}_{N} and θ^{sv}_{n}, however for a polyphase space vehicle with an internal APF generator the amplitude and frequency of the uniaxial APF can be tuned to the situation that the angles θ^{sv}_{N} and θ^{sv}_{n} have the values for a uniaxial artificial APF determined by

$$\sec\theta^{sv}_{N}\cos\theta^{sv}_{n} = i\gamma\gamma^{-1}_{oG}(s^{sv}_{xo}/s^{sv}_{xG}) = \gamma\gamma^{-1}_{oG}(s^{sv}_{xo}/s^{sv}_{mxG}) \qquad s^{sv}_{mxG} \geq s^{sv}_{xo} \qquad (79)$$

where the imaginary number s^{sv}_{xG} and the real number s^{sv}_{mxG} are given by equations (73) and (75) and where $s^{sv}_{mxG} \geq s^{sv}_{xo}$ for the $[hn^{sv}_{o},SN]$ polyphase space vehicle, and in which case the auxiliary vacuum equations obtained from equations (77) through (79) are

$$\cos\theta_{Np}\cos\theta_{np} = v/v_{o} \qquad (80)$$

or

$$\cos\theta_{Dst}\cos\theta_{nDst} = v/v_{o} \qquad (81)$$

which are the equations for the phase angles of the cosmic APF in the $[\ln_{po},SN_{p}]$ or $[\ln^{o}_{Dst},SN_{Dst}]$ vacuum with $v \leq v_{o}$. Equations (79) through (81) give the conditions for superluminal flight. The angles θ^{sv}_{N} and θ^{sv}_{n} are always related by

$$\theta^{sv}_{n} = \theta^{sv}_{N} - \theta^{sv}_{V} = \theta^{sv}_{N} + \left|\theta^{sv}_{V}\right| \qquad (82)$$

where θ^{sv}_{V} = internal phase angle of a volume element within the space vehicle, so that these angles are not independent. Combining equations (79) and (82) gives

$$\sec(\theta^{sv}_{n} + \theta^{sv}_{V})\cos\theta^{sv}_{n} = y_{G} \qquad (83)$$

where

$$y_{G} = \gamma\gamma^{-1}_{oG}(s^{sv}_{xo}/s^{sv}_{mxG}) \qquad 0 \leq y_{G} \leq 1 \qquad s^{sv}_{mxG} \geq s^{sv}_{xo} \qquad (84)$$

Assuming a constant parameter value for $\theta^{sv}_{V} \leq 0$ allows solutions of the phase angle equation (82) and for the tuning equation (83) to be written as[16,17]

$$\tan\theta^{sv}_{n} = -\csc\theta^{sv}_{V}(y^{-1}_{G} - \cos\theta^{sv}_{V}) = \csc\left|\theta^{sv}_{V}\right|(y^{-1}_{G} - \cos\left|\theta^{sv}_{V}\right|) \qquad (85)$$

$$\theta^{sv}_{N} = \theta^{sv}_{n} + \theta^{sv}_{V} = \theta^{sv}_{n} - \left|\theta^{sv}_{V}\right| \qquad (86)$$

Nevertheless, as a crude approximation it can be assumed that θ^{sv}_{N} and θ^{sv}_{n} are independent in equation (79), and this gives the following approximate tuned conditions for superluminal flight

$$\sec\theta_{n}^{sv} = \gamma_{oG} \qquad \sec\theta_{\aleph}^{sv} = i\gamma(s_{xo}^{sv}/s_{xG}^{sv}) = \gamma(s_{xo}^{sv}/s_{mxG}^{sv}) \qquad s_{mxG}^{sv} \geq s_{xo}^{sv} \qquad (87)$$

and with the parameter θ_{V}^{sv} then calculated from equation (82). These equations are only valid if the amplitudes and frequencies of the APFs in the space vehicle, and also the angles $\theta_{\aleph}^{sv}$ and θ_{n}^{sv}, can be tuned so as to satisfy equations (79) through (87) in which the imaginary index i in equations (79) or (87) are also canceled by the imaginary number s_{xG}^{sv} given by equation (73). Equations (79) through (87) show that the engineered APFs within the high-density solid-body polyphase space vehicle can cancel the effects of the singularity of γ_{oG} at $v_o = c$ by requiring

$$\theta_{n}^{sv} = \theta_{\aleph}^{sv} + |\theta_{V}^{sv}| = \pi/2 \qquad \theta_{\aleph}^{sv} = \pi/4 \qquad \theta_{V}^{sv} = -\pi/4 \qquad \varphi_{sxG}^{sv} = \pi/2 \qquad (88)$$

at this point and can cancel the effects of the imaginary index i, and therefore properly tuned polyphase space vehicles can undergo superluminal flight. The condition in equation (88) for a $[hn_{o}^{sv}, SN]$ polyphase space vehicle at the point $v_o = c$ corresponds to an APF/conventional field combination of a particular energy density and frequency.

REFERENCES

1. Hirshfeld, A. and Sinnott, W., Sky Catalogue 2000, Vols. 1&2, Sky Publishing Corp., Cambridge, 1982.

2. Dibon-Smith, R., StarList 2000, John Wiley, New York, 1992.

3. Møller, C., The Theory of Relativity, Oxford University Press, 1955.

4. Aharoni, J., The Special Theory of Relativity, Oxford University Press, 1959.

5. Rindler, W., Introduction to Special Relativity, Clarendon Press, Oxford, 1991.

6. Bergmann, P., Introduction to the Theory of Relativity, Prentice-Hall, New York, 1953.

7. Synge, J., Relativity: The Special Theory, North-Holland, John Wiley, New York, 1965.

8. Pauli, W., Theory of Relativity, Pergamon Press, New York, 1958.

9. Davies, P., The Forces of Nature, Cambridge University Press, New York, 1979.

10. Collins, J., et al, "Lorentz Invariance and Quantum Gravity: An Additional Fine-Tuning Problem?" Phys. Rev. Lett., Vol. 93, Number 19, 5 Nov 2004.

11. Weiss, R. A., Relativistic Thermodynamics, Exposition Press, New York, 1976.

12. Weiss, R. A., "Maxwell's Equations with Broken Internal Symmetries," article in Gauge Theory of Thermodynamics, K&W Publications, Vicksburg, 1989.

13. Weiss, R. A., Clean Fission, K&W Publications, Vicksburg, 1992.

14. Weiss, R. A., Internal Fission, K&W Publications, Vicksburg, 1995.

15. Weiss, R. A., Vacuum Fission, K&W Publications, Vicksburg, 1998.

16. Weiss, R. A., Polyphase Fission and Polyphase Rockets, K&W Publications, Tucson, 2003.

17. Weiss, R. A., Polyphase Fission Nuclear Power, K&W Publications, Tucson, 2004.

A

PHYSICAL NATURE OF SPACE AND TIME COORDINATES

Richard A. Weiss
K&W Research Company
Tucson, Arizona

ABSTRACT. A theory of space and time coordinates in matter and vacuum in the presence of an ambient field is developed that suggests that these coordinates are affected by the presence of the matter and the ambient field and must be represented as complex numbers in an internal space and whose magnitudes and internal phase angles depend on the density of matter and the energy density of the ambient field. These complex number coordinates are referred to as active coordinates and are physical in nature although they are still arbitrary in the sense that they depend on the choice of the origin as in the case of conventional (inactive) coordinates. The basis of this theory of active space and time coordinates is the representation of the complex number space coordinates $\overline{\alpha} = \alpha \exp(j\theta_\alpha)$ in terms of a ratio of a complex number spatial counting number $\overline{\eta}_\alpha$ and a complex number spatial segmentation number $\overline{\pi}_\alpha$ which gives α and θ_α as functions of two fundamental angles $\theta_{\eta\alpha}$ and $\theta_{\pi\alpha}$ where $\alpha = x,y,z; r,\varphi,z$ or ρ,φ,ψ ; and the representation of the complex number time coordinate $\overline{t} = t\exp(\theta_t)$ as a ratio of a complex number temporal counting number $\overline{\eta}_t$ and a complex number temporal segmentation number $\overline{\pi}_t$ which gives t and θ_t as a function of two fundamental phase angles $\theta_{\eta t}$ and $\theta_{\pi t}$. Then the representations of the space and time coordinate magnitudes are $\alpha = \alpha^o C_{\sigma\alpha}(\theta_{\eta\alpha},\theta_{\pi\alpha})$ and $t = t^o C_{\sigma t}(\theta_{\eta t},\theta_{\pi t})$ where α^o and t^o = zero ambient field values of the space and time coordinates, and where $C_{\sigma\alpha}$ and $C_{\sigma t}$ = two-angle space and time coordinate factors respectively where $\sigma = 1$ for a high-density system and where $\sigma = 2, 3$ for a low-density system which is apparently doubly degenerate. The internal phase angles of the complex number space and time coordinates are given by $\theta_\alpha = \theta_{\eta\alpha} - \theta_{\pi\alpha}$ and $\theta_t = \theta_{\eta t} - \theta_{\pi t}$. An ambient field is a field in bulk matter that induces internal phase angles into all physical quantities with the exception of the fundamental constants of nature. For an ambient radiation field, the eight internal phase angles $\theta_{\eta\alpha}, \theta_{\pi\alpha}, \theta_{\eta t}$ and $\theta_{\pi t}$ are calculated in a manner similar to that of a Kerr electro-optic medium in the sense that for for mechanical, electromagnetic or gravitational ambient radiation in matter they are linearly related to the energy density of the ambient radiation field. For a nonradiative ambient field in matter these eight phase angles are more complicated functions of the energy density of an ambient field, and are calculated in terms of the shear, isotropic compression and isotropic expansion deformations of matter that are induced by the ambient field and in terms of the space coordinate factors. The material in this paper will have applications to many fields of physical science including mechanics, atomic and nuclear physics, wave propagation, diffusion, chemical reactions and rocket science.

1. INTRODUCTION. Mathematical coordinates of space and time are universally used in engineering practice and in basic science. The ability to characterize the location in space and time of material bodies lies at the heart of engineering mechanics, astronomy, chemistry and physics. The mathematical nature and theory of coordinates was developed by Fermat, Descartes and others.[1-3] The laws of nature are expressed in terms of the coordinates of particles and fields as in Newtonian mechanics for example, where the velocity and acceleration of bodies depend on the coordinates of the bodies while the potential of Newtonian gravity is described in terms of the location of points in space.[4] Similarly, Maxwell's equations of electromagnetism are expressed in

terms of the space and time coordinate locations of charges and currents and in terms of the space and time coordinates at which points the electromagnetic field is to be calculated.[5] Space and time coordinates are also important parameters for describing the mathematical statements of the principles of the special and general theories of relativity and of quantum mechanics, and in all of these theories the arbitrariness of the choice of coordinate reference frames are stressed, especially in the relativity theories.[6-8] However, recently it has been suggested that in the presence of an ambient field the space and time coordinates are not completely arbitrary in the sense that they are complex numbers in an internal space and that the values of their internal phase angles depend on the energy density of the ambient field.[9-12] An ambient field in bulk matter induces internal phase angles into the space and time coordinates and in all other physical quantities with the exception of the fundamental constants of nature h, c and G. In this paper these complex number space and time coordinates will be referred to as active coordinates.

Space and time coordinates in the presence of ambient fields such as ambient gravitation, electromagnetism, nuclear forces or pressure must be represented as complex numbers in an internal space, so that for cartesian coordinates[9-12]

$$\bar{x} = x\exp(j\theta_x) \qquad \bar{y} = y\exp(j\theta_y) \qquad \bar{z} = z\exp(j\theta_z) \tag{1}$$

while for cylindrical polar coordinates

$$\bar{r} = r\exp(j\theta_r) \qquad \bar{\varphi} = \varphi\exp(j\theta_\varphi) \qquad \bar{z} = z\exp(j\theta_z) \tag{2}$$

and for spherical polar coordinates

$$\bar{\rho} = \rho\exp(j\theta_\rho) \qquad \bar{\varphi} = \varphi\exp(j\theta_\varphi) \qquad \bar{\psi} = \psi\exp(j\theta_\psi) \tag{3}$$

These complex number coordinates can be written in shorthand notation as

$$\bar{\alpha} = \alpha\exp(j\theta_\alpha) \tag{4}$$

where α = x,y,z; r,φ,z or ρ,φ,ψ. The complex number time coordinate is written as

$$\bar{t} = t\exp(j\theta_t)) \tag{5}$$

The measured values of the space and time coordinates are given by[9-12]

$$\alpha_m = \alpha\cos\theta_\alpha \qquad t_m = t\cos\theta_t \tag{6}$$

An ambient field is a field in bulk matter whose forces are complex numbers in an internal space and which are expressed in terms of the complex number space and time coordinates and which also induces the internal phase angles in the space and time coordinates and in other physical quantities of nature within matter such as particle number and particle number density, but the fundamental constants of nature such as h, c and G remain as real numbers. The internal phase angles θ_α and θ_t are thought to be dependent on the energy density of the ambient fields, while the magnitudes α and t are arbitrary in the sense that they depend on the choice of origin of the coordinates.[9-12] The active coordinates are therefore not completely arbitrary because the values of their internal phase angles are in some way related to the energy density of the ambient fields.[9-12] However, in this paper it will be shown that the magnitudes α and t of the complex number space and time coordinates also depend on the energy density of the ambient field and therefore have some physical nature although they are still arbitrary in the sense that they depend on the choice of a reference frame. This paper will determine the dependence of the magnitude α on the internal phase angle θ_α, and the dependence of the magnitude t on the internal phase angle θ_t, by making use of complex number integer functions that are analogous to those that were introduced for the description of γ ray catalyzed internal nuclear fission (γcinf) and internal vacuum fission (γcivf).[10-12]

For a description of internal nuclear fission it has been conjectured that the atomic number, neutron number and atomic mass number of an atomic nucleus in the presence of an ambient γ ray field can be written in the form of complex numbers of the form[10-12]

$$\bar{z} = z\exp(j\theta_z) \qquad \bar{n} = n\exp(j\theta_n) \qquad \bar{a} = a\exp(j\theta_a) \tag{7}$$

where for subactinide nuclei

$$z = Z\cos\theta_z \qquad n = N\cos\theta_n \qquad a = A\cos\theta_a \tag{8}$$

while for actinide and transactinide nuclei

$$z = Z\sec\theta_z \qquad n = N\sec\theta_n \qquad a = A\sec\theta_a \tag{9}$$

where Z, N and A = integer values of the atomic number, neutron number and atomic mass number and where internal phase angles of the atomic number that appears in equations (7) through (9) can be calculated, with t_o = time factor of nuclear matter, by[10-12]

$$\tan\theta_z = (2\mathfrak{E}^\gamma / G_r^\gamma)^{1/2} \qquad G_r^\gamma = \sigma_r^2 G_o A^{-1/3} \qquad G_o = \zeta_o / t_o \tag{10}$$

where $\mathfrak{E}^\gamma$ = energy density of a shearing ambient γ ray field, G_r^γ = resonant dynamic shear modulus of an atomic nucleus, $\sigma_r \approx 3\times10^{-9}$ = resonance parameter for infinite nuclear matter and where G_o, ζ_o = shear modulus and viscosity of nuclear matter for a zero ambient γ ray field. The measured values of the atomic number, neutron number and atomic mass number are given by[10-12]

$$z_m = z\cos\theta_z \qquad n_m = n\cos\theta_n \qquad a_m = a\cos\theta_a \tag{11}$$

Similarly, for the description of vacuum fission, the dimensions of time, space and spacetime in the interior of an atomic nucleus in the presence of an ambient γ ray field are written as complex numbers of the form[10-12]

$$\overline{D}_t = D_t\exp(j\theta_{Dt}) \qquad \overline{D}_s = D_s\exp(j\theta_{Ds}) \qquad \overline{D}_{st} = D_{st}\exp(j\theta_{Dst}) \tag{12}$$

where for low dimensional spacetime[12]

$$D_t = N_{Dt}\cos\theta_{Dt} \qquad D_s = N_{Ds}\cos\theta_{Ds} \qquad D_{st} = N_{Dst}\cos\theta_{Dst} \tag{13}$$

while for high dimensional spacetime[12]

$$D_t = N_{Dt} \sec \theta_{Dt} \qquad D_s = N_{Ds} \sec \theta_{Ds} \qquad D_{st} = N_{Dst} \sec \theta_{Dst} \tag{14}$$

where N_{Dt}, N_{Ds} and N_{Dst} = integer values of the number of time, space and spacetime dimensions that exist within an atomic nucleus under under zero ambient γ ray field conditions and where the internal phase angles of the time dimension that appears in equations (12) through (14) can be calculated, with t_v^o = nuclear vacuum time factor, by[12]

$$\tan \theta_{Dt} = (2\mathfrak{E}^\gamma / G_v^\gamma)^{1/2} \qquad G_v^\gamma = \sigma_v^2 G_v^o N_{Dst}^{-1/3} \qquad G_v^o = \varsigma_v^o / t_v^o \tag{15}$$

where $\mathfrak{E}^\gamma$ = energy density of an ambient γ ray field which is a shearing field, G_v^γ = resonant dynamic shear modulus of a dimensional sphere, σ_v = resonance parameter of the nuclear vacuum and where G_v^o, ς_v^o = shear modulus and viscosity of the vacuum state of nuclear matter for zero γ ray field, with $G_v^o << G_o$ and $\varsigma_v^o << \varsigma_o$. The measured values of the number of time, space and spacetime dimensions are given by[12]

$$D_{tm} = D_t \cos \theta_{Dt} \qquad D_{sm} = D_s \cos \theta_{Ds} \qquad D_{stm} = D_{st} \cos \theta_{Dst} \tag{16}$$

The complex number representations of the nucleon numbers and the dimensions of time, space and spacetime are useful for studies of binary and internal nuclear fission and of binary and internal vacuum fission.[10-12]

It has been shown that for gases undergoing chemical reactions in the presence of an ambient field the molecular number density should be written as[11,12]

$$\bar{n} = n \exp(j\theta_n) \tag{17}$$

$$n = n^o \cos \theta_n \qquad \text{high-density gases} \tag{18}$$

$$n = n^o \sec \theta_n \qquad \text{low-density gases} \tag{19}$$

where n^o = zero ambient field (conventional) value of the molecular number density which is given by

$$n^o = N/V^o \tag{20}$$

where $N = N^o$ = integer number of molecules and V^o = volume at zero ambient field. The measured molecular number density is given by

$$n_m = n \cos \theta_n \tag{21}$$

which is always a real number. The present paper will use the concepts of active space coordinates to establish equations (18) and (19) as general results for matter at high- and low-densities respectively and this is done in Section 6.

Another example of the use of equation (18) is in a theory of thermonuclear reactions in stars.[11] In the presence of ambient gravity the complex number thermonuclear reaction rate of deuterium and tritium producing helium is given by[11]

$$\overline{R} = R\exp(j\theta_R) = \overline{n}_D\overline{n}_T\langle\overline{\sigma v}\rangle \tag{22}$$

where the complex number reaction cross section $\overline{\sigma}$ and the complex number relative velocity $\overline{v}$ of the reacting nuclei in an ambient gravitational field are represented as

$$\overline{\sigma} = \sigma\exp(j\theta_\sigma) \qquad \overline{v} = v\exp(j\theta_v) \tag{23}$$

and where the complex number particle number densities of deuterium and tritium are written as in equation (17) as

$$\overline{n}_D = n_D\exp(j\theta_{nD}) \qquad \overline{n}_T = n_T\exp(j\theta_{nT}) \tag{24}$$

Then the magnitude and internal phase angle of the thermonuclear reaction rate is obtained from the high-density equation (18) combined with equation (22) as [11]

$$R = n_D n_T\langle\sigma v\rangle = n_D^o n_T^o \cos\theta_{nD}\cos\theta_{nT}\langle\sigma v\rangle \tag{25}$$

$$\theta_R = \theta_{nD} + \theta_{nT} + \theta_\sigma + \sigma_v \tag{26}$$

where equation (20) gives the zero ambient field values of the deuterium and tritium particle number densities as

$$n_D^o = N_D / V^o \qquad n_T^o = N_T / V^o \tag{27}$$

where $N_D = N_D^o$ and $N_T = N_T^o$ = integer numbers of deuterium and tritium nuclei. The measured thermonuclear reaction rate is then given by[11]

$$R_m = R\cos\theta_R \tag{28}$$

So that the measured thermonuclear reaction rate is lower than the rate predicted by the conventional calculation by a factor F given by

$$R_m = FR^o \qquad R^o = n_D^o n_T^o\langle\sigma v\rangle \tag{29}$$

where the factor F is given by[11]

$$F = \cos\theta_{nD}\cos\theta_{nT}\cos(\theta_{nD} + \theta_{nT} + \theta_\sigma + \theta_v) \qquad F \le 1 \tag{30}$$

and where R^o = conventional thermonuclear reaction rate.[13] The measured rate of neutrino emission from the sun is found to be about 0.5 of that predicted by the conventional thermonuclear reaction rate R^o.[14] The rate reduction factor F that appears in equation (30) may possibly account for a part of the lower than expected measured neutrino emission from the sun.[11] Thus for the value F = 0.5, and with $\theta_\sigma = 0$ and $\theta_v = 0$, it follows from equation (30) that $\theta_{nD} = \theta_{nT} = 25.91°$ as the values of the phase angles of the particle number densities that are required to explain the measured neutrino deficit without assuming neutrino oscillations and a nonzero neutrino mass. The inclusion of neutrino oscillations and a nonzero neutrino mass would greatly reduce the value of these angles. Note that equations (29) and (30) assume that σ and v are unaffected by the presence of an ambient gravity field so that $\sigma v = \sigma^o v^o$.

In the present paper it will be shown that the complex number active space and time coordinates in matter or in the vacuum in the presence of an ambient field of energy density $\mathfrak{E}$ are represented as in equations (4) and (5) by the ratios

$$\bar{\alpha} = \alpha \exp(j\theta_\alpha) = \bar{\eta}_\alpha L^c_\alpha / \bar{\pi}_\alpha \qquad \bar{t} = t\exp(j\theta_t) = \bar{\eta}_t L^c_t / \bar{\pi}_t \qquad (31)$$

where α = x,y,z; r,φ,z and ρ,φ,ψ; and where the magnitudes of the active space and time coordinates are not completely arbitrary and must be written in the form

$$\alpha = \eta_\alpha L^c_\alpha / \pi_\alpha = \alpha^o C(\theta_{\eta\alpha}, \theta_{\pi\alpha}) \qquad C(\theta_{\eta\alpha} = 0, \theta_{\pi\alpha} = 0) = 1 \qquad (32)$$

$$t = \eta_t L^c_t / \pi_t = t^o C(\theta_{\eta t}, \theta_{\pi t}) \qquad C(\theta_{\eta t} = 0, \theta_{\pi t} = 0) = 1 \qquad (33)$$

where L^c_α and L^c_t = real number constant parameters for space and time coordinates respectively which can be taken as multiples of the Planck length and Planck time, where α^o and t^o = zero ambient field (conventional) values of the space and time coordinates, and where the internal phase angles of the space and time coordinates are given by

$$\theta_\alpha = \theta_{\eta\alpha} - \theta_{\pi\alpha} \qquad \theta_t = \theta_{\eta t} - \theta_{\pi t} \qquad (34)$$

The two-angle space and time coordinate functions $C(\theta_{\eta\alpha},\theta_{\pi\alpha})$ and $C(\theta_{\eta t},\theta_{\pi t})$ depend on eight internal phase angles: six space coordinate phase angles $\theta_{\eta\alpha}$, $\theta_{\pi\alpha}$ and two time coordinate phase angles $\theta_{\eta t}$, $\theta_{\pi t}$ which are the ambient field dependent phase angles of eight complex number integer functions $\bar{\eta}_\alpha, \bar{\pi}_\alpha$ and $\bar{\eta}_t, \bar{\pi}_t$ whose ratios are used to describe the complex number space and time coordinates $\bar{\alpha}$ and $\bar{t}$ as in equation (29), and which are designated as

$$\bar{\eta}_\alpha = \eta_\alpha \exp(j\theta_{\eta\alpha}) \qquad \bar{\pi}_\alpha = \pi_\alpha \exp(j\theta_{\pi\alpha}) \qquad (35)$$

$$\bar{\eta}_t = \eta_t \exp(j\theta_{\eta t}) \qquad \bar{\pi}_t = \pi_t \exp(j\theta_{\pi t}) \qquad (36)$$

where $\bar{\eta}_\alpha$ and $\bar{\pi}_\alpha$ = complex number spatial counting number and complex number spatial segmentation number respectively, $\bar{\eta}_t$ and $\bar{\pi}_t$ = complex number time counting number and complex number time segmentation number respectively, and where three cases are possible for the magnitudes of the complex number space and time integer functions

$\eta_\alpha = N_1 \cos\theta_{\eta\alpha}$	$\pi_\alpha = N_2 \cos\theta_{\pi\alpha}$	high-density matter	(37)
$\eta_\alpha = N_1 \cos\theta_{\eta\alpha}$	$\pi_\alpha = N_2 \sec\theta_{\pi\alpha}$	low-density matter	(38)
$\eta_\alpha = N_1 \sec\theta_{\eta\alpha}$	$\pi_\alpha = N_2 \sec\theta_{\pi\alpha}$	low-density matter	(39)
$\eta_t = N_1 \cos\theta_{\eta t}$	$\pi_t = N_2 \cos\theta_{\pi t}$	high-density matter	(40)
$\eta_t = N_1 \cos\theta_{\eta t}$	$\pi_t = N_2 \sec\theta_{\pi t}$	low-density matter	(41)
$\eta_t = N_1 \sec\theta_{\eta t}$	$\pi_t = N_2 \sec\theta_{\pi t}$	low-density matter	(42)

where N_1 and N_2 are integers and where the low-density case is clearly degenerate.

The zero ambient field space coordinates correspond to $\theta_{\eta\alpha} = 0$, $\theta_{n\alpha} = 0$, $\theta_\alpha = 0$ and $\alpha = \alpha^o = x^o,y^o,z^o$; r^o,φ^o,z^o and ρ^o,φ^o,ψ^o; and the zero ambient field time coordinate corresponds to $\theta_{\eta t} = 0$, $\theta_{nt} = 0$, $\theta_t = 0$ and $t = t^o$. The zero ambient field inactive space and time coordinates depend on the choice of origin and in this sense are completely arbitrary. The two-angle space and time coordinate functions $C(\theta_{\eta\alpha},\theta_{n\alpha})$ and $C(\theta_{\eta t},\theta_{nt})$ depend on the energy density of an ambient field so that the space and time coordinate magnitudes α and t given by equations (32) and (33) depend on the choice of coordinate origin and on the energy density of an ambient field, and therefore they have some physical content and are not simply mathematical constructs as is the case for α^o and t^o. Equation (34) shows that the ambient field dependent internal phase angles θ_α and θ_t of the space and time coordinates are not dependent on the choice of origin of the coordinates. The space and time coordinate functions $C(\theta_{\eta\alpha},\theta_{n\alpha})$ and $C(\theta_{\eta t},\theta_{nt})$ will be denoted as $C_{\sigma\alpha}$ and $C_{\sigma t}$ respectively where $\sigma = 1,2,3$ refers to the three density cases shown in equations (37) through (42). The determination of the coordinate functions $C_{\sigma\alpha}$ and $C_{\sigma t}$ that is accomplished in this paper uses concepts that are analogous to the complex nucleon numbers and complex dimensions of space and time that have been used for nuclear fission and vacuum fission studies, and equations (35) through (42) that are used to describe active coordinates are analogous to equations (7) through (14) that are used in the theory of nuclear and vacuum fission.[10-12]

This paper derives the forms of the two-angle space and time coordinate functions $C_{\sigma\alpha}$ and $C_{\sigma t}$ of equations that appear in equations (32) and (33) and which describe the magnitudes of the complex number space and time coordinates that arise when a physical system is located in ambient fields such as such as ambient gravity and ambient pressure in the case of matter within gravitating planets, stars and galaxies; or γ rays in the case of atomic nuclei that are located in a γ ray catalyzed binary or internal nuclear or vacuum fission reactor. For waves in matter, the phase angles $\theta_{\eta\alpha}$, $\theta_{n\alpha}$, $\theta_{\eta t}$, and θ_{nt} of the space and time counting numbers and segmentation numbers are linearly related to the energy density of an ambient wave field. Equations (31) through (42) can then be used to determine the dependence of the magnitudes α and t and the internal phase angles θ_α and θ_t of the space and time coordinates on the energy density of the ambient field. Section 2 of this paper derives expressions for α, θ_α and for t, θ_t and also for $C_{\sigma\alpha}$ and $C_{\sigma t}$ by representing the complex number space and time coordinates as ratios of complex counting number and segmentation number integer functions as given in equations (31) through (42). Section 3 develops an elastic matter model to quantify the prediction of a linear relationship between the eight internal phase angles $\theta_{\eta\alpha}$, $\theta_{n\alpha}$, $\theta_{\eta t}$ and θ_{nt} and the energy density of an ambient wave field. Sections 4 and 5 consider the effects of combined ambient wave fields, such as ambient electromagnetic waves and ambient gravitational waves, on the active space and time coordinates in a region of space where the ambient fields exist. Section 6 treats the effects of an a general ambient field on the calculation of areas, volumes and particle number densities, and attempts to establish the general validity of the particle number density expressions for high and low matter densities that are given in equations (18) and (19). Section 7 presents an elastic and viscoelastic body model for calculating the internal phase angles of the active space coordinates for a nonradiative ambient field in matter. Finally, Section 8 just touches the subject of inhomogeneous matter. The material in this paper will have applications to many fields of physical science including kinematics and dynamics, wave propagation and diffusion, atomic and nuclear physics, chemical reactions and rocket science. A theory of ambient fields will benefit all of these areas of science and engineering.

2. CHARACTERIZATION OF THE PHYSICAL NATURE OF SPACE AND TIME COORDINATES. The intrinsic energy density $\mathfrak{E}$ of a radiative or nonradiative ambient field can conveniently be written as

$$\mathfrak{E} = \mathfrak{E}^{o} + \mathfrak{E}^{g} + \mathfrak{E}^{\gamma} + \ldots \tag{43}$$

where $\mathfrak{E}^{o} \approx 0$ is the ambient vacuum field energy density, $\mathfrak{E}^{g}$ is the energy density of an ambient gravity field and $\mathfrak{E}^{\gamma}$ is the energy density of an ambient electromagnetic field. This section relates the complex space and time coordinates in the presence of an ambient field of energy density $\mathfrak{E}$ to their values under zero ambient field conditions $\mathfrak{E} = 0$, and determines the forms of the space and time coordinate factors $C_{\sigma\alpha}$ and $C_{\sigma t}$ in terms of the internal phase angles of the complex counting and segmentation numbers that are used to define the space and time coordinates for high and low-density states of matter. As indicated in equations (31) through (33) the complex number active space and time coordinates are written as

$$\overline{\alpha} = \alpha \exp(j\theta_{\alpha}) = \overline{\eta}_{\alpha} L^{c}_{\alpha} / \overline{\imath}_{\alpha} \qquad \overline{t} = t \exp(j\theta_{t}) = \overline{\eta}_{t} L^{c}_{t} / \overline{\imath}_{t} \tag{44}$$

$$\alpha = \alpha^{o} C(\theta_{\eta\alpha}, \theta_{\imath\alpha}) \qquad t = t^{o} C(\theta_{\eta t}, \theta_{\imath t}) \tag{45}$$

$$\theta_{\alpha} = \theta_{\eta\alpha} - \theta_{\imath\alpha} \qquad \theta_{t} = \theta_{\eta t} - \theta_{\imath t} \tag{46}$$

where α = x,y,z; r,φ,z or ρ,φ,ψ are the magnitudes of the space coordinates in the presence of an ambient field, t = magnitude of the time coordinate within an ambient field, $\alpha^{o} = x^{o},y^{o},z^{o}$; r^{o},φ^{o},z^{o} or $\rho^{o},\varphi^{o},\psi^{o}$ are coordinates under zero ambient field conditions, t^{o} = time coordinate under zero ambient field conditions, where $\overline{\eta}_{\alpha}$ = complex number counting number and where $\overline{\imath}_{\alpha}$ = complex number segmentation number, and where L^{c}_{α} and L^{c}_{t} = real number constant parameters for the space and time coordinates respectively. The zero ambient field conditions are specified by

$$\theta_{\eta\alpha} = 0 \qquad \theta_{\imath\alpha} = 0 \qquad \theta_{\alpha} = 0 \tag{47}$$

$$\theta_{\eta t} = 0 \qquad \theta_{\imath t} = 0 \qquad \theta_{t} = 0 \tag{48}$$

$$C(\theta_{\eta\alpha}=0, \theta_{\imath\alpha}=0) = 1 \qquad C(\theta_{\eta t}=0, \theta_{\imath t}=0) = 1 \tag{49}$$

$$\overline{\alpha} = \alpha = \alpha^{o} \qquad \overline{t} = t = t^{o} \tag{50}$$

The values of α^{o} and t^{o} are arbitrary in the sense that they depend on the choice of origin of the space and time coordinates and therefore α and t also contain this arbitrariness which cannot be eliminated. This section determines the possible forms of the space and time coordinate functions $C(\theta_{\eta\alpha},\theta_{\imath\alpha})$ and $C(\theta_{\eta t},\theta_{\imath t})$. Also the values of α^{o} and t^{o} correspond to some group of conventional fields such as a conventional vacuum field, conventional gravity field and a conventional electromagnetic field.

A. Forms of the Space Coordinate Function $C(\theta_{\eta\alpha},\theta_{\pi\alpha})$.

Consider the constant real number spatial parameters L^c_α for the complex number space coordinates $\overline{\alpha}$ in an ambient field in bulk matter. The spatial parameters L^c_α are divided into $\overline{\pi}_\alpha$ parts of length $\Delta\overline{L}_\alpha$ where the complex number spatial segmentation number $\overline{\pi}_\alpha$ is represented as

$$\overline{\pi}_\alpha = \pi_\alpha \exp(j\theta_{\pi\alpha}) \tag{51}$$

so that

$$\Delta\overline{L}_\alpha = \Delta L_\alpha \exp(j\theta_{\Delta L\alpha}) = L^c_\alpha / \overline{\pi}_\alpha = L^c_\alpha / \pi_\alpha \exp(-j\theta_{\pi\alpha}) \tag{52}$$

and

$$\Delta L_\alpha = L^c_\alpha / \pi_\alpha \qquad \theta_{\Delta L\alpha} = -\theta_{\pi\alpha} \tag{53}$$

where $\alpha = x,y,z;\ r,\varphi,z$ or ρ,φ,ψ. Then the complex number space coordinate $\overline{\alpha}$ can be written as

$$\overline{\alpha} = \alpha\exp(j\theta_\alpha) = \overline{\eta}_\alpha \Delta\overline{L}_\alpha = \overline{\eta}_\alpha L^c_\alpha / \overline{\pi}_\alpha = \eta_\alpha L^c_\alpha / \pi_\alpha \exp[j(\theta_{\eta\alpha} - \theta_{\pi\alpha})] \tag{54}$$

where the complex number spatial counting number is represented as

$$\overline{\eta}_\alpha = \eta_\alpha \exp(j\theta_{\eta\alpha}) \tag{55}$$

so that

$$\alpha = \eta_\alpha L^c_\alpha / \pi_\alpha = \eta_\alpha \Delta L_\alpha \qquad \theta_\alpha = \theta_{\eta\alpha} - \theta_{\pi\alpha} \tag{56}$$

where η_α and π_α are not integers in the presence of an ambient field but are integers under zero ambient field conditions.

The internal phase angle $\theta_\alpha = \theta_r$ of the radial coordinate of a gravitating planet and of a Bohr atom in an ambient γ ray field has been found to have negative values.[9,10,12] In this paper it made plausible that $\theta_\alpha \leq 0$ for all radiative and nonradiative ambient fields in matter, so that the internal phase angle relation in equation (56) gives

$$\theta_{\pi\alpha} \geq \theta_{\eta\alpha} \qquad \theta_\alpha \leq 0 \tag{57}$$

Equation (54) shows that the complex number nature of the spatial coordinates arises solely through the complex number counting number $\overline{\eta}_\alpha$ and the complex segmentation number $\overline{\pi}_\alpha$ because L^c_α is chosen to be a real number constant. The measured values of the space coordinates are given by

$$\alpha_m = \alpha\cos\theta_\alpha \tag{58}$$

For the zero ambient field case

$$\eta^o_\alpha = N_1 \qquad \theta^o_{\eta\alpha} = 0 \qquad \pi^o_\alpha = N_2 \qquad \theta^o_{\pi\alpha} = 0 \tag{59}$$

where N_1 and N_2 are integers, and equations (56) and (58) become for this special case of zero ambient field

$$\alpha = \alpha^{o} = N_1 L^{c}_{\alpha} / N_2 = N_1 \Delta L^{o}_{\alpha} \qquad \theta^{o}_{\alpha} = 0 \qquad \alpha^{o}_{m} = \alpha^{o} \tag{60}$$

where

$$\Delta L^{o}_{\alpha} = L^{c}_{\alpha} / N_2 \qquad \theta^{o}_{\Delta L\alpha} = 0 \tag{61}$$

The following inequalities for η_α and π_α and for N_1 and N_2 are always valid in equations (56) and (60)

$$0 \le \eta_\alpha \le \pi_\alpha \qquad 0 \le N_1 \le N_2 \tag{62}$$

which means simply that $0 \le \alpha \le L^{c}_{\alpha}$ and $0 \le \alpha^{o} \le L^{c}_{\alpha}$. The integer N_2 that is associated with the segmentation number is expected to vary inversely with the density of matter, while the integer N_1 is not related to density.

Three cases of the relative sizes of the integers N_1 and N_2 are possible subject to the condition $0 \le N_1 \le N_2$:

Case 1.	N_1 small	N_2 small	high-density matter [hd]	(63)
Case 2.	N_1 small	N_2 large	low-density matter [ld]	(64)
Case 3.	N_1 large	N_2 large	low-density matter [ld]	(65)

where [hd] = high-density matter and [ld] = low-density matter. The case N_1 large and N_2 small violates the integer inequality of equation (62) and is not considered. Following the methods used for nucleon numbers and the dimensions of space and time in atomic nuclei as summarized in equations (8), (9), (13) and (14), the relations between the counting number magnitudes η_α and N_1 and between the segmentation number magnitudes π_α and N_2 for the three cases are determined according to whether N_1 and N_2 is large or small as

Case 1.	$\eta_\alpha = N_1 \cos\theta_{\eta\alpha}$	$\pi_\alpha = N_2 \cos\theta_{\pi\alpha}$	[hd]	(66)
Case 2.	$\eta_\alpha = N_1 \cos\theta_{\eta\alpha}$	$\pi_\alpha = N_2 \sec\theta_{\pi\alpha}$	[ld]	(67)
Case 3.	$\eta_\alpha = N_1 \sec\theta_{\eta\alpha}$	$\pi_\alpha = N_2 \sec\theta_{\pi\alpha}$	[ld]	(68)

Equations (63) through (68) show that the complex number segmentation number is inversely related to the density of the bulk matter in which the coordinate system is located. Combining equations (56), (60) and (66) through (68) gives the active coordinate magnitudes corresponding to an ambient field of energy density $\mathfrak{E}$ as

Case 1. $\qquad \alpha = \alpha^{o} \cos\theta_{\eta\alpha} \sec\theta_{\pi\alpha} \qquad$ [hd] $\qquad$ (69)

Case 2. $\alpha = \alpha^{o} \cos\theta_{\eta\alpha} \cos\theta_{\text{n}\alpha}$ [ld] (70)

Case 3. $\alpha = \alpha^{o} \sec\theta_{\eta\alpha} \cos\theta_{\text{n}\alpha}$ [ld] (71)

where the spatial counting number phase angles $\theta_{\eta\alpha}$ and the spatial segmentation number phase angles $\theta_{\text{n}\alpha}$ are related to the internal phase angles θ_{α} of the active coordinates by the phase angle relation in equation (56), and where $\alpha = x,y,z; r,\varphi,z$ or ρ,φ,ψ. Equations (70) and (71) indicate that the low-density system is doubly degenerate.

Equations (69) through (71) for the three possible cases can be written as

$$\alpha = \alpha^{o} C_{\sigma}(\theta_{\eta\alpha},\theta_{\text{n}\alpha}) \tag{72}$$

where $\sigma = 1,2,3$ corresponds to cases 1, 2 and 3; and where

$C_1(\theta_{\eta\alpha},\theta_{\text{n}\alpha}) = \cos\theta_{\eta\alpha} \sec\theta_{\text{n}\alpha}$ [hd] (73)

$C_2(\theta_{\eta\alpha},\theta_{\text{n}\alpha}) = \cos\theta_{\eta\alpha} \cos\theta_{\text{n}\alpha}$ [ld] (74)

$C_3(\theta_{\eta\alpha},\theta_{\text{n}\alpha}) = \sec\theta_{\eta\alpha} \cos\theta_{\text{n}\alpha}$ [ld] (75)

The measured space coordinates are written as

$$\alpha_m = \alpha \cos\theta_{\alpha} = \alpha^{o} C_{\sigma}(\theta_{\eta\alpha},\theta_{\text{n}\alpha}) \cos(\theta_{\eta\alpha} - \theta_{\text{n}\alpha}) \tag{76}$$

For convenience a shorter notation is introduced

$$C_{\sigma\alpha} = C_{\sigma}(\theta_{\eta\alpha},\theta_{\text{n}\alpha}) \tag{77}$$

where corresponding to an ambient field of energy density $\mathfrak{E}$

$C_{1\alpha} = \cos\theta_{\eta\alpha} \sec\theta_{\text{n}\alpha}$ [hd] (78)

$C_{2\alpha} = \cos\theta_{\eta\alpha} \cos\theta_{\text{n}\alpha}$ [ld] (79)

$C_{3\alpha} = \sec\theta_{\eta\alpha} \cos\theta_{\text{n}\alpha}$ [ld] (80)

which is a less cumbersome notation and refers to an ambient field of energy density $\mathfrak{E}$. The vacuum can be considered as a limiting case of low-density matter, so that equations (79) and (80) are valid for the case of of an ambient field in the vacuum, but the internal phase angles are expected to be very small. For the case when $|\theta_{\text{n}\alpha}| \geq |\theta_{\eta\alpha}|$ and the unlikely case when $|\theta_{\text{n}\alpha}| \leq |\theta_{\eta\alpha}|$ the space coordinate factors $C_{\sigma\alpha}$ that appear in equations (78) through (80) satisfy the following inequalities

$C_{1\alpha} \geq 1$ $C_{2\alpha} \leq 1$ $C_{3\alpha} \leq 1$ $|\theta_{\text{n}\alpha}| \geq |\theta_{\eta\alpha}|$ (81)

$C_{1\alpha} \leq 1$ $C_{2\alpha} \leq 1$ $C_{3\alpha} \geq 1$ $|\theta_{\text{n}\alpha}| \leq |\theta_{\eta\alpha}|$ (82)

The approximation that the angles $\theta_{\eta\alpha}$ and $\theta_{\text{n}\alpha}$ are linearly related to the energy density of an ambient radiation field in bulk matter as in equation (118) of Section 3 and in a more complicated way for nonradiative ambient fields as shown in Section 7, allows the coordinate factors $C_{\sigma\alpha}$ given by equations (78) through (80) to be calculated as functions of the energy density of an ambient field. In terms of this notation equations (72) and (76) can be written with $\sigma = 1$ for [hd] and $\sigma = 2,3$ for [ld] as

$$\alpha = \alpha^o C_{\sigma\alpha} \tag{83}$$

$$\alpha_m = \alpha \cos\theta_\alpha = \alpha^o C_{\sigma\alpha} \cos\theta_\alpha = \alpha^o C_{\sigma\alpha} \cos(\theta_{\eta\alpha} - \theta_{\text{n}\alpha}) \tag{84}$$

which are of use in forthcoming calculations of area, volume and particle number density that are performed in Section 6. Equations (78) through (80), (81) and (83) show that in the presence of an ambient field, $\alpha > \alpha^o$ for high-density matter and $\alpha < \alpha^o$ for low-density matter. The zero ambient field (conventional) values α^o of the coordinates refer to the complete absence of all ambient fields including ambient gravity, and for this case $\theta_{\eta\alpha} = 0$, $\theta_{\text{n}\alpha} = 0$ and $\theta_\alpha = 0$ and therefore

$$C_{\sigma\alpha} = 1 \qquad \alpha = \alpha^o \qquad \alpha_m = \alpha^o \tag{85}$$

but α^o in general corresponds to a multitude of conventional fields such as a conventional vacuum field, a conventional gravity field and a conventional electromagnetic field. For quantized systems such as an atom, equation (83) refers to the Bohr radius ($n = 1$). The higher energy states $n > 1$ are obtained as a factor times the basic relation in equation (83). In fact, equation (83) applies only directly to the fundamental length of any quantum system while other lengths are obtained by additional multiplicative terms.

B. Forms of the Time Coordinate function $C(\theta_{\eta t},\theta_{\text{n}t})$.

The description of the time coordinate function is analogous to that of the space coordinate function. Consider a real number constant L_t^c that is used to measure intervals of the complex number time coordinate $\bar{t}$. In the presence of an ambient field in bulk matter the real number constant L_t^c is divided into $\bar{\text{n}}_t$ parts of length $\Delta\bar{L}_t$ where the complex number time segmentation number is represented as

$$\bar{\text{n}}_t = \text{n}_t \exp(j\theta_{\text{n}t}) \tag{86}$$

and where

$$\Delta\bar{L}_t = \Delta L_t \exp(j\theta_{\Delta Lt}) = L_t^c / \bar{\text{n}}_t = L_t^c / \text{n}_t \exp(-j\theta_{\text{n}t}) \tag{87}$$

and therefore

$$\Delta L_t = L_t^c / \text{n}_t \qquad \theta_{\Delta Lt} = -\theta_{\text{n}t} \tag{88}$$

The complex number time coordinate is then written as

$$\bar{t} = t\exp(j\theta_t) = \bar{\eta}_t \Delta\bar{L}_t = \bar{\eta}_t L_t^c / \bar{\text{n}}_t = \eta_t L_t^c / \text{n}_t \exp[j(\theta_{\eta t} - \theta_{\text{n}t})] \tag{89}$$

where the complex number time counting number is written as

$$\overline{\eta}_t = \eta_t \exp(j\theta_{\eta t}) \tag{90}$$

so that

$$t = \eta_t L_t^c / \pi_t = \eta_t \Delta L_t \qquad \theta_t = \theta_{\eta t} - \theta_{\pi t} \tag{91}$$

where η_t and π_t are not integers when an ambient field is present in matter, and the complex number nature of the time coordinate is $\overline{t}$ due only to the complex number time counting number $\overline{\eta}_t$ and the complex number segmentation number $\overline{\pi}_t$ because the constant L_t^c is a real number in equation (89). In this paper it is suggested that $\theta_t \leq 0$ for all physical systems, and for the same reasons that $\theta_\alpha \leq 0$ was taken in equation (57), so that it follows from the phase angle relation in equation (91) that

$$\theta_{\pi t} \geq \theta_{\eta t} \qquad \theta_t \leq 0 \tag{92}$$

The measured value of the time coordinate is given by

$$t_m = t \cos \theta_t \tag{93}$$

For the zero ambient field case the following equations are valid

$$\eta_t^o = N_1 \qquad \theta_{\eta t}^o = 0 \qquad \pi_t^o = N_2 \qquad \theta_{\pi t}^o = 0 \tag{94}$$

where N_1 and N_2 are integers, and equations (91) and (93) for the zero ambient field case become

$$t = t^o = N_1 L_t^c / N_2 = N_1 \Delta L_t^o \qquad \theta_t^o = 0 \qquad t_m^o = t^o \tag{95}$$

where equation (88) becomes

$$\Delta L_t^o = L_t^c / N_2 \qquad \theta_{\Delta Lt}^o = 0 \tag{96}$$

The parameters η_t, π_t and N_1, N_2 that appear in equations (91) and (95) satisfy the following inequalities

$$0 \leq \eta_t \leq \pi_t \qquad 0 \leq N_1 \leq N_2 \tag{97}$$

or equivalently $0 \leq t \leq L_t^c$ and $0 \leq t^o \leq L_t^c$. And where, as before, the integer N_2 is expected to vary inversely with the density of matter, and the integer N_1 is not expected to depend on the density.

As in the case of the space coordinates, the integers N_1 and N_2 determine three cases according to their relative sizes as shown in equations (63) through (65) for high-density [hd] matter and low-density [ld] matter. Therefore in analogy to the situation of the nucleon numbers and the dimensions of space and time within an atomic nucleus as described by equations (8), (9), (13) and (14), the relations between η_t and N_1 and between π_t and N_2 for these three cases are given by

Case 1.	$\eta_t = N_1 \cos\theta_{\eta t}$	$\mathfrak{n}_t = N_2 \cos\theta_{\mathfrak{n}t}$	[hd]	(98)
Case 2.	$\eta_t = N_1 \cos\theta_{\eta t}$	$\mathfrak{n}_t = N_2 \sec\theta_{\mathfrak{n}t}$	[ld]	(99)
Case 3.	$\eta_t = N_1 \sec\theta_{\eta t}$	$\mathfrak{n}_t = N_2 \sec\theta_{\mathfrak{n}t}$	[ld]	(100)

Then combining equations (91), (95) and (98) through (100) gives the active time coordinate magnitude in an ambient field of energy density $\mathfrak{E}$ as

Case 1.	$t = t^o \cos\theta_{\eta t} \sec\theta_{\mathfrak{n}t}$	[hd]	(101)
Case 2.	$t = t^o \cos\theta_{\eta t} \cos\theta_{\mathfrak{n}t}$	[ld]	(102)
Case 3.	$t = t^o \sec\theta_{\eta t} \cos\theta_{\mathfrak{n}t}$	[ld]	(103)

where the time counting number internal phase angle $\theta_{\eta t}$ and the time segmentation number internal phase angle $\theta_{\mathfrak{n}t}$ are related to the internal phase angle θ_t of the active time coordinate by equation (91). Equations (102) and (103) show that the low-density state is doubly degenerate.

Equations (101) through (103) for the three possible representations of the time coordinate can be written as

$$t = t^o C_\sigma(\theta_{\eta t},\theta_{\mathfrak{n}t}) \tag{104}$$

where $\sigma = 1,2,3$ corresponds to cases 1, 2 and 3 of equation (98) through (100), so that

$C_1(\theta_{\eta t},\theta_{\mathfrak{n}t}) = \cos\theta_{\eta t} \sec\theta_{\mathfrak{n}t}$	[hd]	(105)
$C_2(\theta_{\eta t},\theta_{\mathfrak{n}t}) = \cos\theta_{\eta t} \cos\theta_{\mathfrak{n}t}$	[ld]	(106)
$C_3(\theta_{\eta t},\theta_{\mathfrak{n}t}) = \sec\theta_{\eta t} \cos\theta_{\mathfrak{n}t}$	[ld]	(107)

The measured time coordinate is given by equations (91), (93) and (104) as

$$t_m = t \cos\theta_t = t^o C_\sigma(\theta_{\eta t},\theta_{\mathfrak{n}t}) \cos(\theta_{\eta t} - \theta_{\mathfrak{n}t}) \tag{108}$$

where $\sigma = 1,2,3$. The following shorthand notation is introduced

$$C_{\sigma t} = C_\sigma(\theta_{\eta t},\theta_{\mathfrak{n}t}) \tag{109}$$

where corresponding to an ambient field of energy density $\mathfrak{E}$

$C_{1t} = \cos\theta_{\eta t} \sec\theta_{\mathfrak{n}t}$	[hd]	(110)
$C_{2t} = \cos\theta_{\eta t} \cos\theta_{\mathfrak{n}t}$	[ld]	(111)
$C_{3t} = \sec\theta_{\eta t} \cos\theta_{\mathfrak{n}t}$	[ld]	(112)

Because the vacuum can be considered as a limiting case of low-density matter, equations (111) and (112) are also valid for an ambient field in the vacuum, but the internal phase angles of the counting and segmentation numbers are expected to be very small. For the case when $|\theta_{\imath t}| \geq |\theta_{\eta t}|$ and the unlikely case when $|\theta_{\imath t}| \leq |\theta_{\eta t}|$ the time coordinate factors $C_{\sigma t}$ that appear in equations (110) through (112) satisfy the inequalities

$$C_{1t} \geq 1 \qquad C_{2t} \leq 1 \qquad C_{3t} \leq 1 \qquad |\theta_{\imath t}| \geq |\theta_{\eta t}| \tag{113}$$

$$C_{1t} \leq 1 \qquad C_{2t} \leq 1 \qquad C_{3t} \geq 1 \qquad |\theta_{\imath t}| \leq |\theta_{\eta t}| \tag{114}$$

The approximation that the phase angles $\theta_{\eta t}$ and $\theta_{\imath t}$ are linearly related to the energy density of an ambient wavefield as shown in equation (118) of Section 3 or in a more complicated fashion for an ambient nonradiative field as shown in Section 7, combined with equations (110) through (112) allows the calculation of the time coordinate factors $C_{\sigma t}$ in terms of the energy density. Equations (104) and (108) can be written as

$$t = t^{o}C_{\sigma t} \tag{115}$$

$$t_m = t \cos \theta_t = t^{o}C_{\sigma t} \cos \theta_t = t^{o}C_{\sigma t} \cos(\theta_{\eta t} - \theta_{\imath t}) \tag{116}$$

Equations (110) through (112), (113) and (115) show that $t > t^{o}$ for an ambient field present in high-density matter, while $t < t^{o}$ for an ambient field in low-density matter. The zero ambient field (conventional) value of the time t^{o} corresponds to the absence of all ambient fields in matter including an ambient vacuum field, an ambient gravity field and an ambient electromagnetic field, and all internal phase angles are zero so that $\theta_{\eta t} = 0$, $\theta_{\imath t} = 0$ and $\theta_t = 0$, and therefore

$$C_{\sigma t} = 1 \qquad t = t^{o} \qquad t_m = t^{o} \tag{117}$$

as must be the case for agreement with the conventional case of inactive coordinates. But in general t^{o} corresponds to a group of conventional fields such as a conventional vacuum field, a conventional gravity field and a conventional electromagnetic field.

An important thing to remember is seen from equations (63) through (65) combined with equations (66) through (68) for space coordinates and equations (98) through (100) for the time coordinate, namely that the density of matter in which the coordinates are defined is related to the segmentation integer N_2 in an inverse manner, that is, a small value of N_2 corresponds to high-density while a large value of N_2 corresponds to low-density.

3. APPROXIMATELY LINEAR RELATIONSHIP OF THE INTERNAL PHASE ANGLES OF THE SPACE AND TIME COORDINATES TO THE ENERGY DENSITY OF AN AMBIENT RADIATION FIELD IN MATTER.

It is demonstrated that the internal phase angles of the space and time coordinates within matter that is engulfed in an ambient radiation field of energy density $\mathfrak{E}$, as represented by equation (43), are approximately linearly related to the energy density of the ambient radiation field at least for low energy density of the ambient radiation field.

The phase angles of the complex number counting number and the complex number segmentation number that define active coordinates are found, to first order, to be linearly related to the energy density of the ambient radiation field, and therefore approximately

$$\theta_{\eta\alpha} = f^{o}_{\eta\alpha}\mathfrak{E} \qquad \theta_{\pi\alpha} = f^{o}_{\pi\alpha}\mathfrak{E} \qquad \theta_{\eta t} = f^{o}_{\eta t}\mathfrak{E} \qquad \theta_{\pi t} = f^{o}_{\pi t}\mathfrak{E} \tag{118}$$

and the internal phase angles of the space and time coordinates are then given by equations (56) and (91) approximately as

$$\theta_{\alpha} = \theta_{\eta\alpha} - \theta_{\pi\alpha} = f^{o}_{\alpha}\mathfrak{E} \tag{119}$$

$$\theta_{t} = \theta_{\eta t} - \theta_{\pi t} = f^{o}_{t}\mathfrak{E} \tag{120}$$

where $\theta_{\eta\alpha}$ and $\theta_{\pi\alpha}$ = internal phase angles of the complex number spatial counting numbers $\overline{\eta}_{\alpha}$ and the complex number spatial segmentation numbers $\overline{\pi}_{\alpha}$ as defined by equations (55) and (51) respectively, where $\theta_{\eta t}$ and $\theta_{\pi t}$ = internal phase angles of the complex number time counting number $\overline{\eta}_{t}$ and the complex number time segmentation number $\overline{\pi}_{t}$ as given by equations (90) and (86) respectively, and where $f^{o}_{\eta\alpha}, f^{o}_{\pi\alpha}, f^{o}_{\alpha}$ and $f^{o}_{\eta t}, f^{o}_{\pi t}$ and f^{o}_{t} are frequency dependent material parameters which correspond to zero ambient fields and which may be functions of α^{o} and t^{o}, and which satisfy the relations

$$f^{o}_{\alpha} = f^{o}_{\eta\alpha} - f^{o}_{\pi\alpha} \qquad\qquad f^{o}_{t} = f^{o}_{\eta t} - f^{o}_{\pi t} \tag{121}$$

The angles θ_{α} and θ_{t} are found to have negative or zero values for ambient radiation in matter

$$\theta_{\alpha} \le 0 \qquad\qquad \theta_{t} \le 0 \tag{122}$$

so that equations (119) through (121) give for this case

$$\theta_{\eta\alpha} \le \theta_{\pi\alpha} \qquad \theta_{\eta t} \le \theta_{\pi t} \qquad f^{o}_{\eta\alpha} < f^{o}_{\pi\alpha} \qquad f^{o}_{\eta t} < f^{o}_{\pi t} \tag{123}$$

and so also for ambient radiation in matter

$$f^{o}_{\alpha} < 0 \qquad\qquad f^{o}_{t} < 0 \tag{124}$$

The interesting thing about equations (119) and (120) is that it allows $\theta_{\alpha} \le 0$ and $\theta_{t} \le 0$ which agrees with previous calculations for gravitating bodies such as stars and planets and for Bohr atoms in an ambient γ ray field.[9,10,12] The approximate linear relationship between the internal phase angles of the space and time coordinates and the energy density of the ambient radiation field in matter as given in equations (118) through (120) is somewhat similar to the behavior of the Kerr electro-optic effect of double refraction of light in gases, liquids and solids in the presence of a strong electromagnetic wave field in the sense that the difference of the phase angle between the two beams of light in a Kerr cell is found to be proportional to the square of the electric field strength or in other words proportional to the energy density of the electromagnetic wave field.[15,16]

The following is a simple linear elasticity model for the calculation of the frequency dependent first order material parameters $f^{o}_{\eta\alpha}, f^{o}_{\pi\alpha}, f^{o}_{\alpha}$. The complex number space coordinates for matter in an ambient radiation field are written as

$$\overline{\alpha} = \alpha \exp(j\theta_{\alpha}) = \alpha \exp(jk^{o}_{\alpha}\xi_{\alpha}) \qquad \theta_{\alpha} = k^{o}_{\alpha}\xi_{\alpha} \tag{125}$$

where $\alpha = x,y,z$ and where $k^{o}_{\alpha} = 2\pi / \lambda^{o}_{\alpha}$ = wave number along the α-axis with λ^{o}_{α} = wavelength projection along the α-axis. The displacement ξ_{α} can be written as

$$\xi_{\alpha} = F_r / \mathfrak{k}^{o}_{\alpha} = P_r A^{o} / \mathfrak{k}^{o}_{\alpha} \tag{126}$$

where F_r = ambient radiation force, $\mathfrak{k}^{o}_{\alpha}$ = zero ambient field spring constant, P_r = ambient radiation pressure and A^{o} = zero ambient field area of a surface perpendicular to the energy flux. For ambient radiation along the α-axis the ambient radiation pressure is given by the first order expression

$$P_r = \mathfrak{E}(1 + n^{o} / W^{o}_{\alpha}\, dW^{o}_{\alpha} / dn^{o}) \tag{127}$$

where $\mathfrak{E}$ = energy density of the ambient radiation field, n^{o} = particle number density of the material medium, $W^{o}_{\alpha} = \omega / k^{o}_{\alpha}$ = zero ambient field value of the phase velocity of the waves traveling along the α-axis and where ω = angular frequency of the ambient wave field which is a property of the field and not of the material. Then it follows from equations (118), (119) and (125) through (127) that approximately

$$\theta_{\alpha} = \theta_{\eta\alpha} - \theta_{\pi\alpha} = f^{o}_{\alpha}\mathfrak{E} \qquad \theta_{\eta\alpha} = f^{o}_{\eta\alpha}\mathfrak{E} \qquad \theta_{\pi\alpha} = f^{o}_{\pi\alpha}\mathfrak{E} \tag{128}$$

where

$$f^{o}_{\alpha} = f^{o}_{\eta\alpha} - f^{o}_{\pi\alpha} = (k^{o}_{\alpha} / \mathfrak{k}^{o}_{\alpha}) A^{o}(1 + n^{o} / W^{o}_{\alpha}\, dW^{o}_{\alpha} / dn^{o}) \tag{129}$$

with

$$f^{o}_{\eta\alpha} = (k^{o}_{\alpha} / \mathfrak{k}^{o}_{\eta\alpha}) A^{o}(1 + n^{o} / W^{o}_{\alpha}\, dW^{o}_{\alpha} / dn^{o}) \tag{130}$$

$$f^{o}_{\pi\alpha} = (k^{o}_{\alpha} / \mathfrak{k}^{o}_{\pi\alpha}) A^{o}(1 + n^{o} / W^{o}_{\alpha}\, dW^{o}_{\alpha} / dn^{o}) \tag{131}$$

$$1/\mathfrak{k}^{o}_{\alpha} = 1/\mathfrak{k}^{o}_{\eta\alpha} - 1/\mathfrak{k}^{o}_{\pi\alpha} \qquad \mathfrak{k}^{o}_{\alpha} < 0 \qquad \mathfrak{k}^{o}_{\pi\alpha} < \mathfrak{k}^{o}_{\eta\alpha} \tag{132}$$

and where

$$\mathfrak{k}^{o}_{\eta\alpha} = E_{o} A^{o} / L^{o}_{\alpha} \qquad \mathfrak{k}^{o}_{\pi\alpha} = G_{o} A^{o} / L^{o}_{\alpha} \tag{133}$$

so that

$$1/\mathfrak{k}^{o}_{\alpha} = (L^{o}_{\alpha} / A^{o})(E^{-1}_{o} - G^{-1}_{o}) \tag{134}$$

where for zero ambient fields E_o = Young's modulus, G_o = shear modulus and L^{o}_{α} = characteristic length along the α-axis. Then for one-dimensional waves

$$f^o_{\eta\alpha} = k^o_\alpha L^o_\alpha E_o^{-1}(1 + n^o / W^o_\alpha \, dW^o_\alpha / dn^o) \tag{135}$$

$$f^o_{\pi\alpha} = k^o_\alpha L^o_\alpha G_o^{-1}(1 + n^o / W^o_\alpha \, dW^o_\alpha / dn^o) \tag{136}$$

and therefore

$$f^o_\alpha = k^o_\alpha L^o_\alpha (E_o^{-1} - G_o^{-1})(1 + n^o / W^o_\alpha \, dW^o_\alpha / dn^o) \qquad f^o_\alpha < 0 \tag{137}$$

$$= - k^o_\alpha L^o_\alpha (E_o - G_o) E_o^{-1} G_o^{-1}(1 + n^o / W^o_\alpha \, dW^o_\alpha / dn^o)$$

so that $f^o_\alpha < 0$ because $E_o > G_o$ for an elastic medium. For the case of diffuse ambient radiation in matter as described by spherical polar coordinates, the substitutions $\alpha \to \rho$, $L^o_\alpha \to L^o_\rho/\sqrt{3}$, $W^o_\alpha \to W^o_\rho$, $E_o \to K_o$ and $1 \to 1/3$ in the radiation factor are made in equations (125) through (137) with the following approximate results for three-dimensional ambient waves in elastic materials

$$\theta_\rho = \theta_{\eta\rho} - \theta_{\pi\rho} = f^o_\rho \mathfrak{E} \qquad \theta_{\eta\rho} = f^o_{\eta\rho} \mathfrak{E} \qquad \theta_{\pi\rho} = f^o_{\pi\rho} \mathfrak{E} \tag{138}$$

where

$$f^o_\rho = f^o_{\eta\rho} - f^o_{\pi\rho} = (k^o_\rho / \mathfrak{k}^o_\rho) A^o (1/3 + n^o / W^o_\rho \, dW^o_\rho / dn^o) \tag{139}$$

with

$$f^o_{\eta\rho} = (k^o_\rho / \mathfrak{k}^o_{\eta\rho}) A^o (1/3 + n^o / W^o_\rho \, dW^o_\rho / dn^o) \tag{140}$$

$$f^o_{\pi\rho} = (k^o_\rho / \mathfrak{k}^o_{\pi\rho}) A^o (1/3 + n^o / W^o_\rho \, dW^o_\rho / dn^o) \tag{141}$$

where $k^o_\rho = 2\pi/\lambda^o_\rho = \omega / W^o_\rho$ with $\lambda^o_\rho = \sqrt{3}\,\lambda^o_\alpha$ = zero ambient field value of the wavelength of ambient radiation along a radial path, ω = angular frequency, and where

$$1/\mathfrak{k}^o_\rho = 1/\mathfrak{k}^o_{\eta\rho} - 1/\mathfrak{k}^o_{\pi\rho} \qquad \mathfrak{k}^o_\rho < 0 \qquad \mathfrak{k}^o_{\pi\rho} < \mathfrak{k}^o_{\eta\rho} \tag{142}$$

where

$$\mathfrak{k}^o_{\eta\rho} = K_o A^o / L^o_\rho \qquad \mathfrak{k}^o_{\pi\rho} = G_o A^o / L^o_\rho \tag{143}$$

so that

$$1/\mathfrak{k}^o_\rho = (L^o_\rho / A^o)(K_o^{-1} - G_o^{-1}) \tag{144}$$

Then

$$f^o_{\eta\rho} = k^o_\rho L^o_\rho K_o^{-1}(1/3 + n^o / W^o_\rho \, dW^o_\rho / dn^o) \tag{145}$$

$$f^o_{\pi\rho} = k^o_\rho L^o_\rho G_o^{-1}(1/3 + n^o / W^o_\rho \, dW^o_\rho / dn^o) \tag{146}$$

and therefore

$$f_\rho^o = k_\rho^o L_\rho^o (K_o^{-1} - G_o^{-1})(1/3 + n^o / W_\rho^o \, dW_\rho^o / dn^o) \qquad f_\rho^o < 0 \qquad (147)$$

$$= - k_\rho^o L_\rho^o (K_o - G_o) K_o^{-1} G_o^{-1} (1/3 + n^o / W_\rho^o \, dW_\rho^o / dn^o)$$

where ρ = radial coordinate in spherical polar coordinates, $L_\rho^o = \sqrt{3}\, L_\alpha^o$ = characteristic radial distance in spherical polar coordinates, $W_\rho^o = \omega / k_\rho^o$ = phase velocity of the ambient waves that are traveling along the ρ-axis, ω = angular frequency of the ambient wave field and K_o and G_o = bulk modulus and shear modulus of the elastic medium. It is clear from equation (147) that $f_\rho^o < 0$. Equations (126) through (147) are valid in linear elastic materials whose properties are the conventional values corresponding to the case of zero ambient fields. Finally, the fact that in the presence of an ambient radiation field the material properties of matter are dependent on the energy density of the ambient radiation field shows that the linear relations which appear in equations (128) and (138) are low energy density approximations of the following infinite series

$$\theta_\alpha = f_\alpha^o \mathfrak{E} + f_{\alpha 2}^o \mathfrak{E}^2 + f_{\alpha 3}^o \mathfrak{E}^3 + \ldots \qquad (148)$$

$$\theta_\rho = f_\rho^o \mathfrak{E} + f_{\rho 2}^o \mathfrak{E}^2 + f_{\rho 3}^o \mathfrak{E}^3 + \ldots \qquad (149)$$

which are generally valid. This paper assumes that the low energy approximations in equations (128) and (138) are valid when ambient radiation is treated.

A simple elasticity model for the calculation of the functions $f_{\eta t}^o, f_{\pi t}^o$ and f_t^o is now given. The complex number time coordinate for matter in an ambient radiation field is written as

$$\bar{t} = t \exp(j\theta_t) = t \exp(j\omega\xi_t) \qquad \theta_t = \omega\xi_t \qquad (150)$$

The internal phase angle θ_t of the time coordinate can be rewritten as

$$\theta_t = \omega\xi_\rho / W_\rho^o = (\omega / W_\rho^o)(\theta_\rho / k_\rho^o) = \theta_\rho \qquad (151)$$

where from the analog of equation (125), $\theta_\rho = k_\rho^o \xi_\rho$. Then from equation (120) it follows that

$$\theta_{\eta t} = \theta_{\eta\rho} \qquad \theta_{\pi t} = \theta_{\pi\rho} \qquad (152)$$

and after combining equations (118), (120), (138), (151) and (152)

$$\theta_t = f_t^o \mathfrak{E} \qquad \theta_{\eta t} = f_{\eta t}^o \mathfrak{E} \qquad \theta_{\pi t} = f_{\pi t}^o \mathfrak{E} \qquad (153)$$

where

$$f_t^o = f_\rho^o \qquad f_{\eta t}^o = f_{\eta\rho}^o \qquad f_{\pi t}^o = f_{\pi\rho}^o \qquad (154)$$

where $f_\rho^o, f_{\eta\rho}^o$ and $f_{\pi\rho}^o$ are given by equations (145) through (147).

The energy density of an ambient wave field can be written in the form of equation (43) as

$$\mathfrak{E} = \mathfrak{E}^o + \mathfrak{E}^g + \mathfrak{E}^\gamma = \mathfrak{E}^o + \mathfrak{E}^e \tag{155}$$

where $\mathfrak{E}^o \approx 0$ is the ambient vacuum radiation energy density, $\mathfrak{E}^g$ = energy density of an ambient gravitational wave field, $\mathfrak{E}^\gamma$ = energy density of an ambient electromagnetic wave field such as γ rays, and where $\mathfrak{E}^e$ = external ambient wave field energy density given by

$$\mathfrak{E}^e = \mathfrak{E}^g + \mathfrak{E}^\gamma \tag{156}$$

If it is assumed that the ambient vacuum radiation energy density is comparatively small so that

$$\mathfrak{E}^o << \mathfrak{E}^g \qquad\qquad \mathfrak{E}^o << \mathfrak{E}^\gamma \tag{157}$$

so that approximately

$$\mathfrak{E} = \mathfrak{E}^e = \mathfrak{E}^g + \mathfrak{E}^\gamma \tag{158}$$

Combining the linear relations in equations (118) through (120) with equation (155) shows that the addition rule of the energy density requires that, for ambient radiation, the internal phase angles of the space and time counting and segmentation numbers and of the space and time coordinates add linearly in the manner

$$\theta_{\eta\alpha} = \theta^o_{\eta\alpha} + \theta^g_{\eta\alpha} + \theta^\gamma_{\eta\alpha} \qquad\qquad \theta_{\eta t} = \theta^o_{\eta t} + \theta^g_{\eta t} + \theta^\gamma_{\eta t} \tag{159}$$

$$\theta_{\pi\alpha} = \theta^o_{\pi\alpha} + \theta^g_{\pi\alpha} + \theta^\gamma_{\pi\alpha} \qquad\qquad \theta_{\pi t} = \theta^o_{\pi t} + \theta^g_{\pi t} + \theta^\gamma_{\pi t} \tag{160}$$

$$\theta_{\alpha} = \theta^o_{\alpha} + \theta^g_{\alpha} + \theta^\gamma_{\alpha} \qquad\qquad \theta_{t} = \theta^o_{t} + \theta^g_{t} + \theta^\gamma_{t} \tag{161}$$

where

$$\theta^o_\alpha = \theta^o_{\eta\alpha} - \theta^o_{\pi\alpha} \qquad\qquad \theta^o_t = \theta^o_{\eta t} - \theta^o_{\pi t} \tag{162}$$

$$\theta^g_\alpha = \theta^g_{\eta\alpha} - \theta^g_{\pi\alpha} \qquad\qquad \theta^g_t = \theta^g_{\eta t} - \theta^g_{\pi t} \tag{163}$$

$$\theta^\gamma_\alpha = \theta^\gamma_{\eta\alpha} - \theta^\gamma_{\pi\alpha} \qquad\qquad \theta^\gamma_t = \theta^\gamma_{\eta t} - \theta^\gamma_{\pi t} \tag{164}$$

where

$$\theta^o_{\eta\alpha} = f^o_{\eta\alpha}\mathfrak{E}^o \qquad\qquad \theta^g_{\eta\alpha} = f^o_{\eta\alpha}\mathfrak{E}^g \qquad\qquad \theta^\gamma_{\eta\alpha} = f^o_{\eta\alpha}\mathfrak{E}^\gamma \tag{165}$$

$$\theta^o_{\pi\alpha} = f^o_{\pi\alpha}\mathfrak{E}^o \qquad\qquad \theta^g_{\pi\alpha} = f^o_{\pi\alpha}\mathfrak{E}^g \qquad\qquad \theta^\gamma_{\pi\alpha} = f^o_{\pi\alpha}\mathfrak{E}^\gamma \tag{166}$$

$$\theta^o_{\alpha} = f^o_{\alpha}\mathfrak{E}^o \qquad\qquad \theta^g_{\alpha} = f^o_{\alpha}\mathfrak{E}^g \qquad\qquad \theta^\gamma_{\alpha} = f^o_{\alpha}\mathfrak{E}^\gamma \tag{167}$$

and where

$$\theta^o_{\eta t} = f^o_{\eta t}\mathfrak{E}^o \qquad\qquad \theta^g_{\eta t} = f^o_{\eta t}\mathfrak{E}^g \qquad\qquad \theta^\gamma_{\eta t} = f^o_{\eta t}\mathfrak{E}^\gamma \tag{168}$$

$$\theta^{o}_{\pi t} = f^{o}_{\pi t}\mathfrak{E}^{o} \qquad \theta^{g}_{\pi t} = f^{o}_{\pi t}\mathfrak{E}^{g} \qquad \theta^{\gamma}_{\pi t} = f^{o}_{\pi t}\mathfrak{E}^{\gamma} \tag{169}$$

$$\theta^{o}_{t} = f^{o}_{t}\mathfrak{E}^{o} \qquad \theta^{g}_{t} = f^{o}_{t}\mathfrak{E}^{g} \qquad \theta^{\gamma}_{t} = f^{o}_{t}\mathfrak{E}^{\gamma} \tag{170}$$

where f^{o}_{α} and f^{o}_{t} are given by equation (121). If the ambient vacuum radiation energy density $\mathfrak{E}^{o}$ is relatively small compared to the ambient gravitational wave energy density $\mathfrak{E}^{g}$ and the ambient electromagnetic wave energy density $\mathfrak{E}^{\gamma}$ as in equation (157) it follows from equations (165) through (170) that

$$|\theta^{o}_{\eta\alpha}| << |\theta^{g}_{\eta\alpha}| \quad |\theta^{o}_{\eta\alpha}| << |\theta^{\gamma}_{\eta\alpha}| \quad |\theta^{o}_{\pi\alpha}| << |\theta^{g}_{\pi\alpha}| \quad |\theta^{o}_{\pi\alpha}| << |\theta^{\gamma}_{\pi\alpha}| \tag{171}$$

$$|\theta^{o}_{\eta t}| << |\theta^{g}_{\eta t}| \quad |\theta^{o}_{\eta t}| << |\theta^{\gamma}_{\eta t}| \quad |\theta^{o}_{\pi t}| << |\theta^{g}_{\pi t}| \quad |\theta^{o}_{\pi t}| << |\theta^{\gamma}_{\pi t}| \tag{172}$$

$$|\theta^{o}_{\alpha}| << |\theta^{g}_{\alpha}| \quad |\theta^{o}_{\alpha}| << |\theta^{\gamma}_{\alpha}| \quad |\theta^{o}_{t}| << |\theta^{g}_{t}| \quad |\theta^{o}_{t}| << |\theta^{\gamma}_{t}| \tag{173}$$

so that equations (159) through (161) can be written approximately as

$$\theta_{\eta\alpha} = \theta^{g}_{\eta\alpha} + \theta^{\gamma}_{\eta\alpha} \qquad \theta_{\eta t} = \theta^{g}_{\eta t} + \theta^{\gamma}_{\eta t} \tag{174}$$

$$\theta_{\pi\alpha} = \theta^{g}_{\pi\alpha} + \theta^{\gamma}_{\pi\alpha} \qquad \theta_{\pi t} = \theta^{g}_{\pi t} + \theta^{\gamma}_{\pi t} \tag{175}$$

$$\theta_{\alpha} = \theta^{g}_{\alpha} + \theta^{\gamma}_{\alpha} \qquad \theta_{t} = \theta^{g}_{t} + \theta^{\gamma}_{t} \tag{176}$$

and the internal phase angles associated with the space and time coordinates in the ambient vacuum wave fields can be neglected with respect to the values of these phase angles when ambient gravitational wave fields and ambient electromagnetic wave fields are present.

4. THE THEORY OF SPACE COORDINATES IN MATTER IN THE PRESENCE OF AMBIENT GRAVITATIONAL WAVES AND AMBIENT ELECTROMAGNETIC WAVES. This section describes the case when the ambient field is composed mainly of two distinct ambient radiation fields such as ambient gravitational waves and ambient electromagnetic waves of energy density $\mathfrak{E}^{g} + \mathfrak{E}^{\gamma}$, and determines representations of the magnitudes, phase angles and measured values of the space coordinates for the combined ambient field in terms of the space coordinate magnitudes and internal phase angles that correspond to the component ambient gravitational and ambient electromagnetic wave fields of energy density $\mathfrak{E}^{g}$ and $\mathfrak{E}^{\gamma}$ respectively . In the presence of an ambient radiation field, equation (83) expresses the amplitude α of a complex number space coordinate in terms of the zero ambient field space coordinate amplitude α^{o} and in terms of the internal phase angles $\theta_{\eta\alpha}$ and $\theta_{\pi\alpha}$ that appear in the space coordinate factor $C_{\sigma\alpha}$. Because ambient gravity waves are always present in a laboratory on the earth's surface, the coordinate amplitude α^{o} cannot be measured directly, so that it would be more convenient if the coordinate amplitude α in the presence of a combined ambient gravitational wave and ambient electromagnetic wave field could be expressed in terms of the coordinate amplitude α^{g} for the case when only ambient gravity waves are present or in terms of the coordinate magnitude α^{γ} when an ambient electromagnetic wave field, such as γ rays, is the dominant field, and in terms of the internal phase angles $\theta^{g}_{\eta\alpha}$ and $\theta^{g}_{\pi\alpha}$ of the counting and segmentation numbers when only ambient gravity waves are present and $\theta^{\gamma}_{\eta\alpha}$ and $\theta^{\gamma}_{\pi\alpha}$ when the ambient

electromagnetic wave field is dominant. In this section it is assumed that the zero point ambient vacuum energy can be neglected. Note that each of the coordinate amplitudes α^o, α^g and α^γ correspond to the presence of a multitude of conventional radiation fields including a conventional vacuum radiation field, a conventional gravity radiation field and a conventional electromagnetic radiation field. The space coordinates α^o are just the conventional space coordinates.

Because of the linear relationship between the energy density of an ambient wave field and the internal phase angles of the space coordinates θ_α, counting numbers $\theta_{\eta\alpha}$ and segmentation numbers $\theta_{\iota\alpha}$ for matter, it follows that when more than one ambient field acts these phase angles are the sum of the corresponding phase angles associated with each ambient field. Thus for the case when both an ambient gravitational wave field and an ambient electromagnetic wave field are present it follows from equations (119) and (159) through (161) that after neglecting the zero point phase angles

$$\theta_{\eta\alpha} = \theta^g_{\eta\alpha} + \theta^\gamma_{\eta\alpha} \qquad \theta_{\iota\alpha} = \theta^g_{\iota\alpha} + \theta^\gamma_{\iota\alpha} \tag{177}$$

and

$$\theta_\alpha = \theta_{\eta\alpha} - \theta_{\iota\alpha} = \theta^g_\alpha + \theta^\gamma_\alpha = \theta^g_{\eta\alpha} + \theta^\gamma_{\eta\alpha} - \theta^g_{\iota\alpha} - \theta^\gamma_{\iota\alpha} \tag{178}$$

where

$$\theta^g_\alpha = \theta^g_{\eta\alpha} - \theta^g_{\iota t} \qquad \theta^\gamma_\alpha = \theta^\gamma_{\eta\alpha} - \theta^\gamma_{\iota\alpha} \tag{179}$$

where θ^g_α = internal phase angle of the space coordinates when only ambient gravity waves are present, and θ^γ_α = internal phase angle of the space coordinates when only an ambient electromagnetic wave field is present. Thus in this paper equation (83)

$$\alpha = \alpha^o C_{\sigma\alpha} \tag{180}$$

refers to a combined ambient electromagnetic wave field and an ambient gravitational wave field and in this section equations (77) through (80) have the interpretation

$$C_{\sigma\alpha} = C_\sigma(\theta_{\eta\alpha} = \theta^g_{\eta\alpha} + \theta^\gamma_{\eta\alpha}, \theta_{\iota\alpha} = \theta^g_{\iota\alpha} + \theta^\gamma_{\iota\alpha}) \tag{181}$$

Equations (177) through (181) may be useful, for instance, for the very strong ambient electromagnetic and ambient gravitational wave fields of degenerate stars in which case the internal phase angles of the space and time coordinates are expected to be relatively large.[17] The dynamics of particles and of wave propagation in matter are also expected to be affected by strong ambient wave fields.[18,19]

A. Space Coordinate Factors for the Component Ambient Gravitational Wave Fields and Ambient Electromagnetic Wave Fields.

For the case when only an ambient gravitational wave field is present and the ambient electromagnetic wave field is shut off so that $\theta^\gamma_{\eta\alpha} = 0, \theta^\gamma_{\iota\alpha} = 0$ and $\theta^\gamma_\alpha = 0$, and equation (181) becomes

$$C^g_{\sigma\alpha} = C_\sigma(\theta_{\eta\alpha} = \theta^g_{\eta\alpha}, \theta_{\pi\alpha} = \theta^g_{\pi\alpha}) \tag{182}$$

with $\sigma = 1$ for [hd] and $\sigma = 2,3$ for [ld]; or equivalently after using equations (78) through (80)

$$C^g_{1\alpha} = \cos\theta^g_{\eta\alpha} \sec\theta^g_{\pi\alpha} \qquad \text{[hd]} \tag{183}$$

$$C^g_{2\alpha} = \cos\theta^g_{\eta\alpha} \cos\theta^g_{\pi\alpha} \qquad \text{[ld]} \tag{184}$$

$$C^g_{3\alpha} = \sec\theta^g_{\eta\alpha} \cos\theta^g_{\pi\alpha} \qquad \text{[ld]} \tag{185}$$

which correspond to an ambient gravity wave field of energy density $\mathfrak{E}^g$. Equation (179) shows that, it is possible to have

$$\theta^g_{\pi\alpha} \geq \theta^g_{\eta\alpha} \qquad \theta^g_\alpha \leq 0 \tag{186}$$

and equations (183) through (185) show also that if $|\theta^g_{\pi\alpha}| \geq |\theta^g_{\eta\alpha}|$ then

$$C^g_{1\alpha} \geq 1 \qquad C^g_{2\alpha} \leq 1 \qquad C^g_{3\alpha} \leq 1 \qquad |\theta^g_{\pi\alpha}| \geq |\theta^g_{\eta\alpha}| \tag{187}$$

Then equations (4) and (180) give for the case when only ambient gravity waves are present

$$\alpha^g = \alpha^o C^g_{\sigma\alpha} \qquad \overline{\alpha}^g = \alpha^g \exp(j\theta^g_\alpha) \tag{188}$$

where α^g and θ^g_α = space coordinate magnitude and internal phase angle when only ambient gravity waves are present, and where the space coordinate factors $C^g_{\sigma\alpha}$ with only ambient gravity waves acting are given by equations (183) through (185). From equation (188) it follows that

$$\alpha^o = \alpha^g (C^g_{\sigma\alpha})^{-1} \tag{189}$$

and α^o is calculated in terms of quantities that can be calculated or measured.

Similarly, for the case when only an ambient electromagnetic wave field is present and $\theta^g_{\eta\alpha} = 0$, $\theta^g_{\pi\alpha} = 0$ and $\theta^g_\alpha = 0$, equations (181) and (78) through (80) give the space coordinate factors as

$$C^\gamma_{\sigma\alpha} = C_\sigma(\theta_{\eta\alpha} = \theta^\gamma_{\eta\alpha}, \theta_{\pi\alpha} = \theta^\gamma_{\pi\alpha}) \tag{190}$$

or equivalently with $\sigma = 1$ for [hd] and $\sigma = 2,3$ for [ld]

$$C^\gamma_{1\alpha} = \cos\theta^\gamma_{\eta\alpha} \sec\theta^\gamma_{\pi\alpha} \qquad \text{[hd]} \tag{191}$$

$$C^\gamma_{2\alpha} = \cos\theta^\gamma_{\eta\alpha} \cos\theta^\gamma_{\pi\alpha} \qquad \text{[ld]} \tag{192}$$

$$C^\gamma_{3\alpha} = \sec\theta^\gamma_{\eta\alpha} \cos\theta^\gamma_{\pi\alpha} \qquad \text{[ld]} \tag{193}$$

which corresponds to an energy density of ambient electromagnetic waves $\mathfrak{E}^{\gamma}$, and where in accordance with equation (179) the condition for negative or zero values of the internal phase angles θ^{γ}_{α} of the space coordinates is

$$\theta^{\gamma}_{\pi\alpha} \geq \theta^{\gamma}_{\eta\alpha} \qquad \theta^{\gamma}_{\alpha} \leq 0 \tag{194}$$

while in addition if $|\theta^{\gamma}_{\pi\alpha}| \geq |\theta^{\gamma}_{\eta\alpha}|$ then it follows from equations (191) through (193) that

$$C^{\gamma}_{1\alpha} \geq 1 \qquad C^{\gamma}_{2\alpha} \leq 1 \qquad C^{\gamma}_{3\alpha} \leq 1 \qquad |\theta^{\gamma}_{\pi\alpha}| \geq |\theta^{\gamma}_{\eta\alpha}| \tag{195}$$

When only an ambient electromagnetic wave field is present, such as in outer space far from a star or planet, equations (4) and (180) give

$$\alpha^{\gamma} = \alpha^{o} C^{\gamma}_{\sigma\alpha} \qquad \overline{\alpha}^{\gamma} = \alpha^{\gamma} \exp(j\theta^{\gamma}_{\alpha}) \tag{196}$$

where α^{γ} and θ^{γ}_{α} = space coordinate magnitude and internal phase angle when only an ambient electromagnetic wave field is present, and where the space coordinate factors $C^{\gamma}_{\sigma\alpha}$ for the case when only an ambient electromagnetic wave field is acting are given by equations (191) through (193). Equation (196) gives

$$\alpha^{o} = \alpha^{\gamma} (C^{\gamma}_{\sigma\alpha})^{-1} \tag{197}$$

which can be used to determine α^{o}. Equations (184), (185), (192) and (193) are also valid for the vacuum in the presence of an ambient field because the vacuum can be considered as the low-density limit of matter.

B. Representation of the Magnitudes of Space Coordinates in Matter for a Combined Ambient Gravitational Wave Field and Ambient Electromagnetic Wave Field in Terms of the Magnitudes of the Space Coordinates for the Cases When Only an Ambient Gravity Wave Field or an Ambient Electromagnetic Wave Field is Acting Alone.

Equations (180), (188) and (196) give the following representations of the magnitude α of the complex number space coordinates in matter for the case of a combined ambient gravitational wave and ambient electromagnetic wave field of energy density $\mathfrak{E}^{g} + \mathfrak{E}^{\gamma}$ in terms of the space coordinates α^{o} when zero ambient wave fields are present or in terms of the magnitude α^{g} of the complex number space coordinates when only ambient gravity waves of energy density $\mathfrak{E}^{g}$ are present or in terms of the magnitude α^{γ} of the complex number space coordinates for the case when only ambient electromagnetic waves of energy density $\mathfrak{E}^{\gamma}$ are present

$$\alpha = \alpha^{o} C_{\sigma\alpha} = \alpha^{g} D_{\sigma\alpha} = \alpha^{\gamma} E_{\sigma\alpha} \tag{198}$$

where

$$D_{\sigma\alpha} = C_{\sigma\alpha} / C^{g}_{\sigma\alpha} \qquad E_{\sigma\alpha} = C_{\sigma\alpha} / C^{\gamma}_{\sigma\alpha} \tag{199}$$

where $C_{\sigma\alpha}$, $C^{g}_{\sigma\alpha}$ and $C^{\gamma}_{\sigma\alpha}$ are given by equations (78) through (80), (183) through (185) and (191) through (193) respectively. Equation (198) gives three equivalent representations of the magnitude of the space coordinates in the presence of a combined ambient gravitational and ambient electromagnetic wave field. Four cases of the effects of the relative strengths of the ambient gravitational and ambient electromagnetic wave fields on the magnitudes of the space coordinates will now be considered for a combined ambient gravity wave field and ambient electromagnetic wave field.

Case 1. Strong Ambient Gravity Wave Field and Strong Ambient Electromagnetic Wave Field.

Combining equation (199) with equations (78) through (80), (183) through (185) and (191) through (193) gives $D_{\sigma\alpha}$ and $E_{\sigma\alpha}$ with $\sigma = 1$ for [hd], $\sigma = 2,3$ for [ld], for the general case of a strong ambient gravity wave field and a strong ambient electromagnetic wave field field as

$$D_{1\alpha} = I_\alpha/J_\alpha \qquad D_{2\alpha} = I_\alpha J_\alpha \qquad D_{3\alpha} = J_\alpha/I_\alpha \tag{200}$$

$$E_{1\alpha} = K_\alpha/L_\alpha \qquad E_{2\alpha} = K_\alpha L_\alpha \qquad E_{3\alpha} = L_\alpha/K_\alpha \tag{201}$$

where

$$I_\alpha = \cos\theta^\gamma_{\eta\alpha} - \sin\theta^\gamma_{\eta\alpha}\tan\theta^g_{\eta\alpha} \tag{202}$$

$$J_\alpha = \cos\theta^\gamma_{\pi\alpha} - \sin\theta^\gamma_{\pi\alpha}\tan\theta^g_{\pi\alpha} \tag{203}$$

$$K_\alpha = \cos\theta^g_{\eta\alpha} - \sin\theta^g_{\eta\alpha}\tan\theta^\gamma_{\eta\alpha} \tag{204}$$

$$L_\alpha = \cos\theta^g_{\pi\alpha} - \sin\theta^g_{\pi\alpha}\tan\theta^\gamma_{\pi\alpha} \tag{205}$$

for $\alpha = x,y,z;\ r,\varphi,z$ or ρ,φ,ψ. Then the space coordinate magnitude α is calculated using equation (198). The internal phase angle θ_α of the space coordinates is calculated by equations (177) through (179).

If only ambient gravity waves are present, then $\theta^\gamma_{\eta\alpha} = 0$, $\theta^\gamma_{\pi\alpha} = 0$, $\theta^\gamma_\alpha = 0$ and equations (198) through (205) give $I_\alpha = 1, J_\alpha = 1, K_\alpha = \cos\theta^g_{\eta\alpha}, L_\alpha = \cos\theta^g_{\pi\alpha}$, $D_{\sigma\alpha} = 1$, $C_{\sigma\alpha} = E_{\sigma\alpha} = C^g_{\sigma\alpha}, \alpha^\gamma = \alpha^o$ and $\alpha = \alpha^g$. If only ambient electromagnetic waves exist, then $\theta^g_{\eta\alpha} = 0$, $\theta^g_{\pi\alpha} = 0$, $\theta^g_\alpha = 0$ and equations (198) through (205) give $I_\alpha = \cos\theta^\gamma_{\eta\alpha}$, $J_\alpha = \cos\theta^\gamma_{\pi\alpha}, K_\alpha = 1, L_\alpha = 1$, $E_{\sigma\alpha} = 1$, $C_{\sigma\alpha} = D_{\sigma\alpha} = C^\gamma_{\sigma\alpha}, \alpha^g = \alpha^o$ and $\alpha = \alpha^\gamma$.

Case 2. Strong Ambient Gravity Wave Field and Weak Ambient Electromagnetic Wave Field.

Equations (200) through (205) show that for a strong ambient gravity wave field and a weak ambient electromagnetic wave field with $\theta^\gamma_{\eta\alpha} \approx 0$, $\theta^\gamma_{\pi\alpha} \approx 0$ and $\theta^\gamma_\alpha \approx 0$ the following first order approximations are valid

$$I_\alpha \approx 1 - \theta^\gamma_{\eta\alpha}\tan\theta^g_{\eta\alpha} \tag{206}$$

$$J_\alpha \approx 1 - \theta^\gamma_{\pi\alpha}\tan\theta^g_{\pi\alpha} \tag{207}$$

$$K_\alpha \approx I_\alpha \cos\theta^g_{\eta\alpha} \approx \cos\theta^g_{\eta\alpha} - \theta^\gamma_{\eta\alpha} \sin\theta^g_{\eta\alpha} \tag{208}$$

$$L_\alpha \approx J_\alpha \cos\theta^g_{\pi\alpha} \approx \cos\theta^g_{\pi\alpha} - \theta^\gamma_{\eta\alpha} \sin\theta^g_{\pi\alpha} \tag{209}$$

and with $\sigma = 1$ for [hd] and $\sigma = 2,3$ for [ld]

$$D_{1\alpha} \approx 1 - \theta^\gamma_{\eta\alpha} \tan\theta^g_{\eta\alpha} + \theta^\gamma_{\pi\alpha} \tan\theta^g_{\pi\alpha} \qquad \text{[hd]} \tag{210}$$

$$D_{2\alpha} \approx 1 - \theta^\gamma_{\eta\alpha} \tan\theta^g_{\eta\alpha} - \theta^\gamma_{\pi\alpha} \tan\theta^g_{\pi\alpha} \qquad \text{[ld]} \tag{211}$$

$$D_{3\alpha} \approx 1 + \theta^\gamma_{\eta\alpha} \tan\theta^g_{\eta\alpha} - \theta^\gamma_{\pi\alpha} \tan\theta^g_{\pi\alpha} \qquad \text{[ld]} \tag{212}$$

and $C^\gamma_{\sigma\alpha} \approx 1 +$ second order terms, so that equation (199) gives

$$C_{1\alpha} \approx E_{1\alpha} \approx C^g_{1\alpha} D_{1\alpha} \qquad \text{[hd]} \tag{213}$$

$$C_{2\alpha} \approx E_{2\alpha} \approx C^g_{2\alpha} D_{2\alpha} \qquad \text{[ld]} \tag{214}$$

$$C_{3\alpha} \approx E_{3\alpha} \approx C^g_{3\alpha} D_{3\alpha} \qquad \text{[ld]} \tag{215}$$

or

$$C_{\sigma\alpha} \approx E_{\sigma\alpha} \approx C^g_{\sigma\alpha} D_{\sigma\alpha} \tag{216}$$

where $C^g_{\sigma\alpha}$ is given by equations (183) through (185). Equations (210) through (212) can be written in matrix form as

$$D_\alpha \approx F^g_\alpha \varphi^\gamma_{3\alpha} \tag{217}$$

where

$$F^g_\alpha = \begin{pmatrix} 1 & -\tan\theta^g_{\eta\alpha} & +\tan\theta^g_{\pi\alpha} \\ 1 & -\tan\theta^g_{\eta\alpha} & -\tan\theta^g_{\pi\alpha} \\ 1 & +\tan\theta^g_{\eta\alpha} & -\tan\theta^g_{\pi\alpha} \end{pmatrix} \tag{218}$$

and where

$$D_\alpha = \begin{pmatrix} D_{1\alpha} \\ D_{2\alpha} \\ D_{3\alpha} \end{pmatrix} \qquad \varphi^\gamma_{3\alpha} = \begin{pmatrix} 1 \\ \theta^\gamma_{\eta\alpha} \\ \theta^\gamma_{\pi\alpha} \end{pmatrix} \tag{219}$$

The space coordinate magnitude α is then calculated by equation (198).

For the case of a zero ambient electromagnetic wave field $\theta^\gamma_{\eta\alpha} = 0$, $\theta^\gamma_{\pi\alpha} = 0$ and $\theta^\gamma_\alpha = 0$ it follows that $C^\gamma_{\sigma\alpha} = 1$ and

$$D_{\sigma\alpha} = 1 \qquad C_{\sigma\alpha} = E_{\sigma\alpha} = C^g_{\sigma\alpha} \qquad \alpha^\gamma = \alpha^o \tag{220}$$

and equation (198) gives

$$\alpha = \alpha^g = \alpha^o C^g_{\sigma\alpha} \tag{221}$$

with only an ambient gravity wave field acting.

Case 3. Weak Ambient Gravity Wave Field and Strong Ambient Electromagnetic Wave Field.

Equations (200) through (205) show that for a weak ambient gravity wave field and a strong ambient electromagnetic wave field with $\theta^g_{\eta\alpha} \approx 0$, $\theta^g_{\pi\alpha} \approx 0$ and $\theta^g_\alpha \approx 0$ the following first order approximations are valid

$$I_\alpha \approx K_\alpha \cos\theta^\gamma_{\eta\alpha} \approx \cos\theta^\gamma_{\eta\alpha} - \theta^g_{\eta\alpha} \sin\theta^\gamma_{\eta\alpha} \tag{222}$$

$$J_\alpha \approx L_\alpha \cos\theta^\gamma_{\pi\alpha} \approx \cos\theta^\gamma_{\pi\alpha} - \theta^g_{\pi\alpha} \sin\theta^\gamma_{\pi\alpha} \tag{223}$$

$$K_\alpha \approx 1 - \theta^g_{\eta\alpha} \tan\theta^\gamma_{\eta\alpha} \tag{224}$$

$$L_\alpha \approx 1 - \theta^g_{\pi\alpha} \tan\theta^\gamma_{\pi\alpha} \tag{225}$$

and with $\sigma = 1$ for [hd] and $\sigma = 2,3$ for [ld]

$$E_{1\alpha} \approx 1 - \theta^g_{\eta\alpha} \tan\theta^\gamma_{\eta\alpha} + \theta^g_{\pi\alpha} \tan\theta^\gamma_{\pi\alpha} \qquad [hd] \tag{226}$$

$$E_{2\alpha} \approx 1 - \theta^g_{\eta\alpha} \tan\theta^\gamma_{\eta\alpha} - \theta^g_{\pi\alpha} \tan\theta^\gamma_{\pi\alpha} \qquad [ld] \tag{227}$$

$$E_{3\alpha} \approx 1 + \theta^g_{\eta\alpha} \tan\theta^\gamma_{\eta\alpha} - \theta^g_{\pi\alpha} \tan\theta^\gamma_{\pi\alpha} \qquad [ld] \tag{228}$$

where $C^g_{\sigma\alpha} \approx 1 +$ second order terms, so that equation (199) gives

$$C_{1\alpha} \approx D_{1\alpha} \approx C^\gamma_{1\alpha} E_{1\alpha} \qquad [hd] \tag{229}$$

$$C_{2\alpha} \approx D_{2\alpha} \approx C^\gamma_{2\alpha} E_{2\alpha} \qquad [ld] \tag{230}$$

$$C_{3\alpha} \approx D_{3\alpha} \approx C^\gamma_{3\alpha} E_{3\alpha} \qquad [ld] \tag{231}$$

or

$$C_{\sigma\alpha} \approx D_{\sigma\alpha} \approx C^\gamma_{\sigma\alpha} \tag{232}$$

where $C^\gamma_{\sigma\alpha}$ for $\sigma = 1,2,3$ is given by equations (191) through (193) and where $\alpha = x,y,z$; r,φ,z or ρ,φ,ψ. Equations (226) through (228) can be written in matrix form as

$$E_\alpha \approx F^\gamma_\alpha \varphi^g_{3\alpha} \tag{233}$$

where

$$F^{\gamma}_{\alpha} = \begin{pmatrix} 1 & -\tan\theta^{\gamma}_{\eta\alpha} & +\tan\theta^{\gamma}_{\pi\alpha} \\ 1 & -\tan\theta^{\gamma}_{\eta\alpha} & -\tan\theta^{\gamma}_{\pi\alpha} \\ 1 & +\tan\theta^{\gamma}_{\eta\alpha} & -\tan\theta^{\gamma}_{\pi\alpha} \end{pmatrix} \tag{234}$$

and where

$$E_{\alpha} = \begin{pmatrix} E_{1\alpha} \\ E_{2\alpha} \\ E_{3\alpha} \end{pmatrix} \qquad \varphi^{g}_{3\alpha} = \begin{pmatrix} 1 \\ \theta^{g}_{\eta\alpha} \\ \theta^{g}_{\pi\alpha} \end{pmatrix} \tag{235}$$

Then the space coordinate magnitude α is given by equation (198).

For the case of a zero ambient gravity wave field with $\theta^{g}_{\eta\alpha} = 0$, $\theta^{g}_{\pi\alpha} = 0$ and $\theta^{g}_{\alpha} = 0$ it follows that $C^{g}_{\sigma\alpha} = 1$ and

$$E_{\sigma\alpha} = 1 \qquad C_{\sigma\alpha} = D_{\sigma\alpha} = C^{\gamma}_{\sigma\alpha} \qquad \alpha^{g} = \alpha^{o} \tag{236}$$

and equation (198) gives the space coordinate magnitude as

$$\alpha = \alpha^{\gamma} = \alpha^{o} C^{\gamma}_{\sigma\alpha} \tag{237}$$

which is the case when only an ambient electromagnetic wave field is present.

Case 4. Weak Ambient Gravity Waves and Weak Ambient Electromagnetic Waves.

The case of weak ambient gravity waves and weak ambient electromagnetic waves can be obtained from equations (200) through (205) which give the following second order expressions when $\theta^{g}_{\eta\alpha} \approx 0$, $\theta^{g}_{\pi\alpha} \approx 0$, $\theta^{g}_{\alpha} \approx 0$ and $\theta^{\gamma}_{\eta\alpha} \approx 0$, $\theta^{\gamma}_{\pi\alpha} \approx 0$, $\theta^{\gamma}_{\alpha} \approx 0$

$$I_{\alpha} \approx 1 - 1/2(\theta^{\gamma}_{\eta\alpha})^2 - \theta^{\gamma}_{\eta\alpha}\theta^{g}_{\eta\alpha} \tag{238}$$

$$J_{\alpha} \approx 1 - 1/2(\theta^{\gamma}_{\pi\alpha})^2 - \theta^{\gamma}_{\pi\alpha}\theta^{g}_{\pi\alpha} \tag{239}$$

$$K_{\alpha} \approx 1 - 1/2(\theta^{g}_{\eta\alpha})^2 - \theta^{g}_{\eta\alpha}\theta^{\gamma}_{\eta\alpha} \tag{240}$$

$$L_{\alpha} \approx 1 - 1/2(\theta^{g}_{\pi\alpha})^2 - \theta^{g}_{\pi\alpha}\theta^{\gamma}_{\pi\alpha} \tag{241}$$

Then equations (200) and (201) give with $\sigma = 1$ for [hd] and $\sigma = 2,3$ for [ld]

$$D_{1\alpha} \approx 1 - 1/2(\theta^{\gamma}_{\eta\alpha})^2 + 1/2(\theta^{\gamma}_{\pi\alpha})^2 - \theta^{\gamma}_{\eta\alpha}\theta^{g}_{\eta\alpha} + \theta^{\gamma}_{\pi\alpha}\theta^{g}_{\pi\alpha} \qquad \text{[hd]} \tag{242}$$

$$D_{2\alpha} \approx 1 - 1/2(\theta^{\gamma}_{\eta\alpha})^2 - 1/2(\theta^{\gamma}_{\pi\alpha})^2 - \theta^{\gamma}_{\eta\alpha}\theta^{g}_{\eta\alpha} - \theta^{\gamma}_{\pi\alpha}\theta^{g}_{\pi\alpha} \qquad \text{[ld]} \tag{243}$$

$$D_{3\alpha} \approx 1 + 1/2(\theta^{\gamma}_{\eta\alpha})^2 - 1/2(\theta^{\gamma}_{\pi\alpha})^2 + \theta^{\gamma}_{\eta\alpha}\theta^{g}_{\eta\alpha} - \theta^{\gamma}_{\pi\alpha}\theta^{g}_{\pi\alpha} \qquad [\text{ld}] \qquad (244)$$

$$E_{1\alpha} \approx 1 - 1/2(\theta^{g}_{\eta\alpha})^2 + 1/2(\theta^{g}_{\pi\alpha})^2 - \theta^{g}_{\eta\alpha}\theta^{\gamma}_{\eta\alpha} + \theta^{g}_{\pi\alpha}\theta^{\gamma}_{\pi\alpha} \qquad [\text{hd}] \qquad (245)$$

$$E_{2\alpha} \approx 1 - 1/2(\theta^{g}_{\eta\alpha})^2 - 1/2(\theta^{g}_{\pi\alpha})^2 - \theta^{g}_{\eta\alpha}\theta^{\gamma}_{\eta\alpha} - \theta^{g}_{\pi\alpha}\theta^{\gamma}_{\pi\alpha} \qquad [\text{ld}] \qquad (246)$$

$$E_{3\alpha} \approx 1 + 1/2(\theta^{g}_{\eta\alpha})^2 - 1/2(\theta^{g}_{\pi\alpha})^2 + \theta^{g}_{\eta\alpha}\theta^{\gamma}_{\eta\alpha} - \theta^{g}_{\pi\alpha}\theta^{\gamma}_{\pi\alpha} \qquad [\text{ld}] \qquad (247)$$

Equations (242) through (247) can be written in matrix form as

$$D_{\alpha} \approx W\varphi^{\gamma g}_{5\alpha} \qquad E_{\alpha} \approx W\varphi^{g\gamma}_{5\alpha} \qquad (248)$$

where

$$W = \begin{pmatrix} 1 & -1/2 & +1/2 & -1 & +1 \\ 1 & -1/2 & -1/2 & -1 & -1 \\ 1 & +1/2 & -1/2 & +1 & -1 \end{pmatrix} \qquad (249)$$

$$D_{\alpha} = \begin{pmatrix} D_{1\alpha} \\ D_{2\alpha} \\ D_{3\alpha} \end{pmatrix} \qquad E_{\alpha} = \begin{pmatrix} E_{1\alpha} \\ E_{2\alpha} \\ E_{3\alpha} \end{pmatrix} \qquad (250)$$

$$\varphi^{\gamma g}_{5\alpha} = \begin{pmatrix} 1 \\ (\theta^{\gamma}_{\eta\alpha})^2 \\ (\theta^{\gamma}_{\pi\alpha})^2 \\ \theta^{\gamma}_{\eta\alpha}\theta^{g}_{\eta\alpha} \\ \theta^{\gamma}_{\pi\alpha}\theta^{g}_{\pi\alpha} \end{pmatrix} \qquad \varphi^{g\gamma}_{5\alpha} = \begin{pmatrix} 1 \\ (\theta^{g}_{\eta\alpha})^2 \\ (\theta^{g}_{\pi\alpha})^2 \\ \theta^{g}_{\eta\alpha}\theta^{\gamma}_{\eta\alpha} \\ \theta^{g}_{\pi\alpha}\theta^{\gamma}_{\pi\alpha} \end{pmatrix} \qquad (251)$$

For the case of weak ambient gravity waves and weak ambient electromagnetic waves, equations (78) through (80) give the following second order expressions for $C_{\sigma\alpha}$ with $\theta_{\eta\alpha} \approx 0$ and $\theta_{\pi\alpha} \approx 0$

$$C_{1\alpha} \approx 1 - 1/2\theta^2_{\eta\alpha} + 1/2\theta^2_{\pi\alpha} \qquad [\text{hd}] \qquad (252)$$

$$C_{2\alpha} \approx 1 - 1/2\theta^2_{\eta\alpha} - 1/2\theta^2_{\pi\alpha} \qquad [\text{ld}] \qquad (253)$$

$$C_{3\alpha} \approx 1 + 1/2\theta^2_{\eta\alpha} - 1/2\theta^2_{\pi\alpha} \qquad [\text{ld}] \qquad (254)$$

Combining equation (177) with equations (252) through (254) gives the following second order expressions for $C_{\sigma\alpha}$ with $\sigma = 1$ for [hd] and $\sigma = 1,2$ for [ld]

$$C_{1\alpha} \approx 1 - 1/2(\theta^{g}_{\eta\alpha})^2 + 1/2(\theta^{g}_{\pi\alpha})^2 - 1/2(\theta^{\gamma}_{\eta\alpha})^2 + 1/2(\theta^{\gamma}_{\pi\alpha})^2 - \theta^{g}_{\eta\alpha}\theta^{\gamma}_{\eta\alpha} + \theta^{g}_{\pi\alpha}\theta^{\gamma}_{\pi\alpha} \qquad (255)$$

$$C_{2\alpha} \approx 1 - 1/2(\theta^g_{\eta\alpha})^2 - 1/2(\theta^g_{\pi\alpha})^2 - 1/2(\theta^\gamma_{\eta\alpha})^2 - 1/2(\theta^\gamma_{\pi\alpha})^2 - \theta^g_{\eta\alpha}\theta^\gamma_{\eta\alpha} - \theta^g_{\pi\alpha}\theta^\gamma_{\pi\alpha} \tag{256}$$

$$C_{3\alpha} \approx 1 + 1/2(\theta^g_{\eta\alpha})^2 - 1/2(\theta^g_{\pi\alpha})^2 + 1/2(\theta^\gamma_{\eta\alpha})^2 - 1/2(\theta^\gamma_{\pi\alpha})^2 + \theta^g_{\eta\alpha}\theta^\gamma_{\eta\alpha} - \theta^g_{\pi\alpha}\theta^\gamma_{\pi\alpha} \tag{257}$$

Equations (252) through (254) and (255) through (257) can be written in matrix form as

$$C_\alpha \approx X\varphi_{3\alpha} \approx Y\varphi_{7\alpha} \tag{258}$$

where

$$C_\alpha = \begin{pmatrix} C_{1\alpha} \\ C_{2\alpha} \\ C_{3\alpha} \end{pmatrix} \qquad \varphi_{3\alpha} = \begin{pmatrix} 1 \\ \theta^2_{\eta\alpha} \\ \theta^2_{\pi\alpha} \end{pmatrix} \qquad \varphi_{7\alpha} = \begin{pmatrix} 1 \\ (\theta^g_{\eta\alpha})^2 \\ (\theta^g_{\pi\alpha})^2 \\ (\theta^\gamma_{\eta\alpha})^2 \\ (\theta^\gamma_{\pi\alpha})^2 \\ \theta^g_{\eta\alpha}\theta^\gamma_{\eta\alpha} \\ \theta^g_{\pi\alpha}\theta^\gamma_{\pi\alpha} \end{pmatrix} \tag{259}$$

$$X = \begin{pmatrix} 1 & -1/2 & +1/2 \\ 1 & -1/2 & -1/2 \\ 1 & +1/2 & -1/2 \end{pmatrix} \qquad Y = \begin{pmatrix} 1 & -1/2 & +1/2 & -1/2 & +1/2 & -1 & +1 \\ 1 & -1/2 & -1/2 & -1/2 & -1/2 & -1 & -1 \\ 1 & +1/2 & -1/2 & +1/2 & -1/2 & +1 & -1 \end{pmatrix} \tag{260}$$

Then the space coordinate magnitude is calculated by equation (198).

The case when the ambient electromagnetic wave field is zero has $\theta^\gamma_{\eta\alpha} = 0$, $\theta^\gamma_{\pi\alpha} = 0$ and $\theta^\gamma_\alpha = 0$ so that equations (242) through (244) give $D_{\sigma\alpha} = 1$ or

$$D_{1\alpha} = 1 \qquad D_{2\alpha} = 1 \qquad D_{3\alpha} = 1 \tag{261}$$

and for this case the following second order approximations are obtained from equations (245) through (247) and (255) through (257) with $\sigma = 1$ for [hd] and $\sigma = 2,3$ for [ld]

$$C_{1\alpha} \approx E_{1\alpha} \approx 1 - 1/2(\theta^g_{\eta\alpha})^2 + 1/2(\theta^g_{\pi\alpha})^2 \qquad \text{[hd]} \tag{262}$$

$$C_{2\alpha} \approx E_{2\alpha} \approx 1 - 1/2(\theta^g_{\eta\alpha})^2 - 1/2(\theta^g_{\pi\alpha})^2 \qquad \text{[ld]} \tag{263}$$

$$C_{3\alpha} \approx E_{3\alpha} \approx 1 + 1/2(\theta^g_{\eta\alpha})^2 - 1/2(\theta^g_{\pi\alpha})^2 \qquad \text{[ld]} \tag{264}$$

Equation (198) then gives the space coordinate magnitudes for this special case as

$$\alpha = \alpha^g = \alpha^o E_{\sigma\alpha} \qquad \alpha^\gamma = \alpha^o \tag{265}$$

where $E_{\sigma\alpha}$ is give by equations (262) through (264) for the case of a zero ambient electromagnetic wave field.

For the case when the ambient gravity wave field is zero and $\theta^g_{\eta\alpha} = 0$, $\theta^g_{\pi\alpha} = 0$ and $\theta^g_\alpha = 0$ equations (245) through (247) give with $\sigma = 1$ for [hd] and $\sigma = 2,3$ for [ld]

$$E_{1\alpha} = 1 \qquad E_{2\alpha} = 1 \qquad E_{3\alpha} = 1 \tag{266}$$

and equations (255) through (257) show that the following second order approximations are valid for this special case

$$C_{1\alpha} \approx D_{1\alpha} \approx 1 - 1/2(\theta^\gamma_{\eta\alpha})^2 + 1/2(\theta^\gamma_{\pi\alpha})^2 \qquad \text{[hd]} \tag{267}$$

$$C_{2\alpha} \approx D_{2\alpha} \approx 1 - 1/2(\theta^\gamma_{\eta\alpha})^2 - 1/2(\theta^\gamma_{\pi\alpha})^2 \qquad \text{[ld]} \tag{268}$$

$$C_{3\alpha} \approx D_{3\alpha} \approx 1 + 1/2(\theta^\gamma_{\eta\alpha})^2 - 1/2(\theta^\gamma_{\pi\alpha})^2 \qquad \text{[ld]} \tag{269}$$

Then equation (198) gives for this case

$$\alpha = \alpha^\gamma = \alpha^o D_{\sigma\alpha} \qquad \alpha^g = \alpha^o \tag{270}$$

where $D_{\sigma\alpha}$ for this case of a zero ambient gravity wave field is given by equations (267) through (269).

The case when both an ambient gravity wave field and an ambient electromagnetic wave field are zero gives the following result

$$C_{\sigma\alpha} = D_{\sigma\alpha} = E_{\sigma\alpha} = 1 \qquad \alpha = \alpha^g = \alpha^\gamma = \alpha^o \tag{271}$$

as expected.

C. Representations of the Measured Space Coordinates in Matter for a Combined Ambient Gravitational Wave Field and Ambient Electromagnetic Wave Field in Terms of the Measured Space Coordinates for the Cases When Only an Ambient Gravity Wave Field or an Ambient Electromagnetic Wave Field is Acting Alone.

The measured spatial coordinates in matter when both ambient gravity waves and ambient electromagnetic waves are present are given by equation (84) as

$$\alpha_m = \alpha \cos\theta_\alpha = \alpha^o C_{\sigma\alpha} \cos\theta_\alpha \tag{272}$$

The measured values of the space coordinates under zero ambient wave field conditions are given by

$$\alpha^o_m = \alpha^o \tag{273}$$

The measured values of the space coordinates for the case when only an ambient gravity wave field is acting is given by

$$\alpha^g_m = \alpha^g \cos\theta^g_\alpha = \alpha^o C^g_{\sigma\alpha} \cos\theta^g_\alpha \tag{274}$$

The measured values of the space coordinates when only an ambient electromagnetic wave field is present is given by

$$\alpha^{\gamma}_{m} = \alpha^{\gamma} \cos\theta^{\gamma}_{\alpha} = \alpha^{o} C^{\gamma}_{\sigma\alpha} \cos\theta^{\gamma}_{\alpha} \tag{275}$$

Equations (274) and (275) give

$$\alpha^{o} = \alpha^{g}_{m} (C^{g}_{\sigma\alpha})^{-1} \sec\theta^{g}_{\alpha} = \alpha^{\gamma}_{m} (C^{\gamma}_{\sigma\alpha})^{-1} \sec\theta^{\gamma}_{\alpha} \tag{276}$$

Equations (272) through (275) give the following representations of the measured values of the space coordinates

$$\alpha_{m} = \alpha^{o}_{m} S_{\sigma\alpha} = \alpha^{g}_{m} Q_{\sigma\alpha} = \alpha^{\gamma}_{m} R_{\sigma\alpha} \tag{277}$$

with α^{o}_{m} is given by equation (273) and where

$$S_{\sigma\alpha} = T_{\alpha} C_{\sigma\alpha} \tag{278}$$

$$Q_{\sigma\alpha} = A_{\alpha} D_{\sigma\alpha} \tag{279}$$

$$R_{\sigma\alpha} = B_{\alpha} E_{\sigma\alpha} \tag{280}$$

where $C_{\sigma\alpha}$, $D_{\sigma\alpha}$ and $E_{\sigma\alpha}$ are given by equations (78) through (80) and (199) through (205), and where

$$T_{\alpha} = \cos\theta_{\alpha} \tag{281}$$

$$A_{\alpha} = \cos\theta_{\alpha} \sec\theta^{g}_{\alpha} \tag{282}$$

$$B_{\alpha} = \cos\theta_{\alpha} \sec\theta^{\gamma}_{\alpha} \tag{283}$$

where θ_{α}, θ^{g}_{α} and θ^{γ}_{α} are given by equations (178) and (179). Four special cases of these equations will now be given that are distinguished by the relative strengths of the component ambient gravity wave and ambient electromagnetic wave fields.

Case 1. Strong Ambient Gravity Waves and Strong Ambient Electromagnetic Waves.

Combining equations (178) and (281) through (283) gives for the case of a strong ambient gravity wave field and a strong ambient electromagnetic wave field

$$T_{\alpha} = \cos\theta^{g}_{\alpha} \cos\theta^{\gamma}_{\alpha} - \sin\theta^{g}_{\alpha} \sin\theta^{\gamma}_{\alpha} = \cos\theta^{g}_{\alpha} \cos\theta^{\gamma}_{\alpha} (1 - \tan\theta^{g}_{\alpha} \tan\theta^{\gamma}_{\alpha}) \tag{284}$$

$$A_{\alpha} = \cos\theta^{\gamma}_{\alpha} - \sin\theta^{\gamma}_{\alpha} \tan\theta^{g}_{\alpha} = \cos\theta^{\gamma}_{\alpha} (1 - \tan\theta^{g}_{\alpha} \tan\theta^{\gamma}_{\alpha}) \tag{285}$$

$$B_{\alpha} = \cos\theta^{g}_{\alpha} - \sin\theta^{g}_{\alpha} \tan\theta^{\gamma}_{\alpha} = \cos\theta^{g}_{\alpha} (1 - \tan\theta^{g}_{\alpha} \tan\theta^{\gamma}_{\alpha}) \tag{286}$$

where θ^{g}_{α} and θ^{γ}_{α} are given by equation (179). Then equations (277) through (280) can be used to calculate the measured values of the space coordinates.

Case 2. Strong Ambient Gravity Waves and Weak Ambient Electromagnetic Waves.

For this case $\theta^{\gamma}_{\eta\alpha} \approx 0$, $\theta^{\gamma}_{\pi\alpha} \approx 0$, $\theta^{\gamma}_{\alpha} \approx 0$ and $D_{\sigma\alpha}$ and $E_{\sigma\alpha}$ are given by equations (210) through (215). Then equations (284) through (286) show that the following first order approximations are valid

$$T_{\alpha} \approx B_{\alpha} \approx A_{\alpha} \cos\theta^{g}_{\alpha} \approx \cos\theta^{g}_{\alpha} - \theta^{\gamma}_{\alpha} \sin\theta^{g}_{\alpha} \qquad (287)$$

$$A_{\alpha} \approx 1 - \theta^{\gamma}_{\alpha} \tan\theta^{g}_{\alpha} \qquad (288)$$

$$B_{\alpha} \approx A_{\alpha} \cos\theta^{g}_{\alpha} \approx \cos\theta^{g}_{\alpha} - \theta^{\gamma}_{\alpha} \sin\theta^{g}_{\alpha} \qquad (289)$$

where equation (179) gives

$$\theta^{g}_{\alpha} = \theta^{g}_{\eta\alpha} - \theta^{g}_{\pi\alpha} \qquad \theta^{\gamma}_{\alpha} = \theta^{\gamma}_{\eta\alpha} - \theta^{\gamma}_{\pi\alpha} \qquad (290)$$

and equations (210) through (212), (279), (288) and (290) give the following first order approximations

$$Q_{1\alpha} \approx 1 - a^{g}_{1\alpha}\theta^{\gamma}_{\eta\alpha} + b^{g}_{1\alpha}\theta^{\gamma}_{\pi\alpha} \qquad Q_{1\alpha} \geq 1 \qquad [\text{hd}] \qquad (291)$$

$$Q_{2\alpha} \approx 1 - a^{g}_{1\alpha}\theta^{\gamma}_{\eta\alpha} - d^{g}_{1\alpha}\theta^{\gamma}_{\pi\alpha} \qquad Q_{2\alpha} \leq 1 \qquad [\text{ld}] \qquad (292)$$

$$Q_{3\alpha} \approx 1 + c^{g}_{1\alpha}\theta^{\gamma}_{\eta\alpha} - d^{g}_{1\alpha}\theta^{\gamma}_{\pi\alpha} \qquad Q_{3\alpha} \leq 1 \qquad [\text{ld}] \qquad (293)$$

where $\theta^{\gamma}_{\pi\alpha} \geq \theta^{\gamma}_{\eta\alpha}$ and $\theta^{g}_{\pi\alpha} \geq \theta^{g}_{\eta\alpha}$, and where

$$a^{g}_{1\alpha} = \tan\theta^{g}_{\eta\alpha} - \tan(\theta^{g}_{\pi\alpha} - \theta^{g}_{\eta\alpha}) \qquad (294)$$

$$b^{g}_{1\alpha} = \tan\theta^{g}_{\pi\alpha} - \tan(\theta^{g}_{\pi\alpha} - \theta^{g}_{\eta\alpha}) \qquad (295)$$

$$c^{g}_{1\alpha} = \tan\theta^{g}_{\eta\alpha} + \tan(\theta^{g}_{\pi\alpha} - \theta^{g}_{\eta\alpha}) \qquad (296)$$

$$d^{g}_{1\alpha} = \tan\theta^{g}_{\pi\alpha} + \tan(\theta^{g}_{\pi\alpha} - \theta^{g}_{\eta\alpha}) \qquad (297)$$

which, subject to equation (186), satisfy the following inequalities

$$d^{g}_{1\alpha} \geq c^{g}_{1\alpha} \geq b^{g}_{1\alpha} \geq a^{g}_{1\alpha} \qquad (298)$$

Equations (291) through (293) can be written in matrix form as

$$Q_{\alpha} \approx A^{g}_{\alpha}\varphi^{\gamma}_{3\alpha} \qquad (299)$$

where

$$A^{g}_{\alpha} = \begin{pmatrix} 1 & -a^{g}_{1\alpha} & +b^{g}_{1\alpha} \\ 1 & -a^{g}_{1\alpha} & -d^{g}_{1\alpha} \\ 1 & +c^{g}_{1\alpha} & -d^{g}_{1\alpha} \end{pmatrix} \qquad Q_{\alpha} = \begin{pmatrix} Q_{1\alpha} \\ Q_{2\alpha} \\ Q_{3\alpha} \end{pmatrix} \qquad \varphi^{\gamma}_{3\alpha} = \begin{pmatrix} 1 \\ \theta^{\gamma}_{\eta\alpha} \\ \theta^{\gamma}_{\pi\alpha} \end{pmatrix} \tag{300}$$

which are useful for this first order approximation.

For this case of strong ambient gravity waves and weak ambient electromagnetic waves, equations (213) through (215), (280) and (283) give

$$R_{\sigma\alpha} \approx A_{\alpha} E_{\sigma\alpha} \cos\theta^{g}_{\alpha} \tag{301}$$

where $E_{\sigma\alpha}$ is given by equations (213) through (215), and because equations (213) through (215) show that $C_{\sigma\alpha} \approx E_{\sigma\alpha}$ it follows that for this case

$$S_{\sigma\alpha} \approx B_{\alpha} C_{\sigma\alpha} \approx B_{\alpha} E_{\sigma\alpha} \approx R_{\sigma\alpha} \approx C^{g}_{\sigma\alpha} Q_{\sigma\alpha} \cos\theta^{g}_{\alpha} \tag{302}$$

so that for this case

$$S_{1\alpha} \approx R_{1\alpha} \approx C^{g}_{1\alpha} Q_{1\alpha} \cos\theta^{g}_{\alpha} \qquad [hd] \tag{303}$$

$$S_{2\alpha} \approx R_{2\alpha} \approx C^{g}_{2\alpha} Q_{2\alpha} \cos\theta^{g}_{\alpha} \qquad [ld] \tag{304}$$

$$S_{3\alpha} \approx R_{3\alpha} \approx C^{g}_{3\alpha} Q_{3\alpha} \cos\theta^{g}_{\alpha} \qquad [ld] \tag{305}$$

Equations (301) and (302) give the following first order approximation for the case of strong ambient gravity waves and weak ambient electromagnetic waves

$$A_{\alpha} E_{\sigma\alpha} \approx C^{g}_{\sigma\alpha} Q_{\sigma\alpha} \tag{306}$$

which combined with equation (216) gives equation (279), where $C^{g}_{\sigma\alpha}$ and $Q_{\sigma\alpha}$ are given by equations (183) through (185) and (291) through (293) respectively. Equations (277) through (280) can then be used to calculate the measured values α_m of the space coordinates.

For the case when only ambient gravity waves are present and $\theta^{\gamma}_{\eta\alpha} = 0$, $\theta^{\gamma}_{\pi\alpha} = 0$ and $\theta^{\gamma}_{\alpha} = 0$ it follows that $C^{\gamma}_{\sigma\alpha} = 1$ and

$$Q_{\sigma\alpha} = 1 \qquad A_{\alpha} = 1 \qquad C_{\sigma\alpha} = E_{\sigma\alpha} = C^{g}_{\sigma\alpha} \qquad D_{\sigma\alpha} = 1 \tag{307}$$

and

$$S_{\sigma\alpha} = R_{\sigma\alpha} = C^{g}_{\sigma\alpha} \cos\theta^{g}_{\alpha} \qquad T_{\alpha} = B_{\alpha} = \cos\theta^{g}_{\alpha} \tag{308}$$

and then equation (277) gives

$$\alpha_{m} = \alpha^{g}_{m} = \alpha^{o} C^{g}_{\sigma\alpha} \cos\theta^{g}_{\alpha} \qquad \alpha^{\gamma}_{m} = \alpha^{o} \tag{309}$$

where θ^g_α is given by equation (179) and where $C^g_{\sigma\alpha}$ is given by equations (183) through (185).

Case 3. Weak Ambient Gravity Waves and Strong Ambient Electromagnetic Waves.

For this case $\theta^g_{\eta\alpha} \approx 0$, $\theta^g_{\pi\alpha} \approx 0$, $\theta^g_\alpha \approx 0$ and where $D_{\sigma\alpha}$ and $E_{\sigma\alpha}$ are given by equations (226) through (231). Equations (281) through (283) give the following approximations for this case of weak ambient gravity waves and strong ambient electromagnetic waves

$$T_\alpha \approx A_\alpha \approx B_\alpha \cos\theta^\gamma_\alpha \approx \cos\theta^\gamma_\alpha - \theta^g_\alpha \sin\theta^\gamma_\alpha \tag{310}$$

$$A_\alpha \approx B_\alpha \cos\theta^\gamma_\alpha \approx \cos\theta^\gamma_\alpha - \theta^g_\alpha \sin\theta^\gamma_\alpha \tag{311}$$

$$B_\alpha \approx 1 - \theta^g_\alpha \tan\theta^\gamma_\alpha \tag{312}$$

where equation (179) yields

$$\theta^\gamma_\alpha = \theta^\gamma_{\eta\alpha} - \theta^\gamma_{\pi\alpha} \qquad\qquad \theta^g_\alpha = \theta^g_{\eta\alpha} - \theta^g_{\pi\alpha} \tag{313}$$

while equations (226) through (228), (280), (312) and (313) give the following first order approximations for this case

$$R_{1\alpha} \approx 1 - a^\gamma_{1\alpha}\theta^g_{\eta\alpha} + b^\gamma_{1\alpha}\theta^g_{\pi\alpha} \qquad R_{1\alpha} \geq 1 \qquad [\text{hd}] \tag{314}$$

$$R_{2\alpha} \approx 1 - a^\gamma_{1\alpha}\theta^g_{\eta\alpha} - d^\gamma_{1\alpha}\theta^g_{\pi\alpha} \qquad R_{2\alpha} \leq 1 \qquad [\text{ld}] \tag{315}$$

$$R_{3\alpha} \approx 1 + c^\gamma_{1\alpha}\theta^g_{\eta\alpha} - d^\gamma_{1\alpha}\theta^g_{\pi\alpha} \qquad R_{3\alpha} \leq 1 \qquad [\text{ld}] \tag{316}$$

with $\theta^g_{\pi\alpha} \geq \theta^g_{\eta\alpha}$ and $\theta^\gamma_{\pi\alpha} \geq \theta^\gamma_{\eta\alpha}$, and where

$$a^\gamma_{1\alpha} = \tan\theta^\gamma_{\eta\alpha} - \tan(\theta^\gamma_{\pi\alpha} - \theta^\gamma_{\eta\alpha}) \tag{317}$$

$$b^\gamma_{1\alpha} = \tan\theta^\gamma_{\pi\alpha} - \tan(\theta^\gamma_{\pi\alpha} - \theta^\gamma_{\eta\alpha}) \tag{318}$$

$$c^\gamma_{1\alpha} = \tan\theta^\gamma_{\eta\alpha} + \tan(\theta^\gamma_{\pi\alpha} - \theta^\gamma_{\eta\alpha}) \tag{319}$$

$$d^\gamma_{1\alpha} = \tan\theta^\gamma_{\pi\alpha} + \tan(\theta^\gamma_{\pi\alpha} - \theta^\gamma_{\eta\alpha}) \tag{320}$$

which, according to equation (194), satisfy the inequalities

$$d^\gamma_{1\alpha} \geq c^\gamma_{1\alpha} \geq b^\gamma_{1\alpha} \geq a^\gamma_{1\alpha} \tag{321}$$

Equations (314) through (316) have the following matrix representations

$$R_\alpha = A^\gamma_\alpha \varphi^g_{3\alpha} \tag{322}$$

where

$$A_\alpha^\gamma = \begin{pmatrix} 1 & -a_{1\alpha}^\gamma & +b_{1\alpha}^\gamma \\ 1 & -a_{1\alpha}^\gamma & -d_{1\alpha}^\gamma \\ 1 & +c_{1\alpha}^\gamma & -d_{1\alpha}^\gamma \end{pmatrix} \qquad R_\alpha = \begin{pmatrix} R_{1\alpha} \\ R_{2\alpha} \\ R_{3\alpha} \end{pmatrix} \qquad \varphi_{3\alpha}^g = \begin{pmatrix} 1 \\ \theta_{\eta\alpha}^g \\ \theta_{\pi\alpha}^g \end{pmatrix} \tag{323}$$

which exhibits the characteristic positive and negative signs.

It follows from equations (229) through (231), (279) and (311) that for the case of weak ambient gravity waves and strong ambient electromagnetic waves.

$$Q_{\sigma\alpha} \approx B_\alpha D_{\sigma\alpha} \cos\theta_\alpha^\gamma \tag{324}$$

where $D_{\sigma\alpha}$ is given by equations (229) through (232), and because $C_{\sigma\alpha} \approx D_{\sigma\alpha}$ for this case it follows that after using equation (280)

$$S_{\sigma\alpha} \approx A_\alpha C_{\sigma\alpha} \approx A_\alpha D_{\sigma\alpha} \approx Q_{\sigma\alpha} \approx C_{\sigma\alpha}^\gamma R_{\sigma\alpha} \cos\theta_\alpha^\gamma \tag{325}$$

so that for weak ambient gravity waves and strong ambient electromagnetic waves

$$S_{1\alpha} \approx Q_{1\alpha} \approx C_{1\alpha}^\gamma R_{1\alpha} \cos\theta_\alpha^\gamma \qquad \text{[hd]} \tag{326}$$

$$S_{2\alpha} \approx Q_{2\alpha} \approx C_{2\alpha}^\gamma R_{2\alpha} \cos\theta_\alpha^\gamma \qquad \text{[ld]} \tag{327}$$

$$S_{3\alpha} \approx Q_{3\alpha} \approx C_{3\alpha}^\gamma R_{3\alpha} \cos\theta_\alpha^\gamma \qquad \text{[ld]} \tag{328}$$

where $C_{\sigma\alpha}^\gamma$ is given by equations (191) through (193) and where $R_{\sigma\alpha}$, for $\sigma = 1,2,3$ and $\alpha = x,y,z$; r,φ,z or ρ,φ,ψ, is given by equations (314) through (316). Equations (324) and (325) give for the case of weak ambient gravity waves and strong ambient electromagnetic waves

$$B_\alpha D_{\sigma\alpha} \approx C_{\sigma\alpha}^\gamma R_{\sigma\alpha} \tag{329}$$

which combined with equation (232) gives equation (280). The measured values α_m of the space coordinates for the case of weak ambient gravity waves and strong ambient electromagnetic waves can then be calculated using equations (277) through (280).

When only ambient electromagnetic waves are present and $\theta_{\eta\alpha}^g = 0$, $\theta_{\pi\alpha}^g = 0$ and $\theta_\alpha^g = 0$, then $C_{\sigma\alpha}^g = 1$ and

$$R_{\sigma\alpha} = 1 \qquad B_\alpha = 1 \qquad C_{\sigma\alpha} = D_{\sigma\alpha} = C_{\sigma\alpha}^\gamma \qquad E_{\sigma\alpha} = 1 \tag{330}$$

and

$$S_{\sigma\alpha} = Q_{\sigma\alpha} = C_{\sigma\alpha}^\gamma \cos\theta_\alpha^\gamma \qquad T_\alpha = A_\alpha = \cos\theta_\alpha^\gamma \tag{331}$$

and

$$\alpha_m = \alpha_m^\gamma = \alpha^o C_{\sigma\alpha}^\gamma \cos\theta_\alpha^\gamma \qquad \alpha_m^g = \alpha^o \tag{332}$$

where θ_α^γ is given by equation (313) and where $C_{\sigma\alpha}^\gamma$ is given by equations (191) through (193).

Case 4. Weak Ambient Gravity Waves and Weak Ambient Electromagnetic Waves.

The case of weak ambient gravity waves and weak ambient electromagnetic waves is obtained from equations (279) and (242) through (244) and from equations (280) and (245) through (247). Equations (281) through (286) for this case give the following second order approximations with $\theta_\alpha^g \approx 0$ and $\theta_\alpha^\gamma \approx 0$

$$T_\alpha \approx 1 - 1/2(\theta_\alpha^\gamma)^2 - 1/2(\theta_\alpha^g)^2 - \theta_\alpha^\gamma \theta_\alpha^g \tag{333}$$

$$A_\alpha \approx 1 - 1/2(\theta_\alpha^\gamma)^2 - \theta_\alpha^\gamma \theta_\alpha^g \tag{334}$$

$$B_\alpha \approx 1 - 1/2(\theta_\alpha^g)^2 - \theta_\alpha^g \theta_\alpha^\gamma \tag{335}$$

Combining equation (179) with equations (333) through (335) gives the following second order approximations with $\theta_{\eta\alpha}^g \approx 0$, $\theta_{\pi\alpha}^g \approx 0$, $\theta_{\eta\alpha}^\gamma \approx 0$ and $\theta_{\pi\alpha}^\gamma \approx 0$

$$\begin{aligned} T_\alpha \approx{}& 1 - 1/2(\theta_{\eta\alpha}^g)^2 - 1/2(\theta_{\pi\alpha}^g)^2 - 1/2(\theta_{\eta\alpha}^\gamma)^2 - 1/2(\theta_{\pi\alpha}^\gamma)^2 + \theta_{\eta\alpha}^g \theta_{\pi\alpha}^g + \theta_{\eta\alpha}^\gamma \theta_{\pi\alpha}^\gamma \\ & - \theta_{\eta\alpha}^g \theta_{\eta\alpha}^\gamma + \theta_{\eta\alpha}^\gamma \theta_{\pi\alpha}^g + \theta_{\eta\alpha}^g \theta_{\pi\alpha}^\gamma - \theta_{\pi\alpha}^g \theta_{\pi\alpha}^\gamma \end{aligned} \tag{336}$$

$$A_\alpha \approx 1 - 1/2(\theta_{\eta\alpha}^\gamma)^2 - 1/2(\theta_{\pi\alpha}^\gamma)^2 + \theta_{\eta\alpha}^\gamma \theta_{\pi\alpha}^\gamma - \theta_{\eta\alpha}^g \theta_{\eta\alpha}^\gamma + \theta_{\eta\alpha}^\gamma \theta_{\pi\alpha}^g + \theta_{\eta\alpha}^g \theta_{\pi\alpha}^\gamma - \theta_{\pi\alpha}^g \theta_{\pi\alpha}^\gamma \tag{337}$$

$$B_\alpha \approx 1 - 1/2(\theta_{\eta\alpha}^g)^2 - 1/2(\theta_{\pi\alpha}^g)^2 + \theta_{\eta\alpha}^g \theta_{\pi\alpha}^g - \theta_{\eta\alpha}^g \theta_{\eta\alpha}^\gamma + \theta_{\eta\alpha}^\gamma \theta_{\pi\alpha}^g + \theta_{\eta\alpha}^g \theta_{\pi\alpha}^\gamma - \theta_{\pi\alpha}^g \theta_{\pi\alpha}^\gamma \tag{338}$$

For the case of weak ambient gravity waves and weak ambient electromagnetic waves equations (278) through (280) show that the functions T_α, A_α and B_α given by equations (336) through (338) allow the calculation of the functions $S_{\sigma\alpha}$, $Q_{\sigma\alpha}$ and $R_{\sigma\alpha}$ which enter the calculation of α_m through equation (277).

For small angles equations (255) through (257), (278) and (336) give the following second order approximations for $S_{\sigma\alpha}$

$$S_{1\alpha} \approx 1 - (\theta_{\eta\alpha}^g)^2 - (\theta_{\eta\alpha}^\gamma)^2 - 2\theta_{\eta\alpha}^g \theta_{\eta\alpha}^\gamma + \theta_{\eta\alpha}^g \theta_{\pi\alpha}^g + \theta_{\eta\alpha}^\gamma \theta_{\pi\alpha}^\gamma + \theta_{\eta\alpha}^\gamma \theta_{\pi\alpha}^g + \theta_{\eta\alpha}^g \theta_{\pi\alpha}^\gamma \quad [\text{hd}] \tag{339}$$

$$\begin{aligned} S_{2\alpha} \approx{}& 1 - (\theta_{\eta\alpha}^g)^2 - (\theta_{\pi\alpha}^g)^2 - (\theta_{\eta\alpha}^\gamma)^2 - (\theta_{\pi\alpha}^\gamma)^2 - 2\theta_{\eta\alpha}^g \theta_{\eta\alpha}^\gamma - 2\theta_{\pi\alpha}^g \theta_{\pi\alpha}^\gamma + \theta_{\eta\alpha}^g \theta_{\pi\alpha}^g \quad [\text{ld}] \\ & + \theta_{\eta\alpha}^\gamma \theta_{\pi\alpha}^\gamma + \theta_{\eta\alpha}^\gamma \theta_{\pi\alpha}^g + \theta_{\eta\alpha}^g \theta_{\pi\alpha}^\gamma \end{aligned} \tag{340}$$

$$S_{3\alpha} \approx 1 - (\theta_{\pi\alpha}^g)^2 - (\theta_{\pi\alpha}^\gamma)^2 - 2\theta_{\pi\alpha}^g \theta_{\pi\alpha}^\gamma + \theta_{\eta\alpha}^g \theta_{\pi\alpha}^g + \theta_{\eta\alpha}^\gamma \theta_{\pi\alpha}^\gamma + \theta_{\eta\alpha}^\gamma \theta_{\pi\alpha}^g + \theta_{\eta\alpha}^g \theta_{\pi\alpha}^\gamma \quad [\text{ld}] \tag{341}$$

Equations (242) through (244), (279) and (334) or (337) give the following second order approximations for $Q_{\sigma\alpha}$

$$Q_{1\alpha} \approx 1 - 1/2(\theta^{\gamma}_{\eta\alpha})^2 + 1/2(\theta^{\gamma}_{\pi\alpha})^2 - \theta^{\gamma}_{\eta\alpha}\theta^{g}_{\eta\alpha} + \theta^{\gamma}_{\pi\alpha}\theta^{g}_{\pi\alpha} - 1/2(\theta^{\gamma}_{\alpha})^2 - \theta^{\gamma}_{\alpha}\theta^{g}_{\alpha} \quad \text{[hd]} \qquad (342)$$

$$\approx 1 - (\theta^{\gamma}_{\eta\alpha})^2 - 2\theta^{\gamma}_{\eta\alpha}\theta^{g}_{\eta\alpha} + \theta^{\gamma}_{\eta\alpha}\theta^{\gamma}_{\pi\alpha} + \theta^{\gamma}_{\eta\alpha}\theta^{g}_{\pi\alpha} + \theta^{g}_{\eta\alpha}\theta^{\gamma}_{\pi\alpha}$$

$$Q_{2\alpha} \approx 1 - 1/2(\theta^{\gamma}_{\eta\alpha})^2 - 1/2(\theta^{\gamma}_{\pi\alpha})^2 - \theta^{\gamma}_{\eta\alpha}\theta^{g}_{\eta\alpha} - \theta^{\gamma}_{\pi\alpha}\theta^{g}_{\pi\alpha} - 1/2(\theta^{\gamma}_{\alpha})^2 - \theta^{\gamma}_{\alpha}\theta^{g}_{\alpha} \quad \text{[ld]} \qquad (343)$$

$$\approx 1 - (\theta^{\gamma}_{\eta\alpha})^2 - (\theta^{\gamma}_{\pi\alpha})^2 - 2\theta^{\gamma}_{\eta\alpha}\theta^{g}_{\eta\alpha} - 2\theta^{\gamma}_{\pi\alpha}\theta^{g}_{\pi\alpha} + \theta^{\gamma}_{\eta\alpha}\theta^{\gamma}_{\pi\alpha} + \theta^{\gamma}_{\eta\alpha}\theta^{g}_{\pi\alpha} + \theta^{g}_{\eta\alpha}\theta^{\gamma}_{\pi\alpha}$$

$$Q_{3\alpha} \approx 1 + 1/2(\theta^{\gamma}_{\eta\alpha})^2 - 1/2(\theta^{\gamma}_{\pi\alpha})^2 + \theta^{\gamma}_{\eta\alpha}\theta^{g}_{\eta\alpha} - \theta^{\gamma}_{\pi\alpha}\theta^{g}_{\pi\alpha} - 1/2(\theta^{\gamma}_{\alpha})^2 - \theta^{\gamma}_{\alpha}\theta^{g}_{\alpha} \quad \text{[ld]} \qquad (344)$$

$$\approx 1 - (\theta^{\gamma}_{\pi\alpha})^2 - 2\theta^{\gamma}_{\pi\alpha}\theta^{g}_{\pi\alpha} + \theta^{\gamma}_{\eta\alpha}\theta^{\gamma}_{\pi\alpha} + \theta^{\gamma}_{\eta\alpha}\theta^{g}_{\pi\alpha} + \theta^{g}_{\eta\alpha}\theta^{\gamma}_{\pi\alpha}$$

with the help of equation (179). Second order approximations for $R_{\sigma\alpha}$ are obtained from equations (245) through (247), (280) and (335) or (338) as follows

$$R_{1\alpha} \approx 1 - 1/2(\theta^{g}_{\eta\alpha})^2 + 1/2(\theta^{g}_{\pi\alpha})^2 - \theta^{g}_{\eta\alpha}\theta^{\gamma}_{\eta\alpha} + \theta^{g}_{\pi\alpha}\theta^{\gamma}_{\pi\alpha} - 1/2(\theta^{g}_{\alpha})^2 - \theta^{g}_{\alpha}\theta^{\gamma}_{\alpha} \quad \text{[hd]} \qquad (345)$$

$$\approx 1 - (\theta^{g}_{\eta\alpha})^2 - 2\theta^{g}_{\eta\alpha}\theta^{\gamma}_{\eta\alpha} + \theta^{g}_{\eta\alpha}\theta^{g}_{\pi\alpha} + \theta^{g}_{\eta\alpha}\theta^{\gamma}_{\pi\alpha} + \theta^{\gamma}_{\eta\alpha}\theta^{g}_{\pi\alpha}$$

$$R_{2\alpha} \approx 1 - 1/2(\theta^{g}_{\eta\alpha})^2 - 1/2(\theta^{g}_{\pi\alpha})^2 - \theta^{g}_{\eta\alpha}\theta^{\gamma}_{\eta\alpha} - \theta^{g}_{\pi\alpha}\theta^{\gamma}_{\pi\alpha} - 1/2(\theta^{g}_{\alpha})^2 - \theta^{g}_{\alpha}\theta^{\gamma}_{\alpha} \quad \text{[ld]} \qquad (346)$$

$$\approx 1 - (\theta^{g}_{\eta\alpha})^2 - (\theta^{g}_{\pi\alpha})^2 - 2\theta^{g}_{\eta\alpha}\theta^{\gamma}_{\eta\alpha} - 2\theta^{g}_{\pi\alpha}\theta^{\gamma}_{\pi\alpha} + \theta^{g}_{\eta\alpha}\theta^{g}_{\pi\alpha} + \theta^{g}_{\eta\alpha}\theta^{\gamma}_{\pi\alpha} + \theta^{\gamma}_{\eta\alpha}\theta^{g}_{\pi\alpha}$$

$$R_{3\alpha} \approx 1 + 1/2(\theta^{g}_{\eta\alpha})^2 - 1/2(\theta^{g}_{\pi\alpha})^2 + \theta^{g}_{\eta\alpha}\theta^{\gamma}_{\eta\alpha} - \theta^{g}_{\pi\alpha}\theta^{\gamma}_{\pi\alpha} - 1/2(\theta^{g}_{\alpha})^2 - \theta^{g}_{\alpha}\theta^{\gamma}_{\alpha} \quad \text{[ld]} \qquad (347)$$

$$\approx 1 - (\theta^{g}_{\pi\alpha})^2 - 2\theta^{g}_{\pi\alpha}\theta^{\gamma}_{\pi\alpha} + \theta^{g}_{\eta\alpha}\theta^{g}_{\pi\alpha} + \theta^{g}_{\eta\alpha}\theta^{\gamma}_{\pi\alpha} + \theta^{\gamma}_{\eta\alpha}\theta^{g}_{\pi\alpha}$$

after using equation (179).

The quantities $S_{\sigma\alpha}$, $Q_{\sigma\alpha}$ and $R_{\sigma\alpha}$ in equations (339) through (347) that also appear in the calculation of α_m in equation (277), can be written in matrix form as

$$S_{\alpha} \approx Z_m \varphi_{11\alpha} \qquad Q_{\alpha} \approx W_m \varphi^{\gamma g}_{8\alpha} \qquad R_{\alpha} \approx W_m \varphi^{g\gamma}_{8\alpha} \qquad (348)$$

where Z_m is given by

$$Z_m = \begin{pmatrix} 1 & -1 & 0 & -1 & 0 & -2 & 0 & 1 & 1 & 1 & 1 \\ 1 & -1 & -1 & -1 & -1 & -2 & -2 & 1 & 1 & 1 & 1 \\ 1 & 0 & -1 & 0 & -1 & 0 & -2 & 1 & 1 & 1 & 1 \end{pmatrix} \qquad (349)$$

while W_m is given by

$$W_m = \begin{pmatrix} 1 & -1 & 0 & -2 & 0 & 1 & 1 & 1 \\ 1 & -1 & -1 & -2 & -2 & 1 & 1 & 1 \\ 1 & 0 & -1 & 0 & -2 & 1 & 1 & 1 \end{pmatrix} \tag{350}$$

and where

$$S_\alpha = \begin{pmatrix} S_{1\alpha} \\ S_{2\alpha} \\ S_{3\alpha} \end{pmatrix} \qquad Q_\alpha = \begin{pmatrix} Q_{1\alpha} \\ Q_{2\alpha} \\ Q_{3\alpha} \end{pmatrix} \qquad R_\alpha = \begin{pmatrix} R_{1\alpha} \\ R_{2\alpha} \\ R_{3\alpha} \end{pmatrix} \tag{351}$$

and where

$$\varphi_{11\alpha} = \begin{pmatrix} 1 \\ (\theta^g_{\eta\alpha})^2 \\ (\theta^g_{\pi\alpha})^2 \\ (\theta^\gamma_{\eta\alpha})^2 \\ (\theta^\gamma_{\pi\alpha})^2 \\ \theta^g_{\eta\alpha}\theta^\gamma_{\eta\alpha} \\ \theta^g_{\pi\alpha}\theta^\gamma_{\pi\alpha} \\ \theta^g_{\eta\alpha}\theta^g_{\pi\alpha} \\ \theta^\gamma_{\eta\alpha}\theta^\gamma_{\pi\alpha} \\ \theta^\gamma_{\eta\alpha}\theta^g_{\pi\alpha} \\ \theta^g_{\eta\alpha}\theta^\gamma_{\pi\alpha} \end{pmatrix} \qquad \varphi^{\gamma g}_{8\alpha} = \begin{pmatrix} 1 \\ (\theta^\gamma_{\eta\alpha})^2 \\ (\theta^\gamma_{\pi\alpha})^2 \\ \theta^\gamma_{\eta\alpha}\theta^g_{\eta\alpha} \\ \theta^\gamma_{\pi\alpha}\theta^g_{\pi\alpha} \\ \theta^\gamma_{\eta\alpha}\theta^\gamma_{\pi\alpha} \\ \theta^\gamma_{\eta\alpha}\theta^g_{\pi\alpha} \\ \theta^g_{\eta\alpha}\theta^\gamma_{\pi\alpha} \end{pmatrix} \qquad \varphi^{g\gamma}_{8\alpha} = \begin{pmatrix} 1 \\ (\theta^g_{\eta\alpha})^2 \\ (\theta^g_{\pi\alpha})^2 \\ \theta^g_{\eta\alpha}\theta^\gamma_{\eta\alpha} \\ \theta^g_{\pi\alpha}\theta^\gamma_{\pi\alpha} \\ \theta^g_{\eta\alpha}\theta^g_{\pi\alpha} \\ \theta^g_{\eta\alpha}\theta^\gamma_{\pi\alpha} \\ \theta^\gamma_{\eta\alpha}\theta^g_{\pi\alpha} \end{pmatrix} \tag{352}$$

The measured space coordinates α_m are then calculated from equations (277) through (280).

For the special case when the ambient electromagnetic wave field is zero and $\theta^\gamma_{\eta\alpha} = 0$, $\theta^\gamma_{\pi\alpha} = 0$ and $\theta^\gamma_\alpha = 0$ it follows from equation (342) through (344) that $Q_{\sigma\alpha} = 1$ with $\sigma = 1$ for [hd] and $\sigma = 2,3$ for [ld], or

$$Q_{1\alpha} = 1 \qquad Q_{2\alpha} = 1 \qquad Q_{3\alpha} = 1 \tag{353}$$

while equations (336) through (338) give $A_\alpha = 1$ and

$$T_\alpha = B_\alpha \approx 1 - 1/2(\theta^g_{\eta\alpha})^2 - 1/2(\theta^g_{\pi\alpha})^2 + \theta^g_{\eta\alpha}\theta^g_{\pi\alpha} \tag{354}$$

and equations (339) through (341) and (345) through (347) give $S_{\sigma\alpha} = R_{\sigma\alpha}$ to second order approximation as

$$S_{1\alpha} = R_{1\alpha} \approx 1 - (\theta^g_{\eta\alpha})^2 + \theta^g_{\eta\alpha}\theta^g_{\pi\alpha} \qquad \text{[hd]} \tag{355}$$

$$S_{2\alpha} = R_{2\alpha} \approx 1 - (\theta^g_{\eta\alpha})^2 - (\theta^g_{\pi\alpha})^2 + \theta^g_{\eta\alpha}\theta^g_{\pi\alpha} \qquad [ld] \qquad (356)$$

$$S_{3\alpha} = R_{3\alpha} \approx 1 - (\theta^g_{\pi\alpha})^2 + \theta^g_{\eta\alpha}\theta^g_{\pi\alpha} \qquad [ld] \qquad (357)$$

Then equation (277) gives for this special case with $\sigma = 1,2,3$

$$\alpha_m = \alpha^g_m = \alpha^o S_{\sigma\alpha} = \alpha^o R_{\sigma\alpha} \qquad \alpha^\gamma_m = \alpha^o \qquad (358)$$

where $S_{\sigma\alpha} = R_{\sigma\alpha}$ is given by equations (355) through (357) for the special case of zero ambient electromagnetic wave fields and weak ambient gravitational waves.

For the case when the ambient gravity wave field is zero and $\theta^g_{\eta\alpha} = 0$, $\theta^g_{\pi\alpha} = 0$ and $\theta^g_\alpha = 0$ it follows from equations (345) through (347) that $R_{\sigma\alpha} = 1$ with $\sigma = 1$ for [hd] and $\sigma = 2,3$ for [ld], or

$$R_{1\alpha} = 1 \qquad R_{2\alpha} = 1 \qquad R_{3\alpha} = 1 \qquad (359)$$

while equations (336) through (338) give $B_\alpha = 1$ and

$$T_\alpha = A_\alpha \approx 1 - 1/2(\theta^\gamma_{\eta\alpha})^2 - 1/2(\theta^\gamma_{\pi\alpha})^2 + \theta^\gamma_{\eta\alpha}\theta^\gamma_{\pi\alpha} \qquad (360)$$

and equations (342) through (344) give $S_{\sigma\alpha} = Q_{\sigma\alpha}$ to second order as

$$S_{1\alpha} = Q_{1\alpha} \approx 1 - (\theta^\gamma_{\eta\alpha})^2 + \theta^\gamma_{\eta\alpha}\theta^\gamma_{\pi\alpha} \qquad [hd] \qquad (361)$$

$$S_{2\alpha} = Q_{2\alpha} \approx 1 - (\theta^\gamma_{\eta\alpha})^2 - (\theta^\gamma_{\pi\alpha})^2 + \theta^\gamma_{\eta\alpha}\theta^\gamma_{\pi\alpha} \qquad [ld] \qquad (362)$$

$$S_{3\alpha} = Q_{3\alpha} \approx 1 - (\theta^\gamma_{\pi\alpha})^2 + \theta^\gamma_{\eta\alpha}\theta^\gamma_{\pi\alpha} \qquad [ld] \qquad (363)$$

For this special case equation (277) gives for $\sigma = 1,2,3$

$$\alpha_m = \alpha^\gamma_m = \alpha^o S_{\sigma\alpha} = \alpha^o Q_{\sigma\alpha} \qquad \alpha^g_m = \alpha^o \qquad (364)$$

where $S_{\sigma\alpha} = Q_{\sigma\alpha}$ is given by equations (361) through (363) for this special case of zero ambient gravity wave fields and weak ambient electromagnetic waves.

Consider now the case when both the ambient gravity wave fields and the ambient electromagnetic wave fields are zero which gives

$$T_\alpha = 1 \quad A_\alpha = 1 \quad B_\alpha = 1 \quad S_{\sigma\alpha} = 1 \quad Q_{\sigma\alpha} = 1 \quad R_{\sigma\alpha} = 1 \quad \alpha_m = \alpha^o \qquad (365)$$

as must be the case. Finally, it should be pointed out that γ and g can represent any two dominant wave ambient fields and not just ambient electromagnetic waves and ambient gravitational waves, as for example seismic waves, acoustic waves, electro-acoustic waves and plasma waves.

5. THE THEORY OF TIME COORDINATES IN MATTER IN THE PRESENCE OF AMBIENT GRAVITATIONAL WAVES AND AMBIENT ELECTROMAGNETIC WAVES. Equation (115) gives the amplitude t of the complex number active time coordinate in the presence of ambient gravity waves and ambient electromagnetic waves in terms of the conventional time coordinate t^o that corresponds to zero ambient fields. However, t^o cannot be measured in a laboratory on the earth because ambient gravitational waves are always present. This section shows how to express the magnitude and phase angle of the active time coordinate for a combined ambient gravity wave field and ambient electromagnetic wave field of energy density $\mathfrak{E}^g + \mathfrak{E}^\gamma$ in terms of the time coordinate magnitude t^g when only ambient gravity waves of energy density $\mathfrak{E}^g$ are present or in terms of the time coordinate magnitude t^γ when an ambient electromagnetic wave field of energy density $\mathfrak{E}^\gamma$ is dominant and in terms of the internal phase angles $\theta^g_{\eta t}$ and $\theta^g_{\pi t}$ of the time counting and segmentation numbers when only ambient gravity waves are present and $\theta^\gamma_{\eta t}$ and $\theta^\gamma_{\pi t}$ when the ambient electromagnetic wave field is dominant. Note that each coordinate amplitude t^o, t^g and t^γ corresponds to the presence of a multitude of conventional radiation fields including a conventional vacuum radiation field, a conventional gravity radiation field and a conventional electromagnetic radiation field.

When ambient gravity waves and ambient electromagnetic waves of energy density are both present in matter as indicated in equations (155) and (159) through (161) and after neglecting the zero point energy contributions, it follows from the linearity of the internal phase angles and the energy density of the ambient wave fields, as shown in equation (153), that

$$\theta_{\eta t} = \theta^g_{\eta t} + \theta^\gamma_{\eta t} \qquad \theta_{\pi t} = \theta^g_{\pi t} + \theta^\gamma_{\pi t} \tag{366}$$

$$\theta_t = \theta_{\eta t} - \theta_{\pi t} = \theta^g_t + \theta^\gamma_t = \theta^g_{\eta t} + \theta^\gamma_{\eta t} - \theta^g_{\pi t} - \theta^\gamma_{\pi t} \tag{367}$$

and

$$\theta^g_t = \theta^g_{\eta t} - \theta^g_{\pi t} \qquad \theta^\gamma_t = \theta^\gamma_{\eta t} - \theta^\gamma_{\pi t} \tag{368}$$

where θ^g_t = internal phase angle of the time coordinate when only ambient gravity waves are present and θ^γ_t = internal phase angle of the time coordinate when only ambient electromagnetic waves are present. Therefore in this section equation (115)

$$t = t^o C_{\sigma t} \tag{369}$$

corresponds to the case of combined ambient gravity waves and ambient electromagnetic waves of energy density $\mathfrak{E}^g + \mathfrak{E}^\gamma$. In this section equations (109) through (112) signify

$$C_{\sigma t} = C_\sigma(\theta_{\eta t} = \theta^g_{\eta t} + \theta^\gamma_{\eta t}, \theta_{\pi t} = \theta^g_{\pi t} + \theta^\gamma_{\pi t}) \tag{370}$$

which is the general case when ambient gravity waves and ambient electromagnetic waves have comparable energy densities so that neither can be neglected in the calculations.

A. Time Coordinate Factors for the Component Ambient Gravity Waves and Ambient Electromagnetic Waves.

When only ambient gravity waves are present and the ambient electromagnetic waves have been turned off and $\theta^{\gamma}_{\eta t} = 0, \theta^{\gamma}_{\pi t} = 0$ and $\theta^{\gamma}_{t} = 0$ then equation (370) becomes

$$C^{g}_{\sigma t} = C_{\sigma}(\theta_{\eta t} = \theta^{g}_{\eta t}, \theta_{\pi t} = \theta^{g}_{\pi t}) \qquad (371)$$

or with $\sigma = 1$ for [hd] and $\sigma = 2,3$ for [ld] equations (110) through (112) become

$$C^{g}_{1t} = \cos\theta^{g}_{\eta t} \sec\theta^{g}_{\pi t} \qquad [\text{hd}] \qquad (372)$$

$$C^{g}_{2t} = \cos\theta^{g}_{\eta t} \cos\theta^{g}_{\pi t} \qquad [\text{ld}] \qquad (373)$$

$$C^{g}_{3t} = \sec\theta^{g}_{\eta t} \cos\theta^{g}_{\pi t} \qquad [\text{ld}] \qquad (374)$$

which corresponds to ambient gravity waves with an energy density of $\mathfrak{E}^{g}$ where according to equation (368) the case for ambient gravitational waves can be written as

$$\theta^{g}_{\pi t} \geq \theta^{g}_{\eta t} \qquad \theta^{g}_{t} \leq 0 \qquad (375)$$

and equations (372) through (374) show that if $|\theta^{g}_{\pi t}| \geq |\theta^{g}_{\eta t}|$ then

$$C^{g}_{1t} \geq 1 \qquad C^{g}_{2t} \leq 1 \qquad C^{g}_{3t} \leq 1 \qquad |\theta^{g}_{\pi t}| \geq |\theta^{g}_{\eta t}| \qquad (376)$$

When only ambient gravity waves are present equation (369) becomes

$$t^{g} = t^{o} C^{g}_{\sigma t} \qquad (377)$$

where t^{g} = time coordinate magnitude for the case when only ambient gravity waves are present and where the time coordinate factor is given by equations (372) through (374). Equation (377) allows the calculation of t^{o} as

$$t^{o} = t^{g} (C^{g}_{\sigma t})^{-1} \qquad (378)$$

and can be used to calculate t^{o} if t^{g} and $C^{g}_{\sigma t}$ are known.

For the case when only ambient electromagnetic waves are present (or at least when the ambient electromagnetic wave field is totally dominant over the ambient gravity wave field) with $\theta^{g}_{\eta t} = 0, \theta^{g}_{\pi t} = 0$ and $\theta^{g}_{t} = 0$, then equation (370) becomes

$$C^{\gamma}_{\sigma t} = C_{\sigma}(\theta_{\eta t} = \theta^{\gamma}_{\eta t}, \theta_{\pi t} = \theta^{\gamma}_{\pi t}) \qquad (379)$$

while with $\sigma = 1$ for [hd] and $\sigma = 2,3$ for [ld] equations (105) through (107) give

$$C^{\gamma}_{1t} = \cos\theta^{\gamma}_{\eta t} \sec\theta^{\gamma}_{\pi t} \qquad [\text{hd}] \qquad (380)$$

$$C^{\gamma}_{2t} = \cos\theta^{\gamma}_{\eta t} \cos\theta^{\gamma}_{\pi t} \qquad [\text{ld}] \qquad (381)$$

$$C^{\gamma}_{3t} = \sec\theta^{\gamma}_{\eta t} \cos\theta^{\gamma}_{\pi t} \qquad [1d] \qquad (382)$$

which corresponds to an ambient electromagnetic wave field of energy density $\mathfrak{E}^{\gamma}$, where according to equation (368) the condition for negative or zero value for the internal phase angle θ^{γ}_{t} of the time coordinate is given by

$$\theta^{\gamma}_{\pi t} \geq \theta^{\gamma}_{\eta t} \qquad \theta^{\gamma}_{t} \leq 0 \qquad (383)$$

while equations (380) through (382) show that if $|\theta^{\gamma}_{\pi t}| \geq |\theta^{\gamma}_{\eta t}|$ then

$$C^{\gamma}_{1t} \geq 1 \qquad C^{\gamma}_{2t} \leq 1 \qquad C^{\gamma}_{3t} \leq 1 \qquad |\theta^{\gamma}_{\pi t}| \geq |\theta^{\gamma}_{\eta t}| \qquad (384)$$

For the case when only ambient electromagnetic waves are present equation (369) becomes

$$t^{\gamma} = t^{o} C^{\gamma}_{\sigma t} \qquad (385)$$

where t^{γ} = time coordinate magnitude for the situation when only ambient electromagnetic waves are present, and where the time coordinate factors $C^{\gamma}_{\sigma t}$ are given by equations (380) through (382). From equation (385) it follows that

$$t^{o} = t^{\gamma} (C^{\gamma}_{\sigma t})^{-1} \qquad (386)$$

which allows the calculation of t^{o} if t^{γ} and $C^{\gamma}_{\sigma t}$ are known quantities.

B. Time Coordinate Magnitude Representations in Matter for Combined Ambient Gravity Waves and Ambient Electromagnetic Waves in Terms of the Magnitudes of the Time Coordinates for the Cases When Only Ambient Gravity Waves or Ambient Electromagnetic Waves are Acting Alone.

The representations of the magnitude of the time coordinate in matter for the case of a combined ambient gravitational waves and ambient electromagnetic waves in terms of the conventional time coordinate t^{o} when no ambient wave fields are acting or in terms of the magnitude t^{g} of the complex number active time coordinate when only ambient gravity waves are present or in terms of the magnitude t^{γ} of the complex number active time coordinate when only ambient electromagnetic waves are present are obtained from equations (369), (377) and (385) as

$$t = t^{o} C_{\sigma t} = t^{g} D_{\sigma t} = t^{\gamma} E_{\sigma t} \qquad (387)$$

where

$$D_{\sigma t} = C_{\sigma t} / C^{g}_{\sigma t} \qquad E_{\sigma t} = C_{\sigma t} / C^{\gamma}_{\sigma t} \qquad (388)$$

where $C_{\sigma t}$, $C^{g}_{\sigma t}$ and $C^{\gamma}_{\sigma t}$ are given by equations (110) through (112), (372) through (374) and (380) through (382) respectively. In a similar way to equation (198) for space coordinate magnitudes, equation (387) describes three equivalent representations of the time coordinate in the presence of ambient combined ambient gravitational and ambient electromagnetic waves. Depending on the relative strengths of the component ambient gravity waves and ambient electromagnetic waves, four cases of the effects of the combined ambient field on the magnitude of the active time coordinate will now be delineated.

Case 1. Strong Ambient Gravity Waves and Strong Ambient Electromagnetic Waves.

Combining equations (388) with equations (110) through (112), (372) through (374) and (380) through (382) gives $D_{\sigma t}$ and $E_{\sigma t}$ with $\sigma = 1$ for [hd] and $\sigma = 2,3$ for [ld] and for the general case of strong ambient gravity waves and strong ambient electromagnetic waves as

$$D_{1t} = I_t/J_t \qquad D_{2t} = I_t J_t \qquad D_{3t} = J_t/I_t \tag{389}$$

$$E_{1t} = K_t/L_t \qquad E_{2t} = K_t L_t \qquad E_{3t} = L_t/K_t \tag{390}$$

Where

$$I_t = \cos\theta^{\gamma}_{\eta t} - \sin\theta^{\gamma}_{\eta t}\tan\theta^{g}_{\eta t} \tag{391}$$

$$J_t = \cos\theta^{\gamma}_{\pi t} - \sin\theta^{\gamma}_{\pi t}\tan\theta^{g}_{\pi t} \tag{392}$$

$$K_t = \cos\theta^{g}_{\eta t} - \sin\theta^{g}_{\eta t}\tan\theta^{\gamma}_{\eta t} \tag{393}$$

$$L_t = \cos\theta^{g}_{\pi t} - \sin\theta^{g}_{\pi t}\tan\theta^{\gamma}_{\pi t} \tag{394}$$

subject to equations (366) through (368). Then equation (387) can be used to calculate the magnitude t of the active time coordinate. The internal phase angle θ_t of the active time coordinate is calculated by equation (367).

If only ambient gravity waves are present, then $\theta^{\gamma}_{\eta t} = 0$, $\theta^{\gamma}_{\pi t} = 0$, $\theta^{\gamma}_{t} = 0$ and equations (387) through (394) give $I_t = 1, J_t = 1, K_t = \cos\theta^{g}_{\eta t}, L_t = \cos\theta^{g}_{\pi t}$, $D_{\sigma t} = 1$, $C_{\sigma t} = E_{\sigma t} = C^{g}_{\sigma t}, t^{\gamma} = t^{o}$ and $t = t^{g}$. If only ambient electromagnetic waves exist, then $\theta^{g}_{\eta t} = 0$, $\theta^{g}_{\pi t} = 0$, $\theta^{g}_{t} = 0$ and equations (387) through (394) give $I_t = \cos\theta^{\gamma}_{\eta t}$, $J_t = \cos\theta^{\gamma}_{\pi t}, K_t = 1, L_t = 1$, $E_{\sigma t} = 1$, $C_{\sigma t} = D_{\sigma t} = C^{\gamma}_{\sigma t}, t^{g} = t^{o}$ and $t = t^{\gamma}$.

Case 2. Strong Ambient Gravity Waves and Weak Ambient Electromagnetic Waves.

Equations (389) through (394) show that for the case of strong ambient gravity waves and weak ambient electromagnetic waves with $\theta^{\gamma}_{\eta t} \approx 0, \theta^{\gamma}_{\pi t} \approx 0$ and $\theta^{\gamma}_{t} \approx 0$, the following first order approximations are valid

$$I_t \approx 1 - \theta^{\gamma}_{\eta t}\tan\theta^{g}_{\eta t} \tag{395}$$

$$J_t \approx 1 - \theta^{\gamma}_{\pi t}\tan\theta^{g}_{\pi t} \tag{396}$$

$$K_t \approx I_t \cos\theta^g_{\eta t} \approx \cos\theta^g_{\eta t} - \theta^\gamma_{\eta t} \sin\theta^g_{\eta t} \tag{397}$$

$$L_t \approx J_t \cos\theta^g_{\pi t} \approx \cos\theta^g_{\pi t} - \theta^\gamma_{\pi t} \sin\theta^g_{\pi t} \tag{398}$$

and with $\sigma = 1$ for [hd] and $\sigma = 2,3$ for [ld]

$$D_{1t} \approx 1 - \theta^\gamma_{\eta t} \tan\theta^g_{\eta t} + \theta^\gamma_{\pi t} \tan\theta^g_{\pi t} \qquad \text{[hd]} \tag{399}$$

$$D_{2t} \approx 1 - \theta^\gamma_{\eta t} \tan\theta^g_{\eta t} - \theta^\gamma_{\pi t} \tan\theta^g_{\pi t} \qquad \text{[ld]} \tag{400}$$

$$D_{3t} \approx 1 + \theta^\gamma_{\eta t} \tan\theta^g_{\eta t} - \theta^\gamma_{\pi t} \tan\theta^g_{\pi t} \qquad \text{[ld]} \tag{401}$$

and because $C^\gamma_{\sigma t} \approx 1 +$ second order terms, it follows that equation (388) gives the first order approximation

$$C_{1t} \approx E_{1t} \approx C^g_{1t} D_{1t} \qquad \text{[hd]} \tag{402}$$

$$C_{2t} \approx E_{2t} \approx C^g_{2t} D_{2t} \qquad \text{[ld]} \tag{403}$$

$$C_{3t} \approx E_{3t} \approx C^g_{3t} D_{3t} \qquad \text{[ld]} \tag{404}$$

or in general

$$C_{\sigma t} \approx E_{\sigma t} \approx C^g_{\sigma t} D_{\sigma t} \tag{405}$$

where $C^g_{\sigma t}$ is given by equations (372) through (374). Equations (399) through (401) can be written in matrix form as

$$D_t = F^g_t \varphi^\gamma_{3t} \tag{406}$$

where

$$F^g_t = \begin{pmatrix} 1 & -\tan\theta^g_{\eta t} & +\tan\theta^g_{\pi t} \\ 1 & -\tan\theta^g_{\eta t} & -\tan\theta^g_{\pi t} \\ 1 & +\tan\theta^g_{\eta t} & -\tan\theta^g_{\pi t} \end{pmatrix} \tag{407}$$

and where

$$D_t = \begin{pmatrix} D_{1t} \\ D_{2t} \\ D_{3t} \end{pmatrix} \qquad \varphi^\gamma_{3t} = \begin{pmatrix} 1 \\ \theta^\gamma_{\eta t} \\ \theta^\gamma_{\pi t} \end{pmatrix} \tag{408}$$

Then equation (387) can be used to calculate the magnitude t of the time coordinate.

The case of a zero ambient electromagnetic wave field with $\theta^\gamma_{\eta t} = 0, \theta^\gamma_{\pi t} = 0$ and $\theta^\gamma_t = 0$ gives

$$C^{\gamma}_{\sigma t} = 1 \qquad D_{\sigma t} = 1 \qquad C_{\sigma t} = E_{\sigma t} = C^{g}_{\sigma t} \qquad t^{\gamma} = t^{o} \tag{409}$$

and equation (387) gives

$$t = t^{g} = t^{o} C^{g}_{\sigma t} \tag{410}$$

with only ambient gravity waves present.

Case 3. Weak Ambient Gravity Waves and Strong Ambient Electromagnetic Waves.

Equations (389) through (394) indicate that for the case of weak ambient gravity waves and strong ambient electromagnetic waves with $\theta^{g}_{\eta t} \approx 0, \theta^{g}_{\pi t} \approx 0$ and $\theta^{g}_{t} \approx 0$ give the following first order approximations

$$I_{t} \approx K_{t} \cos\theta^{\gamma}_{\eta t} \approx \cos\theta^{\gamma}_{\eta t} - \theta^{g}_{\eta t} \sin\theta^{\gamma}_{\eta t} \tag{411}$$

$$J_{t} \approx L_{t} \cos\theta^{\gamma}_{\pi t} \approx \cos\theta^{\gamma}_{\pi t} - \theta^{g}_{\pi t} \sin\theta^{\gamma}_{\pi t} \tag{412}$$

$$K_{t} \approx 1 - \theta^{g}_{\eta t} \tan\theta^{\gamma}_{\eta t} \tag{413}$$

$$L_{t} \approx 1 - \theta^{g}_{\pi t} \tan\theta^{\gamma}_{\pi t} \tag{414}$$

and with $\sigma = 1$ for [hd] and $\sigma = 2,3$ for [ld]

$$E_{1t} \approx 1 - \theta^{g}_{\eta t} \tan\theta^{\gamma}_{\eta t} + \theta^{g}_{\pi t} \tan\theta^{\gamma}_{\pi t} \qquad \text{[hd]} \tag{415}$$

$$E_{2t} \approx 1 - \theta^{g}_{\eta t} \tan\theta^{\gamma}_{\eta t} - \theta^{g}_{\pi t} \tan\theta^{\gamma}_{\pi t} \qquad \text{[ld]} \tag{416}$$

$$E_{3t} \approx 1 + \theta^{g}_{\eta t} \tan\theta^{\gamma}_{\eta t} - \theta^{g}_{\pi t} \tan\theta^{\gamma}_{\pi t} \qquad \text{[ld]} \tag{417}$$

and because $C^{g}_{\sigma t} \approx 1 +$ second order terms, it follows from equation (388) that

$$C_{1t} \approx D_{1t} \approx C^{\gamma}_{1t} E_{1t} \qquad \text{[hd]} \tag{418}$$

$$C_{2t} \approx D_{2t} \approx C^{\gamma}_{2t} E_{2t} \qquad \text{[ld]} \tag{419}$$

$$C_{3t} \approx D_{3t} \approx C^{\gamma}_{3t} E_{3t} \qquad \text{[ld]} \tag{420}$$

or in short notation

$$C_{\sigma t} \approx D_{\sigma t} \approx C^{\gamma}_{\sigma t} E_{\sigma t} \tag{421}$$

where $C^{\gamma}_{\sigma t}$ for $\sigma = 1,2,3$ is given by equations (380) through (382). Equations (415) through (417) can be written in matrix form in the following way

$$E_{t} \approx F^{\gamma}_{t} \varphi^{g}_{3t} \tag{422}$$

where

$$F_t^\gamma = \begin{pmatrix} 1 & -\tan\theta_{\eta t}^\gamma & +\tan\theta_{\pi t}^\gamma \\ 1 & -\tan\theta_{\eta t}^\gamma & -\tan\theta_{\pi t}^\gamma \\ 1 & +\tan\theta_{\eta t}^\gamma & -\tan\theta_{\pi t}^\gamma \end{pmatrix} \qquad (423)$$

and where

$$E_t = \begin{pmatrix} E_{1t} \\ E_{2t} \\ E_{3t} \end{pmatrix} \qquad \varphi_{3t}^g = \begin{pmatrix} 1 \\ \theta_{\eta t}^g \\ \theta_{\pi t}^g \end{pmatrix} \qquad (424)$$

Equation (387) can then be used to calculate the magnitude t of the complex number active time coordinate.

The case of a zero ambient gravity wave field with $\theta_{\eta t}^g = 0, \theta_{\pi t}^g = 0$ and $\theta_t^g = 0$ gives

$$C_{\sigma t}^g = 1 \qquad E_{\sigma t} = 1 \qquad C_{\sigma t} = D_{\sigma t} = C_{\sigma t}^\gamma \qquad t^g = t^o \qquad (425)$$

and equation (365) gives

$$t = t^\gamma = t^o C_{\sigma t}^\gamma \qquad (426)$$

for the case when only ambient electromagnetic waves present. The case of zero ambient gravity waves can occur at great distances from an ambient gravitational wave source.

Case 4. Weak Ambient Gravity Waves and Weak Ambient Electromagnetic Waves.

The case of weak ambient gravity waves and weak ambient electromagnetic waves is obtained from equations (389) through (394) for the case when $\theta_{\eta t}^g \approx 0, \theta_{\pi t}^g \approx 0, \theta_t^g \approx 0$ and $\theta_{\eta t}^\gamma \approx 0, \theta_{\pi t}^\gamma \approx 0, \theta_t^\gamma \approx 0$, and yields the following second order approximations

$$I_t \approx 1 - 1/2(\theta_{\eta t}^\gamma)^2 - \theta_{\eta t}^\gamma \theta_{\eta t}^g \qquad (427)$$

$$J_t \approx 1 - 1/2(\theta_{\pi t}^\gamma)^2 - \theta_{\pi t}^\gamma \theta_{\pi t}^g \qquad (428)$$

$$K_t \approx 1 - 1/2(\theta_{\eta t}^g)^2 - \theta_{\eta t}^g \theta_{\eta t}^\gamma \qquad (429)$$

$$L_t \approx 1 - 1/2(\theta_{\pi t}^g)^2 - \theta_{\pi t}^g \theta_{\pi t}^\gamma \qquad (430)$$

Equations (389) and (390) then give with $\sigma = 1$ for [hd] and $\sigma = 2,3$ for [ld]

$$D_{1t} \approx 1 - 1/2(\theta^{\gamma}_{\eta t})^2 + 1/2(\theta^{\gamma}_{\pi t})^2 - \theta^{\gamma}_{\eta t}\theta^{g}_{\eta t} + \theta^{\gamma}_{\pi t}\theta^{g}_{\pi t} \qquad \text{[hd]} \qquad (431)$$

$$D_{2t} \approx 1 - 1/2(\theta^{\gamma}_{\eta t})^2 - 1/2(\theta^{\gamma}_{\pi t})^2 - \theta^{\gamma}_{\eta t}\theta^{g}_{\eta t} - \theta^{\gamma}_{\pi t}\theta^{g}_{\pi t} \qquad \text{[ld]} \qquad (432)$$

$$D_{3t} \approx 1 + 1/2(\theta^{\gamma}_{\eta t})^2 - 1/2(\theta^{\gamma}_{\pi t})^2 + \theta^{\gamma}_{\eta t}\theta^{g}_{\eta t} - \theta^{\gamma}_{\pi t}\theta^{g}_{\pi t} \qquad \text{[ld]} \qquad (433)$$

$$E_{1t} \approx 1 - 1/2(\theta^{g}_{\eta t})^2 + 1/2(\theta^{g}_{\pi t})^2 - \theta^{g}_{\eta t}\theta^{\gamma}_{\eta t} + \theta^{g}_{\pi t}\theta^{\gamma}_{\pi t} \qquad \text{[hd]} \qquad (434)$$

$$E_{2t} \approx 1 - 1/2(\theta^{g}_{\eta t})^2 - 1/2(\theta^{g}_{\pi t})^2 - \theta^{g}_{\eta t}\theta^{\gamma}_{\eta t} - \theta^{g}_{\pi t}\theta^{\gamma}_{\pi t} \qquad \text{[ld]} \qquad (435)$$

$$E_{3t} \approx 1 + 1/2(\theta^{g}_{\eta t})^2 - 1/2(\theta^{g}_{\pi t})^2 + \theta^{g}_{\eta t}\theta^{\gamma}_{\eta t} - \theta^{g}_{\pi t}\theta^{\gamma}_{\pi t} \qquad \text{[ld]} \qquad (436)$$

The matrix form of equations (431) through (436) is written as

$$D_t \approx W\varphi^{\gamma g}_{5t} \qquad E_t \approx W\varphi^{g\gamma}_{5t} \qquad (437)$$

where the matrix W is given by equation (249) as

$$W = \begin{pmatrix} 1 & -1/2 & +1/2 & -1 & +1 \\ 1 & -1/2 & -1/2 & -1 & -1 \\ 1 & +1/2 & -1/2 & +1 & -1 \end{pmatrix} \qquad (438)$$

and where

$$D_t = \begin{pmatrix} D_{1t} \\ D_{2t} \\ D_{3t} \end{pmatrix} \qquad E_t = \begin{pmatrix} E_{1t} \\ E_{2t} \\ E_{3t} \end{pmatrix} \qquad (439)$$

$$\varphi^{\gamma g}_{5t} = \begin{pmatrix} 1 \\ (\theta^{\gamma}_{\eta t})^2 \\ (\theta^{\gamma}_{\pi t})^2 \\ \theta^{\gamma}_{\eta t}\theta^{g}_{\eta t} \\ \theta^{\gamma}_{\pi t}\theta^{g}_{\pi t} \end{pmatrix} \qquad \varphi^{g\gamma}_{5t} = \begin{pmatrix} 1 \\ (\theta^{g}_{\eta t})^2 \\ (\theta^{g}_{\pi t})^2 \\ \theta^{g}_{\eta t}\theta^{\gamma}_{\eta t} \\ \theta^{g}_{\pi t}\theta^{\gamma}_{\pi t} \end{pmatrix} \qquad (440)$$

Equations (110) through (112) for the case of weak ambient gravity waves and weak ambient electromagnetic waves with $\theta_{\eta t} \approx 0$ and $\theta_{\pi t} \approx 0$ give the following second order approximations for $C_{\sigma t}$

$$C_{1t} \approx 1 - 1/2\theta^2_{\eta t} + 1/2\theta^2_{\pi t} \qquad \text{[hd]} \qquad (441)$$

$$C_{2t} \approx 1 - 1/2\theta^2_{\eta t} - 1/2\theta^2_{\pi t} \qquad \text{[ld]} \qquad (442)$$

$$C_{3t} \approx 1 + 1/2\theta^2_{\eta t} - 1/2\theta^2_{\pi t} \qquad \text{[ld]} \qquad (443)$$

Combining equation (366) with equations (441) through (443), and with σ = 1 for [hd] and σ = 2,3 for [ld], gives the following second order approximations for $C_{\sigma t}$

$$C_{1t} \approx 1 - 1/2(\theta^g_{\eta t})^2 + 1/2(\theta^g_{\pi t})^2 - 1/2(\theta^\gamma_{\eta t})^2 + 1/2(\theta^\gamma_{\pi t})^2 - \theta^g_{\eta t}\theta^\gamma_{\eta t} + \theta^g_{\pi t}\theta^\gamma_{\pi t} \tag{444}$$

$$C_{2t} \approx 1 - 1/2(\theta^g_{\eta t})^2 - 1/2(\theta^g_{\pi t})^2 - 1/2(\theta^\gamma_{\eta t})^2 - 1/2(\theta^\gamma_{\pi t})^2 - \theta^g_{\eta t}\theta^\gamma_{\eta t} - \theta^g_{\pi t}\theta^\gamma_{\pi t} \tag{445}$$

$$C_{3t} \approx 1 + 1/2(\theta^g_{\eta t})^2 - 1/2(\theta^g_{\pi t})^2 + 1/2(\theta^\gamma_{\eta t})^2 - 1/2(\theta^\gamma_{\pi t})^2 + \theta^g_{\eta t}\theta^\gamma_{\eta t} - \theta^g_{\pi t}\theta^\gamma_{\pi t} \tag{446}$$

Equations (441) through (443) and (444) through (446) are written in matrix form as

$$C_t \approx X\varphi_{3t} \approx Y\varphi_{7t} \tag{447}$$

where

$$C_t = \begin{pmatrix} C_{1t} \\ C_{2t} \\ C_{3t} \end{pmatrix} \qquad \varphi_{3t} = \begin{pmatrix} 1 \\ \theta^2_{\eta t} \\ \theta^2_{\pi t} \end{pmatrix} \qquad \varphi_{7t} = \begin{pmatrix} 1 \\ (\theta^g_{\eta t})^2 \\ (\theta^g_{\pi t})^2 \\ (\theta^\gamma_{\eta t})^2 \\ (\theta^\gamma_{\pi t})^2 \\ \theta^g_{\eta t}\theta^\gamma_{\eta t} \\ \theta^g_{\pi t}\theta^\gamma_{\pi t} \end{pmatrix} \tag{448}$$

and where X and Y are given in equation (260) as

$$X = \begin{pmatrix} 1 & -1/2 & +1/2 \\ 1 & -1/2 & -1/2 \\ 1 & +1/2 & -1/2 \end{pmatrix} \qquad Y = \begin{pmatrix} 1 & -1/2 & +1/2 & -1/2 & +1/2 & -1 & +1 \\ 1 & -1/2 & -1/2 & -1/2 & -1/2 & -1 & -1 \\ 1 & +1/2 & -1/2 & +1/2 & -1/2 & +1 & -1 \end{pmatrix} \tag{449}$$

Equation (387) can then be used to calculate the time magnitude t.

The case when the ambient electromagnetic wave field is set to zero with $\theta^\gamma_{\eta t} = 0, \theta^\gamma_{\pi t} = 0$ and $\theta^\gamma_t = 0$ combined with equations (431) through (433) gives $D_{\sigma t} = 1$ or

$$D_{1t} = 1 \qquad D_{2t} = 1 \qquad D_{3t} = 1 \tag{450}$$

and equations (434) through (436) and (444) through (446) with σ = 1 for [hd] and σ = 2,3 for [ld] give the following second order approximations for this case

$$C_{1t} \approx E_{1t} \approx 1 - 1/2(\theta^g_{\eta t})^2 + 1/2(\theta^g_{\pi t})^2 \qquad \text{[hd]} \tag{451}$$

$$C_{2t} \approx E_{2t} \approx 1 - 1/2(\theta^g_{\eta t})^2 - 1/2(\theta^g_{\pi t})^2 \qquad \text{[ld]} \tag{452}$$

$$C_{3t} \approx E_{3t} \approx 1 + 1/2(\theta^{g}_{\eta t})^2 - 1/2(\theta^{g}_{\pi t})^2 \qquad [ld] \qquad (453)$$

Then equation (387) gives the magnitude t of the complex number active time coordinate for this special case when only ambient gravity waves are present as

$$t = t^{g} = t^{o}E_{\sigma t} \qquad t^{\gamma} = t^{o} \qquad (454)$$

where $E_{\sigma t}$ is given by equations (451) through (453) for the case of zero ambient electromagnetic wave field.

For the case when the ambient gravity wave field is zero with $\theta^{g}_{\eta t} = 0, \theta^{g}_{\pi t} = 0$ and $\theta^{g}_{t} = 0$ equations (434) through (436) give $E_{\sigma t} = 1$ or

$$E_{1t} = 1 \qquad E_{2t} = 1 \qquad E_{3t} = 1 \qquad (455)$$

while equations (431) through (433) and (444) through (446) with $\sigma = 1$ for [hd] and $\sigma =$ 2,3 for [ld] give the second order approximations

$$C_{1t} \approx D_{1t} \approx 1 - 1/2(\theta^{\gamma}_{\eta t})^2 + 1/2(\theta^{\gamma}_{\pi t})^2 \qquad [hd] \qquad (456)$$

$$C_{2t} \approx D_{2t} \approx 1 - 1/2(\theta^{\gamma}_{\eta t})^2 - 1/2(\theta^{\gamma}_{\pi t})^2 \qquad [ld] \qquad (457)$$

$$C_{3t} \approx D_{3t} \approx 1 + 1/2(\theta^{\gamma}_{\eta t})^2 - 1/2(\theta^{\gamma}_{\pi t})^2 \qquad [ld] \qquad (458)$$

Equation (387) then gives the time coordinate magnitude for this case when only ambient electromagnetic waves are present as

$$t = t^{\gamma} = t^{o}D_{\sigma t} \qquad t^{g} = t^{o} \qquad (459)$$

where $D_{\sigma t}$ for this case is given by equations (456) through (458) for the case of a zero ambient gravity wave field. The case when both the ambient gravity wave field and the ambient electromagnetic wave field are zero and all internal phase angles of the time coordinates are zero is described by

$$C_{\sigma t} = D_{\sigma t} = E_{\sigma t} = 1 \qquad t = t^{g} = t^{\gamma} = t^{o} \qquad (460)$$

as expected.

C. Representations of the Measured Time Coordinate in Matter for Combined Ambient Gravitational and Ambient Electromagnetic Waves in Terms of the Measured Time Coordinates for the Cases When Ambient Gravity Waves or Ambient Electromagnetic Waves are Acting Alone.

The measured time coordinate in matter when both ambient gravity waves and ambient electromagnetic waves are present is given by equation (116) as

$$t_m = t\cos\theta_t = t^{o}C_{\sigma t}\cos\theta_t \qquad (461)$$

The measured value of the time coordinate for the case of a zero ambient wave field is

$$t_m^o = t^o \tag{462}$$

The measured value of the time coordinate for the situation when only ambient gravity waves are present is given by

$$t_m^g = t^g \cos\theta_t^g = t^o C_{\sigma t}^g \cos\theta_t^g \tag{463}$$

The measured valued of the time coordinated for the case when only an ambient electromagnetic wave field is acting is given as

$$t_m^\gamma = t^\gamma \cos\theta_t^\gamma = t^o C_{\sigma t}^\gamma \cos\theta_t^\gamma \tag{464}$$

Then equations (463) and (464) give

$$t^o = t_m^g (C_{\sigma t}^g)^{-1} \sec\theta_t^g = t_m^\gamma (C_{\sigma t}^\gamma)^{-1} \sec\theta_t^\gamma \tag{465}$$

Equations (461) through (464) give the following three representations of the measured time coordinates

$$t_m = t_m^o S_{\sigma t} = t_m^g Q_{\sigma t} = t_m^\gamma R_{\sigma t} \tag{466}$$

with t_m^o given by equation (462) and where

$$S_{\sigma t} = T_t C_{\sigma t} \tag{467}$$

$$Q_{\sigma t} = A_t D_{\sigma t} \tag{468}$$

$$R_{\sigma t} = B_t E_{\sigma t} \tag{469}$$

where $C_{\sigma t}$, $D_{\sigma t}$ and $E_{\sigma t}$ are given by equations (110) through (112) and (388) through (394), and where

$$T_t = \cos\theta_t \tag{470}$$

$$A_t = \cos\theta_t \sec\theta_t^g \tag{471}$$

$$B_t = \cos\theta_t \sec\theta_t^\gamma \tag{472}$$

where θ_t, θ_t^g and θ_t^γ are given by equations (367) and (368).

Case 1. Strong Ambient Gravity Waves and Strong Ambient Electromagnetic Waves.

Combining equation (367) with equations (470) through (472) gives the strong ambient gravity wave and strong ambient electromagnetic wave case as

$$T_t = \cos\theta_t^g \cos\theta_t^\gamma - \sin\theta_t^g \sin\theta_t^\gamma = \cos\theta_t^g \cos\theta_t^\gamma (1 - \tan\theta_t^g \tan\theta_t^\gamma) \tag{473}$$

$$A_t = \cos\theta_t^\gamma - \sin\theta_t^\gamma \tan\theta_t^g = \cos\theta_t^\gamma (1 - \tan\theta_t^g \tan\theta_t^\gamma) \tag{474}$$

$$B_t = \cos\theta_t^g - \sin\theta_t^g \tan\theta_t^\gamma = \cos\theta_t^g (1 - \tan\theta_t^g \tan\theta_t^\gamma) \tag{475}$$

where θ_t^g and θ_t^γ are given by equation (368). Then equations (466) through (469) are used to calculate the measured time coordinate t_m for the case of strong ambient gravity waves and strong ambient electromagnetic waves. In general, the inequality conditions, $\theta_t^g < 0$ and $\theta_t^\gamma < 0$, are valid for ambient waves in matter.

Case 2. Strong Ambient Gravity Waves and Weak Ambient Electromagnetic Waves.

The case of a weak ambient electromagnetic waves corresponds to the conditions, $\theta_{\eta t}^\gamma \approx 0, \theta_{\pi t}^\gamma \approx 0$ and $\theta_t^\gamma \approx 0$, and where $D_{\sigma t}$ and $E_{\sigma t}$ given by equations (399) through (404). Equations (473) through (475) show that for the case of strong ambient gravity waves and weak ambient electromagnetic waves the following first order approximations are valid

$$T_t \approx B_t \approx A_t \cos\theta_t^g \approx \cos\theta_t^g - \theta_t^\gamma \sin\theta_t^g \tag{476}$$

$$A_t \approx 1 - \theta_t^\gamma \tan\theta_t^g \tag{477}$$

$$B_t \approx A_t \cos\theta_t^g \approx \cos\theta_t^g - \theta_t^\gamma \sin\theta_t^g \tag{478}$$

where equation (368) gives

$$\theta_t^g = \theta_{\eta t}^g - \theta_{\pi t}^g \qquad \theta_t^\gamma = \theta_{\eta t}^\gamma - \theta_{\pi t}^\gamma \tag{479}$$

and equations (399) through (401), (468), (477) and (479) give the first order approximations for $Q_{\sigma t}$ with $\sigma = 1$ for [hd] and $\sigma = 2,3$ for [ld]

$$Q_{1t} \approx 1 - a_{1t}^g \theta_{\eta t}^\gamma + b_{1t}^g \theta_{\pi t}^\gamma \qquad Q_{1t} \geq 1 \qquad [\text{hd}] \tag{480}$$

$$Q_{2t} \approx 1 - a_{1t}^g \theta_{\eta t}^\gamma - d_{1t}^g \theta_{\pi t}^\gamma \qquad Q_{2t} \leq 1 \qquad [\text{ld}] \tag{481}$$

$$Q_{3t} \approx 1 + c_{1t}^g \theta_{\eta t}^\gamma - d_{1t}^g \theta_{\pi t}^\gamma \qquad Q_{3t} \leq 1 \qquad [\text{ld}] \tag{482}$$

where $\theta_{\pi t}^\gamma \geq \theta_{\eta t}^\gamma$ and $\theta_{\pi t}^g \geq \theta_{\eta t}^g$, and where

$$a_{1t}^g = \tan\theta_{\eta t}^g - \tan(\theta_{\pi t}^g - \theta_{\eta t}^g) \tag{483}$$

$$b_{1t}^g = \tan\theta_{\pi t}^g - \tan(\theta_{\pi t}^g - \theta_{\eta t}^g) \tag{484}$$

$$c_{1t}^g = \tan\theta_{\eta t}^g + \tan(\theta_{\pi t}^g - \theta_{\eta t}^g) \tag{485}$$

$$d_{1t}^g = \tan\theta_{\pi t}^g + \tan(\theta_{\pi t}^g - \theta_{\eta t}^g) \tag{486}$$

which satisfy the following inequalities that are obtained by inspection after using equation (375)

$$d_{1t}^g \geq c_{1t}^g \geq b_{1t}^g \geq a_{1t}^g \tag{487}$$

Equations (480) through (482) are written in matrix notation as

$$Q_t \approx A_t^g \varphi_t^\gamma \tag{488}$$

where

$$A_t^g = \begin{pmatrix} 1 & -a_{1t}^g & +b_{1t}^g \\ 1 & -a_{1t}^g & -d_{1t}^g \\ 1 & +c_{1t}^g & -d_{1t}^g \end{pmatrix} \qquad Q_t = \begin{pmatrix} Q_{1t} \\ Q_{2t} \\ Q_{3t} \end{pmatrix} \qquad \varphi_t^\gamma = \begin{pmatrix} 1 \\ \theta_{\eta t}^\gamma \\ \theta_{\pi t}^\gamma \end{pmatrix} \tag{489}$$

which are valid for strong ambient gravity waves and weak ambient electromagnetic waves.

For this case of strong ambient gravity waves and weak ambient electromagnetic waves, equations (402) through (404), (469) and (475) give to first order

$$R_{\sigma t} \approx A_t E_{\sigma t} \cos\theta_t^g \tag{490}$$

where $E_{\sigma t}$ is given by equations (402) through (404) which also show that $C_{\sigma t} \approx E_{\sigma t}$ for this case, so that it follows from equations (467) through (469) and (476) that

$$S_{\sigma t} \approx B_t C_{\sigma t} \approx B_t E_{\sigma t} \approx R_{\sigma t} \approx C_{\sigma t}^g Q_{\sigma t} \cos\theta_t^g \tag{491}$$

so that for strong ambient gravity waves and weak ambient electromagnetic waves in matter with $\sigma = 1$ for [hd] and $\sigma = 2,3$ for [ld]

$$S_{1t} \approx R_{1t} \approx C_{1t}^g Q_{1t} \cos\theta_t^g \qquad [\text{hd}] \tag{492}$$

$$S_{2t} \approx R_{2t} \approx C_{2t}^g Q_{2t} \cos\theta_t^g \qquad [\text{ld}] \tag{493}$$

$$S_{3t} \approx R_{3t} \approx C_{3t}^g Q_{3t} \cos\theta_t^g \qquad [\text{ld}] \tag{494}$$

and equations (490) and (492) through (494) give the following first order approximation for the case of strong ambient gravity waves and weak ambient electromagnetic waves

$$A_t E_{\sigma t} \approx C_{\sigma t}^g Q_{\sigma t} \tag{495}$$

which combined with equation (405) gives equation (468), where $C_{\sigma t}^g$ and $Q_{\sigma t}$ are given by equations (372) through (374) and (480) through (482) respectively. The measured value t_m of the complex number active coordinate can then be obtained from equations (466) through (469).

The case when only ambient gravity waves are present with the conditions $\theta_{\eta t}^\gamma = 0, \theta_{\pi t}^\gamma = 0$ and $\theta_t^\gamma = 0$, requires that $C_{\sigma t}^\gamma = 1$ and

$$Q_{\sigma t} = 1 \qquad A_t = 1 \qquad C_{\sigma t} = E_{\sigma t} = C^g_{\sigma t} \qquad D_{\sigma t} = 1 \tag{496}$$

$$S_{\sigma t} = R_{\sigma t} = C^g_{\sigma t} \cos\theta^g_t \qquad T_t = B_t = \cos\theta^g_t \tag{497}$$

$$t_m = t^g_m = t^o C^g_{\sigma t} \cos\theta^g_t \qquad t^\gamma_m = t^o \tag{498}$$

where θ^g_t is given by equation (479) and where $C^g_{\sigma t}$ for $\sigma = 1,2,3$ is given by equations (372) through (374). In general, $\theta^g_t < 0$ for ambient gravitational waves.

Case 3. Weak Ambient Gravity Waves and Strong Ambient Electromagnetic Waves.

In this case $\theta^g_{\eta t} \approx 0, \theta^g_{\pi t} \approx 0$ and $\theta^g_t \approx 0$ and where $D_{\sigma t}$ and $E_{\sigma t}$ are given by equations (415) through (420). Equations (470) through (472) give, for the case of weak ambient gravity waves and strong ambient electromagnetic waves, the following first order expressions

$$T_t \approx A_t \approx B_t \cos\theta^\gamma_t \approx \cos\theta^\gamma_t - \theta^g_t \sin\theta^\gamma_t \tag{499}$$

$$A_t \approx B_t \cos\theta^\gamma_t \approx \cos\theta^\gamma_t - \theta^g_t \sin\theta^\gamma_t \tag{500}$$

$$B_t \approx 1 - \theta^g_t \tan\theta^\gamma_t \tag{501}$$

where equation (368) gives

$$\theta^\gamma_t = \theta^\gamma_{\eta t} - \theta^\gamma_{\pi t} \qquad \theta^g_t = \theta^g_{\eta t} - \theta^g_{\pi t} \tag{502}$$

while equations (415) through (417), (469), (501) and (502) give the following first order approximations for $R_{\sigma t}$ in this case

$$R_{1t} \approx 1 - a^\gamma_{1t}\theta^g_{\eta t} + b^\gamma_{1t}\theta^g_{\pi t} \qquad R_{1t} \geq 1 \qquad [\text{hd}] \tag{503}$$

$$R_{2t} \approx 1 - a^\gamma_{1t}\theta^g_{\eta t} - d^\gamma_{1t}\theta^g_{\pi t} \qquad R_{2t} \leq 1 \qquad [\text{ld}] \tag{504}$$

$$R_{3t} \approx 1 + c^\gamma_{1t}\theta^g_{\eta t} - d^\gamma_{1t}\theta^g_{\pi t} \qquad R_{3t} \leq 1 \qquad [\text{ld}] \tag{505}$$

with $\theta^g_{\pi t} \geq \theta^g_{\eta t}$ and $\theta^\gamma_{\pi t} \geq \theta^\gamma_{\eta t}$, and where

$$a^\gamma_{1t} = \tan\theta^\gamma_{\eta t} - \tan(\theta^\gamma_{\pi t} - \theta^\gamma_{\eta t}) \tag{506}$$

$$b^\gamma_{1t} = \tan\theta^\gamma_{\pi t} - \tan(\theta^\gamma_{\pi t} - \theta^\gamma_{\eta t}) \tag{507}$$

$$c^\gamma_{1t} = \tan\theta^\gamma_{\eta t} + \tan(\theta^\gamma_{\pi t} - \theta^\gamma_{\eta t}) \tag{508}$$

$$d^\gamma_{1t} = \tan\theta^\gamma_{\pi t} + \tan(\theta^\gamma_{\pi t} - \theta^\gamma_{\eta t}) \tag{509}$$

which, according to equation (383), satisfy the inequalities

$$d^\gamma_{1t} \geq c^\gamma_{1t} \geq b^\gamma_{1t} \geq a^\gamma_{1t} \tag{510}$$

Equations (503) through (505) have the following matrix representations

$$R_t \approx A_t^\gamma \varphi_t^g \tag{511}$$

where

$$A_t^\gamma = \begin{pmatrix} 1 & -a_{1t}^\gamma & +b_{1t}^\gamma \\ 1 & -a_{1t}^\gamma & -d_{1t}^\gamma \\ 1 & +c_{1t}^\gamma & -d_{1t}^\gamma \end{pmatrix} \qquad R_t = \begin{pmatrix} R_{1t} \\ R_{2t} \\ R_{3t} \end{pmatrix} \qquad \varphi_t^g = \begin{pmatrix} 1 \\ \theta_{\eta t}^g \\ \theta_{\pi t}^g \end{pmatrix} \tag{512}$$

which hold for the case of weak ambient gravity waves and strong ambient electromagnetic waves.

It follows from equations (418) through (420), (468) and (500) that

$$Q_{\sigma t} = B_t D_{\sigma t} \cos\theta_t^\gamma \tag{513}$$

where $D_{\sigma t} \approx C_{\sigma t}$ is given by equations (418) through (421), and it follows to first order that

$$S_{\sigma t} \approx A_t C_{\sigma t} \approx A_t D_{\sigma t} \approx Q_{\sigma t} \approx C_{\sigma t}^\gamma R_{\sigma t} \cos\theta_t^\gamma \tag{514}$$

so that for weak ambient gravity waves and strong ambient electromagnetic waves the following first order approximations are valid

$$S_{1t} \approx Q_{1t} \approx C_{1t}^\gamma R_{1t} \cos\theta_t^\gamma \qquad [hd] \tag{515}$$

$$S_{2t} \approx Q_{2t} \approx C_{2t}^\gamma R_{2t} \cos\theta_t^\gamma \qquad [ld] \tag{516}$$

$$S_{3t} \approx Q_{3t} \approx C_{3t}^\gamma R_{3t} \cos\theta_t^\gamma \qquad [ld] \tag{517}$$

and equations (513) and (515) through (517) give the following first order approximations for this case of weak ambient gravity waves and strong ambient electromagnetic waves

$$B_t D_{\sigma t} \approx C_{\sigma t}^\gamma R_{\sigma t} \tag{518}$$

where $C_{\sigma t}^\gamma$ is given by equations (380) through (382), and where $R_{\sigma t}$ is given by equations (503) through (505). The measured value t_m of the time coordinate can then be calculated from equations (466) through (469).

When only ambient electromagnetic waves are present with $\theta_{\eta t}^g = 0$, $\theta_{\pi t}^g = 0$ and $\theta_t^g = 0$ then $C_{\sigma t}^g = 1$ and

$$R_{\sigma t} = 1 \qquad B_t = 1 \qquad C_{\sigma t} = D_{\sigma t} = C_{\sigma t}^\gamma \qquad E_{\sigma t} = 1 \tag{519}$$

$$S_{\sigma t} = Q_{\sigma t} = C_{\sigma t}^\gamma \cos\theta_t^\gamma \qquad T_t = A_t = \cos\theta_t^\gamma \tag{520}$$

$$t_m = t_m^\gamma = t^o C_{\sigma t}^\gamma \cos\theta_t^\gamma \qquad t_m^g = t^o \tag{521}$$

where θ_t^γ is given by equation (502) and where $C_{\sigma t}^\gamma$ is given by equations (380) through (382). In general, $\theta_t^\gamma < 0$ for ambient electromagnetic waves.

Case 4. Weak Ambient Gravity Waves and Weak Ambient Electromagnetic Waves.

The case of weak ambient gravity waves and weak ambient electromagnetic waves is obtained from equation (468) combined with equations (431) through (433), and equation (469) combined with equations (434) through (436). For this case, equations (473) through (475) become the following second order approximations

$$T_t \approx 1 - 1/2(\theta_t^\gamma)^2 - 1/2(\theta_t^g)^2 - \theta_t^\gamma\theta_t^g \tag{522}$$

$$A_t \approx 1 - 1/2(\theta_t^\gamma)^2 - \theta_t^\gamma\theta_t^g \tag{523}$$

$$B_t \approx 1 - 1/2(\theta_t^g)^2 - \theta_t^\gamma\theta_t^g \tag{524}$$

Combining equation (368) with equations (522) through (524) gives

$$T_t \approx 1 - 1/2(\theta_{\eta t}^g)^2 - 1/2(\theta_{\iota t}^g)^2 - 1/2(\theta_{\eta t}^\gamma)^2 - 1/2(\theta_{\iota t}^\gamma)^2 + \theta_{\eta t}^g\theta_{\iota t}^g + \theta_{\eta t}^\gamma\theta_{\iota t}^\gamma - \theta_{\eta t}^g\theta_{\eta t}^\gamma + \theta_{\eta t}^\gamma\theta_{\iota t}^g + \theta_{\eta t}^g\theta_{\iota t}^\gamma - \theta_{\iota t}^g\theta_{\iota t}^\gamma \tag{525}$$

$$A_t \approx 1 - 1/2(\theta_{\eta t}^\gamma)^2 - 1/2(\theta_{\iota t}^\gamma)^2 + \theta_{\eta t}^\gamma\theta_{\iota t}^\gamma - \theta_{\eta t}^g\theta_{\eta t}^\gamma + \theta_{\eta t}^\gamma\theta_{\iota t}^g + \theta_{\eta t}^g\theta_{\iota t}^\gamma - \theta_{\iota t}^g\theta_{\iota t}^\gamma \tag{526}$$

$$B_t \approx 1 - 1/2(\theta_{\eta t}^g)^2 - 1/2(\theta_{\iota t}^g)^2 + \theta_{\eta t}^g\theta_{\iota t}^g - \theta_{\eta t}^g\theta_{\eta t}^\gamma + \theta_{\eta t}^\gamma\theta_{\iota t}^g + \theta_{\eta t}^g\theta_{\iota t}^\gamma - \theta_{\iota t}^g\theta_{\iota t}^\gamma \tag{527}$$

Equations (467) through (469) show that the functions T_t, A_t and B_t given by the second order approximation equations (525) through (527) facilitate the calculation of the functions $S_{\sigma t}$, $Q_{\sigma t}$ and $R_{\sigma t}$ which allow the calculation of the measured time t_m for the case of weak ambient gravity waves and weak ambient electromagnetic waves using equation (470).

Equations (444) through (446), (467) and (525) give the second order small angle approximation of $S_{\sigma t}$ as

$$S_{1t} \approx 1 - (\theta_{\eta t}^g)^2 - (\theta_{\eta t}^\gamma)^2 - 2\theta_{\eta t}^g\theta_{\eta t}^\gamma + \theta_{\eta t}^g\theta_{\iota t}^g + \theta_{\eta t}^\gamma\theta_{\iota t}^\gamma + \theta_{\eta t}^\gamma\theta_{\iota t}^g + \theta_{\eta t}^g\theta_{\iota t}^\gamma \qquad [\text{hd}] \tag{528}$$

$$S_{2t} \approx 1 - (\theta_{\eta t}^g)^2 - (\theta_{\iota t}^g)^2 - (\theta_{\eta t}^\gamma)^2 - (\theta_{\iota t}^\gamma)^2 - 2\theta_{\eta t}^g\theta_{\eta t}^\gamma - 2\theta_{\iota t}^g\theta_{\iota t}^\gamma + \theta_{\eta t}^g\theta_{\iota t}^g + \theta_{\eta t}^\gamma\theta_{\iota t}^\gamma + \theta_{\eta t}^\gamma\theta_{\iota t}^g + \theta_{\eta t}^g\theta_{\iota t}^\gamma \qquad [\text{ld}] \tag{529}$$

$$S_{3t} \approx 1-(\theta^{g}_{\pi t})^2-(\theta^{\gamma}_{\pi t})^2-2\theta^{\gamma}_{\pi t}\theta^{g}_{\pi t}+\theta^{g}_{\eta t}\theta^{g}_{\pi t}+\theta^{\gamma}_{\eta t}\theta^{\gamma}_{\pi t}+\theta^{\gamma}_{\eta t}\theta^{g}_{\pi t}+\theta^{g}_{\eta t}\theta^{\gamma}_{\pi t} \quad [\text{ld}] \qquad (530)$$

Equations (468), (431) through (433) and (523) or (526) give the following second order approximations for $Q_{\sigma t}$

$$Q_{1t} \approx 1-1/2(\theta^{\gamma}_{\eta t})^2+1/2(\theta^{\gamma}_{\pi t})^2-\theta^{\gamma}_{\eta t}\theta^{g}_{\eta t}+\theta^{\gamma}_{\pi t}\theta^{g}_{\pi t}-1/2(\theta^{\gamma}_{t})^2-\theta^{\gamma}_{t}\theta^{g}_{t} \quad [\text{hd}] \qquad (531)$$

$$\approx 1-(\theta^{\gamma}_{\eta t})^2-2\theta^{\gamma}_{\eta t}\theta^{g}_{\eta t}+\theta^{\gamma}_{\eta t}\theta^{\gamma}_{\pi t}+\theta^{\gamma}_{\eta t}\theta^{g}_{\pi t}+\theta^{g}_{\eta t}\theta^{\gamma}_{\pi t}$$

$$Q_{2t} \approx 1-1/2(\theta^{\gamma}_{\eta t})^2-1/2(\theta^{\gamma}_{\pi t})^2-\theta^{\gamma}_{\eta t}\theta^{g}_{\eta t}-\theta^{\gamma}_{\pi t}\theta^{g}_{\pi t}-1/2(\theta^{\gamma}_{t})^2-\theta^{\gamma}_{t}\theta^{g}_{t} \quad [\text{ld}] \qquad (532)$$

$$\approx 1-(\theta^{\gamma}_{\eta t})^2-(\theta^{\gamma}_{\pi t})^2-2\theta^{\gamma}_{\eta t}\theta^{g}_{\eta t}-2\theta^{\gamma}_{\pi t}\theta^{g}_{\pi t}+\theta^{\gamma}_{\eta t}\theta^{\gamma}_{\pi t}+\theta^{\gamma}_{\eta t}\theta^{g}_{\pi t}+\theta^{g}_{\eta t}\theta^{\gamma}_{\pi t}$$

$$Q_{3t} \approx 1+1/2(\theta^{\gamma}_{\eta t})^2-1/2(\theta^{\gamma}_{\pi t})^2+\theta^{\gamma}_{\eta t}\theta^{g}_{\eta t}-\theta^{\gamma}_{\pi t}\theta^{g}_{\pi t}-1/2(\theta^{\gamma}_{t})^2-\theta^{\gamma}_{t}\theta^{g}_{t} \quad [\text{hd}] \qquad (533)$$

$$\approx 1-(\theta^{\gamma}_{\pi t})^2-2\theta^{\gamma}_{\pi t}\theta^{g}_{\pi t}+\theta^{\gamma}_{\eta t}\theta^{\gamma}_{\pi t}+\theta^{\gamma}_{\eta t}\theta^{g}_{\pi t}+\theta^{g}_{\eta t}\theta^{\gamma}_{\pi t}$$

where equation (368) was used in these calculations. Equations (469), (434) through (436) and (524) or (527) give the second order approximations for $R_{\sigma t}$ as

$$R_{1t} \approx 1-1/2(\theta^{g}_{\eta t})^2+1/2(\theta^{g}_{\pi t})^2-\theta^{g}_{\eta t}\theta^{\gamma}_{\eta t}+\theta^{g}_{\pi t}\theta^{\gamma}_{\pi t}-1/2(\theta^{g}_{t})^2-\theta^{g}_{t}\theta^{\gamma}_{t} \quad [\text{hd}] \qquad (534)$$

$$\approx 1-(\theta^{g}_{\eta t})^2-2\theta^{g}_{\eta t}\theta^{\gamma}_{\eta t}+\theta^{g}_{\eta t}\theta^{g}_{\pi t}+\theta^{g}_{\eta t}\theta^{\gamma}_{\pi t}+\theta^{\gamma}_{\eta t}\theta^{g}_{\pi t}$$

$$R_{2t} \approx 1-1/2(\theta^{g}_{\eta t})^2-1/2(\theta^{g}_{\pi t})^2-\theta^{g}_{\eta t}\theta^{\gamma}_{\eta t}-\theta^{g}_{\pi t}\theta^{\gamma}_{\pi t}-1/2(\theta^{g}_{t})^2-\theta^{g}_{t}\theta^{\gamma}_{t} \quad [\text{ld}] \qquad (535)$$

$$\approx 1-(\theta^{g}_{\eta t})^2-(\theta^{g}_{\pi t})^2-2\theta^{g}_{\eta t}\theta^{\gamma}_{\eta t}-2\theta^{g}_{\pi t}\theta^{\gamma}_{\pi t}+\theta^{g}_{\eta t}\theta^{g}_{\pi t}+\theta^{g}_{\eta t}\theta^{\gamma}_{\pi t}+\theta^{\gamma}_{\eta t}\theta^{g}_{\pi t}$$

$$R_{3t} \approx 1+1/2(\theta^{g}_{\eta t})^2-1/2(\theta^{g}_{\pi t})^2+\theta^{g}_{\eta t}\theta^{\gamma}_{\eta t}-\theta^{g}_{\pi t}\theta^{\gamma}_{\pi t}-1/2(\theta^{g}_{t})^2-\theta^{g}_{t}\theta^{\gamma}_{t} \quad [\text{ld}] \qquad (536)$$

$$\approx 1-(\theta^{g}_{\pi t})^2-2\theta^{g}_{\pi t}\theta^{\gamma}_{\pi t}+\theta^{g}_{\eta t}\theta^{g}_{\pi t}+\theta^{g}_{\eta t}\theta^{\gamma}_{\pi t}+\theta^{\gamma}_{\eta t}\theta^{g}_{\pi t}$$

after using equation (368).

The expressions for $S_{\sigma t}$, $Q_{\sigma t}$ and $R_{\sigma t}$ given by equations (528) through (536) can be written in matrix form as

$$S_t = Z_m \varphi_{11t} \qquad Q_t = W_m \varphi^{\gamma g}_{8t} \qquad R_t = W_m \varphi^{g\gamma}_{8t} \qquad (537)$$

where Z_m is given by equation (349) as

$$Z_m = \begin{pmatrix} 1 & -1 & 0 & -1 & 0 & -2 & 0 & 1 & 1 & 1 & 1 \\ 1 & -1 & -1 & -1 & -1 & -2 & -2 & 1 & 1 & 1 & 1 \\ 1 & 0 & -1 & 0 & -1 & 0 & -2 & 1 & 1 & 1 & 1 \end{pmatrix} \qquad (538)$$

while W_m is given by equation (350) as

$$W_m = \begin{pmatrix} 1 & -1 & 0 & -2 & 0 & 1 & 1 & 1 \\ 1 & -1 & -1 & -2 & -2 & 1 & 1 & 1 \\ 1 & 0 & -1 & 0 & -2 & 1 & 1 & 1 \end{pmatrix} \qquad (539)$$

and where

$$S_t = \begin{pmatrix} S_{1t} \\ S_{2t} \\ S_{3t} \end{pmatrix} \qquad Q_t = \begin{pmatrix} Q_{1t} \\ Q_{2t} \\ Q_{3t} \end{pmatrix} \qquad R_t = \begin{pmatrix} R_{1t} \\ R_{2t} \\ R_{3t} \end{pmatrix} \qquad (540)$$

and

$$\varphi_{11t} = \begin{pmatrix} 1 \\ (\theta^g_{\eta t})^2 \\ (\theta^g_{\pi t})^2 \\ (\theta^\gamma_{\eta t})^2 \\ (\theta^\gamma_{\pi t})^2 \\ \theta^g_{\eta t}\theta^\gamma_{\eta t} \\ \theta^g_{\pi t}\theta^\gamma_{\pi t} \\ \theta^g_{\eta t}\theta^g_{\pi t} \\ \theta^\gamma_{\eta t}\theta^\gamma_{\pi t} \\ \theta^\gamma_{\eta t}\theta^g_{\pi t} \\ \theta^g_{\eta t}\theta^\gamma_{\pi t} \end{pmatrix} \qquad \varphi^{\gamma g}_{8t} = \begin{pmatrix} 1 \\ (\theta^\gamma_{\eta t})^2 \\ (\theta^\gamma_{\pi t})^2 \\ \theta^\gamma_{\eta t}\theta^g_{\eta t} \\ \theta^\gamma_{\pi t}\theta^g_{\pi t} \\ \theta^\gamma_{\eta t}\theta^\gamma_{\pi t} \\ \theta^\gamma_{\eta t}\theta^g_{\pi t} \\ \theta^g_{\eta t}\theta^\gamma_{\pi t} \end{pmatrix} \qquad \varphi^{g\gamma}_{8t} = \begin{pmatrix} 1 \\ (\theta^g_{\eta t})^2 \\ (\theta^g_{\pi t})^2 \\ \theta^g_{\eta t}\theta^\gamma_{\eta t} \\ \theta^g_{\pi t}\theta^\gamma_{\pi t} \\ \theta^g_{\eta t}\theta^g_{\pi t} \\ \theta^g_{\eta t}\theta^\gamma_{\pi t} \\ \theta^\gamma_{\eta t}\theta^g_{\pi t} \end{pmatrix} \qquad (541)$$

Then the measured time coordinate t_m can be calculated from equations (466) through (469).

The case when the ambient electromagnetic wave field is zero and $\theta^\gamma_{\eta t} = 0$, $\theta^\gamma_{\pi t} = 0$ and $\theta^\gamma_t = 0$ can be deduced from equations (531) through (533) which gives $Q_{\sigma t} = 1$ with $\sigma = 1$ for [hd] and $\sigma = 2,3$ for [ld], or written out

$$Q_{1t} = 0 \qquad Q_{2t} = 0 \qquad Q_{3t} = 0 \qquad (542)$$

while equations (525) through (527) give $A_t = 1$ and

$$T_t = B_t \approx 1 - 1/2(\theta^g_{\eta t})^2 - 1/2(\theta^g_{\pi t})^2 + \theta^g_{\eta t}\theta^g_{\pi t} \qquad (543)$$

and for this case equations (528) through (530) and (534) through (536) give $S_{\sigma t} = R_{\sigma t}$ and the second order approximations

$$S_{1t} = R_{1t} \approx 1 - (\theta^g_{\eta t})^2 + \theta^g_{\eta t}\theta^g_{\pi t} \qquad [hd] \qquad (544)$$

$$S_{2t} = R_{2t} \approx 1 - (\theta^g_{\eta t})^2 - (\theta^g_{\pi t})^2 + \theta^g_{\eta t}\theta^g_{\pi t} \qquad [ld] \qquad (545)$$

$$S_{3t} = R_{3t} \approx 1 - (\theta^g_{\pi t})^2 + \theta^g_{\eta t}\theta^g_{\pi t} \qquad [ld] \qquad (546)$$

Equation (466) gives for this special case of zero ambient electromagnetic wave field and with $\sigma = 1,2,3$

$$t_m = t^g_m = t^o R_{\sigma t} = t^o S_{\sigma t} \qquad t^\gamma_m = t^o \qquad (547)$$

where for this special case $S_{\sigma t} = R_{\sigma t}$ is given by equations (544) through (546).

For the case when the ambient gravity wave field is zero and $\theta^g_{\eta t} = 0, \theta^g_{\pi t} = 0$ and $\theta^g_t = 0$, equations (534) through (536) give $R_{\sigma t} = 1$ with $\sigma = 1$ for [hd] and $\sigma = 2,3$ for [ld], or

$$R_{1t} = 1 \qquad R_{2t} = 1 \qquad R_{3t} = 1 \qquad (548)$$

and where equations (525) through (527) give $B_t = 1$ and

$$T_t = A_t \approx 1 - 1/2(\theta^\gamma_{\eta t})^2 - 1/2(\theta^\gamma_{\pi t})^2 + \theta^\gamma_{\eta t}\theta^\gamma_{\pi t} \qquad (549)$$

and equations (528) through (533) give $S_{\sigma t} = Q_{\sigma t}$ which to a second order approximation is written as

$$S_{1t} = Q_{1t} \approx 1 - (\theta^\gamma_{\eta t})^2 + \theta^\gamma_{\eta t}\theta^\gamma_{\pi t} \qquad [hd] \qquad (550)$$

$$S_{2t} = Q_{2t} \approx 1 - (\theta^\gamma_{\eta t})^2 - (\theta^\gamma_{\pi t})^2 + \theta^\gamma_{\eta t}\theta^\gamma_{\pi t} \qquad [ld] \qquad (551)$$

$$S_{3t} = Q_{3t} \approx 1 - (\theta^\gamma_{\pi t})^2 + \theta^\gamma_{\eta t}\theta^\gamma_{\pi t} \qquad [ld] \qquad (552)$$

Equation (466) gives the measured time coordinate for this special case of a zero ambient gravity wave field as

$$t_m = t^\gamma_m = t^o Q_{\sigma t} = t^o S_{\sigma t} \qquad t^g_m = t^o \qquad (553)$$

where $S_{\sigma t} = Q_{\sigma t}$ is given for this special case by equations (550) through (552).

For the case when both the ambient gravitational wave field and the ambient electromagnetic wave field are zero it follows that

$$T_t = 1 \quad A_t = 1 \quad B_t = 1 \quad S_{\sigma t} = 1 \quad Q_{\sigma t} = 1 \quad R_{\sigma t} = 1 \quad t_m = t^o \qquad (554)$$

as must be the case. Finally, it should be noted that γ and g can also represent any two dominant ambient wave fields present in matter. An extension of the analysis done in this section and the previous one should be done to include the effects of three or four radiative ambient field effects.

6. AREA, VOLUME AND PARTICLE NUMBER DENSITY OF MATTER LOCATED IN AN AMBIENT FIELD. As a simple use of space coordinate factors this section considers the representation of area, volume and particle number density of matter in the presence of a combined ambient gravitational and ambient electromagnetic field of either a radiative or nonradiative type. This section justifies the expressions for particle number density of matter in an ambient field that are given in equations (17) through (19) because these formulas play an important role in the theory of thermonuclear reactions, chemical reactions, polyphase rocket design and polyphase nuclear fission.[11,12,21-24]

A. Geometrical Areas in the Presence of an Ambient Field.

In the presence of an ambient field areas are written as complex numbers in an internal space and the area perpendicular to the α axis is written as

$$\overline{A}_\alpha = A_\alpha \exp(j\theta_{A\alpha}) \tag{555}$$

The measured area is given by

$$A_{\alpha m} = A_\alpha \cos\theta_{A\alpha} \tag{556}$$

where A_α and $\theta_{A\alpha}$ = magnitude and internal phase angle of the complex number area. When only an ambient gravity field or an ambient electromagnetic field is present the complex number areas are written as

$$\overline{A}^g_\alpha = A^g_\alpha \exp(j\theta^g_{A\alpha}) \tag{557}$$

$$\overline{A}^\gamma_\alpha = A^\gamma_\alpha \exp(j\theta^\gamma_{A\alpha}) \tag{558}$$

Consider now the representation of complex number areas in three common coordinate systems.

Cartesian Coordinates

In Cartesian coordinates the complex number areas of rectangles in an ambient field are obtained from equation (1) as

$$\overline{A}_x = \overline{y}\,\overline{z} \qquad A_x = yz \qquad \theta_{Ax} = \theta_y + \theta_z \tag{559}$$

$$\overline{A}_y = \overline{z}\,\overline{x} \qquad A_y = zx \qquad \theta_{Ay} = \theta_z + \theta_x \tag{560}$$

$$\overline{A}_z = \overline{x}\,\overline{y} \qquad A_z = xy \qquad \theta_{Az} = \theta_x + \theta_y \tag{561}$$

The magnitudes of the complex number areas in the presence of an ambient field can be related to the areas under zero ambient field conditions by combining equation (83) with equations (559) through (561) which gives

$$A_x = A^o_x C_{\sigma y} C_{\sigma z} \qquad \theta_{Ax} = \theta_y + \theta_z \tag{562}$$

$$A_y = A_y^o C_{\sigma z} C_{\sigma x} \qquad \theta_{Ay} = \theta_z + \theta_x \tag{563}$$

$$A_z = A_z^o C_{\sigma x} C_{\sigma y} \qquad \theta_{Az} = \theta_x + \theta_y \tag{564}$$

where the zero ambient field areas are written as

$$A_x^o = y_o z_o \qquad A_y^o = z_o x_o \qquad A_z^o = x_o y_o \tag{565}$$

and where $C_{\sigma\alpha}$ are the space coordinate factors given by equations (78) through (80) for $\alpha = x, y, z$ and with $\sigma = 1$ for [hd] and $\sigma = 2,3$ for [ld], and where θ_x, θ_y and θ_z are represented as in equation (56). The measured areas in the presence of an ambient field are then obtained from equations (556) and (562) through (564). The zero ambient field areas A_α^o are not in general measurable quantities because ambient gravity cannot be shut off, so that it is more logical to refer areas in the presence of a combined ambient electromagnetic and ambient gravitational field to their values when only an ambient gravitational field is acting or to the case when the dominant ambient field is an ambient electromagnetic field. For the case when only an ambient gravitational field is acting the magnitudes and internal phase angles of the areas given in equations (562) through (564) become

$$A_x^g = A_x^o C_{\sigma y}^g C_{\sigma z}^g \qquad \theta_{Ax}^g = \theta_y^g + \theta_z^g \tag{566}$$

$$A_y^g = A_y^o C_{\sigma z}^g C_{\sigma x}^g \qquad \theta_{Ay}^g = \theta_z^g + \theta_x^g \tag{567}$$

$$A_z^g = A_z^o C_{\sigma x}^g C_{\sigma y}^g \qquad \theta_{Az}^g = \theta_x^g + \theta_y^g \tag{568}$$

where the space coordinate factors $C_{\sigma\alpha}^g$ are defined by equations (183) through (185). For the case of an ambient electromagnetic field acting alone equations (562) through (564) become

$$A_x^\gamma = A_x^o C_{\sigma y}^\gamma C_{\sigma z}^\gamma \qquad \theta_{Ax}^\gamma = \theta_y^\gamma + \theta_z^\gamma \tag{569}$$

$$A_y^\gamma = A_y^o C_{\sigma z}^\gamma C_{\sigma x}^\gamma \qquad \theta_{Ay}^\gamma = \theta_z^\gamma + \theta_x^\gamma \tag{570}$$

$$A_z^\gamma = A_z^o C_{\sigma x}^\gamma C_{\sigma y}^\gamma \qquad \theta_{Az}^\gamma = \theta_x^\gamma + \theta_y^\gamma \tag{571}$$

where the space coordinate factors $C_{\sigma\alpha}^\gamma$ are defined by equations (191) through (193). Combining equations (562) through (564) with equations (566) through (571) allows the magnitudes of the areas to be written as

$$A_x = A_x^o C_{\sigma y} C_{\sigma z} = A_x^g D_{\sigma y} D_{\sigma z} = A_x^\gamma E_{\sigma y} E_{\sigma z} \tag{572}$$

$$A_y = A_y^o C_{\sigma z} C_{\sigma x} = A_y^g D_{\sigma z} D_{\sigma x} = A_y^\gamma E_{\sigma z} E_{\sigma x} \tag{573}$$

$$A_z = A_z^o C_{\sigma x} C_{\sigma y} = A_z^g D_{\sigma x} D_{\sigma y} = A_z^\gamma E_{\sigma x} E_{\sigma y} \tag{574}$$

where $D_{\sigma\alpha}$ and $E_{\sigma\alpha}$ are defined by equation (198) and (199) which can be used directly to obtain equations (572) through (574).

Cylindrical Polar Coordinates

In the presence of a combined ambient electromagnetic and ambient gravitational field the complex number area of a circle, which can be located within matter, is given by

$$\overline{A}_r = \pi\overline{r}^2 \qquad A_r = \pi r^2 \qquad \theta_{Ar} = 2\theta_r \tag{575}$$

The magnitude of the complex number area in the presence of an ambient field can be related to the area under zero ambient field conditions by combining equation (83) with equation (575), which gives

$$A_r = A_r^o C_{\sigma r}^2 \qquad \theta_{Ar} = 2\theta_r \tag{576}$$

where the area of a circle under zero ambient field conditions is given by

$$A_r^o = \pi r_o^2 \tag{577}$$

and where $C_{\sigma r}$ are the space coordinate factors defined by equations (78) through (80) with $\sigma = 1$ for [hd] and $\sigma = 2,3$ for [ld], and where the angle θ_r is given by equation (56). When only an ambient gravitational field or only an ambient electromagnetic field is present equation (576) becomes

$$A_r^g = A_r^o (C_{\sigma r}^g)^2 \qquad \theta_{Ar}^g = 2\theta_r^g \tag{578}$$

$$A_r^\gamma = A_r^o (C_{\sigma r}^\gamma)^2 \qquad \theta_{Ar}^\gamma = 2\theta_r^\gamma \tag{579}$$

where the space coordinate factors $C_{\sigma r}^g$ and $C_{\sigma r}^\gamma$ are defined by equations (183) through (185) and (191) through (193) respectively. Combining equations (576) through (579) gives

$$A_r = A_r^o C_{\sigma r}^2 = A_r^g D_{\sigma r}^2 = A_r^\gamma E_{\sigma r}^2 \tag{580}$$

where $D_{\sigma r}$ and $E_{\sigma r}$ are defined by equations (198) and (199) which can be used to obtain equation (580) directly.

Spherical Polar Coordinates

The complex number area of a sphere that is located in matter of high or low-density and which is immersed in an ambient field is given by

$$\overline{A}_\rho = 4\pi\overline{\rho}^2 \qquad A_\rho = 4\pi\rho^2 \qquad \theta_{A\rho} = 2\theta_\rho \tag{581}$$

The magnitude of the complex number of a sphere that is located in an ambient field can be related to the zero ambient field value of the area by using equations (83) and (581) with the result that

$$A_\rho = A_\rho^o C_{\sigma\rho}^2 \qquad \theta_{A\rho} = 2\theta_\rho \tag{582}$$

where

$$A_\rho^o = 4\pi\rho_o^2 \qquad (583)$$

where A_ρ^o = zero ambient field value of the area of a sphere, and where the space coordinate factors $C_{\sigma\rho}$ are given by equations (78) through (80) with $\sigma = 1$ for [hd] and $\sigma = 2,3$ for [ld], and where θ_ρ can be represented in terms of the internal phase angles of the counting number and segmentation number using equation (56). When only an ambient gravitational field or an ambient electromagnetic field is present equation (582) becomes

$$A_\rho^g = A_\rho^o (C_{\sigma\rho}^g)^2 \qquad \theta_{A\rho}^g = 2\theta_\rho^g \qquad (584)$$

$$A_\rho^\gamma = A_\rho^o (C_{\sigma\rho}^\gamma)^2 \qquad \theta_{A\rho}^\gamma = 2\theta_\rho^\gamma \qquad (585)$$

where the space coordinate factors $C_{\sigma\rho}^g$ and $C_{\sigma\rho}^\gamma$ are defined by equations (183) through (185) and (191) through (193) respectively. Finally, combining equations (582) through (585) gives

$$A_\rho = A_\rho^o C_{\sigma\rho}^2 = A_\rho^g D_{\sigma\rho}^2 = A_\rho^\gamma E_{\sigma\rho}^2 \qquad (586)$$

where $D_{\sigma\rho}$ and $E_{\sigma\rho}$ are given by equations (198) and (199) which can be used to directly obtain equation (586).

B. The Representation of Volume in the Presence of an Ambient Field.

When a volume element exists within an ambient field it must be represented by a complex number in internal space as follows

$$\overline{V} = V \exp(j\theta_V) \qquad (587)$$

and the measured volume is just the real part given by

$$V_m = V \cos\theta_V \qquad (588)$$

When only an ambient gravity field or an ambient electromagnetic field is present equation (587) becomes

$$\overline{V}^g = V^g \exp(j\theta_V^g) \qquad (589)$$

$$\overline{V}^\gamma = V^\gamma \exp(j\theta_V^\gamma) \qquad (590)$$

When all ambient fields are turned off

$$\overline{V} = V = V^o \qquad \theta_V = 0 \qquad (591)$$

Consider now the representation of geometric volumes in matter that is located in an ambient field for the three common coordinate systems: Cartesian coordinates, cylindrical polar coordinates and spherical polar coordinates.

Cartesian Coordinates

The complex number volume of a rectangular parallelepiped of matter in an ambient field is obtained by using equation (1) as

$$\overline{V} = \overline{x}\,\overline{y}\,\overline{z} \qquad V = xyz \qquad \theta_V = \theta_x + \theta_y + \theta_z \tag{592}$$

The magnitude of the complex number volume can be related to the volume under zero ambient field conditions by equations (83) and (592) with the result that

$$V = V^o C_{\sigma x} C_{\sigma y} C_{\sigma z} \qquad \theta_V = \theta_x + \theta_y + \theta_z \tag{593}$$

where the volume under zero ambient field conditions is given by

$$V^o = x^o y^o z^o \tag{594}$$

and where $C_{\sigma\alpha}$ = coordinate factors given by equations (78) through (80) for $\alpha = x, y, z$ and with $\sigma = 1$ for [hd] and $\sigma = 2,3$ for [ld], and where θ_x, θ_y and θ_z are represented in terms of the internal phase angles of the counting and segmentation numbers by equation (56). For the cases when only an ambient gravitational field or only an ambient electromagnetic field is present the volumes are given respectively by

$$V^g = V^o C^g_{\sigma x} C^g_{\sigma y} C^g_{\sigma z} \qquad \theta^g_V = \theta^g_x + \theta^g_y + \theta^g_z \tag{595}$$

$$V^\gamma = V^o C^\gamma_{\sigma x} C^\gamma_{\sigma y} C^\gamma_{\sigma z} \qquad \theta^\gamma_V = \theta^\gamma_x + \theta^\gamma_y + \theta^\gamma_z \tag{596}$$

where the space coordinate factors $C^g_{\sigma\alpha}$ and $C^\gamma_{\sigma\alpha}$ are given by equations (183) through (185) and (191) through (193) respectively. Combining equations (593) through (596) gives

$$V = V^o C_{\sigma x} C_{\sigma y} C_{\sigma z} = V^g D_{\sigma x} D_{\sigma y} D_{\sigma z} = V^\gamma E_{\sigma x} E_{\sigma y} E_{\sigma z} \tag{597}$$

where $D_{\sigma\alpha}$ and $E_{\sigma\alpha}$ are given by equations (198) and (199) for $\alpha = x,y,z$ which can also be used to obtain equation (597) more directly.

Cylindrical Polar Coordinates

The complex number volume of a cylinder located in matter in an ambient field is given by

$$\overline{V} = \pi \overline{r}^2 \overline{z} \qquad V = \pi r^2 z \qquad \theta_V = 2\theta_r + \theta_z \tag{598}$$

The magnitude of the complex number volume can be related to the volume under zero ambient field conditions by combining equations (83) and (598) which gives

$$V = V^o C^2_{\sigma r} C_{\sigma z} \qquad \theta_V = 2\theta_r + \theta_z \tag{599}$$

where the volume of a cylinder under zero ambient field conditions is given by

$$V^{o} = \pi r_{o}^{2} z_{o} \tag{600}$$

and where the space coordinate factors $C_{\sigma r}$ and $C_{\sigma z}$ are defined by equations (78) through (80) with $\sigma = 1$ for [hd] and $\sigma = 2,3$ for [ld], and where the internal phase angles θ_r and θ_z are given by equation (56) in terms of the internal phase angles of the complex counting and segmentation numbers. When only an ambient gravitational field or an ambient electromagnetic field is present the volumes are given respectively by

$$V^{g} = V^{o}(C_{\sigma r}^{g})^{2} C_{\sigma z}^{g} \qquad \theta_{V}^{g} = 2\theta_{r}^{g} + \theta_{z}^{g} \tag{601}$$

$$V^{\gamma} = V^{o}(C_{\sigma r}^{\gamma})^{2} C_{\sigma z}^{\gamma} \qquad \theta_{V} = 2\theta_{r}^{\gamma} + \theta_{z}^{\gamma} \tag{602}$$

where the space coordinate factors $C_{\sigma\alpha}^{g}$ and $C_{\sigma\alpha}^{\gamma}$ are given by equations (183) through (185) and (191) through (193) respectively. Combining equations (599) through (602) gives the volume magnitude as

$$V = V^{o} C_{\sigma r}^{2} C_{\sigma z} = V^{g} D_{\sigma r}^{2} D_{\sigma z} = V^{\gamma} E_{\sigma r}^{2} E_{\sigma z} \tag{603}$$

where $D_{\sigma\alpha}$ and $E_{\sigma\alpha}$ for $\alpha = r, z$ and with $\sigma = 1$ for [hd] and $\sigma = 2,3$ for [ld] are given by equations (198) and (199) which can be used to obtain equation (603) more directly.

Spherical Polar Coordinates

In the presence of an ambient field the complex number volume of a sphere is given by

$$\overline{V} = 4/3\pi\overline{\rho}^{3} \qquad V = 4/3\pi\rho^{3} \qquad \theta_{V} = 3\theta_{\rho} \tag{604}$$

The magnitude of the complex number volume of a sphere located in matter which is surrounded by an ambient field can be related to the volume of the sphere under zero ambient field conditions by combining equations (83) and (604) and obtaining the result

$$V = V^{o} C_{\sigma\rho}^{3} \qquad \theta_{V} = 3\theta_{\rho} \tag{605}$$

where the zero ambient field volume of a sphere is given by

$$V^{o} = 4/3\pi\rho_{o}^{3} \tag{606}$$

and where the space coordinate factors $C_{\sigma\rho}$ are given by equations (78) through (80) with $\sigma = 1$ for [hd] and $\sigma = 2,3$ for [ld], and where equation (56) relates θ_ρ to the internal phase angles of the complex counting and segmentation numbers. For the case when only an ambient gravitational field or only an ambient electromagnetic field is acting, the magnitudes of the spherical volumes are obtained from equation (605) and are written respectively as

$$V^{g} = V^{o}(C_{\sigma\rho}^{g})^{3} \qquad \theta_{V}^{g} = 3\theta_{\rho}^{g} \tag{607}$$

$$V^{\gamma} = V^{o}(C_{\sigma\rho}^{\gamma})^{3} \qquad \theta_{V}^{\gamma} = 3\theta_{\rho}^{\gamma} \tag{608}$$

where the space coordinate factors $C^{g}_{\sigma\rho}$ and $C^{\gamma}_{\sigma\rho}$ are defined by equations (183) through (185) and (191) through (193) respectively. Combining equations (605) through (608) give the relations

$$V = V^{o}C^{3}_{\sigma\rho} = V^{g}D^{3}_{\sigma\rho} = V^{\gamma}E^{3}_{\sigma\rho} \tag{609}$$

where $D_{\sigma\rho}$ and $E_{\sigma\rho}$ are given by equation (199). Additional practical expressions for the volume magnitude V are derived in subsection E based on the concepts of complex particle number and complex particle number density which are introduced in subsections C and D.

C. General Expressions for Particle Number Density of Large and Small Amounts of Matter in an Ambient Field.

The basic equation that describes the complex number average particle number density is written as

$$\overline{n} = n\exp(j\theta_{n}) = \overline{\mathfrak{N}}/\overline{V} \tag{610}$$

where the complex number volume $\overline{V}$ is defined by equations (587) through (609), and where in the presence of an ambient field the complex particle number is given by [10-12]

$$\overline{\mathfrak{N}} = \mathfrak{N}\exp(j\theta_{\mathfrak{N}}) \tag{611}$$

$$\mathfrak{N} = N\cos\theta_{\mathfrak{N}} \qquad [SN] \tag{612}$$

$$\mathfrak{N} = N\sec\theta_{\mathfrak{N}} \qquad [LN] \tag{613}$$

where N = integer number of particles, $\mathfrak{N}$ = noninteger number of particles and where [SN] = small number of particles and [LN] = large number of particles. In analogy to equation (10), the internal phase angle of the complex number of particles is given for a combined static elastic and viscoelastic system without resonances, with t_o = time factor for the continuous matter, as

$$\tan\theta_{\mathfrak{N}} = (2\mathfrak{E}_{s}/G_{o})^{1/2} \qquad G_{o} = \varsigma_{o}/t_{o} \qquad \varsigma_{o} = \text{viscosity of continuous matter} \tag{614}$$

where $\mathfrak{E}_{s}$ = ambient shear field energy density and G_o = shear modulus of a material medium corresponding to zero ambient field. The replacement $G_{o} \rightarrow G_{r}$, where G_{r} = an effective dynamical shear modulus, allows a simple treatment of oscillatory systems at a resonance frequency, for example. Combining equations (610) through (613) gives the magnitude and internal phase angle of the particle number density as

$$n = \mathfrak{N}/V \qquad \theta_{n} = \theta_{\mathfrak{N}} - \theta_{V} = \theta_{\mathfrak{N}} + |\theta_{V}| \tag{615}$$

$$= N/V\cos\theta_{\mathfrak{N}} \qquad [SN] \tag{616}$$

$$= N/V\sec\theta_{\mathfrak{N}} \qquad [LN] \tag{617}$$

where V and θ_V = magnitude and internal phase angle of the complex number volume of particles as described in equations (587) through (609), and where for a system of particles $\theta_V \leq 0$ because $\theta_\alpha \leq 0$ for α = x,y,z; r,φ,z or ρ,φ,ψ as mentioned in the previous sections. When only an ambient gravitational field or only an ambient electromagnetic field is acting, equations (610) through (617) become respectively

$$\bar{n}^g = n^g \exp(j\theta_n^g) = \bar{\mathcal{N}}^g / \bar{V}^g \tag{618}$$

$$\bar{n}^\gamma = n^\gamma \exp(j\theta_n^\gamma) = \bar{\mathcal{N}}^\gamma / \bar{V}^\gamma \tag{619}$$

where $\bar{V}^g$ and $\bar{V}^\gamma$ are given by equations (589), (595) and (590), (596) respectively, and where

$$\bar{\mathcal{N}}^g = \mathcal{N}^g \exp(j\theta_{\mathcal{N}}^g) \tag{620}$$

$$\bar{\mathcal{N}}^\gamma = \mathcal{N}^\gamma \exp(j\theta_{\mathcal{N}}^\gamma) \tag{621}$$

$$\mathcal{N}^g = N\cos\theta_{\mathcal{N}}^g \qquad \text{[SN]} \tag{622}$$

$$\mathcal{N}^g = N\sec\theta_{\mathcal{N}}^g \qquad \text{[LN]} \tag{623}$$

$$\mathcal{N}^\gamma = N\cos\theta_{\mathcal{N}}^\gamma \qquad \text{[SN]} \tag{624}$$

$$\mathcal{N}^\gamma = N\sec\theta_{\mathcal{N}}^\gamma \qquad \text{[LN]} \tag{625}$$

with N = integer number of particles. Combining equations (589) and (590) with equations (618) through (625) gives the magnitudes and internal phase angles for the corresponding particle number densities as

$$n^g = \mathcal{N}^g / V^g \qquad \theta_n^g = \theta_{\mathcal{N}}^g - \theta_V^g = \theta_{\mathcal{N}}^g + |\theta_V^g| \tag{626}$$

$$= N / V^g \cos\theta_{\mathcal{N}}^g \qquad \text{[SN]} \tag{627}$$

$$= N / V^g \sec\theta_{\mathcal{N}}^g \qquad \text{[LN]} \tag{628}$$

$$n^\gamma = \mathcal{N}^\gamma / V^\gamma \qquad \theta_n^\gamma = \theta_{\mathcal{N}}^\gamma - \theta_V^\gamma = \theta_{\mathcal{N}}^\gamma + |\theta_V^\gamma| \tag{629}$$

$$= N / V^\gamma \cos\theta_{\mathcal{N}}^\gamma \qquad \text{[SN]} \tag{630}$$

$$= N / V^\gamma \sec\theta_{\mathcal{N}}^\gamma \qquad \text{[LN]} \tag{631}$$

For the case of zero ambient field

$$\theta_{\mathcal{N}} = 0 \qquad \theta_V = 0 \qquad V = V^o \tag{632}$$

and the particle number density is given by

$$n^o = N / V^o \qquad \theta_n = 0 \tag{633}$$

where V^{o} = zero ambient field volume given by equations (594), (600) or (606).

It should be pointed out that the particle number density expressions that appear in equations (615) through (617), (626) through (628) and (629) through (631) for large and small numbers of particles of matter are not complete because the magnitude V of the complex number volume has not yet been expressed in terms of the zero ambient field volume V^{o}. This will be done in terms of the space coordinate factors for high and low-density matter in subsections D and E.

D. Expressions for High and Low Particle Number Density of Matter Consisting of Large and Small Numbers of Particles in an Ambient Field.

Now the forms of the particle number density of matter in an ambient field are related to the zero ambient field value of the particle number density and are evaluated for high [hd] and low [ld] matter densities and for large [LN] and small [SN] numbers of particles and for Cartesian coordinates, cylindrical polar coordinates and spherical polar coordinates.

Cartesian Coordinates

The particle number density for matter in an ambient field can be calculated in Cartesian coordinates by using equations (593) and (615) through (617) and give the particle number density magnitudes as

$$n = \mathcal{N}/V = n^{o} C_{1x}^{-1} C_{1y}^{-1} C_{1z}^{-1} \cos\theta_{\mathcal{N}} \le n^{o} \qquad [\text{hd}, \text{SN}] \tag{634}$$

$$n = \mathcal{N}/V = n^{o} C_{1x}^{-1} C_{1y}^{-1} C_{1z}^{-1} \sec\theta_{\mathcal{N}} \le n^{o} \qquad [\text{hd}, \text{LN}] \tag{635}$$

$$n = \mathcal{N}/V = n^{o} C_{2x}^{-1} C_{2y}^{-1} C_{2z}^{-1} \cos\theta_{\mathcal{N}} \ge n^{o} \qquad [\text{ld}, \text{SN}] \tag{636}$$

$$n = \mathcal{N}/V = n^{o} C_{2x}^{-1} C_{2y}^{-1} C_{2z}^{-1} \sec\theta_{\mathcal{N}} \ge n^{o} \qquad [\text{ld}, \text{LN}] \tag{637}$$

$$n = \mathcal{N}/V = n^{o} C_{3x}^{-1} C_{3y}^{-1} C_{3z}^{-1} \cos\theta_{\mathcal{N}} \ge n^{o} \qquad [\text{ld}, \text{SN}] \tag{638}$$

$$n = \mathcal{N}/V = n^{o} C_{3x}^{-1} C_{3y}^{-1} C_{3z}^{-1} \sec\theta_{\mathcal{N}} \ge n^{o} \qquad [\text{ld}, \text{LN}] \tag{639}$$

and the phase angle as

$$\theta_n = \theta_{\mathcal{N}} - \theta_x - \theta_y - \theta_z = \theta_{\mathcal{N}} + |\theta_x| + |\theta_y| + |\theta_z| \qquad \theta_n \ge \theta_{\mathcal{N}} \tag{640}$$

where n^{o} = zero ambient field particle number density given by equation (594) as

$$n^{o} = N / V^{o} \tag{641}$$

and where the space coordinate factors $C_{\sigma\alpha}$ are given by equations (78) through (80) for $\alpha = x,y,z$ and where $\sigma = 1$ for [hd] and $\sigma = 2,3$ for [ld], and finally where $\theta_\alpha \le 0$ for systems such as, for example, the earth's atmosphere and its extension into interplanetary space. When only an ambient gravity field is present the particle number densities expressed in terms of Cartesian coordinates as given in equations (634) through (640) become after using equation (595)

$$n^g = \mathcal{N}^g / V^g = n^o (C^g_{1x})^{-1} (C^g_{1y})^{-1} (C^g_{1z})^{-1} \cos\theta^g_{\mathcal{N}} \le n^o \qquad [hd, SN] \qquad (642)$$

$$n^g = \mathcal{N}^g / V^g = n^o (C^g_{1x})^{-1} (C^g_{1y})^{-1} (C^g_{1z})^{-1} \sec\theta^g_{\mathcal{N}} \le n^o \qquad [hd, LN] \qquad (643)$$

$$n^g = \mathcal{N}^g / V^g = n^o (C^g_{2x})^{-1} (C^g_{2y})^{-1} (C^g_{2z})^{-1} \cos\theta^g_{\mathcal{N}} \ge n^o \qquad [ld, SN] \qquad (644)$$

$$n^g = \mathcal{N}^g / V^g = n^o (C^g_{2x})^{-1} (C^g_{2y})^{-1} (C^g_{2z})^{-1} \sec\theta^g_{\mathcal{N}} \ge n^o \qquad [ld, LN] \qquad (645)$$

$$n^g = \mathcal{N}^g / V^g = n^o (C^g_{3x})^{-1} (C^g_{3y})^{-1} (C^g_{3z})^{-1} \cos\theta^g_{\mathcal{N}} \ge n^o \qquad [ld, SN] \qquad (646)$$

$$n^g = \mathcal{N}^g / V^g = n^o (C^g_{3x})^{-1} (C^g_{3y})^{-1} (C^g_{3z})^{-1} \sec\theta^g_{\mathcal{N}} \ge n^o \qquad [ld, LN] \qquad (647)$$

$$\theta^g_n = \theta^g_{\mathcal{N}} - \theta^g_x - \theta^g_y - \theta^g_z = \theta^g_{\mathcal{N}} + |\theta^g_x| + |\theta^g_y| + |\theta^g_z| \qquad \theta^g_n \ge \theta^g_{\mathcal{N}} \qquad (648)$$

where $C^g_{\sigma\alpha}$ for $\alpha = x,y,z$ is given by equations (183) through (185). For the case of an ambient electromagnetic field acting alone, equations (634) through (640) become after using equation (596)

$$n^\gamma = \mathcal{N}^\gamma / V^\gamma = n^o (C^\gamma_{1x})^{-1} (C^\gamma_{1y})^{-1} (C^\gamma_{1z})^{-1} \cos\theta^\gamma_{\mathcal{N}} \le n^o \qquad [hd, SN] \qquad (649)$$

$$n^\gamma = \mathcal{N}^\gamma / V^\gamma = n^o (C^\gamma_{1x})^{-1} (C^\gamma_{1y})^{-1} (C^\gamma_{1z})^{-1} \sec\theta^\gamma_{\mathcal{N}} \le n^o \qquad [hd, LN] \qquad (650)$$

$$n^\gamma = \mathcal{N}^\gamma / V^\gamma = n^o (C^\gamma_{2x})^{-1} (C^\gamma_{2y})^{-1} (C^\gamma_{2z})^{-1} \cos\theta^\gamma_{\mathcal{N}} \ge n^o \qquad [ld, SN] \qquad (651)$$

$$n^\gamma = \mathcal{N}^\gamma / V^\gamma = n^o (C^\gamma_{2x})^{-1} (C^\gamma_{2y})^{-1} (C^\gamma_{2z})^{-1} \sec\theta^\gamma_{\mathcal{N}} \ge n^o \qquad [ld, LN] \qquad (652)$$

$$n^\gamma = \mathcal{N}^\gamma / V^\gamma = n^o (C^\gamma_{3x})^{-1} (C^\gamma_{3y})^{-1} (C^\gamma_{3z})^{-1} \cos\theta^\gamma_{\mathcal{N}} \ge n^o \qquad [ld, SN] \qquad (653)$$

$$n^\gamma = \mathcal{N}^\gamma / V^\gamma = n^o (C^\gamma_{3x})^{-1} (C^\gamma_{3y})^{-1} (C^\gamma_{3z})^{-1} \sec\theta^\gamma_{\mathcal{N}} \ge n^o \qquad [ld, LN] \qquad (654)$$

$$\theta^\gamma_n = \theta^\gamma_{\mathcal{N}} - \theta^\gamma_x - \theta^\gamma_y - \theta^\gamma_z = \theta^\gamma_{\mathcal{N}} + |\theta^\gamma_x| + |\theta^\gamma_y| + |\theta^\gamma_z| \qquad \theta^\gamma_n \ge \theta^\gamma_{\mathcal{N}} \qquad (655)$$

where the space coordinate factors $C^\gamma_{\sigma\alpha}$ for $\alpha = x,y,z$ are given by equations (191) through (193).

Equations (634) through (654) can be used to express the particle number density n for matter in a rectangular volume located in an ambient field in terms of n^o, n^g or n^γ. Thus, equations (634), (635), (642), (643), (649) and (650) give for the [hd] case

$$n = n^o C_{1x}^{-1} C_{1y}^{-1} C_{1z}^{-1} \cos\theta_{\mathcal{N}} \le n^o \qquad [hd, SN] \qquad (656)$$

$$= n^g D_{1x}^{-1} D_{1y}^{-1} D_{1z}^{-1} \cos\theta_{\mathcal{N}} \sec\theta_{\mathcal{N}}^g \le n^g \qquad (657)$$

$$= n^\gamma E_{1x}^{-1} E_{1y}^{-1} E_{1z}^{-1} \cos\theta_{\mathcal{N}} \sec\theta_{\mathcal{N}}^\gamma \le n^\gamma \qquad (658)$$

$$n = n^o C_{1x}^{-1} C_{1y}^{-1} C_{1z}^{-1} \sec\theta_{\mathcal{N}} \le n^o \qquad [hd, LN] \qquad (659)$$

$$= n^g D_{1x}^{-1} D_{1y}^{-1} D_{1z}^{-1} \sec\theta_{\mathcal{N}} \cos\theta_{\mathcal{N}}^g \le n^g \qquad (660)$$

$$= n^\gamma E_{1x}^{-1} E_{1y}^{-1} E_{1z}^{-1} \sec\theta_{\mathcal{N}} \cos\theta_{\mathcal{N}}^\gamma \le n^\gamma \qquad (661)$$

while equations (636), (637), (644), (645), (651) and (652) give the first set of the two degenerate [ld] representations of the particle number density magnitudes as

$$n = n^o C_{2x}^{-1} C_{2y}^{-1} C_{2z}^{-1} \cos\theta_{\mathcal{N}} \ge n^o \qquad [ld, SN] \qquad (662)$$

$$= n^g D_{2x}^{-1} D_{2y}^{-1} D_{2z}^{-1} \cos\theta_{\mathcal{N}} \sec\theta_{\mathcal{N}}^g \ge n^g \qquad (663)$$

$$= n^\gamma E_{2x}^{-1} E_{2y}^{-1} E_{2z}^{-1} \cos\theta_{\mathcal{N}} \sec\theta_{\mathcal{N}}^\gamma \ge n^\gamma \qquad (664)$$

$$n = n^o C_{2x}^{-1} C_{2y}^{-1} C_{2z}^{-1} \sec\theta_{\mathcal{N}} \ge n^o \qquad [ld, LN] \qquad (665)$$

$$= n^g D_{2x}^{-1} D_{2y}^{-1} D_{2z}^{-1} \sec\theta_{\mathcal{N}} \cos\theta_{\mathcal{N}}^g \ge n^g \qquad (666)$$

$$= n^\gamma E_{2x}^{-1} E_{2y}^{-1} E_{2z}^{-1} \sec\theta_{\mathcal{N}} \cos\theta_{\mathcal{N}}^\gamma \ge n^\gamma \qquad (667)$$

and equations (638), (639), (646), (647), (653) and (654) give the second set of [ld] representations of the particle density magnitudes as

$$n = n^o C_{3x}^{-1} C_{3y}^{-1} C_{3z}^{-1} \cos\theta_{\mathcal{N}} \ge n^o \qquad [ld, SN] \qquad (668)$$

$$= n^g D_{3x}^{-1} D_{3y}^{-1} D_{3z}^{-1} \cos\theta_{\mathcal{N}} \sec\theta_{\mathcal{N}}^g \ge n^g \qquad (669)$$

$$= n^\gamma E_{3x}^{-1} E_{3y}^{-1} E_{3z}^{-1} \cos\theta_{\mathcal{N}} \sec\theta_{\mathcal{N}}^\gamma \ge n^\gamma \qquad (670)$$

$$n = n^o C_{3x}^{-1} C_{3y}^{-1} C_{3z}^{-1} \sec\theta_{\mathcal{N}} \ge n^o \qquad [ld, LN] \qquad (671)$$

$$= n^g D_{3x}^{-1} D_{3y}^{-1} D_{3z}^{-1} \sec\theta_{\mathcal{N}} \cos\theta_{\mathcal{N}}^g \ge n^g \qquad (672)$$

$$= n^\gamma E_{3x}^{-1} E_{3y}^{-1} E_{3z}^{-1} \sec\theta_{\mathcal{N}} \cos\theta_{\mathcal{N}}^\gamma \ge n^\gamma \qquad (673)$$

where equations (198) and (199) give

$$D_{\sigma x} = C_{\sigma x} / C_{\sigma x}^g \qquad D_{\sigma y} = C_{\sigma y} / C_{\sigma y}^g \qquad D_{\sigma z} = C_{\sigma z} / C_{\sigma z}^g \qquad (674)$$

$$E_{\sigma x} = C_{\sigma x} / C^{\gamma}_{\sigma x} \qquad E_{\sigma y} = C_{\sigma y} / C^{\gamma}_{\sigma y} \qquad E_{\sigma z} = C_{\sigma z} / C^{\gamma}_{\sigma z} \tag{675}$$

where $\sigma = 1$ for [hd] and $\sigma = 2,3$ for the degenerate [ld] case.

Cylindrical Polar Coordinates

For the case of a cylindrical volume of matter in an ambient field, equations (594) and (615) through (617) can be used to calculate the particle number density magnitudes as

$$n = \mathcal{N}/V = n^{\circ} C_{1r}^{-2} C_{1z}^{-1} \cos\theta_{\mathcal{N}} \le n^{\circ} \qquad [\mathrm{hd, SN}] \tag{676}$$

$$n = \mathcal{N}/V = n^{\circ} C_{1r}^{-2} C_{1z}^{-1} \sec\theta_{\mathcal{N}} \le n^{\circ} \qquad [\mathrm{hd, LN}] \tag{677}$$

$$n = \mathcal{N}/V = n^{\circ} C_{2r}^{-2} C_{2z}^{-1} \cos\theta_{\mathcal{N}} \ge n^{\circ} \qquad [\mathrm{ld, SN}] \tag{678}$$

$$n = \mathcal{N}/V = n^{\circ} C_{2r}^{-2} C_{2z}^{-1} \sec\theta_{\mathcal{N}} \ge n^{\circ} \qquad [\mathrm{ld, LN}] \tag{679}$$

$$n = \mathcal{N}/V = n^{\circ} C_{3r}^{-2} C_{3z}^{-1} \cos\theta_{\mathcal{N}} \ge n^{\circ} \qquad [\mathrm{ld, SN}] \tag{680}$$

$$n = \mathcal{N}/V = n^{\circ} C_{3r}^{-2} C_{3z}^{-1} \sec\theta_{\mathcal{N}} \ge n^{\circ} \qquad [\mathrm{ld, LN}] \tag{681}$$

and the particle number internal phase angle as

$$\theta_n = \theta_{\mathcal{N}} - 2\theta_r - \theta_z = \theta_{\mathcal{N}} + 2|\theta_r| + |\theta_z| \qquad \theta_n \ge \theta_{\mathcal{N}} \tag{682}$$

where the zero ambient field particle number density is given by equations (600) and (633) as

$$n^{\circ} = N / V^{\circ} \tag{683}$$

and where the space coordinate factors $C_{\sigma\alpha}$ are given by equations (78) through (80) for $\alpha = r, z$ and where $\sigma = 1$ for [hd] and $\sigma = 2,3$ for [ld], and where $\theta_{\alpha} \le 0$ for systems such as solids, liquids and gases. Equations (676) through (681) are useful for the study polyphase rockets.[21] For a cylindrical volume of matter in the presence of ambient gravity only, equations (676) through (682) become

$$n^{g} = \mathcal{N}^{g} / V^{g} = n^{\circ} (C_{1r}^{g})^{-2} (C_{1z}^{g})^{-1} \cos\theta_{\mathcal{N}}^{g} \le n^{\circ} \qquad [\mathrm{hd, SN}] \tag{684}$$

$$n^{g} = \mathcal{N}^{g} / V^{g} = n^{\circ} (C_{1r}^{g})^{-2} (C_{1z}^{g})^{-1} \sec\theta_{\mathcal{N}}^{g} \le n^{\circ} \qquad [\mathrm{hd, LN}] \tag{685}$$

$$n^{g} = \mathcal{N}^{g} / V^{g} = n^{\circ} (C_{2r}^{g})^{-2} (C_{2z}^{g})^{-1} \cos\theta_{\mathcal{N}}^{g} \ge n^{\circ} \qquad [\mathrm{ld, SN}] \tag{686}$$

$$n^{g} = \mathcal{N}^{g} / V^{g} = n^{\circ} (C_{2r}^{g})^{-2} (C_{2z}^{g})^{-1} \sec\theta_{\mathcal{N}}^{g} \ge n^{\circ} \qquad [\mathrm{ld, LN}] \tag{687}$$

$$n^g = \mathfrak{N}^g / V^g = n^o (C^g_{3r})^{-2} (C^g_{3z})^{-1} \cos\theta^g_{\mathfrak{N}} \geq n^o \qquad [\mathrm{ld,SN}] \qquad (688)$$

$$n^g = \mathfrak{N}^g / V^g = n^o (C^g_{3r})^{-2} (C^g_{3z})^{-1} \sec\theta^g_{\mathfrak{N}} \geq n^o \qquad [\mathrm{ld,LN}] \qquad (689)$$

$$\theta^g_n = \theta^g_{\mathfrak{N}} - 2\theta^g_r - \theta^g_z = \theta^g_{\mathfrak{N}} + 2|\theta^g_r| + |\theta^g_z| \qquad \theta^g_n \geq \theta^g_{\mathfrak{N}} \qquad (690)$$

where the space coordinate factors $C^g_{\sigma\alpha}$ are given by equations (183) through (185) for $\alpha = r, z$ and with $\sigma = 1$ for [hd] and $\sigma = 2,3$ for [ld]. When only an ambient electromagnetic field is present, equations (676) through (682) become for a cylindrical volume of matter

$$n^\gamma = \mathfrak{N}^\gamma / V^\gamma = n^o (C^\gamma_{1r})^{-2} (C^\gamma_{1z})^{-1} \cos\theta^\gamma_{\mathfrak{N}} \leq n^o \qquad [\mathrm{hd,SN}] \qquad (691)$$

$$n^\gamma = \mathfrak{N}^\gamma / V^\gamma = n^o (C^\gamma_{1r})^{-2} (C^\gamma_{1z})^{-1} \sec\theta^\gamma_{\mathfrak{N}} \leq n^o \qquad [\mathrm{hd,LN}] \qquad (692)$$

$$n^\gamma = \mathfrak{N}^\gamma / V^\gamma = n^o (C^\gamma_{2r})^{-2} (C^\gamma_{2z})^{-1} \cos\theta^\gamma_{\mathfrak{N}} \geq n^o \qquad [\mathrm{ld,SN}] \qquad (693)$$

$$n^\gamma = \mathfrak{N}^\gamma / V^\gamma = n^o (C^\gamma_{2r})^{-2} (C^\gamma_{2z})^{-1} \sec\theta^\gamma_{\mathfrak{N}} \geq n^o \qquad [\mathrm{ld,LN}] \qquad (694)$$

$$n^\gamma = \mathfrak{N}^\gamma / V^\gamma = n^o (C^\gamma_{3r})^{-2} (C^\gamma_{3z})^{-1} \cos\theta^\gamma_{\mathfrak{N}} \geq n^o \qquad [\mathrm{ld,SN}] \qquad (695)$$

$$n^\gamma = \mathfrak{N}^\gamma / V^\gamma = n^o (C^\gamma_{3r})^{-2} (C^\gamma_{3z})^{-1} \sec\theta^\gamma_{\mathfrak{N}} \geq n^o \qquad [\mathrm{ld,LN}] \qquad (696)$$

$$\theta^\gamma_n = \theta^\gamma_{\mathfrak{N}} - 2\theta^\gamma_r - \theta^\gamma_z = \theta^\gamma_{\mathfrak{N}} + 2|\theta^\gamma_r| + |\theta^\gamma_z| \qquad \theta^\gamma_n \geq \theta^\gamma_{\mathfrak{N}} \qquad (697)$$

where the space coordinate factors $C^\gamma_{\sigma\alpha}$ for $\alpha = r,z$ are given by equations (191) through (193).

For a cylindrical volume element of matter located in an ambient field, equations (676) through (696) can be used to express the particle number density magnitude n in terms of n^o, n^g or n^γ. In particular, equations (676), (677), (684), (685), (691) and (692) give the results for the [hd] case as

$$n = n^o C^{-2}_{1r} C^{-1}_{1z} \cos\theta_{\mathfrak{N}} \leq n^o \qquad [\mathrm{hd,SN}] \qquad (698)$$

$$= n^g D^{-2}_{1r} D^{-1}_{1z} \cos\theta_{\mathfrak{N}} \sec\theta^g_{\mathfrak{N}} \leq n^g \qquad (699)$$

$$= n^\gamma E^{-2}_{1r} E^{-1}_{1z} \cos\theta_{\mathfrak{N}} \sec\theta^\gamma_{\mathfrak{N}} \leq n^\gamma \qquad (700)$$

$$n = n^o C^{-2}_{1r} C^{-1}_{1z} \sec\theta_{\mathfrak{N}} \leq n^o \qquad [\mathrm{hd,LN}] \qquad (701)$$

$$= n^g D^{-2}_{1r} D^{-1}_{1z} \sec\theta_{\mathfrak{N}} \cos\theta^g_{\mathfrak{N}} \leq n^g \qquad (702)$$

$$= n^\gamma E^{-2}_{1r} E^{-1}_{1z} \sec\theta_{\mathfrak{N}} \cos\theta^\gamma_{\mathfrak{N}} \leq n^\gamma \qquad (703)$$

while equations (678), (679), (686), (687), (693) and (694) give the first set of the two degenerate sets of [ld] equations as

$$n = n^{o} C_{2r}^{-2} C_{2z}^{-1} \cos\theta_{\aleph} \geq n^{o} \qquad [ld, SN] \qquad (704)$$

$$= n^{g} D_{2r}^{-2} D_{2z}^{-1} \cos\theta_{\aleph} \sec\theta_{\aleph}^{g} \geq n^{g} \qquad (705)$$

$$= n^{\gamma} E_{2r}^{-2} E_{2z}^{-1} \cos\theta_{\aleph} \sec\theta_{\aleph}^{\gamma} \geq n^{\gamma} \qquad (706)$$

$$n = n^{o} C_{2r}^{-2} C_{2z}^{-1} \sec\theta_{\aleph} \geq n^{o} \qquad [ld, LN] \qquad (707)$$

$$= n^{g} D_{2r}^{-2} D_{2z}^{-1} \sec\theta_{\aleph} \cos\theta_{\aleph}^{g} \geq n^{g} \qquad (708)$$

$$= n^{\gamma} E_{2r}^{-2} E_{2z}^{-1} \sec\theta_{\aleph} \cos\theta_{\aleph}^{\gamma} \geq n^{\gamma} \qquad (709)$$

and equations (680), (681), (688), (689), (695) and (696) give the second set of [ld] equations as

$$n = n^{o} C_{3r}^{-2} C_{3z}^{-1} \cos\theta_{\aleph} \geq n^{o} \qquad [ld, SN] \qquad (710)$$

$$= n^{g} D_{3r}^{-2} D_{3z}^{-1} \cos\theta_{\aleph} \sec\theta_{\aleph}^{g} \geq n^{g} \qquad (711)$$

$$= n^{\gamma} E_{3r}^{-2} E_{3z}^{-1} \cos\theta_{\aleph} \sec\theta_{\aleph}^{\gamma} \geq n^{\gamma} \qquad (712)$$

$$n = n^{o} C_{3r}^{-2} C_{3z}^{-1} \sec\theta_{\aleph} \geq n^{o} \qquad [ld, LN] \qquad (713)$$

$$= n^{g} D_{3r}^{-2} D_{3z}^{-1} \sec\theta_{\aleph} \cos\theta_{\aleph}^{g} \geq n^{g} \qquad (714)$$

$$= n^{\gamma} E_{3r}^{-2} E_{3z}^{-1} \sec\theta_{\aleph} \cos\theta_{\aleph}^{\gamma} \geq n^{\gamma} \qquad (715)$$

where equations (198) and (199) give for cylindrical polar coordinates

$$D_{\sigma r} = C_{\sigma r} / C_{\sigma r}^{g} \qquad D_{\sigma z} = C_{\sigma z} / C_{\sigma z}^{g} \qquad (716)$$

$$E_{\sigma r} = C_{\sigma r} / C_{\sigma r}^{\gamma} \qquad E_{\sigma z} = C_{\sigma z} / C_{\sigma z}^{\gamma} \qquad (717)$$

where $D_{\sigma\alpha}$ and $E_{\sigma\alpha}$ are calculated in Section 4.

Spherical Polar Coordinates

For the case of a spherical volume of matter in an ambient field, equations (605) and (615) through (617) are available to calculate the magnitudes and internal phase angle of the particle number density as

$$n = \aleph/V = n^{o} C_{1\rho}^{-3} \cos\theta_{\aleph} \leq n^{o} \qquad [hd, SN] \qquad (718)$$

$$n = \aleph/V = n^{o} C_{1\rho}^{-3} \sec\theta_{\aleph} \le n^{o} \qquad [hd, LN] \qquad (719)$$

$$n = \aleph/V = n^{o} C_{2\rho}^{-3} \cos\theta_{\aleph} \ge n^{o} \qquad [ld, SN] \qquad (720)$$

$$n = \aleph/V = n^{o} C_{2\rho}^{-3} \sec\theta_{\aleph} \ge n^{o} \qquad [ld, LN] \qquad (721)$$

$$n = \aleph/V = n^{o} C_{3\rho}^{-3} \cos\theta_{\aleph} \ge n^{o} \qquad [ld, SN] \qquad (722)$$

$$n = \aleph/V = n^{o} C_{3\rho}^{-3} \sec\theta_{\aleph} \ge n^{o} \qquad [ld, LN] \qquad (723)$$

$$\theta_{n} = \theta_{\aleph} - 3\theta_{\rho} = \theta_{\aleph} + 3|\theta_{\rho}| \qquad \theta_{n} \ge \theta_{\aleph} \qquad (724)$$

where the zero ambient field particle number density is given by equations (606) and (633) as

$$n^{o} = N / V^{o} \qquad (725)$$

and where the space coordinate factors $C_{\sigma\alpha}$ are given by equations (78) through (80) with $\alpha = \rho$ and where $\sigma = 1$ for [hd] and $\sigma = 2,3$ for the degenerate [ld] case, and where $\theta_{\rho} \le 0$ for systems of matter such as planets, stars, and galaxies. For the case of a spherical volume element of matter, equations (718) through (724) become for an ambient gravity field

$$n^{g} = \aleph^{g} / V^{g} = n^{o} (C_{1\rho}^{g})^{-3} \cos\theta_{\aleph}^{g} \le n^{o} \qquad [hd, SN] \qquad (726)$$

$$n^{g} = \aleph^{g} / V^{g} = n^{o} (C_{1\rho}^{g})^{-3} \sec\theta_{\aleph}^{g} \le n^{o} \qquad [hd, LN] \qquad (727)$$

$$n^{g} = \aleph^{g} / V^{g} = n^{o} (C_{2\rho}^{g})^{-3} \cos\theta_{\aleph}^{g} \ge n^{o} \qquad [ld, SN] \qquad (728)$$

$$n^{g} = \aleph^{g} / V^{g} = n^{o} (C_{2\rho}^{g})^{-3} \sec\theta_{\aleph}^{g} \ge n^{o} \qquad [ld, LN] \qquad (729)$$

$$n^{g} = \aleph^{g} / V^{g} = n^{o} (C_{3\rho}^{g})^{-3} \cos\theta_{\aleph}^{g} \ge n^{o} \qquad [ld, SN] \qquad (730)$$

$$n^{g} = \aleph^{g} / V^{g} = n^{o} (C_{3\rho}^{g})^{-3} \sec\theta_{\aleph}^{g} \ge n^{o} \qquad [ld, LN] \qquad (731)$$

$$\theta_{n}^{g} = \theta_{\aleph}^{g} - 3\theta_{\rho}^{g} = \theta_{\aleph}^{g} + 3|\theta_{\rho}^{g}| \qquad \theta_{n}^{g} \ge \theta_{\aleph}^{g} \qquad (732)$$

where the space coordinate factors $C_{\sigma\alpha}^{g}$ are given by equations (183) through (185) for $\alpha = \rho$ and with $\sigma = 1$ for [hd] and $\sigma = 2,3$ for the possibly doubly degenerate [ld] state. For an ambient electromagnetic field acting alone on a spherical volume element of matter, equations (718) through (724) become

$$n^\gamma = \mathcal{N}^\gamma / V^\gamma = n^o (C^\gamma_{1\rho})^{-3} \cos\theta^\gamma_{\mathcal{N}} \le n^o \qquad [hd, SN] \qquad (733)$$

$$n^\gamma = \mathcal{N}^\gamma / V^\gamma = n^o (C^\gamma_{1\rho})^{-3} \sec\theta^\gamma_{\mathcal{N}} \le n^o \qquad [hd, LN] \qquad (734)$$

$$n^\gamma = \mathcal{N}^\gamma / V^\gamma = n^o (C^\gamma_{2\rho})^{-3} \cos\theta^\gamma_{\mathcal{N}} \ge n^o \qquad [ld, SN] \qquad (735)$$

$$n^\gamma = \mathcal{N}^\gamma / V^\gamma = n^o (C^\gamma_{2\rho})^{-3} \sec\theta^\gamma_{\mathcal{N}} \ge n^o \qquad [ld, LN] \qquad (736)$$

$$n^\gamma = \mathcal{N}^\gamma / V^\gamma = n^o (C^\gamma_{3\rho})^{-3} \cos\theta^\gamma_{\mathcal{N}} \ge n^o \qquad [ld, SN] \qquad (737)$$

$$n^\gamma = \mathcal{N}^\gamma / V^\gamma = n^o (C^\gamma_{3\rho})^{-3} \sec\theta^\gamma_{\mathcal{N}} \ge n^o \qquad [ld, LN] \qquad (738)$$

$$\theta^\gamma_n = \theta^\gamma_{\mathcal{N}} - 3\theta^\gamma_\rho = \theta^\gamma_{\mathcal{N}} + 3\left|\theta^\gamma_\rho\right| \qquad \theta^\gamma_n \ge \theta^\gamma_{\mathcal{N}} \qquad (739)$$

where the space coordinate factors $C^\gamma_{\sigma\alpha}$ for $\alpha = \rho$ are given by equations (191) through (193).

For a spherical volume of matter that is located in an ambient field, equations (718) through (738) can be utilized to represent the particle number density n for a combined ambient field in terms of n^o, n^g or n^γ. Thus, equations (718), (719), (726), (727), (733) and (734) give for the [hd] case

$$n = n^o C^{-3}_{1\rho} \cos\theta_{\mathcal{N}} \le n^o \qquad [hd, SN] \qquad (740)$$

$$= n^g D^{-3}_{1\rho} \cos\theta_{\mathcal{N}} \sec\theta^g_{\mathcal{N}} \le n^g \qquad (741)$$

$$= n^\gamma E^{-3}_{1\rho} \cos\theta_{\mathcal{N}} \sec\theta^\gamma_{\mathcal{N}} \le n^\gamma \qquad (742)$$

$$n = n^o C^{-3}_{1\rho} \sec\theta_{\mathcal{N}} \le n^o \qquad [hd, LN] \qquad (743)$$

$$= n^g D^{-3}_{1\rho} \sec\theta_{\mathcal{N}} \cos\theta^g_{\mathcal{N}} \le n^g \qquad (744)$$

$$= n^\gamma E^{-3}_{1\rho} \sec\theta_{\mathcal{N}} \cos\theta^\gamma_{\mathcal{N}} \le n^\gamma \qquad (745)$$

while equations (720), (721), (728), (729), (735) and (736) give the first set of the two degenerate classes of the [ld] equations as

$$n = n^o C^{-3}_{2\rho} \cos\theta_{\mathcal{N}} \ge n^o \qquad [ld, SN] \qquad (746)$$

$$= n^g D^{-3}_{2\rho} \cos\theta_{\mathcal{N}} \sec\theta^g_{\mathcal{N}} \ge n^g \qquad (747)$$

$$= n^\gamma E^{-3}_{2\rho} \cos\theta_{\mathcal{N}} \sec\theta^\gamma_{\mathcal{N}} \ge n^\gamma \qquad (748)$$

$$n = n^{o} C_{2\rho}^{-3} \sec\theta_{\mathcal{N}} \geq n^{o} \qquad [\text{ld, LN}] \tag{749}$$

$$= n^{g} D_{2\rho}^{-3} \sec\theta_{\mathcal{N}} \cos\theta_{\mathcal{N}}^{g} \geq n^{g} \tag{750}$$

$$= n^{\gamma} E_{2\rho}^{-3} \sec\theta_{\mathcal{N}} \cos\theta_{\mathcal{N}}^{\gamma} \geq n^{\gamma} \tag{751}$$

and where equations (722), (723), (730), (731), (737) and (738) produce the second set of the doubly degenerate [ld] equations as

$$n = n^{o} C_{3\rho}^{-3} \cos\theta_{\mathcal{N}} \geq n^{o} \qquad [\text{ld, SN}] \tag{752}$$

$$= n^{g} D_{3\rho}^{-3} \cos\theta_{\mathcal{N}} \sec\theta_{\mathcal{N}}^{g} \geq n^{g} \tag{753}$$

$$= n^{\gamma} E_{3\rho}^{-3} \cos\theta_{\mathcal{N}} \sec\theta_{\mathcal{N}}^{\gamma} \geq n^{\gamma} \tag{754}$$

$$n = n^{o} C_{3\rho}^{-3} \sec\theta_{\mathcal{N}} \geq n^{o} \qquad [\text{ld, LN}] \tag{755}$$

$$= n^{g} D_{3\rho}^{-3} \sec\theta_{\mathcal{N}} \cos\theta_{\mathcal{N}}^{g} \geq n^{g} \tag{756}$$

$$= n^{\gamma} E_{3\rho}^{-3} \sec\theta_{\mathcal{N}} \cos\theta_{\mathcal{N}}^{\gamma} \geq n^{\gamma} \tag{757}$$

where for spherical polar coordinates, equations (198) and (199) give

$$D_{\sigma\rho} = C_{\sigma\rho} / C_{\sigma\rho}^{g} \qquad \sigma = 1,2,3 \tag{758}$$

$$E_{\sigma\rho} = C_{\sigma\rho} / C_{\sigma\rho}^{\gamma} \qquad \sigma = 1,2,3 \tag{759}$$

where $D_{\sigma\rho}$ and $E_{\sigma\rho}$ are calculated in Section 4.

E. Functional Relationships Between n, V, $\theta_{\mathcal{N}}$ and θ_n.

This section shows, among other things, that the complex number particle number density of matter that is located in an ambient field can always be written in the form of equations (17) through (19) as

$$\bar{n} = n \exp(j\theta_n) \tag{760}$$

where the internal phase angle θ_n is given by equation (615) while the magnitude n is always given by[11,12]

$$n = \mathcal{N}/V = n^{o} \cos\theta_n \qquad [\text{hd}] \tag{761}$$

$$n = \mathcal{N}/V = n^{o} \sec\theta_n \qquad [\text{ld}] \tag{762}$$

where $\mathcal{N}$ = noninteger values of the particle number of matter in the presence of an ambient field as defined by equations (612) and (613), V = volume of matter in the presence of an ambient field, n^{o} = zero ambient field particle number density given by equation (633) as

$$n^{o} = N/V^{o} \tag{763}$$

where N = integer number of particles of matter and V^o = zero ambient field volume of matter. The measured values of the particle number density are given by

$$n_m = n\cos\theta_n \tag{764}$$

$$= n^o \cos^2\theta_n \qquad [hd] \tag{765}$$

$$= n^o \qquad [ld] \tag{766}$$

In this section the volume V will be calculated in terms of V^o, θ_n and $\theta_{\mathcal{N}}$ in such a way that equations (761) and (762) are universally valid.

A comparison of equations (615) through (617) with equations (761) and (762) gives

$$\cos\theta_n = n/n^o = (\mathcal{N}/V)/n^o = (V^o/V)\cos\theta_{\mathcal{N}} \qquad [hd, SN] \tag{767}$$

$$\cos\theta_n = n/n^o = (\mathcal{N}/V)/n^o = (V^o/V)\sec\theta_{\mathcal{N}} \qquad [hd, LN] \tag{768}$$

$$\sec\theta_n = n/n^o = (\mathcal{N}/V)/n^o = (V^o/V)\cos\theta_{\mathcal{N}} \qquad [ld, SN] \tag{769}$$

$$\sec\theta_n = n/n^o = (\mathcal{N}/V)/n^o = (V^o/V)\sec\theta_{\mathcal{N}} \qquad [ld, LN] \tag{770}$$

where $\mathcal{N}$ = noninteger number of particles in volume V of matter which is located in an ambient field, n = particle number density of matter in the presence of an ambient field, and where n^o and V^o = particle number density and volume for zero ambient field conditions as related by equation (763). Equations (767) through (770) can also be used to determine the volume V as

$$V/V^o = \cos\theta_{\mathcal{N}} \sec\theta_n \geq 1 \qquad [hd, SN] \tag{771}$$

$$V/V^o = \sec\theta_{\mathcal{N}} \sec\theta_n \geq 1 \qquad [hd, LN] \tag{772}$$

$$V/V^o = \cos\theta_{\mathcal{N}} \cos\theta_n \leq 1 \qquad [ld, SN] \tag{773}$$

$$V/V^o = \sec\theta_{\mathcal{N}} \cos\theta_n \leq 1 \qquad [ld, LN] \tag{774}$$

Equations (771) through (774) allow the calculation of the volume element V in terms of the energy density of the ambient field. Without knowledge of the above analysis the arbitrary assumption $V = V^o$ is made and the only way equations (761) and (762) can be correct is to have the $\cos\theta_{\mathcal{N}}$ and $\sec\theta_{\mathcal{N}}$ terms reversed in equations (612) and (613) and $\theta_n = \theta_{\mathcal{N}}$ as was done in the chemical reaction paper of Reference 11, however this is not necessary and equations (761) and (762) are completely compatible with equations (612) and (613) provided that equations (771) through (774) are valid with θ_n and $\theta_{\mathcal{N}}$ related by the space coordinate factors. Equations (760) through (774) are useful for the theory of equilibrium distributions and chemical reactions in gases.[11,22]

F. Relationship Between $C_{\sigma\alpha}$, $\theta_{\mathbb{N}}$ and θ_n.

This subsection determines the relationship between the space coordinate factors $C_{\sigma\alpha}$ and the internal phase angles $\theta_{\mathbb{N}}$ and θ_n of the particle number and particle number density of matter in an ambient field, and shows that for the doubly degenerate low-density homogeneous matter case there are two possible sets of counting number and segmentation number internal phase angles $\theta_{\eta\alpha}$ and $\theta_{\mathrm{n}\alpha}$ for specified values of the internal phase angles $\theta_{\mathbb{N}}$ and θ_n because $C_{2\alpha} = C_{3\alpha}$ for the [ld,SN] and [ld,LN] cases. Consider now the relationships between $C_{\sigma\alpha}$, $\theta_{\mathbb{N}}$ and θ_n for the three basic coordinate systems.

Cartesian Coordinates

Comparing equations (767) through (770) with equations (656), (659), (662), (665), (668) and (671) gives

$$V/V^{\circ} = C_{1x}C_{1y}C_{1z} = \cos\theta_{\mathbb{N}} \sec\theta_n \geq 1 \qquad \text{[hd,SN]} \qquad (775)$$

$$V/V^{\circ} = C_{1x}C_{1y}C_{1z} = \sec\theta_{\mathbb{N}} \sec\theta_n \geq 1 \qquad \text{[hd,LN]} \qquad (776)$$

$$V/V^{\circ} = C_{2x}C_{2y}C_{2z} = \cos\theta_{\mathbb{N}} \cos\theta_n \leq 1 \qquad \text{[ld,SN]} \qquad (777)$$

$$V/V^{0} = C_{2x}C_{2y}C_{2z} = \sec\theta_{\mathbb{N}} \cos\theta_n \leq 1 \qquad \text{[ld,LN]} \qquad (778)$$

$$V/V^{\circ} = C_{3x}C_{3y}C_{3z} = \cos\theta_{\mathbb{N}} \cos\theta_n \leq 1 \qquad \text{[ld,SN]} \qquad (779)$$

$$V/V^{\circ} = C_{3x}C_{3y}C_{3z} = \sec\theta_{\mathbb{N}} \cos\theta_n \leq 1 \qquad \text{[ld,LN]} \qquad (780)$$

where the space coordinate factors $C_{\sigma\alpha}$ for $\sigma = 1,2,3$ and $\alpha = x,y,z$ are given by equations (78) through (80) in terms of the space coordinate counting number and segmentation number angle pairs $\theta_{\eta\alpha}$ and $\theta_{\mathrm{n}\alpha}$.

Cylindrical Polar Coordinates

A comparison of equations (767) through (770) with equations (698), (701), (704), (707), (710) and (713) give

$$V/V^{\circ} = C_{1r}^{2}C_{1z} = \cos\theta_{\mathbb{N}} \sec\theta_n \geq 1 \qquad \text{[hd,SN]} \qquad (781)$$

$$V/V^{\circ} = C_{1r}^{2}C_{1z} = \sec\theta_{\mathbb{N}} \sec\theta_n \geq 1 \qquad \text{[hd,LN]} \qquad (782)$$

$$V/V^{\circ} = C_{2r}^{2}C_{2z} = \cos\theta_{\mathbb{N}} \cos\theta_n \leq 1 \qquad \text{[ld,SN]} \qquad (783)$$

$$V/V^o = C_{2r}^2 C_{2z} = \sec\theta_{\mathcal{N}} \cos\theta_n \leq 1 \qquad [\text{ld, LN}] \qquad (784)$$

$$V/V^o = C_{3r}^2 C_{3z} = \cos\theta_{\mathcal{N}} \cos\theta_n \leq 1 \qquad [\text{ld, SN}] \qquad (785)$$

$$V/V^o = C_{3r}^2 C_{3z} = \sec\theta_{\mathcal{N}} \cos\theta_n \leq 1 \qquad [\text{ld, LN}] \qquad (786)$$

where $C_{\sigma\alpha}$ for $\sigma = 1,2,3$ and $\alpha = r,z$ are given by equations (78) through (80) in terms of the angle pairs $\theta_{\eta\alpha}$ and $\theta_{\text{n}\alpha}$.

Spherical Polar Coordinates

Comparing equations (767) through (770) with equations (740), (743), (746), (749), (752) and (755) or with equations (718) through (723) and (605) gives

$$V/V^o = C_{1\rho}^3 = \cos\theta_{\mathcal{N}} \sec\theta_n \geq 1 \qquad [\text{hd, SN}] \qquad (787)$$

$$V/V^o = C_{1\rho}^3 = \sec\theta_{\mathcal{N}} \sec\theta_n \geq 1 \qquad [\text{hd, LN}] \qquad (788)$$

$$V/V^o = C_{2\rho}^3 = \cos\theta_{\mathcal{N}} \cos\theta_n \leq 1 \qquad [\text{ld, SN}] \qquad (789)$$

$$V/V^o = C_{2\rho}^3 = \sec\theta_{\mathcal{N}} \cos\theta_n \leq 1 \qquad [\text{ld, LN}] \qquad (790)$$

$$V/V^o = C_{3\rho}^3 = \cos\theta_{\mathcal{N}} \cos\theta_n \leq 1 \qquad [\text{ld, SN}] \qquad (791)$$

$$V/V^o = C_{3\rho}^3 = \sec\theta_{\mathcal{N}} \cos\theta_n \leq 1 \qquad [\text{ld, LN}] \qquad (792)$$

where $C_{\sigma\alpha}$ for $\sigma = 1,2,3$ and $\alpha = \rho$ is given by equations (78) through (80) in terms of the pairs of angles $\theta_{\eta\rho}$ and $\theta_{\text{n}\rho}$ that are associated with the complex number counting and segmentation numbers. Similarly, using equations (582) and (787) through (792) gives the values of the areas perpendicular to the radial coordinates as

$$A_\rho / A_\rho^o = C_{1\rho}^2 = \cos^{2/3}\theta_{\mathcal{N}} \sec^{2/3}\theta_n \geq 1 \qquad [\text{hd, SN}] \qquad (793)$$

$$A_\rho / A_\rho^o = C_{1\rho}^2 = \sec^{2/3}\theta_{\mathcal{N}} \sec^{2/3}\theta_n \geq 1 \qquad [\text{hd, LN}] \qquad (794)$$

$$A_\rho / A_\rho^o = C_{2\rho}^2 = \cos^{2/3}\theta_{\mathcal{N}} \cos^{2/3}\theta_n \leq 1 \qquad [\text{ld, SN}] \qquad (795)$$

$$A_\rho / A_\rho^o = C_{2\rho}^2 = \sec^{2/3}\theta_{\mathcal{N}} \cos^{2/3}\theta_n \leq 1 \qquad [\text{ld, LN}] \qquad (796)$$

$$A_\rho / A_\rho^o = C_{3\rho}^2 = \cos^{2/3}\theta_{\mathcal{N}} \cos^{2/3}\theta_n \leq 1 \qquad [\text{ld, SN}] \qquad (797)$$

$$A_\rho / A_\rho^o = C_{3\rho}^2 = \sec^{2/3}\theta_{\mathcal{N}} \cos^{2/3}\theta_n \leq 1 \qquad [\text{ld, LN}] \qquad (798)$$

which are valid for spherical surface elements. These area expressions are useful to describe thermonuclear reaction cross sections which occur in strong ambient gravitational fields or intense ambient laser beams.

Combining equations (78) through (80) with equations (787) through (792) gives

$$C_{1\rho}^3 = \cos^3\theta_{\eta\rho}\sec^3\theta_{\pi\rho} = \cos\theta_{\mathcal{N}}\sec\theta_{n} \geq 1 \qquad [\text{hd},\text{SN}] \qquad (799)$$

$$C_{1\rho}^3 = \cos^3\theta_{\eta\rho}\sec^3\theta_{\pi\rho} = \sec\theta_{\mathcal{N}}\sec\theta_{n} \geq 1 \qquad [\text{hd},\text{LN}] \qquad (800)$$

$$C_{2\rho}^3 = \cos^3\theta_{\eta\rho}\cos^3\theta_{\pi\rho} = \cos\theta_{\mathcal{N}}\cos\theta_{n} \leq 1 \qquad [\text{ld},\text{SN}] \qquad (801)$$

$$C_{2\rho}^3 = \cos^3\theta_{\eta\rho}\cos^3\theta_{\pi\rho} = \sec\theta_{\mathcal{N}}\cos\theta_{n} \leq 1 \qquad [\text{ld},\text{LN}] \qquad (802)$$

$$C_{3\rho}^3 = \sec^3\theta_{\eta\rho}\cos^3\theta_{\pi\rho} = \cos\theta_{\mathcal{N}}\cos\theta_{n} \leq 1 \qquad [\text{ld},\text{SN}] \qquad (803)$$

$$C_{3\rho}^3 = \sec^3\theta_{\eta\rho}\cos^3\theta_{\pi\rho} = \sec\theta_{\mathcal{N}}\cos\theta_{n} \leq 1 \qquad [\text{ld},\text{LN}] \qquad (804)$$

for which in general $\theta_{\pi\rho} \geq \theta_{\eta\rho}$ and $\theta_n \geq \theta_{\mathcal{N}}$. Equations (799) through (804) give the space coordinate factors for spherical polar coordinates in homogeneous matter in an ambient field as

$$C_{1\rho} = \cos\theta_{\eta\rho}\sec\theta_{\pi\rho} = \cos^{1/3}\theta_{\mathcal{N}}\sec^{1/3}\theta_{n} \geq 1 \qquad [\text{hd},\text{SN}] \qquad (805)$$

$$C_{1\rho} = \cos\theta_{\eta\rho}\sec\theta_{\pi\rho} = \sec^{1/3}\theta_{\mathcal{N}}\sec^{1/3}\theta_{n} \geq 1 \qquad [\text{hd},\text{LN}] \qquad (806)$$

$$C_{2\rho} = \cos\theta_{\eta\rho}\cos\theta_{\pi\rho} = \cos^{1/3}\theta_{\mathcal{N}}\cos^{1/3}\theta_{n} \leq 1 \qquad [\text{ld},\text{SN}] \qquad (807)$$

$$C_{2\rho} = \cos\theta_{\eta\rho}\cos\theta_{\pi\rho} = \sec^{1/3}\theta_{\mathcal{N}}\cos^{1/3}\theta_{n} \leq 1 \qquad [\text{ld},\text{LN}] \qquad (808)$$

$$C_{3\rho} = \sec\theta_{\eta\rho}\cos\theta_{\pi\rho} = \cos^{1/3}\theta_{\mathcal{N}}\cos^{1/3}\theta_{n} \leq 1 \qquad [\text{ld},\text{SN}] \qquad (809)$$

$$C_{3\rho} = \sec\theta_{\eta\rho}\cos\theta_{\pi\rho} = \sec^{1/3}\theta_{\mathcal{N}}\cos^{1/3}\theta_{n} \leq 1 \qquad [\text{ld},\text{LN}] \qquad (810)$$

Equations (805), (807) and (810) have the same separable solutions

$$\cos\theta_{\eta\rho} = \cos^{1/3}\theta_{\mathcal{N}} \qquad\qquad \cos\theta_{\pi\rho} = \cos^{1/3}\theta_{n} \qquad (811)$$

Note that equation (811) agrees with the previously determined results that $\theta_{\pi\rho} \geq \theta_{\eta\rho}$ and $\theta_n \geq \theta_{\mathcal{N}}$. Equations (806), (808) and (809) do not have separable solutions. The values of the coordinate factor $C_{1\rho}$ for the high-density states [hd,SN] and [hd,LN] given by equations (805) and (806) are used to describe the radii of subactinide nuclei and actinide and transactinide nuclei respectively.[23]

If the zero ambient field volume element V^o is taken to be the same for all coordinate systems, it follows from equations (593), (599) and (605) combined with equations (771) through (774) that in the presence of an ambient field

$$V / V^o = C_{1x} C_{1y} C_{1z} = C_{1r}^2 C_{1z} = C_{1\rho}^3 = \cos\theta_{\mathcal{N}} \sec\theta_n \geq 1 \qquad [\text{hd,SN}] \qquad (812)$$

$$V / V^o = C_{1x} C_{1y} C_{1z} = C_{1r}^2 C_{1z} = C_{1\rho}^3 = \sec\theta_{\mathcal{N}} \sec\theta_n \geq 1 \qquad [\text{hd,LN}] \qquad (813)$$

$$V / V^o = C_{2x} C_{2y} C_{2z} = C_{2r}^2 C_{2z} = C_{2\rho}^3 = \cos\theta_{\mathcal{N}} \cos\theta_n \leq 1 \qquad [\text{ld,SN}] \qquad (814)$$

$$V / V^o = C_{2x} C_{2y} C_{2z} = C_{2r}^2 C_{2z} = C_{2\rho}^3 = \sec\theta_{\mathcal{N}} \cos\theta_n \leq 1 \qquad [\text{ld,LN}] \qquad (815)$$

$$V / V^o = C_{3x} C_{3y} C_{3z} = C_{3r}^2 C_{3z} = C_{3\rho}^3 = \cos\theta_{\mathcal{N}} \cos\theta_n \leq 1 \qquad [\text{ld,SN}] \qquad (816)$$

$$V / V^o = C_{3x} C_{3y} C_{3z} = C_{3r}^2 C_{3z} = C_{3\rho}^3 = \sec\theta_{\mathcal{N}} \cos\theta_n \leq 1 \qquad [\text{ld,LN}] \qquad (817)$$

Homologous equations are valid for the cases when only an ambient gravity field or an ambient electromagnetic field such as ambient γ ray field is present by adding a superscript g or γ respectively to each quantity in equations (812) through (817).

G. Degeneracy of the Active Coordinates of the Low-density States of Matter.

Equations (807) and (809) for the [ld,SN] state give two independent sets of pairs of angles $\theta_{\eta\rho}$, $\theta_{\text{n}\rho}$ for each pair of angles $\theta_{\mathcal{N}}$, θ_n according to the occurrence of either $C_{2\rho}$ or $C_{3\rho}$ so that the [ld,SN] state is doubly in regard to active coordinates. Similarly, equations (808) and (810) for the [ld,LN] state give two independent solutions for the pairs of angles $\theta_{\eta\rho}$, $\theta_{\text{n}\rho}$ for a given choice of angles $\theta_{\mathcal{N}}$ and θ_n because of the occurrence of $C_{2\rho}$ and $C_{3\rho}$ in the left hand side of these equations, and therefore [ld,LN] is also doubly degenerate. The double degeneracy of the active coordinates that appears in equations (775) through (817) is described most easily for spherical polar coordinates by observing that for specified values of $\theta_{\mathcal{N}}$ and θ_n the following equations are valid

$$C_{2\rho} = C_{3\rho} = \cos^{1/3}\theta_{\mathcal{N}} \cos^{1/3}\theta_n \qquad [\text{ld,SN}] \qquad (818)$$

$$C_{2\rho} = C_{3\rho} = \sec^{1/3}\theta_{\mathcal{N}} \cos^{1/3}\theta_n \qquad [\text{ld,LN}] \qquad (819)$$

so that for each case of the [ld] state $C_{2\rho}$ and $C_{3\rho}$ have the same values although $C_{2\rho}$ and $C_{3\rho}$ are different functions of the angles $\theta_{\eta\rho}$ and $\theta_{\text{n}\rho}$ as given by equations (79) and (80). Therefore the small and large number forms of the [ld] case given in equations (818) and (819) are associated with two possible solutions relating $\theta_{\eta\rho}$ and $\theta_{\text{n}\rho}$

$$\theta_{\eta\rho} = f_2(\theta_{\text{n}\rho}, \theta_{\mathcal{N}}, \theta_n) \qquad \theta_{\eta\rho} = f_3(\theta_{\text{n}\rho}, \theta_{\mathcal{N}}, \theta_n) \qquad [\text{ld,SN}] \qquad (820)$$

$$\theta_{\eta\rho} = g_2(\theta_{\text{n}\rho}, \theta_{\mathcal{N}}, \theta_n) \qquad \theta_{\eta\rho} = g_3(\theta_{\text{n}\rho}, \theta_{\mathcal{N}}, \theta_n) \qquad [\text{ld,LN}] \qquad (821)$$

for specified values of $\theta_{\mathfrak{N}}$ and θ_n. The degeneracy associated with the [ld,SN] case is used in the description of Moseley's x-ray emission law for low Z atoms in an ambient electromagnetic field.[20]

H. Determination of the Phase Angles $\theta_{\eta\rho}$ and $\theta_{\text{n}\rho}$ of the Complex Counting and Segmentation Numbers.

For spherical polar coordinates consider the case where the internal phase angle θ_ρ of the complex number active radial coordinate $\overline{\rho}$ is, through experiment, a known function of the energy density of a combined ambient gravitational and ambient electromagnetic field. Then equations (31), (32) and (34) give the active radial coordinate as

$$\overline{\rho} = \rho \exp(j\theta_\rho) \tag{822}$$

$$\rho = \rho^{\circ} C_{\sigma\rho}(\theta_{\eta\rho}, \theta_{\text{n}\rho}) \qquad \theta_\rho = \theta_{\eta\rho} - \theta_{\text{n}\rho} \tag{823}$$

where θ_ρ is a known function of the ambient fields

$$\theta_\rho = \theta_\rho(\mathfrak{E}^g, \mathfrak{E}^\gamma) \tag{824}$$

Combining the phase angle relation in equation (823) with equations (805) through (810) gives the following equations for the internal phase angle $\theta_{\text{n}\rho}$ of the complex number segmentation number of the radial coordinate

$$\cos(\theta_{\text{n}\rho} + \theta_\rho)\sec\theta_{\text{n}\rho} = u \qquad [\text{hd,SN}] \tag{825}$$

$$\cos(\theta_{\text{n}\rho} + \theta_\rho)\sec\theta_{\text{n}\rho} = w \qquad [\text{hd,LN}] \tag{826}$$

$$\cos(\theta_{\text{n}\rho} + \theta_\rho)\cos\theta_{\text{n}\rho} = y \qquad [\text{ld,SN}] \tag{827}$$

$$\cos(\theta_{\text{n}\rho} + \theta_\rho)\cos\theta_{\text{n}\rho} = z \qquad [\text{ld,LN}] \tag{828}$$

$$\sec(\theta_{\text{n}\rho} + \theta_\rho)\cos\theta_{\text{n}\rho} = y \qquad [\text{ld,SN}] \tag{829}$$

$$\sec(\theta_{\text{n}\rho} + \theta_\rho)\cos\theta_{\text{n}\rho} = z \qquad [\text{ld,LN}] \tag{830}$$

where u, w, y and z are known functions of $\theta_{\mathfrak{N}}$ and θ_n given by

$$u = \cos^{1/3}\theta_{\mathfrak{N}}\sec^{1/3}\theta_n \qquad 1 \le u \le \infty \qquad [\text{hd,SN}] \tag{831}$$

$$w = \sec^{1/3}\theta_{\mathfrak{N}}\sec^{1/3}\theta_n \qquad 1 \le w \le \infty \qquad [\text{hd,LN}] \tag{832}$$

$$y = \cos^{1/3}\theta_{\mathfrak{N}}\cos^{1/3}\theta_n \qquad 0 \le y \le 1 \qquad [\text{ld,SN}] \tag{833}$$

$$z = \sec^{1/3}\theta_{\mathfrak{N}}\cos^{1/3}\theta_n \qquad 0 \le z \le 1 \qquad [\text{ld,LN}] \tag{834}$$

Equations (827), (829) and (828), (830) express the degeneracy that occurs in the [ld,SN] and [ld,LN] cases as shown in equations (818) and (819). For the case when θ_ρ is a known quantity, equations (825) through (830) combined with the phase angle relation in equation (823) are easily solved for $\theta_{\text{n}\rho}$ and $\theta_{\eta\rho}$ by using trigonometric identities.[22,23] The angle $\theta_{\mathbb{N}}$ is known from equation (614), the angle θ_n is obtained from equations (615), (640), (682) or (724) in terms of the internal phase angle of the space coordinates. For an ambient radiation field in matter the internal phase angles of the space coordinates, such as θ_ρ, $\theta_{\text{n}\rho}$ and $\theta_{\eta\rho}$, for example, are linearly related to the energy density of the ambient wave field as in equations (118) through (149) and equations (825) through (830) are not needed to determine these angles. This is not the case, however, for nonradiative ambient fields in matter. The determination of the internal phase angles of the space coordinates for the case of a nonradiative ambient field in matter is treated in Section 7.

7. COMBINED ELASTIC AND VISCOELASTIC BODY MODEL FOR THE DETERMINATION OF THE INTERNAL PHASE ANGLES OF THE SPATIAL COORDINATES FOR A NONRADIATIVE AMBIENT FIELD IN MATTER. The internal phase angle θ_ρ is required to utilize equations (825) through (830) for the determination of the internal phase angles $\theta_{\text{n}\rho}$ and $\theta_{\eta\rho}$ of the complex segmentation and counting numbers for the case of nonradiative ambient fields because these angles are not simply related to the energy density of the ambient nonradiative field. This section determines the internal phase angle θ_ρ for high-density and low-density matter by first using the theory of elasticity to calculate the internal phase angle θ_V of a volume element of matter in the presence of an nonradiative ambient field and then using equations (592), (598) and (604) to calculate the internal phase angles θ_α of the spatial coordinates by assuming they are all equal. The general expression for the internal phase angle of a volume element is obtained from equation (615) that is written as

$$\theta_V = \theta_{\mathbb{N}} - \theta_n \leq 0 \qquad \theta_n \geq \theta_{\mathbb{N}} \tag{835}$$

and where for an arbitrary ambient field in three dimensions

$$\theta_x + \theta_y + \theta_z = 2\theta_r + \theta_z = 3\theta_\rho = \theta_V = \theta_{\mathbb{N}} - \theta_n \leq 0 \tag{836}$$

$$\theta_t = \theta_\rho = 1/3\theta_V = 1/3(\theta_{\mathbb{N}} - \theta_n) \leq 0 \tag{837}$$

while for a homogeneous and isotropic ambient field in homogeneous and isotropic matter, equation (836) can be written as

$$\theta_x = \theta_y = \theta_z = \theta_r = \theta_\rho = 1/3\theta_V = 1/3(\theta_{\mathbb{N}} - \theta_n) \leq 0 \tag{838}$$

and where for a one-dimensional uniaxial case of an ambient field

$$\theta_x = \theta_V \leq 0 \tag{839}$$

and finally where $\theta_{\mathbb{N}}$ is given by equation (614) and where θ_n will now be calculated from the theory of elasticity for high-density and low-density matter in nonradiative ambient fields.

A. High-density Matter, [hd,SN] and [hd,LN].

Consider first the case of a static elastic system. For high-density matter (gas, liquid or solids) the nonradiative ambient field must be one of isotropic tension (T) because the application of this field lowers the particle number density of matter as indicated in equation (761). For [hd] elastic matter the energy density $\mathfrak{E}_T$ of the matter is one of isotropic tension and is written as

$$\mathfrak{E}_T = 1/2K_o(n/n^o - 1)^2 \qquad \text{[hd]} \qquad (840)$$

where K_o = constant bulk modulus for high-density matter at zero nonradiative ambient field, n = particle number density in the expanded state of matter due to the presence of an ambient nonradiative isotropic tension field and n^o = conventional particle number density that corresponds to the case of zero nonradiative ambient fields. Combining equations (761) and (840) gives the isotropic tension energy density of matter as

$$\mathfrak{E}_T = 1/2K_o(\cos\theta_n - 1)^2 = 1/2K_o(1-\cos\theta_n)^2 \qquad \text{[hd]} \qquad (841)$$

or equivalently

$$\cos\theta_n = 1-(2\mathfrak{E}_T/K_o)^{1/2} \qquad \text{[hd]} \qquad (842)$$

subject to the condition $\mathfrak{E}_T/K_o \le 1/2$. As before in equation (614), it is assumed that a nonradiative ambient shear field in [hd] matter is also present and which for static elasticity is described by the equation

$$\mathfrak{E}_{ST} = 1/2G_o\tan^2\theta_{\mathcal{N}} \qquad \tan\theta_{\mathcal{N}} = (2\mathfrak{E}_{ST}/G_o)^{1/2} \qquad G_o = \varsigma_o/t_o \qquad \text{[hd]} \qquad (843)$$

where G_o, ζ_o and t_o = constant shear modulus, viscosity and time factor for [hd] matter at zero ambient field conditions. The total energy density for the [hd] matter case is given by

$$\mathfrak{E} = \mathfrak{E}_T + \mathfrak{E}_{ST} \qquad \text{[hd]} \qquad (844)$$

and, in order to obtain a simple relation between the angles $\theta_{\mathcal{N}}$ and θ_n, this total energy density is assumed to be approximately distributed among the isotropic tension and shear energy density components by the rule

$$\mathfrak{E}_T = K_o(K_o+G_o)^{-1}\mathfrak{E} \qquad \mathfrak{E}_{ST} = G_o(K_o+G_o)^{-1}\mathfrak{E} \qquad \text{[hd]} \qquad (845)$$

Equation (844) is a linear elastic approximation of a more general equation that includes a thermal energy density component and is written as

$$\mathfrak{E} = \mathfrak{E}_T + \mathfrak{E}_{ST} + \mathfrak{E}_{thermal} \qquad \text{[hd]} \qquad (846)$$

where $\mathfrak{E}_{thermal}$ = thermal energy component which is neglected in this paper but which must be included if both $G_o = 0$ and $K_o = 0$. Combining equations (842) and (843) with equation (845) gives

$$\cos\theta_n = 1 - [2\mathfrak{E}(K_o + G_o)^{-1}]^{1/2} \qquad \mathfrak{E}(K_o + G_o)^{-1} \le 1/2 \quad [hd] \qquad (847)$$

$$\tan\theta_{\mathfrak{N}} = [2\mathfrak{E}(K_o + G_o)^{-1}]^{1/2} \qquad \mathfrak{E}(K_o + G_o)^{-1} \le 1/2 \quad [hd] \qquad (848)$$

which show that for a nonradiative ambient field in [hd] matter

$$\tan\theta_{\mathfrak{N}} = 1 - \cos\theta_n \qquad \theta_n \ge \theta_{\mathfrak{N}} \qquad 0 \le \theta_n \le \pi/2 \qquad 0 \le \theta_{\mathfrak{N}} \le \pi/4 \qquad [hd] \quad (849)$$

For use in the representation of the spatial coordinate factor $C_{1\rho}$ in terms of the angle θ_n alone, it is useful to rewrite equation (849) as

$$\sec\theta_{\mathfrak{N}} = [1 + (1 - \cos\theta_n)^2]^{1/2} \qquad \cos\theta_{\mathfrak{N}} = [1 + (1 - \cos\theta_n)^2]^{-1/2} \quad [hd] \quad (850)$$

Equations (805), (825), (831) and (850) give the [hd,SN] spatial coordinate factor $C_{1\rho}$ in terms of θ_n as

$$C_{1\rho} = u = \sec^{1/3}\theta_n[1 + (1 - \cos\theta_n)^2]^{-1/6} \qquad [hd,SN] \qquad (851)$$

Equations (806), (826), (832) and (850) give the spatial coordinate factor $C_{1\rho}$ for the [hd,LN] case as

$$C_{1\rho} = w = \sec^{1/3}\theta_n[1 + (1 - \cos\theta_n)^2]^{1/6} \qquad [hd,LN] \qquad (852)$$

where in both of the above cases $C_{1\rho} \ge 1$ for $0 \le \theta_n \le \pi/2$. The internal phase angle $\theta_V \le 0$ of a volume element of high-density matter in nonradiative ambient fields is obtained from equations (835), (842), (843), (847) and (848) as

$$\theta_V = \tan^{-1}(2\mathfrak{E}_{ST}/G_o)^{1/2} - \cos^{-1}[1 - (2\mathfrak{E}_T/K_o)^{1/2}] \quad [hd] \qquad (853)$$

$$= \tan^{-1}(w_D^{1/2}) - \cos^{-1}(1 - w_D^{1/2})$$

where w_D is given by

$$w_D = 2\mathfrak{E}(K_o + G_o)^{-1} \qquad (854)$$

where $\mathfrak{E}$ = total energy density of the nonradiative ambient field. The internal phase angles of the space coordinates are then obtained from equation (838). Thus for a [hd] system in nonradiative ambient fields, this procedure determines the quantities $\theta_n(\mathfrak{E}_T)$, $\theta_{\mathfrak{N}}(\mathfrak{E}_{ST})$, $C_{1\rho}(\mathfrak{E}_{ST},\mathfrak{E}_T)$, $\theta_V(\mathfrak{E}_{ST},\mathfrak{E}_T)$ and $\theta_\rho(\mathfrak{E}_{ST},\mathfrak{E}_T)$ or equivalently after using equation (845) these internal phase angles can be expressed in terms of the total energy density of the nonradiative ambient fields in the manner $\theta_n(\mathfrak{E})$, $\theta_{\mathfrak{N}}(\mathfrak{E})$, $C_{1\rho}(\mathfrak{E})$, $\theta_V(\mathfrak{E})$ and $\theta_\rho(\mathfrak{E})$. And then for a [hd] system the phase angle relation in equation (823) can be combined with equations (825) and (826) and these equations can be solved for the internal phase angles $\theta_{\eta\rho}(\mathfrak{E})$ and $\theta_{n\rho}(\mathfrak{E})$ of the complex counting and segmentation numbers.

Now consider the case of a time dependent [hd] elastic system in which the average kinetic energy is equal to the average potential energy so that equation (844) must be changed to

$$\mathfrak{E} = 2\mathfrak{E}_{T} + 2\mathfrak{E}_{ST} \qquad [hd] \qquad (855)$$

and equations (845), (847) and (848) must be replaced by

$$\mathfrak{E}_{T} = 1/2K_o(K_o + G_o)^{-1}\mathfrak{E} \qquad \mathfrak{E}_{ST} = 1/2G_o(K_o + G_o)^{-1}\mathfrak{E} \qquad [hd] \qquad (856)$$

$$\cos\theta_n = 1 - [\mathfrak{E}(K_o + G_o)^{-1}]^{1/2} \qquad \mathfrak{E}(K_o + G_o)^{-1} \le 1 \qquad [hd] \qquad (857)$$

$$\tan\theta_{\mathfrak{N}} = [\mathfrak{E}(K_o + G_o)^{-1}]^{1/2} \qquad \mathfrak{E}(K_o + G_o)^{-1} \le 1 \qquad [hd] \qquad (858)$$

where a thermal energy density component has been neglected in equation (855). Then equations (849) through (852) remain the same, but equation (853) becomes

$$\theta_V = \tan^{-1}(2\mathfrak{E}_{ST}/G_o)^{1/2} - \cos^{-1}[1 - (2\mathfrak{E}_{T}/K_o)^{1/2}] \qquad [hd] \qquad (859)$$

$$= \tan^{-1}(w_{Dt}^{1/2}) - \cos^{-1}(1 - w_{Dt}^{1/2})$$

where w_{Dt} is given by

$$w_{Dt} = \mathfrak{E}(K_o + G_o)^{-1} \qquad (860)$$

where $\mathfrak{E}$ = total energy density of nonradiative ambient fields given by equation (855) .

B. Low-density Matter, [ld,SN] and [ld,LN].

Consider first a static elastic system. For low-density matter (which can be a gas, liquid or solid) the application of a nonradiative ambient field must be associated with a compression energy density of the matter because the application of this ambient field induces an increase of the particle number density as indicated in equation (762), and the isotropic compression energy density of the [ld] elastic matter is written as

$$\mathfrak{E}_{C} = 1/2K_o(n/n^o - 1)^2 \qquad [ld] \qquad (861)$$

where K_o = constant bulk modulus for low-density matter at zero ambient field, n = particle number density in the compressed state of matter due to the presence of an ambient compressive nonradiative field and n^o = conventional particle number density. Combining equations (762) and (861) gives the isotropic compression energy density of the [ld] matter due to the application of a nonradiative ambient field as

$$\mathfrak{E}_{C} = 1/2K_o(\sec\theta_n - 1)^2 \qquad [ld] \qquad (862)$$

or equivalently

$$\sec\theta_n = 1 + (2\mathfrak{E}_{C}/K_o)^{1/2} \qquad [ld] \qquad (863)$$

where $0 \le \theta_n \le \pi/2$. As before in equations (614), it is assumed that a nonradiative ambient shear field is also present in [ld] matter and is of the form

$$\mathfrak{E}_{SC} = 1/2G_o\tan^2\theta_{\mathfrak{N}} \qquad \tan\theta_{\mathfrak{N}} = (2\mathfrak{E}_{SC}/G_o)^{1/2} \qquad G_o = \varsigma_o/t_o \qquad [ld] \qquad (864)$$

where G_o, ζ_o and t_o = constant shear modulus, viscosity and time factor for [ld] matter at zero ambient field. The total static elastic energy density for [ld] matter is given by

$$\mathfrak{E} = \mathfrak{E}_C + \mathfrak{E}_{SC} \qquad \text{[ld]} \qquad (865)$$

and, in order to obtain a simple relation between the angles $\theta_{\mathfrak{N}}$ and θ_n for low-density matter, this total energy density is assumed to be approximately distributed among the isotropic compression and shear energy density components by the rule

$$\mathfrak{E}_C = K_o(K_o + G_o)^{-1}\mathfrak{E} \qquad \mathfrak{E}_{SC} = G_o(K_o + G_o)^{-1}\mathfrak{E} \qquad \text{[ld]} \qquad (866)$$

so that equation (865) is automatically satisfied. A thermal energy component energy density has been neglected in equation (865). Combining equations (863) and (864) with equation (866) gives

$$\sec\theta_n = 1 + [2\mathfrak{E}(K_o + G_o)^{-1}]^{1/2} \qquad \text{[ld]} \qquad (867)$$

$$\tan\theta_{\mathfrak{N}} = [2\mathfrak{E}(K_o + G_o)^{-1}]^{1/2} \qquad \text{[ld]} \qquad (868)$$

which show that for a nonradiative ambient field in [ld] matter the angles $\theta_{\mathfrak{N}}$ and θ_n are related by

$$\tan\theta_{\mathfrak{N}} = \sec\theta_n - 1 \qquad \theta_n \geq \theta_{\mathfrak{N}} \qquad 0 \leq \theta_n \leq \pi/2 \qquad 0 \leq \theta_{\mathfrak{N}} \leq \pi/2 \qquad \text{[ld]} \qquad (869)$$

For use in the representation of the [ld] spatial coordinate factors $C_{\beta\rho}$ with $\beta = 2,3$ in terms of the angle θ_n alone, it is useful to rewrite equation (869) as

$$\sec\theta_{\mathfrak{N}} = [1 + (\sec\theta_n - 1)^2]^{1/2} \qquad \cos\theta_{\mathfrak{N}} = [1 + (\sec\theta_n - 1)^2]^{-1/2} \qquad \text{[ld]} \qquad (870)$$

Combining equations (807), (809) or (827), (829) and (833) with (870) gives the [ld,SN] spatial coordinate factors $C_{\beta\rho}$ with $\beta = 2,3$ in terms of θ_n as

$$C_{\beta\rho} = y = \cos^{1/3}\theta_n[1 + (\sec\theta_n - 1)^2]^{-1/6} \qquad \text{[ld,SN]} \qquad (871)$$

Equations (808), (810) or (828), (830) and (834) with (870) give the spatial coordinate factors $C_{\beta\rho}$ with $\beta = 2,3$ in terms of θ_n for the [ld,LN] case as

$$C_{\beta\rho} = z = \cos^{1/3}\theta_n[1 + (\sec\theta_n - 1)^2]^{1/6} \qquad \text{[ld,LN]} \qquad (872)$$

where in both the [ld,SN] and [ld,LN] cases $C_{\beta\rho} \leq 1$ for $0 \leq \theta_n \leq \pi/2$. The internal phase angle $\theta_V \leq 0$ of a volume element of low-density matter in nonradiative ambient fields is obtained from equations (835), (863), (864), (867) and (868) as

$$\theta_V = \tan^{-1}(2\mathfrak{E}_{SC}/G_o)^{1/2} - \sec^{-1}[1 + (2\mathfrak{E}_C/K_o)^{1/2}] \qquad \text{[ld]} \qquad (873)$$

$$= \tan^{-1}(w_D^{1/2}) - \sec^{-1}(1 + w_D^{1/2})$$

where w_D is given by

$$w_D = 2\mathfrak{E}(K_o + G_o)^{-1} \tag{874}$$

where $\mathfrak{E}$ = total energy density of the ambient field. The internal phase angles of the space coordinates are then obtained from equation (838). Thus for a [ld] system in nonradiative ambient fields, this procedure determines $\theta_n(\mathfrak{E}_C)$, $\theta_{\aleph}(\mathfrak{E}_{SC})$, $C_{\beta\rho}(\mathfrak{E}_{SC},\mathfrak{E}_C)$, $\theta_V(\mathfrak{E}_{SC},\mathfrak{E}_C)$ and $\theta_\rho(\mathfrak{E}_{SC},\mathfrak{E}_C)$ or equivalently after using equation (866) these internal phase angles can be expressed in terms of the total energy density of the nonradiative ambient fields in the manner $\theta_n(\mathfrak{E})$, $\theta_{\aleph}(\mathfrak{E})$, $C_{\beta\rho}(\mathfrak{E})$, $\theta_V(\mathfrak{E})$ and $\theta_\rho(\mathfrak{E})$. Then for a [ld] system the phase angle relation in equation (823) can be combined with equations (827) through (830) and these equations can be solved for the internal phase angles $\theta_{\eta\rho}(\mathfrak{E})$ and $\theta_{n\rho}(\mathfrak{E})$ of the complex counting and segmentation numbers.

Consider now the case of a time dependent [ld] elastic system in which the average kinetic energy is equal to the average potential energy so that equation (865) must be changed to

$$\mathfrak{E} = 2\mathfrak{E}_C + 2\mathfrak{E}_{SC} \qquad \text{[ld]} \tag{875}$$

and equations (866), (867) and (868) must be replaced by

$$\mathfrak{E}_C = 1/2K_o(K_o + G_o)^{-1}\mathfrak{E} \qquad \mathfrak{E}_{SC} = 1/2G_o(K_o + G_o)^{-1}\mathfrak{E} \qquad \text{[ld]} \tag{876}$$

$$\sec\theta_n = 1 + [\mathfrak{E}(K_o + G_o)^{-1}]^{1/2} \qquad \text{[ld]} \tag{877}$$

$$\tan\theta_{\aleph} = [\mathfrak{E}(K_o + G_o)^{-1}]^{1/2} \qquad \text{[ld]} \tag{878}$$

A thermal energy component energy density has been neglected in equation (875). Then equations (869) through (872) remain the same, but equation (873) becomes

$$\theta_V = \tan^{-1}(2\mathfrak{E}_{SC}/G_o)^{1/2} - \sec^{-1}[1 + (2\mathfrak{E}_C/K_o)^{1/2}] \qquad \text{[ld]} \tag{879}$$

$$= \tan^{-1}(w_{Dt}^{1/2}) - \sec^{-1}(1 + w_{Dt}^{1/2})$$

where w_{Dt} is given by

$$w_{Dt} = \mathfrak{E}(K_o + G_o)^{-1} \tag{880}$$

where $\mathfrak{E}$ = total energy density of the ambient fields in [ld] matter as in equation (875).

Combining equations (835), (838), (849) and (869) gives the internal phase angles of the volume and space coordinates α = x, y, z, r and ρ in terms of the angle θ_n as

$$\theta_V = \tan^{-1}(1 - \cos\theta_n) - \theta_n \qquad \theta_\alpha = 1/3[\tan^{-1}(1 - \cos\theta_n) - \theta_n] \qquad \text{[hd]} \tag{881}$$

$$\theta_V = \tan^{-1}(\sec\theta_n - 1) - \theta_n \qquad \theta_\alpha = 1/3[\tan^{-1}(\sec\theta_n - 1) - \theta_n] \qquad \text{[ld]} \tag{882}$$

The small internal phase angle limiting form of both equations (849) and (869) is

$$\theta_{\mathcal{N}} \approx \theta_n^2/2 \qquad \theta_n \geq \theta_{\mathcal{N}} \qquad \text{[hd] and [ld]} \qquad (883)$$

and equations (835) and (838) or equations (881) and (882) give

$$\theta_V = \theta_{\mathcal{N}} - \theta_n \approx -\theta_n(1-\theta_n/2) \approx -\theta_n \qquad \text{[hd] and [ld]} \qquad (884)$$

$$\approx -\sqrt{\theta_{\mathcal{N}}}(\sqrt{2}-\sqrt{\theta_{\mathcal{N}}}) \approx -\sqrt{2\theta_{\mathcal{N}}} \qquad \text{[hd] and [ld]} \qquad (885)$$

$$\theta_\alpha = 1/3(\theta_{\mathcal{N}} - \theta_n) \approx -1/3\theta_n(1-\theta_n/2) \approx -1/3\theta_n \qquad \text{[hd] and [ld]} \qquad (886)$$

$$\approx -1/3\sqrt{\theta_{\mathcal{N}}}(\sqrt{2}-\sqrt{\theta_{\mathcal{N}}}) \approx -1/3\sqrt{2\theta_{\mathcal{N}}} \qquad \text{[hd] and [ld]} \qquad (887)$$

which are useful for quick order of magnitude calculations.

A combined static elastic and viscoelastic medium calculation of the internal phase angles of the space and time coordinates has been presented. A more realistic dynamical calculation of these internal phase angles is more complicated and has not been accomplished. For polyphase rockets, a simple approach to this calculation may possibly be the substitution of effective, dynamical (oscillatory), frequency-dependent elastic and viscoelastic moduli for the static elastic moduli in the manner $K_e^o \rightarrow K_e^r$, $G_e^o \rightarrow G_e^r$, $K_c^o \rightarrow K_c^r$, $G_c^o \rightarrow G_c^r$, $K_b^o \rightarrow K_b^r$ and $G_b^o \rightarrow G_b^r$ where $K_e^r, G_e^r, K_c^r, G_c^r$, K_b^r and G_b^r correspond to characteristic resonance frequencies of the exhaust gases (e), combustion chamber gases (c) and rocket body (b).[21] This approach is also followed in the theory of two polyphase fission processes, (γcinf) and (γcivf), wherein the internal phase angles θ_z and θ_{Dt} of the atomic number and the time dimension are calculated in terms of an effective dynamic frequency-dependent resonant shear moduli G_r^γ and G_v^γ and which in turn are used to calculate the effective magnetic fields (intensities) of the shearing ambient γ ray fields that are required for (γcinf) and (γcivf).[10-12,23,24]

<u>8. ACTIVE COORDINATES FOR INHOMOGENEOUS MATTER.</u> In general, matter is distributed inhomogeneously in space and time. For example, planets, stars and galaxies have inhomogeneous distributions of matter in the radial and azimuthal directions. The radial variation of density in these systems is due to the inverse square law of the gravitational forces that bind the particles of these systems. The azimuthal variation is due to complex forces associated with the diffusion and flow of matter. For example, on a macroscopic scale the earth has an inhomogeneous ambient gravity field that increases in an approximately linear fashion from the center to a point on the surface and then decreases with distance from the center in an approximately inverse square manner up through the atmosphere and into the rarified gases of the partial vacuum of interplanetary space. The pressure within the earth decreases from its large maximum value at the center of the earth to the much smaller value of the pressure of the atmosphere at the earth surface. The density of matter at the center of the earth is only a few times greater than the density of rocks at the surface.

The generalizations of equations (83) and (115) to the case of inhomogeneous matter are

$$d\alpha = C_{\sigma\alpha} d\alpha^o \qquad \alpha = \int C_{\sigma\alpha} d\alpha^o \qquad (888)$$

$$dt = C_{\sigma t} dt^{o} \qquad t = \int C_{\sigma t} dt^{o} \tag{889}$$

where the space coordinate factors $C_{\sigma\alpha}$ are given by equations (78) through (80) and the time coordinate factors $C_{\sigma t}$ are given by equations (110) through (112), and where $\sigma = 1$ for high-density matter such as solids while $\sigma = 2,3$ for low-density matter such as gases. Consider an inhomogeneous distribution of matter such as in the earth, which consists of a solid body part and an atmosphere whose density decreases with height and which eventually approaches the density of the rarified gases of interplanetary space. For this case the complex number radial distance of a body, such as a satellite or a rocket, in the atmosphere above the earth's surface can be written as

$$\overline{\rho} = \overline{R}_E + \overline{z} \tag{890}$$

where $\overline{\rho}$, $\overline{R}_E$ and $\overline{z}$ = complex number radial distance of body from center of the earth, complex number radius of the earth and complex number distance of the body above the earth's surface respectively, which an be represented as

$$\overline{\rho} = r\exp(j\theta_\rho) \qquad \overline{R}_E = R_E \exp(j\theta_{RE}) \qquad \overline{z} = z\exp(j\theta_z) \tag{891}$$

Then it follows simply that

$$\rho\cos\theta_\rho = I \qquad \rho\sin\theta_\rho = J \tag{892}$$

where

$$I = R_E \cos\theta_{RE} + z\cos\theta_z \qquad J = R_E \sin\theta_{RE} + z\sin\theta_z \tag{893}$$

The values of the magnitude ρ and internal phase angle θ_ρ are then given by

$$\rho = (I^2 + J^2)^{1/2} = [R_E^2 + z^2 + 2zR_E \cos(\theta_{RE} - \theta_z)]^{1/2} \tag{894}$$

$$\tan\theta_\rho = J/I = (R_E \sin\theta_{RE} + z\sin\theta_z)(R_E \cos\theta_{RE} + z\cos\theta_Z)^{-1} \tag{895}$$

The approximation $\theta_{RE} \approx \theta_z$ gives the following approximate values for ρ and θ_ρ

$$\rho \approx R_E + z \qquad \theta_\rho \approx \theta_{RE} \approx \theta_z \tag{896}$$

Consider now some approximations obtained from equation (888) by assuming that $C_{1\rho}$ is a slowly varying function of the radial distance from the center of the earth, and $C_{\beta\rho}$ for $\beta = 2,3$ is a slowly varying function of the distance through the atmosphere above the earth's surface. Then equation (888) gives the approximations

$$R_E = \int_o^{R_E^o} C_{1\rho}^g d\rho^o \approx R_E^o C_{1\rho}^g (R_E^o) \tag{897}$$

$$z = \int_o^{z^o} C_{\beta z}^g dz^o \approx z^o C_{\beta z}^g (z^o) \tag{898}$$

Combining equations (896) through (898) gives the approximation

$$\rho \approx R_E^o C_{1\rho}^g (R_E^o) + z^o C_{\beta z}^g (z^o) \qquad (899)$$

The corresponding radial distance to a body in the earth's atmosphere assuming inactive coordinates is given by

$$\rho^o = R_E^o + z^o \qquad (900)$$

An effective space coordinate factor C_ρ^g for a point above the surface of the earth can be defined in the following way

$$\rho = \rho^o C_\rho^g \qquad (901)$$

which combined with equations (899) and (900) gives an expression for C_ρ^g as

$$C_\rho^g = \rho/\rho^o \approx (R_E^o + z^o)^{-1}[R_E^o C_{1\rho}^g (R_E^o) + z^o C_{\beta z}^g (z^o)] \qquad 0 \le z^o \le \infty \qquad (902)$$

which is an approximation valid for the inhomogeneous solid earth-atmosphere-space system. For large values of z^o the low-density matter coordinate factor $C_{\beta z}^g (z^o)$ can describe the near vacuum of interstellar space.

CONCLUSION. The nature of space and time has not yet been discovered and although the concepts of space and time are ancient, new and surprising discoveries have yet to be made.[25-28] An ambient field is a field that induces internal phase angles in all physical quantities except the fundamental constants of nature (which include at least h, c and G). Space and time coordinates in matter located in an ambient field are represented as complex numbers in an internal space and have magnitudes and phase angles which depend on the energy density of the ambient field. The magnitudes of the space and time coordinates are still arbitrary in the sense that they depend on the choice of origin of the coordinates as well as on the local energy density of the ambient field, but the coordinate phase angles depend only on the local energy density of the ambient field. The effect of the ambient field on the magnitudes of the complex number active space and time coordinates is described by the space and time coordinate factors $C_{\sigma\alpha}$ and $C_{\sigma t}$ respectively, with $\alpha = x,y,z; r,\varphi,z$ and ρ,φ,ψ and $\sigma = 1$ for high-density matter and $\sigma =$ 2,3 for low-density matter, which multiplicatively relate the coordinate magnitudes to their zero ambient field real number inactive values. The concept of complex number active space and time coordinates arises from the use of complex counting numbers and complex segmentation numbers to describe coordinates in matter in the presence of ambient fields. The real number constants L_α^c and L_t^c that appear in equation (31) can be taken to be multiples (generally very large multiples) of the Planck length and Planck time respectively because an ambient field is a field that induces internal phase angles in all physical quantities except the fundamental constants of nature which include (h, c and G). The coordinate factors and the coordinate phase angles are relatively simple functions of the internal phase angles of the complex counting and segmentation numbers and have been delineated for high and low-density matter. In particular the coordinate factors turn out to be products of cosine and secant functions of the phase angles of the complex counting and segmentation numbers, while the coordinate phase angles are represented as the differences between the phase angles of the counting numbers and the segmentation numbers, and this allows for the possibility that the phase angles of the space and time coordinates can be negative as is the case for radiative and nonradiative ambient fields in matter as for example for atomic nuclei and Bohr atoms in an ambient electromagnetic field and for gravitating planets, stars and galaxies within an ambient gravitational field.[9,10,12,20] The internal phase angles of the complex number space and time coordinates are calculated for radiative and nonradiative ambient fields in terms of the shear modulus and bulk modulus modulus of elastic bulk matter. The cosine and secant terms which enter the description of the coordinate counting and segmentation numbers are analogous to the cosine and secant terms which enter the descriptions of the complex number nucleon numbers that arise in atomic nuclei that are located in an ambient field of γ rays and the complex number of molecules that enter the calculations of chemical reactions in the presence of ambient electromagnetic or ambient gravitational fields.[10-12] All of these systems have some parameter which has a critical value below which a cosine term appears in the magnitude of a complex number of objects and above which a secant term appears in the magnitude of the complex number of objects. If stellar gravitational collapse is considered and also cosmological problems, it may be that all systems have a transition parameter and the cosine and secant terms appear in all systems. In any case, equations (83), (115), (612), (613), (761), (762) and (771) through (774) are generally true for homogeneous matter in the presence of an ambient field. Modern physics and engineering, and in particular nuclear physics and rocket engineering and science, do not incorporate the concepts of active space and time coordinates.[29,30] The results of this paper may perhaps have scientific and engineering applications to polyphase fission nuclear reactors and to the design concepts of polyphase rockets.

REFERENCES

1. Jourdain, P.E.B., "The Nature of Mathematics," article in The World of Mathematics, p. 4, Simon and Schuster, New York, 1956.

2. Kline, M., Mathematics in Western Culture, Oxford University Press, New York, 1953.

3. Coolidge, J.L., A History of Geometrical Methods, Oxford University Press, New York, 1947.

4. Corben, H.C. and Stehle, P., Classical Mechanics, John Wiley, New York, 1957.

5. Stratton, J.A., Electromagnetic Theory, McGraw-Hill, New York, 1941.

6. Rindler, W., Introduction to Special Relativity, Clarendon Press, Oxford, 1991.

7. Misner, C.W., Thorne, K.S. and Wheeler, J.A., Gravitation, Freeman, San Francisco, 1970.

8. Bohm, D., Quantum Theory, Prentice-Hall, New York, 1951.

9. Weiss, R.A., Gauge Theory of Thermodynamics, reprint, K&W Publications, Vicksburg, 1989.

10. Weiss, R.A., Clean Fission, K&W Publications, Vicksburg, 1992.

11. Weiss, R.A., Internal Fission, K&W Publications, Vicksburg, 1995.

12. Weiss, R.A., Vacuum Fission, K&W Publications, Vicksburg, 1998.

13. Harwit, M., Astrophysical Concepts, Springer-Verlag, New York, 1988.

14. Davis, R., "The Search for Solar Neutrinos," article in Frontiers In Cosmic Physics, edited by Mendell, R.B. and Mincer, A.I., Annals of the New York Academy of Sciences, p. 209, 1992.

15. Jenkins, F.A. and White, H.E., Fundamentals of Optics, McGraw-Hill, New York, 1957.

16. Whittaker, E., A History of the Theories of Aether and Electricity, Vols. 1&2, Philosophical Library, New York, 1951.

17. Chanmugam, G., "Magnetic Fields of Degenerate Stars," article in Annual Reviews of Astronomy and Astrophysics, Vol. 30, p. 143, 1992.

18. Weiss, R.A., "Kinematics and Dynamics of Bodies in Ambient Fields," article in Polyphase Fission and Polyphase Rockets, p.95, K&W Publications, Tucson, 2003.

19. Weiss, R.A., "Wave Propagation and Diffusion in Matter in the Presence of an Ambient Field," article in Polyphase Fission and Polyphase Rockets, p.159, K&W Publications, Tucson, 2003.

20. Weiss, R.A., "Moseley's Law for Atoms in an Ambient γ Ray Field," article in Polyphase Fission and Polyphase Rockets, p.307, K&W Publications, Tucson, 2003.

21. Weiss, R.A., "Polyphase Rockets," article in Polyphase Fission and Polyphase Rockets, p.495, K&W Publications, Tucson, 2003.

22. Weiss, R.A., "Equilibrium Distributions and Chemical Reactions of Gases in Ambient Fields," article in Polyphase Fission and Polyphase Rockets, p.343, K&W Publications, Tucson, 2003.

23. Weiss, R.A., "Radius, Binding Energy and Polyphase Fission Parameters of an Atomic Nucleus in an Ambient γ Ray Field," article in Polyphase Fission and Polyphase Rockets, p.387, K&W Publications, Tucson, 2003.

24. Weiss, R.A., "Dynamics of Polyphase Fission Reactors," article in Polyphase Fission and Polyphase Rockets, p.449, K&W Publications, Tucson, 2003.

25. Whitrow, G.J., Time in History, Oxford, New york, 1989.

26. Margenau, H., The Nature of Physical Reality, McGraw-Hill, New York, 1950.

27. Landsberg, P.T., The Enigma of Time, Adam Hilger, Bristol, 1982.

28. Szamosi, G., The Twin Dimensions, McGraw-Hill, New York, 1986.

29. Pais, A., Inward Bound, Oxford Univerity Press, New York, 1986.

30. Sutton, G. P. and Biblarz, O., Rocket Propulsion Elements, John Wiley, New York, 2001.

B

POLYPHASE ROCKETS

Richard A. Weiss
K&W Research Company
Tucson, Arizona

ABSTRACT. This paper applies the theory of complex number active space and time coordinates and their density dependent coordinate factors to study the liftoff behavior of a polyphase rocket. The design concept of a polyphase rocket is introduced and consists of a chemical rocket that has six ambient energy fields of different intensities and phases applied to six rocket component systems including: (1) the exhaust gas aperture walls, (2) the combustion chamber foramen walls, (3) the walls of the personnel compartment, (4) the solid or liquid fuel and other solid or liquid systems that constitute the main body of a rocket, (5) the exhaust gases in the exhaust tube, and (6) the combustion gases in the combustion chamber. For practical purpose these ambient fields will be ambient electromagnetic wave fields. Because the first four systems are high-density solid or liquid systems while the latter two are low-density gaseous systems, it is shown that the application of ambient energy fields leads to enhancement of the rockets thrust and an increase in the liftoff parameter of the polyphase rocket. An ambient field is a field in bulk matter that induces internal phase angles into all physical quantities with the exception of the fundamental constants of nature (including h, c and G). The application of ambient energy fields such as ambient electromagnetic wave fields to the rocket systems requires that the dynamics of the rocket motion and the internal dynamics of the rocket systems be described by complex number active coordinates and by geometrical and dynamical quantities that are represented by complex numbers in an internal space. The space and time coordinate factors that are associated with active coordinates enter the calculation of the physical parameters of the six rocket systems and also relate the internal phase angles of the complex number of molecules to the internal phase angles of the complex molecular number densities of these systems. A complex number liftoff equation is developed and solved using approximations that allow analytical solutions. It is shown that the application of ambient energy fields leads to liftoff enhancement so that liftoff can be achieved in underpowered chemical rockets for which liftoff is impossible under zero ambient field conditions. Polyphase rockets may be of value for heavy-lift rockets.

1. INTRODUCTION. No other field of engineering or science has such a history of dreams and imagination as does rocket flight from the Earth to other celestial bodies.[1-7] The history of flight itself is closely entwined with the ideas of extraterrestrial travel. Dreams finally touched reality in the early part of the twentieth century when small liquid propellant rockets began to be developed.[1-7] Engineering and scientific evolution over the past century has brought chemical propellant rockets to the point that, now at the beginning of the third millennium, heavy payloads can be lifted into orbit.[1-7] Chemical rocket engines are very inefficient for delivering large payloads into space because most of the weight of a rocket at liftoff is due to the fuel. The large sustained thrust needed to lift heavy payloads requires a huge weight of chemical fuel. Alternative propulsion systems such as nuclear rocket engines and electromagnetic gun propulsion have been considered but with little practical success.[6] New ideas for heavy payload delivery into space are needed.

The question arises as to whether a low thrust chemical rocket with insufficient thrust to achieve liftoff can have its thrust enhanced by the application of ambient fields so that the effective thrust is sufficient for liftoff. If something like this were possible it would lead to an increased efficiency because the amount of fuel needed by the rocket would be greatly reduced provided that the weight of the equipment that generates the ambient fields does not weigh a great deal. It is suggested in this paper that the possibility of an enhancement of rocket liftoff capability requires that ambient fields of various kinds be applied to six rocket component systems: (1) the exhaust aperture walls, (2) the combustion chamber foramen walls, (3) the walls of the personnel chamber, (4) the solid or liquid main components of the rocket body, (5) the exhaust gases in the exhaust tube and (6) the combustion product gases in the combustion chamber, which are respectively four high-density systems consisting of solids or liquids and two low-density systems consisting of gases. An ambient field is a field in bulk matter that induces internal phase angles into all physical quantities with the exception of the fundamental constants of nature. In the presence of ambient fields in matter the geometrical and dynamical parameters of a rocket must be represented as complex numbers in an internal space. In addition, the number an number density of molecules of matter that are located in ambient fields are complex numbers that assume special forms for large and small systems and for high and low-density systems respectively, so that the six large rocket component systems mentioned above have unique and specific internal phase angles associated with their physical properties. The enhancement of the rocket thrust and liftoff capability requires the application of ambient fields to the six basic components of a rocket and this requires that the space and time coordinates used to describe the rocket components must also be represented as complex numbers in an internal space. In this description of a polyphase rocket it should be remembered that the rocket component systems are macroscopic in size and will be treated as large systems of molecules and atoms in various high-density and low-density states in the presence of ambient fields.

The ambient fields applied to the six rocket component systems can take the form of electromagnetic fields, magnetic fields, electric fields, laser fields, ultrasonic fields, stress and strain fields and any other kinds of fields that are appropriate for the materials in the six rocket components systems. For example, an ambient electromagnetic wave field such as ambient microwaves can be applied to the exhaust gases in the exhaust tube and to the combustion product gases in the combustion chamber because an electromagnetic wave field can be sustained within gases, but an electromagnetic wave field cannot be sustained in the walls of the exhaust tube and the combustion chamber tube if they are metallic because electromagnetic waves attenuate and dissipate rapidly in metals due to resistance heating resulting in metal fatigue. If these walls were made of ceramic materials then ambient electromagnetic wave fields can be used. It may be possible in the future to apply electromagnetic waves to the solid or liquid fuels that constitute the main portion of the body weight of a chemical rocket at liftoff. In this paper the energy density of the ambient field that is applied to a component system of a polyphase rocket is denoted by $\mathfrak{E}^{\gamma}$ where γ is an index which denotes the type of ambient field that is applied which can range from γ rays to sound waves. Thus the energy density of the ambient fields applied to the six rocket component systems mentioned above are denoted as $\mathfrak{E}^{\gamma a}$ = energy density of the ambient field in the high-density exhaust aperture walls, $\mathfrak{E}^{\gamma f}$ = energy density of the ambient field in the high-density combustion chamber foramen walls, $\mathfrak{E}^{\gamma p}$ = energy density of the ambient field in the high-density walls of the personnel compartment, $\mathfrak{E}^{\gamma b}$ = energy density of the ambient field in the high-density solid or liquid components of the rocket body, $\mathfrak{E}^{\gamma e}$ = energy density of the ambient field in the low-density exhaust gases in the exhaust tube and $\mathfrak{E}^{\gamma c}$ = energy density of the ambient field in the low-density combustion product gases in the combustion chamber.

It has been suggested that in matter in the presence of ambient fields, the space and time coordinates in matter or vacuum must be represented as complex numbers in an internal space and are of the form[8-11]

$$\bar{\alpha} = \alpha \exp(j\theta_\alpha) \qquad \bar{t} = t\exp(j\theta_t) \tag{1}$$

where α = x,y,z for Cartesian coordinates, α = r,φ,z for cylindrical polar coordinates, and where $\alpha = \rho,\varphi,\psi$ for spherical polar coordinates. The space and time coordinate magnitudes α and t are arbitrary in the sense that the origin of the space and time coordinates can be chosen arbitrarily, however, the internal phase angles of the space and time coordinates depend on the energy density of the ambient field. The measured values of the space and time coordinates are obtained as the real parts of equation (1) as[8-11]

$$\alpha_m = \alpha \cos\theta_\alpha \qquad t_m = t\cos\theta_t \tag{2}$$

For a body accelerating in the z direction the complex number particle speed is written as[8-10]

$$\bar{v} = v\exp(j\theta_v) = d\bar{z}/d\bar{t} \tag{3}$$

and the complex number acceleration is written as

$$\bar{a} = a\exp(j\theta_a) = d\bar{v}/d\bar{t} = d^2\bar{z}/d\bar{t}^2 \tag{4}$$

which for simplicity is taken to be constant. Assuming zero initial conditions for the displacement and speed allows the first integral of these equations to be written as[8-10]

$$\bar{v} = \bar{a}\bar{t} \qquad v = at \qquad \theta_v = \theta_a + \theta_t \tag{5}$$

$$\bar{z} = 1/2\bar{a}\bar{t}^2 \qquad z = 1/2at^2 \qquad \theta_z = \theta_a + 2\theta_t \tag{6}$$

Equations (5) and (6) can also be written as

$$\bar{v}^2 = 2\bar{a}\bar{z} \qquad v^2 = 2az \qquad 2\theta_v = \theta_a + \theta_z \tag{7}$$

These formulas will be used in Section 6 to describe the motion of exhaust gas molecules in the rocket exhaust tube.

The magnitudes of the complex number space coordinates of equation (1) are related to their zero ambient field values by[12]

$$\alpha = \alpha^{\circ} C_{\sigma\alpha} \tag{8}$$

while the internal phase angles of the complex number space coordinates are

$$\theta_\alpha = \theta_{\eta\alpha} - \theta_{\pi\alpha} \tag{9}$$

where the dimensionless space coordinate factors $C_{\sigma\alpha}$ for $\sigma = 1,2,3$ and $\alpha = x,y,z;\ r,\varphi,z$ or ρ,φ,ψ are given for high-density [hd] matter and low-density [ld] matter by[12]

$$C_{1\alpha} = \cos\theta_{\eta\alpha}\sec\theta_{\pi\alpha} \qquad \text{[hd]} \tag{10}$$

$$C_{2\alpha} = \cos\theta_{\eta\alpha}\cos\theta_{\pi\alpha} \qquad \text{[ld]} \tag{11}$$

$$C_{3\alpha} = \sec\theta_{\eta\alpha}\cos\theta_{\pi\alpha} \qquad \text{[ld]} \tag{12}$$

where $\theta_{\eta\alpha}$ and $\theta_{\pi\alpha}$ = internal phase angles of the complex number spatial counting and segmentation numbers respectively. For ambient wave fields in matter, the angles $\theta_{\eta\alpha}$ and $\theta_{\pi\alpha}$ are, to first order, linearly related to the energy density of the ambient wave fields in the manner[12]

$$\theta_{\eta\alpha} = f^{o}_{\eta\alpha}\mathfrak{E} \qquad \theta_{\pi\alpha} = f^{o}_{\pi\alpha}\mathfrak{E} \tag{13}$$

where $f^{o}_{\eta\alpha}$ and $f^{o}_{\pi\alpha}$ are frequency dependent parameters that depend on the shear modulus and bulk modulus of the matter, and $\mathfrak{E}$ = energy density of the ambient wave fields.[12] For nonradiative ambient fields in matter the angles $\theta_{\eta\alpha}$ and $\theta_{\pi\alpha}$ are not linearly related to the energy densities of the ambient fields but are calculated in a more complicated fashion in terms of the internal phase angles of the particle number and particle number densities which themselves are related to the shear modulus and bulk modulus of the materials under consideration.[12] Equations (11) and (12) show that [ld] matter exhibits doubly degenerate space coordinate factors. The following inequalities are valid

$$C_{1\alpha} \geq 1 \qquad C_{2\alpha} \leq 1 \qquad C_{3\alpha} \leq 1 \tag{14}$$

under the assumption that $|\theta_{\eta\alpha}| \leq |\theta_{\pi\alpha}|$ and therefore it follows from equation (7a)

$$\alpha \geq \alpha_o \qquad \text{[hd]} \tag{15}$$

$$\alpha \leq \alpha_o \qquad \text{[ld]} \tag{16}$$

for all coordinates. Also from equation (9) it follows that if $\theta_{\eta\alpha} \leq \theta_{\pi\alpha}$ then

$$\theta_\alpha \leq 0 \tag{17}$$

which is true for bound systems and for waves in matter.

Cylindrical polar coordinates are of particular importance for rocket design calculations because all main component systems can be approximately modeled as solid cylinders or cylindrical shells. When ambient fields are present in a cylindrical volume element of matter in which a cylindrical coordinate system is placed whose z-axis is along the cylinder axis, the complex number radial and axial coordinates are written as[12]

$$\bar{r} = r\exp(j\theta_r) \qquad \bar{z} = z\exp(j\theta_z) \tag{18}$$

$$r = r^o C_{\sigma r} \qquad z = z^o C_{\sigma z} \tag{19}$$

$$\theta_r = \theta_{\eta r} - \theta_{\iota r} \qquad \theta_z = \theta_{\eta z} - \theta_{\iota z} \tag{20}$$

where r^o and z^o = radial and axial coordinates under zero ambient field conditions, and where $C_{\sigma r}$ and $C_{\sigma z}$ = spatial coordinate factors given by equations (10) through (12) for σ = 1,2,3 and for α = r,z and where equations (14) and (19) show that

$$r \geq r^o \qquad z \geq z^o \qquad [hd] \tag{21}$$

$$r \leq r^o \qquad z \leq z^o \qquad [ld] \tag{22}$$

and where equation (17) shows that if $\theta_{\eta r} \leq \theta_{\iota r}$ and $\theta_{\eta z} \leq \theta_{\iota z}$ then[12]

$$\theta_r \leq 0 \qquad \theta_z \leq 0 \tag{23}$$

For ambient wave fields, the values of the internal phase angles of the coordinates are, to first order, linearly related to the total energy density of the ambient wave fields present in the manner

$$\theta_{\eta r} = f^o_{\eta r}\mathfrak{E} \qquad \theta_{\iota r} = f^o_{\iota r}\mathfrak{E} \qquad \theta_{\eta z} = f^o_{\eta z}\mathfrak{E} \qquad \theta_{\iota z} = f^o_{\iota z}\mathfrak{E} \tag{24}$$

where $f^o_{\eta r}, f^o_{\iota r}, f^o_{\eta z}$ and $f^o_{\iota z}$ are frequency dependent material parameters and $\mathfrak{E}$ = total energy density of the ambient wave fields.[12] The area of the end of a vertical cylinder volume element is given by

$$\bar{A} = A\exp(j\theta_A) = \pi\bar{r}^2 \tag{25}$$

$$A = \pi r^2 = A^o C^2_{\sigma r} \qquad A^o = \pi(r^o)^2 \tag{26}$$

$$\theta_A = 2\theta_r \leq 0 \tag{27}$$

The measured area is given by

$$A_m = A\cos\theta_A = A^o C^2_{\sigma r}\cos(2\theta_r) \tag{28}$$

The complex volume of a cylindrical volume element is written as

$$\bar{V} = V\exp(j\theta_V) = \pi\bar{r}^2\bar{z} \tag{29}$$

$$V = \pi r^2 z = V^o C^2_{\sigma r} C_{\sigma z} \qquad V^o = A^o z^o = \pi(r^o)^2 z^o \tag{30}$$

$$\theta_V = 2\theta_r + \theta_z \leq 0 \tag{31}$$

The measured volume element is given by

$$V_m = V\cos\theta_V = V^o C^2_{\sigma r} C_{\sigma z}\cos(2\theta_r + \theta_z) \tag{32}$$

The fact that the internal phase angles of the coordinates, areas and volumes of the various bulk components of a rocket are negative plays a role in the development of the liftoff equations of a polyphase rocket is treated in Sections 4, 6,8 and 9.

A polyphase piston engine is described in Section 2. In addition to gravity, which acts on all components of a piston engine, there are three ambient fields that are applied to the three basic components of a piston engine- the piston head (h), the piston body (b) and the working fluid (w). An ambient field of energy density $\mathfrak{E}^{\gamma h}$ is applied to the piston head upon which the working fluid is acting, an ambient field of energy density $\mathfrak{E}^{\gamma b}$ is applied to the body of the piston, and an ambient field of energy density $\mathfrak{E}^{\gamma w}$ is applied to the working fluid; in addition there are natural ambient gravitational field energy densities $\mathfrak{E}^{gh}$, $\mathfrak{E}^{gb}$ and $\mathfrak{E}^{gw}$ associated with the head, body and working fluid of the piston engine. In order to calculate the values of the internal phase angles of the space and time coordinates within the components of a polyphase piston engine, the ambient fields, including gravity, are assumed to behave like ambient wave fields. The total energy densities of the ambient fields acting in the engine components are written as

$$\mathfrak{E}^{h} = \mathfrak{E}^{gh} + \mathfrak{E}^{\gamma h} \qquad \text{piston head} \qquad (33)$$

$$\mathfrak{E}^{b} = \mathfrak{E}^{gb} + \mathfrak{E}^{\gamma b} \qquad \text{piston body} \qquad (34)$$

$$\mathfrak{E}^{w} = \mathfrak{E}^{gw} + \mathfrak{E}^{\gamma w} \qquad \text{working fluid} \qquad (35)$$

The complex number space coordinates in each component system are written as

$$\overline{\alpha}_h = \alpha_h \exp(j\theta^h_\alpha) \qquad \alpha_h = \alpha^o_h C^h_{1\alpha} \qquad \theta^h_\alpha = \theta^h_{\eta\alpha} - \theta^h_{\pi\alpha} \qquad \text{piston head} \qquad (36)$$

$$\overline{\alpha}_b = \alpha_b \exp(j\theta^b_\alpha) \qquad \alpha_b = \alpha^o_b C^b_{1\alpha} \qquad \theta^b_\alpha = \theta^b_{\eta\alpha} - \theta^b_{\pi\alpha} \qquad \text{piston body} \qquad (37)$$

$$\overline{\alpha}_w = \alpha_w \exp(j\theta^w_\alpha) \qquad \alpha_w = \alpha^o_w C^w_{\beta\alpha} \qquad \theta^w_\alpha = \theta^w_{\eta\alpha} - \theta^w_{\pi\alpha} \qquad \text{working fluid} \qquad (38)$$

where $\beta = 2,3$. For a polyphase piston engine the values of the space coordinate factors $C^h_{1\alpha}, C^b_{1\alpha}$ and $C^w_{\beta\alpha}$ with $\beta = 2,3$ and the internal phase angles $\theta^h_\alpha, \theta^b_\alpha$ and θ^w_α of the cylindrical coordinates $\alpha = r,z$ of the piston engine components depend on the internal phase angles of the complex spatial counting and segmentation numbers that enter into equations (9) through (12) and which for ambient wave fields are written for each of the components of a polyphase piston engine as a linear sum

$$\theta^h_{\eta\alpha} = \theta^{gh}_{\eta\alpha} + \theta^{\gamma h}_{\eta\alpha} \qquad \theta^h_{\pi\alpha} = \theta^{gh}_{\pi\alpha} + \theta^{\gamma h}_{\pi\alpha} \qquad \text{piston head} \qquad (39)$$

$$\theta^b_{\eta\alpha} = \theta^{gb}_{\eta\alpha} + \theta^{\gamma b}_{\eta\alpha} \qquad \theta^b_{\pi\alpha} = \theta^{gb}_{\pi\alpha} + \theta^{\gamma b}_{\pi\alpha} \qquad \text{piston body} \qquad (40)$$

$$\theta^w_{\eta\alpha} = \theta^{gw}_{\eta\alpha} + \theta^{\gamma w}_{\eta\alpha} \qquad \theta^w_{\pi\alpha} = \theta^{gw}_{\pi\alpha} + \theta^{\gamma w}_{\pi\alpha} \qquad \text{working fluid} \qquad (41)$$

For ambient wave fields, these internal phase angles of the complex space coordinate counting and segmentation numbers are calculated approximately for the components of a polyphase piston engine by using equation (13) as follows[12]

$$\theta^{gh}_{\eta\alpha} = f^{oh}_{\eta\alpha}\mathfrak{E}^{gh} \quad \theta^{gh}_{\pi\alpha} = f^{oh}_{\pi\alpha}\mathfrak{E}^{gh} \quad \theta^{\gamma h}_{\eta\alpha} = f^{oh}_{\eta\alpha}\mathfrak{E}^{\gamma h} \quad \theta^{\gamma h}_{\pi\alpha} = f^{oh}_{\pi\alpha}\mathfrak{E}^{\gamma h} \tag{42}$$

$$\theta^{gb}_{\eta\alpha} = f^{ob}_{\eta\alpha}\mathfrak{E}^{gb} \quad \theta^{gb}_{\pi\alpha} = f^{ob}_{\pi\alpha}\mathfrak{E}^{gb} \quad \theta^{\gamma b}_{\eta\alpha} = f^{ob}_{\eta\alpha}\mathfrak{E}^{\gamma b} \quad \theta^{\gamma b}_{\pi\alpha} = f^{ob}_{\pi\alpha}\mathfrak{E}^{\gamma b} \tag{43}$$

$$\theta^{gw}_{\eta\alpha} = f^{ow}_{\eta\alpha}\mathfrak{E}^{gw} \quad \theta^{gw}_{\pi\alpha} = f^{ow}_{\pi\alpha}\mathfrak{E}^{gw} \quad \theta^{\gamma w}_{\eta\alpha} = f^{ow}_{\eta\alpha}\mathfrak{E}^{\gamma w} \quad \theta^{\gamma w}_{\pi\alpha} = f^{ow}_{\pi\alpha}\mathfrak{E}^{\gamma w} \tag{44}$$

where the f^{o} – parameters are frequency dependent and material dependent coefficients that do not depend on the energy density of the ambient fields.[12] Then for ambient wave fields the internal phase angles of the counting and segmentation numbers in equations (39) through (41) are obtained after using equations (33) through (35) and written in the approximate form of equation (13) as[12]

$$\theta^{h}_{\eta\alpha} = f^{oh}_{\eta\alpha}\mathfrak{E}^{h} \qquad \theta^{h}_{\pi\alpha} = f^{oh}_{\pi\alpha}\mathfrak{E}^{h} \tag{45}$$

$$\theta^{b}_{\eta\alpha} = f^{ob}_{\eta\alpha}\mathfrak{E}^{b} \qquad \theta^{b}_{\pi\alpha} = f^{ob}_{\pi\alpha}\mathfrak{E}^{b} \tag{46}$$

$$\theta^{w}_{\eta\alpha} = f^{ow}_{\eta\alpha}\mathfrak{E}^{w} \qquad \theta^{w}_{\pi\alpha} = f^{ow}_{\pi\alpha}\mathfrak{E}^{w} \tag{47}$$

where $\mathfrak{E}^{h}$, $\mathfrak{E}^{b}$ and $\mathfrak{E}^{w}$ are given by equations (33) through (35). Then equation (9) gives the internal phase angles of the space coordinates in each component of the polyphase piston engine for the applied ambient wave fields as[12]

$$\theta^{h}_{\alpha} = f^{oh}_{\alpha}\mathfrak{E}^{h} \qquad \theta^{b}_{\alpha} = f^{ob}_{\alpha}\mathfrak{E}^{b} \qquad \theta^{w}_{\alpha} = f^{ow}_{\alpha}\mathfrak{E}^{w} \tag{48}$$

where

$$f^{oh}_{\alpha} = f^{oh}_{\eta\alpha} - f^{oh}_{\pi\alpha} \qquad f^{ob}_{\alpha} = f^{ob}_{\eta\alpha} - f^{ob}_{\pi\alpha} \qquad f^{ow}_{\alpha} = f^{ow}_{\eta\alpha} - f^{ow}_{\pi\alpha} \tag{49}$$

For a polyphase piston engine the space coordinate factors $C^{h}_{1\alpha}$, $C^{b}_{1\alpha}$ and $C^{w}_{\beta\alpha}$ with β = 2,3 that appear in equations (10) through (12) and the space coordinate internal phase angles $\theta^{h}_{\alpha}, \theta^{b}_{\alpha}$ and θ^{w}_{α} with α = r,z of equation (20), corresponding to the ambient fields in equation (33) through (35) and the internal phase angles of the complex counting and segmentation numbers that appear equations (39) through (41), are given for the high-density piston head and body as

$$C^{h}_{1\alpha} = \cos\theta^{h}_{\eta\alpha}\sec\theta^{h}_{\pi\alpha} \qquad \theta^{h}_{\alpha} = \theta^{h}_{\eta\alpha} - \theta^{h}_{\pi\alpha} \qquad \text{piston head} \tag{50}$$

$$C^{b}_{1\alpha} = \cos\theta^{b}_{\eta\alpha}\sec\theta^{b}_{\pi\alpha} \qquad \theta^{b}_{\alpha} = \theta^{b}_{\eta\alpha} - \theta^{b}_{\pi\alpha} \qquad \text{piston body} \tag{51}$$

and for the doubly degenerate low-density working fluid

$$C^{w}_{2\alpha} = \cos\theta^{w}_{\eta\alpha}\cos\theta^{w}_{\pi\alpha} \qquad \theta^{w}_{\alpha} = \theta^{w}_{\eta\alpha} - \theta^{w}_{\pi\alpha} \qquad \text{working fluid} \tag{52}$$

$$C^{w}_{3\alpha} = \sec\theta^{h}_{\eta\alpha}\cos\theta^{w}_{\pi\alpha} \qquad \theta^{w}_{\alpha} = \theta^{w}_{\eta\alpha} - \theta^{w}_{\pi\alpha} \qquad \text{working fluid} \tag{53}$$

where α = r,z. The coordinate and volume magnitudes given by equations (19) and (30) can be written for the components of a polyphase piston engine as

$$r_h = r^{o}_{h}C^{h}_{1r} \qquad z_h = z^{o}_{h}C^{h}_{1z} \qquad V_h = V^{o}_{h}(C^{h}_{1r})^2 C^{h}_{1z} \qquad \text{piston head} \tag{54}$$

$$r_b = r_b^o C_{1r}^b \qquad z_b = z_b^o C_{1z}^b \qquad V_b = V_b^o (C_{1r}^b)^2 C_{1z}^b \qquad \text{piston body} \tag{55}$$

$$r_w = r_w^o C_{2r}^w \qquad z_w = z_w^o C_{2z}^w \qquad V_w = V_w^o (C_{2r}^w)^2 C_{2z}^w \qquad \text{working fluid} \tag{56}$$

$$r_w = r_w^o C_{3r}^w \qquad z_w = z_w^o C_{3z}^w \qquad V_w = V_w^o (C_{3r}^w)^2 C_{3z}^w \qquad \text{working fluid} \tag{57}$$

where $C_{1r}^h, C_{1z}^h, C_{1r}^b, C_{1z}^b, C_{2r}^w, C_{2z}^w, C_{3r}^w$ and C_{3z}^w are given by equations (50) through (53) with $\alpha = r,z$.

For a polyphase rocket as described in Section 3, the internal phase angles of the space and time coordinates and in fact of all physical quantities except the constants of nature (including at least h, c and G which are real numbers) are due to the presence of ambient gravity fields and other applied ambient fields throughout six component systems of the polyphase rocket. In particular, an ambient field of energy density $\mathfrak{E}^{\gamma a}$ is applied to the high-density rocket exhaust aperture system (nozzle), an ambient field of energy density $\mathfrak{E}^{\gamma f}$ is applied to the high-density combustion chamber foramen system, an ambient field of energy density $\mathfrak{E}^{\gamma p}$ is applied to the high-density walls of the personnel compartment, an ambient field of energy density $\mathfrak{E}^{\gamma b}$ is applied to the high-density liquid or solid body component of the rocket body, an ambient field $\mathfrak{E}^{\gamma e}$ is applied to the low-density exhaust gases in the nozzle and an ambient field of energy density $\mathfrak{E}^{\gamma c}$ is applied to the low-density gases of the combustion chamber; and in addition there are the six energy densities that are associated with the natural ambient gravitational fields $\mathfrak{E}^{ga}, \mathfrak{E}^{gf}, \mathfrak{E}^{gp}, \mathfrak{E}^{gb}, \mathfrak{E}^{ge}$ and $\mathfrak{E}^{gc}$ acting on the exhaust aperture, combustion chamber foramen system, personnel chamber walls, rocket body, exhaust gases and the combustion chamber gases respectively. All of the ambient fields, including gravity, are treated as ambient wave fields. The total energy densities of the ambient wave fields acting on the six components of a polyphase rocket are written as

$$\mathfrak{E}^a = \mathfrak{E}^{ga} + \mathfrak{E}^{\gamma a} \qquad \text{exhaust aperture system} \tag{58}$$

$$\mathfrak{E}^f = \mathfrak{E}^{gf} + \mathfrak{E}^{\gamma f} \qquad \text{combustion chamber foramen system} \tag{59}$$

$$\mathfrak{E}^p = \mathfrak{E}^{gp} + \mathfrak{E}^{\gamma p} \qquad \text{walls of the personnel compartment} \tag{60}$$

$$\mathfrak{E}^b = \mathfrak{E}^{gb} + \mathfrak{E}^{\gamma b} \qquad \text{rocket body} \tag{61}$$

$$\mathfrak{E}^e = \mathfrak{E}^{ge} + \mathfrak{E}^{\gamma e} \qquad \text{exhaust gases} \tag{62}$$

$$\mathfrak{E}^c = \mathfrak{E}^{gc} + \mathfrak{E}^{\gamma c} \qquad \text{combustion chamber gases} \tag{63}$$

The complex number space coordinates in each component system are written as

$$\overline{\alpha}_a = \alpha_a \exp(j\theta_\alpha^a) \qquad \alpha_a = \alpha_a^o C_{1\alpha}^a \qquad \theta_\alpha^a = \theta_{\eta\alpha}^a - \theta_{\pi\alpha}^a \qquad \text{exhaust aperture system} \tag{64}$$

$$\overline{\alpha}_f = \alpha_f \exp(j\theta_\alpha^f) \qquad \alpha_f = \alpha_f^o C_{1\alpha}^f \qquad \theta_\alpha^f = \theta_{\eta\alpha}^f - \theta_{\pi\alpha}^f \qquad \text{combustion chamber foramen system} \tag{65}$$

$$\overline{\alpha}_p = \alpha_p \exp(j\theta_\alpha^p) \qquad \alpha_p = \alpha_p^o C_{1\alpha}^p \qquad \theta_\alpha^p = \theta_{\eta\alpha}^p - \theta_{\pi\alpha}^p \qquad \text{personnel compartment wall} \tag{66}$$

$$\overline{\alpha}_b = \alpha_b \exp(j\theta^b_\alpha) \qquad \alpha_b = \alpha^o_b C^b_{1\alpha} \qquad \theta^b_\alpha = \theta^b_{\eta\alpha} - \theta^b_{\pi\alpha} \qquad \text{rocket body} \tag{67}$$

$$\overline{\alpha}_e = \alpha_e \exp(j\theta^e_\alpha) \qquad \alpha_e = \alpha^o_e C^e_{\beta\alpha} \qquad \theta^e_\alpha = \theta^e_{\eta\alpha} - \theta^e_{\pi\alpha} \qquad \text{exhaust gases} \tag{68}$$

$$\overline{\alpha}_c = \alpha_c \exp(j\theta^c_\alpha) \qquad \alpha_c = \alpha^o_c C^c_{\beta\alpha} \qquad \theta^c_\alpha = \theta^c_{\eta\alpha} - \theta^c_{\pi\alpha} \qquad \text{combustion chamber gases} \tag{69}$$

where $\alpha = r,z$ and $\beta = 2,3$. Thus there are a total of six applied energy densities for a polyphase rocket, and there are apparently six space coordinate factors $C^a_{1\alpha}, C^f_{1\alpha}, C^p_{1\alpha}$, $C^b_{1\alpha}, C^e_{\beta\alpha}$ and $C^c_{\beta\alpha}$ (but actually eight if the degeneracy of the low-density systems are considered which have $\beta = 2,3$) and six space coordinate internal phase angles $\theta^a_\alpha, \theta^f_\alpha$, $\theta^p_\alpha, \theta^b_\alpha, \theta^e_\alpha$ and θ^c_α of the cylindrical coordinates $\alpha = r,z$ of the rocket components and which depend on the internal phase angles of the complex number spatial coordinate counting and segmentation numbers that enter equations (9) through (12) and equations (64) through (69) and which for ambient wave fields are written for the six component systems of a polyphase rocket as[12]

$$\theta^a_{\eta\alpha} = \theta^{ga}_{\eta\alpha} + \theta^{\gamma a}_{\eta\alpha} \qquad \theta^a_{\pi\alpha} = \theta^{ga}_{\pi\alpha} + \theta^{\gamma a}_{\pi\alpha} \qquad \text{exhaust aperture system} \tag{70}$$

$$\theta^f_{\eta\alpha} = \theta^{gf}_{\eta\alpha} + \theta^{\gamma f}_{\eta\alpha} \qquad \theta^f_{\pi\alpha} = \theta^{gf}_{\pi\alpha} + \theta^{\gamma f}_{\pi\alpha} \qquad \text{combustion chamber foramen system} \tag{71}$$

$$\theta^p_{\eta\alpha} = \theta^{gp}_{\eta\alpha} + \theta^{\gamma p}_{\eta\alpha} \qquad \theta^p_{\pi\alpha} = \theta^{gp}_{\pi\alpha} + \theta^{\gamma p}_{\pi\alpha} \qquad \text{personnel compartment walls} \tag{72}$$

$$\theta^b_{\eta\alpha} = \theta^{gb}_{\eta\alpha} + \theta^{\gamma b}_{\eta\alpha} \qquad \theta^b_{\pi\alpha} = \theta^{gb}_{\pi\alpha} + \theta^{\gamma b}_{\pi\alpha} \qquad \text{rocket body} \tag{73}$$

$$\theta^e_{\eta\alpha} = \theta^{ge}_{\eta\alpha} + \theta^{\gamma e}_{\eta\alpha} \qquad \theta^e_{\pi\alpha} = \theta^{ge}_{\pi\alpha} + \theta^{\gamma e}_{\pi\alpha} \qquad \text{exhaust gases} \tag{74}$$

$$\theta^c_{\eta\alpha} = \theta^{gc}_{\eta\alpha} + \theta^{\gamma c}_{\eta\alpha} \qquad \theta^c_{\pi\alpha} = \theta^{gc}_{\pi\alpha} + \theta^{\gamma c}_{\pi\alpha} \qquad \text{combustion chamber gases} \tag{75}$$

For ambient wave fields, these internal phase angles of the complex space coordinate counting and segmentation numbers are calculated for the components of a polyphase rocket by using the approximate equation (13) as follows[12]

$$\theta^{ga}_{\eta\alpha} = f^{oa}_{\eta\alpha}\mathfrak{E}^{ga} \quad \theta^{ga}_{\pi\alpha} = f^{oa}_{\pi\alpha}\mathfrak{E}^{ga} \quad \theta^{\gamma a}_{\eta\alpha} = f^{oa}_{\eta\alpha}\mathfrak{E}^{\gamma a} \quad \theta^{\gamma a}_{\pi\alpha} = f^{oa}_{\pi\alpha}\mathfrak{E}^{\gamma a} \tag{76}$$

$$\theta^{gf}_{\eta\alpha} = f^{of}_{\eta\alpha}\mathfrak{E}^{gf} \quad \theta^{gf}_{\pi\alpha} = f^{of}_{\pi\alpha}\mathfrak{E}^{gf} \quad \theta^{\gamma f}_{\eta\alpha} = f^{of}_{\eta\alpha}\mathfrak{E}^{\gamma f} \quad \theta^{\gamma f}_{\pi\alpha} = f^{of}_{\pi\alpha}\mathfrak{E}^{\gamma f} \tag{77}$$

$$\theta^{gp}_{\eta\alpha} = f^{op}_{\eta\alpha}\mathfrak{E}^{gp} \quad \theta^{gp}_{\pi\alpha} = f^{op}_{\pi\alpha}\mathfrak{E}^{gp} \quad \theta^{\gamma p}_{\eta\alpha} = f^{op}_{\eta\alpha}\mathfrak{E}^{\gamma p} \quad \theta^{\gamma p}_{\pi\alpha} = f^{op}_{\pi\alpha}\mathfrak{E}^{\gamma p} \tag{78}$$

$$\theta^{gb}_{\eta\alpha} = f^{ob}_{\eta\alpha}\mathfrak{E}^{gb} \quad \theta^{gb}_{\pi\alpha} = f^{ob}_{\pi\alpha}\mathfrak{E}^{gb} \quad \theta^{\gamma b}_{\eta\alpha} = f^{ob}_{\eta\alpha}\mathfrak{E}^{\gamma b} \quad \theta^{\gamma b}_{\pi\alpha} = f^{ob}_{\pi\alpha}\mathfrak{E}^{\gamma b} \tag{79}$$

$$\theta^{ge}_{\eta\alpha} = f^{oe}_{\eta\alpha}\mathfrak{E}^{ge} \quad \theta^{ge}_{\pi\alpha} = f^{oe}_{\pi\alpha}\mathfrak{E}^{ge} \quad \theta^{\gamma e}_{\eta\alpha} = f^{oe}_{\eta\alpha}\mathfrak{E}^{\gamma e} \quad \theta^{\gamma e}_{\pi\alpha} = f^{oe}_{\pi\alpha}\mathfrak{E}^{\gamma e} \tag{80}$$

$$\theta^{gc}_{\eta\alpha} = f^{oc}_{\eta\alpha}\mathfrak{E}^{gc} \quad \theta^{gc}_{\pi\alpha} = f^{oc}_{\pi\alpha}\mathfrak{E}^{gc} \quad \theta^{\gamma c}_{\eta\alpha} = f^{oc}_{\eta\alpha}\mathfrak{E}^{\gamma c} \quad \theta^{\gamma c}_{\pi\alpha} = f^{oc}_{\pi\alpha}\mathfrak{E}^{\gamma c} \tag{81}$$

where the f^o – functions are frequency dependent material parameters. For the case of ambient wave fields, the internal phase angles are obtained from equations (70) through (81) and equations (58) through (63) in the approximately linear form of equation (13) as

$$\theta^{a}_{\eta\alpha} = f^{oa}_{\eta\alpha}\mathfrak{E}^{a} \qquad \theta^{a}_{n\alpha} = f^{oa}_{n\alpha}\mathfrak{E}^{a} \qquad \text{exhaust gas aperture system} \tag{82}$$

$$\theta^{f}_{\eta\alpha} = f^{of}_{\eta\alpha}\mathfrak{E}^{f} \qquad \theta^{f}_{n\alpha} = f^{of}_{n\alpha}\mathfrak{E}^{f} \qquad \text{combustion chamber foramen system} \tag{83}$$

$$\theta^{p}_{\eta\alpha} = f^{op}_{\eta\alpha}\mathfrak{E}^{p} \qquad \theta^{p}_{n\alpha} = f^{op}_{n\alpha}\mathfrak{E}^{p} \qquad \text{personnel compartment walls} \tag{84}$$

$$\theta^{b}_{\eta\alpha} = f^{ob}_{\eta\alpha}\mathfrak{E}^{b} \qquad \theta^{b}_{n\alpha} = f^{ob}_{n\alpha}\mathfrak{E}^{b} \qquad \text{rocket body} \tag{85}$$

$$\theta^{e}_{\eta\alpha} = f^{oe}_{\eta\alpha}\mathfrak{E}^{e} \qquad \theta^{e}_{n\alpha} = f^{oe}_{n\alpha}\mathfrak{E}^{e} \qquad \text{exhaust gases} \tag{86}$$

$$\theta^{c}_{\eta\alpha} = f^{oc}_{\eta\alpha}\mathfrak{E}^{c} \qquad \theta^{c}_{n\alpha} = f^{oc}_{n\alpha}\mathfrak{E}^{c} \qquad \text{combustion chamber gases} \tag{87}$$

where $\mathfrak{E}^{a}$, $\mathfrak{E}^{f}$, $\mathfrak{E}^{p}$, $\mathfrak{E}^{b}$, $\mathfrak{E}^{e}$ and $\mathfrak{E}^{c}$ are given by equations (58) through (63). For ambient wave fields, equations (9), (13) and (82) through (87) give the internal phase angles of the space coordinates for each component of a polyphase rocket approximately as

$$\theta^{a}_{\alpha} = f^{oa}_{\alpha}\mathfrak{E}^{a} \qquad \theta^{f}_{\alpha} = f^{of}_{\alpha}\mathfrak{E}^{f} \qquad \theta^{p}_{\alpha} = f^{op}_{\alpha}\mathfrak{E}^{p} \tag{88}$$

$$\theta^{b}_{\alpha} = f^{ob}_{\alpha}\mathfrak{E}^{b} \qquad \theta^{e}_{\alpha} = f^{oe}_{\alpha}\mathfrak{E}^{e} \qquad \theta^{c}_{\alpha} = f^{oc}_{\alpha}\mathfrak{E}^{c} \tag{89}$$

where[12]

$$f^{oa}_{\alpha} = f^{oa}_{\eta\alpha} - f^{oa}_{n\alpha} \qquad f^{of}_{\alpha} = f^{of}_{\eta\alpha} - f^{of}_{n\alpha} \qquad f^{op}_{\alpha} = f^{op}_{\eta\alpha} - f^{op}_{n\alpha} \tag{90}$$

$$f^{ob}_{\alpha} = f^{ob}_{\eta\alpha} - f^{ob}_{n\alpha} \qquad f^{oe}_{\alpha} = f^{oe}_{\eta\alpha} - f^{oe}_{n\alpha} \qquad f^{oc}_{\alpha} = f^{oc}_{\eta\alpha} - f^{oc}_{n\alpha} \tag{91}$$

The space coordinate factors $C_{\sigma\alpha}$ and the internal phase angles θ_{α} of the space coordinates as given by equations (10) through (12) and (9) respecively, and which correspond to the combined energy densities of equation (58) through (63) and to the internal phase angles of the counting and segmentation numbers as given by equations (70) through (75), for the six component systems of a polyphase rocket can now be given for $\alpha = r,z$. The results for the high-density systems are

$$C^{a}_{1\alpha} = \cos\theta^{a}_{\eta\alpha}\sec\theta^{a}_{n\alpha} \qquad \theta^{a}_{\alpha} = \theta^{a}_{\eta\alpha} - \theta^{a}_{n\alpha} \qquad \text{exhaust aperture system} \tag{92}$$

$$C^{f}_{1\alpha} = \cos\theta^{f}_{\eta\alpha}\sec\theta^{f}_{n\alpha} \qquad \theta^{f}_{\alpha} = \theta^{f}_{\eta\alpha} - \theta^{f}_{n\alpha} \qquad \text{combustion chamber foramen system} \tag{93}$$

$$C^{p}_{1\alpha} = \cos\theta^{p}_{\eta\alpha}\sec\theta^{p}_{n\alpha} \qquad \theta^{p}_{\alpha} = \theta^{p}_{\eta\alpha} - \theta^{p}_{n\alpha} \qquad \text{personnel compartment walls} \tag{94}$$

$$C^{b}_{1\alpha} = \cos\theta^{b}_{\eta\alpha}\sec\theta^{b}_{n\alpha} \qquad \theta^{b}_{\alpha} = \theta^{b}_{\eta\alpha} - \theta^{b}_{n\alpha} \qquad \text{rocket body} \tag{95}$$

and for the degenerate low-density systems

$$C^{e}_{2\alpha} = \cos\theta^{e}_{\eta\alpha}\cos\theta^{e}_{n\alpha} \qquad \theta^{e}_{\alpha} = \theta^{e}_{\eta\alpha} - \theta^{e}_{n\alpha} \qquad \text{exhaust gases} \tag{96}$$

$$C^{e}_{3\alpha} = \sec\theta^{e}_{\eta\alpha}\cos\theta^{e}_{n\alpha} \qquad \theta^{e}_{\alpha} = \theta^{e}_{\eta\alpha} - \theta^{e}_{n\alpha} \qquad \text{exhaust gases} \tag{97}$$

$$C^{c}_{2\alpha} = \cos\theta^{c}_{\eta\alpha}\cos\theta^{c}_{n\alpha} \qquad \theta^{c}_{\alpha} = \theta^{c}_{\eta\alpha} - \theta^{c}_{n\alpha} \qquad \text{combustion chamber gases} \tag{98}$$

$$C^{c}_{3\alpha} = \sec\theta^{c}_{\eta\alpha}\cos\theta^{c}_{\pi\alpha} \qquad \theta^{c}_{\alpha} = \theta^{c}_{\eta\alpha} - \theta^{c}_{\pi\alpha} \qquad \text{combustion chamber gases} \tag{99}$$

which can be used in equations (64) through (69). Equations (19) and (30) applied to the six components of a polyphase rocket can be written as

$$r_a = r^o_a C^a_{1r} \qquad z_a = z^o_a C^a_{1z} \qquad V_a = V^o_a (C^a_{1r})^2 C^a_{1z} \qquad \text{exhaust aperture system} \tag{100}$$

$$r_f = r^o_f C^f_{1r} \qquad z_f = z^o_f C^f_{1z} \qquad V_f = V^o_f (C^f_{1r})^2 C^f_{1z} \qquad \text{combustion chamber foramen system} \tag{101}$$

$$r_p = r^o_p C^p_{1r} \qquad z_p = z^o_p C^p_{1z} \qquad V_p = V^o_p (C^p_{1r})^2 C^p_{1z} \qquad \text{personnel compartment walls} \tag{102}$$

$$r_b = r^o_b C^b_{1r} \qquad z_b = z^o_b C^b_{1z} \qquad V_b = V^o_b (C^b_{1r})^2 C^b_{1z} \qquad \text{rocket body} \tag{103}$$

$$r_e = r^o_e C^w_{2e} \qquad z_e = z^o_e C^e_{2z} \qquad V_e = V^o_e (C^e_{2r})^2 C^e_{2z} \qquad \text{exhaust gases} \tag{104}$$

$$r_e = r^o_e C^e_{3r} \qquad z_e = z^o_e C^e_{3z} \qquad V_e = V^o_e (C^e_{3r})^2 C^e_{3z} \qquad \text{exhaust gases} \tag{105}$$

$$r_c = r^o_c C^c_{2r} \qquad z_c = z^o_c C^c_{2z} \qquad V_c = V^o_c (C^c_{2r})^2 C^c_{2z} \qquad \text{combustion chamber gases} \tag{106}$$

$$r_c = r^o_c C^c_{3r} \qquad z_c = z^o_c C^c_{3z} \qquad V_c = V^o_c (C^c_{3r})^2 C^c_{3z} \qquad \text{combustion chamber gases} \tag{107}$$

where $C^a_{1r}, C^a_{1z}, C^f_{1r}, C^f_{1z}, C^p_{1r}, C^p_{1z}, C^b_{1r}, C^b_{1z}, C^e_{2r}, C^e_{2z}, C^e_{3r}, C^e_{3z}, C^c_{2r}, C^c_{2z}, C^c_{3r}$ and C^c_{3z} are given by equations (92) through (99) with $\alpha = r,z$.

For the special case when all artificially applied ambient fields are zero and only ambient gravity is acting approximately equally throughout the rocket with its natural ambient field component in matter, the space coordinate factors $C_{\sigma\alpha}$ and the internal phase angles θ_α of the coordinates are written with $\alpha = r,z$ as

$$C^{ag}_{1\alpha} = \cos\theta^{ag}_{\eta\alpha}\sec\theta^{ag}_{\pi\alpha} \qquad \theta^{ag}_{\alpha} = \theta^{ag}_{\eta\alpha} - \theta^{ag}_{\pi\alpha} \qquad \text{exhaust aperture system} \tag{108}$$

$$C^{fg}_{1\alpha} = \cos\theta^{fg}_{\eta\alpha}\sec\theta^{fg}_{\pi\alpha} \qquad \theta^{fg}_{\alpha} = \theta^{fg}_{\eta\alpha} - \theta^{fg}_{\pi\alpha} \qquad \text{combustion chamber foramen system} \tag{109}$$

$$C^{pg}_{1\alpha} = \cos\theta^{pg}_{\eta\alpha}\sec\theta^{pg}_{\pi\alpha} \qquad \theta^{pg}_{\alpha} = \theta^{pg}_{\eta\alpha} - \theta^{pg}_{\pi\alpha} \qquad \text{personnel compartment walls} \tag{110}$$

$$C^{bg}_{1\alpha} = \cos\theta^{bg}_{\eta\alpha}\sec\theta^{bg}_{\pi\alpha} \qquad \theta^{bg}_{\alpha} = \theta^{bg}_{\eta\alpha} - \theta^{bg}_{\pi\alpha} \qquad \text{rocket body} \tag{111}$$

and for the degenerate low-density systems

$$C^{eg}_{2\alpha} = \cos\theta^{eg}_{\eta\alpha}\cos\theta^{eg}_{\pi\alpha} \qquad \theta^{eg}_{\alpha} = \theta^{eg}_{\eta\alpha} - \theta^{eg}_{\pi\alpha} \qquad \text{exhaust gases} \tag{112}$$

$$C^{eg}_{3\alpha} = \sec\theta^{eg}_{\eta\alpha}\cos\theta^{eg}_{\pi\alpha} \qquad \theta^{eg}_{\alpha} = \theta^{eg}_{\eta\alpha} - \theta^{eg}_{\pi\alpha} \qquad \text{exhaust gases} \tag{113}$$

$$C^{cg}_{2\alpha} = \cos\theta^{cg}_{\eta\alpha}\cos\theta^{cg}_{\pi\alpha} \qquad \theta^{cg}_{\alpha} = \theta^{cg}_{\eta\alpha} - \theta^{cg}_{\pi\alpha} \qquad \text{combustion chamber gases} \tag{114}$$

$$C_{3\alpha}^{cg} = \sec\theta_{\eta\alpha}^{cg} \cos\theta_{\pi\alpha}^{cg} \qquad \theta_{\alpha}^{cg} = \theta_{\eta\alpha}^{cg} - \theta_{\pi\alpha}^{cg} \qquad \text{combustion chamber gases} \tag{115}$$

For the case when only an ambient gravitational field is acting, equations (19) and (30) become

$$r^g = r^o C_{\sigma r}^g \qquad z^g = z^o C_{\sigma z} \qquad V^g = V^o (C_{\sigma r}^g)^2 C_{\sigma z}^g \tag{116}$$

The magnitudes of the complex number coordinates and of a cylindrical volume as given by equations (19) and (30) for the general case of a natural ambient gravity field plus an arbitrary ambient field are expressed in terms of their zero ambient field values. However, by combining equations (19), (30) and (116) the expressions for the magnitudes of the complex number coordinates and volume can also be expressed in terms of their values when only an ambient gravity field is present as follows

$$r = r^g D_{\sigma r} \qquad z = z^g D_{\sigma z} \qquad V = V^g D_{\sigma r}^2 D_{\sigma z} \tag{117}$$

where[12]

$$D_{\sigma r} = C_{\sigma r} / C_{\sigma r}^g \qquad D_{\sigma z} = C_{\sigma z} / C_{\sigma z}^g \tag{118}$$

for $\sigma = 1,2,3$. For the case of the polyphase rocket systems, equation (116) becomes

$$r_a^g = r_a^o C_{1r}^{ag} \qquad z_a^g = z_a^o C_{1z}^{ag} \qquad V_a^g = V_a^o (C_{1r}^{ag})^2 C_{1z}^{ag} \qquad \text{exhaust aperture system} \tag{119}$$

$$r_f^g = r_f^o C_{1r}^{fg} \qquad z_f^g = z_f^o C_{1z}^{fg} \qquad V_f^g = V_f^o (C_{1r}^{fg})^2 C_{1z}^{fg} \qquad \text{combustion chamber foramen system} \tag{120}$$

$$r_p^g = r_p^o C_{1r}^{pg} \qquad z_p^g = z_p^o C_{1z}^{pg} \qquad V_p^g = V_p^o (C_{1r}^{pg})^2 C_{1z}^{pg} \qquad \text{personnel compartment wall} \tag{121}$$

$$r_b^g = r_b^o C_{1r}^{bg} \qquad z_b^g = z_b^o C_{1z}^{bg} \qquad V_b^g = V_b^o (C_{1r}^{bg})^2 C_{1z}^{bg} \qquad \text{rocket body} \tag{122}$$

$$r_e^g = r_e^o C_{2r}^{eg} \qquad z_e^g = z_e^o C_{2z}^{eg} \qquad V_e^g = V_e^o (C_{2r}^{eg})^2 C_{2z}^{eg} \qquad \text{exhaust gases} \tag{123}$$

$$r_e^g = r_e^o C_{3r}^{eg} \qquad z_e^g = z_e^o C_{3z}^{eg} \qquad V_e^g = V_e^o (C_{3r}^{eg})^2 C_{3z}^{eg} \qquad \text{exhaust gases} \tag{124}$$

$$r_c^g = r_c^o C_{2r}^{cg} \qquad z_c^g = z_c^o C_{2z}^{cg} \qquad V_c^g = V_c^o (C_{2r}^{cg})^2 C_{2z}^{cg} \qquad \text{combustion chamber gases} \tag{125}$$

$$r_c^g = r_c^o C_{3r}^{cg} \qquad z_c^g = z_c^o C_{3z}^{cg} \qquad V_c^g = V_c^o (C_{3r}^{cg})^2 C_{3z}^{cg} \qquad \text{combustion chamber gases} \tag{126}$$

while equation (117) becomes

$$r_a = r_a^g D_{1r}^a \qquad z_a = z_a^g D_{1z}^a \qquad V_a = V_a^g (D_{1r}^a)^2 D_{1z}^a \qquad \text{exhaust aperture system} \tag{127}$$

$$r_f = r_f^g D_{1r}^f \qquad z_f = z_f^g D_{1z}^f \qquad V_f = V_f^g (D_{1r}^f)^2 D_{1z}^f \qquad \text{combustion chamber foramen system} \tag{128}$$

$r_p = r_p^g D_{1r}^p \qquad z_p = z_p^g D_{1z}^p \qquad V_p = V_p^g (D_{1r}^p)^2 D_{1z}^p$ personnel compartment wall (129)

$r_b = r_b^g D_{1r}^b \qquad z_b = z_b^g D_{1z}^b \qquad V_b = V_b^g (D_{1r}^b)^2 D_{1z}^b$ rocket body (130)

$r_e = r_e^g D_{2r}^e \qquad z_e = z_e^g D_{2z}^e \qquad V_e = V_e^g (D_{2r}^e)^2 D_{2z}^e$ exhaust gases (131)

$r_e = r_e^g D_{3r}^e \qquad z_e = z_e^g D_{3z}^e \qquad V_e = V_e^g (D_{3r}^e)^2 D_{3z}^e$ exhaust gases (132)

$r_c = r_c^g D_{2r}^c \qquad z_c = z_c^g D_{2z}^c \qquad V_c = V_c^g (D_{2r}^c)^2 D_{2z}^c$ combustion chamber gases (133)

$r_c = r_c^g D_{3r}^c \qquad z_c = z_c^g D_{3z}^c \qquad V_c = V_c^g (D_{3r}^c)^2 D_{3z}^c$ combustion chamber gases (134)

and where equation (118) becomes

$D_{1r}^a = C_{1r}^a / C_{1r}^{ag} \qquad D_{1z}^a = C_{1z}^a / C_{1z}^{ag}$ exhaust aperture system (135)

$D_{1r}^f = C_{1r}^f / C_{1r}^{fg} \qquad D_{1z}^f = C_{1z}^f / C_{1z}^{fg}$ combustion chamber foramen system (136)

$D_{1r}^p = C_{1r}^p / C_{1r}^{pg} \qquad D_{1z}^p = C_{1z}^p / C_{1z}^{pg}$ personnel compartment walls (137)

$D_{1r}^b = C_{1r}^b / C_{1r}^{bg} \qquad D_{1z}^b = C_{1z}^b / C_{1z}^{bg}$ rocket body (138)

$D_{2r}^e = C_{2r}^e / C_{2r}^{eg} \qquad D_{2z}^e = C_{2z}^e / C_{2z}^{eg}$ exhaust gases (139)

$D_{3r}^e = C_{3r}^e / C_{3r}^{eg} \qquad D_{3z}^e = C_{3z}^e / C_{3z}^{eg}$ exhaust gases (140)

$D_{2r}^c = C_{2r}^c / C_{2r}^{cg} \qquad D_{2z}^c = C_{2z}^c / C_{2z}^{cg}$ combustion chamber gases (141)

$D_{3r}^c = C_{3r}^c / C_{3r}^{cg} \qquad D_{3z}^c = C_{3z}^c / C_{3z}^{cg}$ combustion chamber gases (142)

so that each component system of a polyphase rocket has its own set of space coordinate factors which scales the spatial coordinates within the system.

In the presence of an ambient field the time coordinate is given by a complex number of the form[12]

$$\bar{t} = t \exp(j\theta_t) \tag{143}$$

where the magnitude of the complex number time coordinate is related to its zero ambient field value by[12]

$$t = t^{o} C_{\sigma t} \tag{144}$$

and the internal phase angle of the complex number time coordinate is given by[12]

$$\theta_t = \theta_{\eta t} - \theta_{\pi t} \tag{145}$$

where the dimensionless time coordinate factors $C_{\sigma t}$ for $\sigma = 1,2,3$ are given for high-density [hd] matter and low-density [ld] matter by[12]

$$C_{1t} = \cos\theta_{\eta t} \sec\theta_{\pi t} \quad \text{[hd]} \tag{146}$$

$$C_{2t} = \cos\theta_{\eta t} \cos\theta_{\pi t} \quad \text{[ld]} \tag{147}$$

$$C_{3t} = \sec\theta_{\eta t} \cos\theta_{\pi t} \quad \text{[ld]} \tag{148}$$

where $\theta_{\eta t}$ and $\theta_{\pi t}$ = internal phase angles of the complex number temporal counting and segmentation numbers respectively, and which for ambient wave fields in matter are related to the total energy density of the ambient wave fields by the approximations[12]

$$\theta_{\eta t} = f^{o}_{\eta t}\mathfrak{E} \qquad \theta_{\pi t} = f^{o}_{\pi t}\mathfrak{E} \tag{149}$$

where $f^{o}_{\eta t}$ and $f^{o}_{\pi t}$ are frequency dependent material parameters.[12] The time coordinate factors are doubly degenerate for [ld] matter as can be seen from equations (147) and (148). If $|\theta_{\eta t}| \le |\theta_{\pi t}|$ it follows that[12]

$$C_{1t} \ge 1 \qquad C_{2t} \le 1 \qquad C_{3t} \le 1 \tag{150}$$

and where equations (144) through (150) show that[12]

$$t \ge t_o \quad \text{[hd]} \tag{151}$$

$$t \le t_o \quad \text{[ld]} \tag{152}$$

and where if $\theta_{\eta t} \le \theta_{\pi t}$ then[12]

$$\theta_t \le 0 \tag{153}$$

which apparently is valid for bound systems and radiation in matter.

Equations (33) through (35) show that for a polyphase piston engine, three energy densities of the ambient gravity fields along with three energy densities of the applied ambient fields are associated with the piston head, piston body and the working fluid, for a total of six energy densities. The complex number time coordinates for each of these three components are written as

$$\bar{t}_h = t_h \exp(j\theta^h_t) \qquad t_h = t^o_h C^h_{1t} \qquad \theta^h_t = \theta^h_{\eta t} - \theta^h_{\pi t} \qquad \text{piston head} \tag{154}$$

$$\bar{t}_b = t_b \exp(j\theta^b_t) \qquad t_b = t^o_b C^b_{1t} \qquad \theta^b_t = \theta^b_{\eta t} - \theta^b_{\pi t} \qquad \text{piston body} \tag{155}$$

$$\bar{t}_w = t_w \exp(j\theta^w_t) \qquad t_w = t^o_w C^w_{\beta t} \qquad \theta^w_t = \theta^w_{\eta t} - \theta^w_{\pi t} \qquad \text{working fluid} \tag{156}$$

where $\beta = 2,3$. The internal phase angles of the complex time coordinate counting and segmentation numbers that are required for the time coordinate factors C^h_{1t}, C^b_{1t} and $C^w_{\beta t}$

with $\beta = 2,3$ and for the internal phase angles $\theta^{h}_{t}, \theta^{b}_{t}$ and θ^{w}_{t}, as in equations (145) through (148), are for ambient wave fields written for each of the components of the polyphase piston engine as a linear sum as follows

$$\theta^{h}_{\eta t} = \theta^{gh}_{\eta t} + \theta^{\gamma h}_{\eta t} \qquad \theta^{h}_{\pi t} = \theta^{gh}_{\pi t} + \theta^{\gamma h}_{\pi t} \qquad \text{piston head} \qquad (157)$$

$$\theta^{b}_{\eta t} = \theta^{gb}_{\eta t} + \theta^{\gamma b}_{\eta t} \qquad \theta^{b}_{\pi t} = \theta^{gb}_{\pi t} + \theta^{\gamma b}_{\pi t} \qquad \text{piston body} \qquad (158)$$

$$\theta^{w}_{\eta t} = \theta^{gw}_{\eta t} + \theta^{\gamma w}_{\eta t} \qquad \theta^{w}_{\pi t} = \theta^{gw}_{\pi t} + \theta^{\gamma w}_{\pi t} \qquad \text{working fluid} \qquad (159)$$

These internal phase angles of the complex time coordinate counting and segmentation numbers are calculated for the components of a polyphase piston engine by using the approximate equation (149) for ambient wave fields as follows[12]

$$\theta^{gh}_{\eta t} = f^{oh}_{\eta t}\mathfrak{E}^{gh} \qquad \theta^{gh}_{\pi t} = f^{oh}_{\pi t}\mathfrak{E}^{gh} \qquad \theta^{\gamma h}_{\eta t} = f^{oh}_{\eta t}\mathfrak{E}^{\gamma h} \qquad \theta^{\gamma h}_{\pi t} = f^{oh}_{\pi t}\mathfrak{E}^{\gamma h} \qquad (160)$$

$$\theta^{gb}_{\eta t} = f^{ob}_{\eta t}\mathfrak{E}^{gb} \qquad \theta^{gb}_{\pi t} = f^{ob}_{\pi t}\mathfrak{E}^{gb} \qquad \theta^{\gamma b}_{\eta t} = f^{ob}_{\eta t}\mathfrak{E}^{\gamma b} \qquad \theta^{\gamma b}_{\pi t} = f^{ob}_{\pi t}\mathfrak{E}^{\gamma b} \qquad (161)$$

$$\theta^{gw}_{\eta t} = f^{ow}_{\eta t}\mathfrak{E}^{gw} \qquad \theta^{gw}_{\pi t} = f^{ow}_{\pi t}\mathfrak{E}^{gw} \qquad \theta^{\gamma w}_{\eta t} = f^{ow}_{\eta t}\mathfrak{E}^{\gamma w} \qquad \theta^{\gamma w}_{\pi t} = f^{ow}_{\pi t}\mathfrak{E}^{\gamma w} \qquad (162)$$

where the f^{o} – parameters are frequency dependent material coefficients. Then for ambient wave fields, the internal phase angles of the counting and segmentation numbers in equations (157) through (159) are obtained after using equations (33) through (35) and written in the form of equation (149) as[12]

$$\theta^{h}_{\eta t} = f^{oh}_{\eta t}\mathfrak{E}^{h} \qquad \theta^{h}_{\pi t} = f^{oh}_{\pi t}\mathfrak{E}^{h} \qquad (163)$$

$$\theta^{b}_{\eta t} = f^{ob}_{\eta t}\mathfrak{E}^{b} \qquad \theta^{b}_{\pi t} = f^{ob}_{\pi t}\mathfrak{E}^{b} \qquad (164)$$

$$\theta^{w}_{\eta t} = f^{ow}_{\eta t}\mathfrak{E}^{w} \qquad \theta^{w}_{\pi t} = f^{ow}_{\pi t}\mathfrak{E}^{w} \qquad (165)$$

where $\mathfrak{E}^{h}, \mathfrak{E}^{b}$ and $\mathfrak{E}^{w}$ are given by equations (33) through (35). Then for ambient wave fields equation (145) gives the internal phase angle for the time coordinate in each component of the polyphase piston engine as

$$\theta^{h}_{t} = f^{oh}_{t}\mathfrak{E}^{h} \qquad \theta^{b}_{t} = f^{ob}_{t}\mathfrak{E}^{b} \qquad \theta^{w}_{t} = f^{ow}_{t}\mathfrak{E}^{w} \qquad (166)$$

where[12]

$$f^{oh}_{t} = f^{oh}_{\eta t} - f^{oh}_{\pi t} \qquad f^{ob}_{t} = f^{ob}_{\eta t} - f^{ob}_{\pi t} \qquad f^{ow}_{t} = f^{ow}_{\eta t} - f^{ow}_{\pi t} \qquad (167)$$

For a polyphase piston engine the time coordinate factors C^{h}_{1t}, C^{b}_{1t} and $C^{w}_{\beta t}$ with $\beta = 2,3$ and the internal phase angles $\theta^{h}_{t}, \theta^{b}_{t}$ and θ^{w}_{t} are given for the high-density piston head and piston body as[12]

$$C^{h}_{1t} = \cos\theta^{h}_{\eta t}\sec\theta^{h}_{\pi t} \qquad \theta^{h}_{t} = \theta^{h}_{\eta t} - \theta^{h}_{\pi t} \qquad \text{piston head} \qquad (168)$$

$$C^{b}_{1t} = \cos\theta^{b}_{\eta t}\sec\theta^{b}_{\pi t} \qquad \theta^{b}_{t} = \theta^{b}_{\eta t} - \theta^{b}_{\pi t} \qquad \text{piston body} \qquad (169)$$

and for the doubly degenerate low-density working fluid

$$C^{w}_{2t} = \cos\theta^{w}_{\eta t}\cos\theta^{w}_{\iota t} \qquad \theta^{w}_{t} = \theta^{w}_{\eta t} - \theta^{w}_{\iota t} \qquad \text{working fluid} \qquad (170)$$

$$C^{w}_{3t} = \sec\theta^{h}_{\eta t}\cos\theta^{w}_{\iota t} \qquad \theta^{w}_{t} = \theta^{w}_{\eta t} - \theta^{w}_{\iota t} \qquad \text{working fluid} \qquad (171)$$

which are used in equations (154) through (156).

For a polyphase rocket, whose ambient energy density distribution over the various components of the rocket are described by equations (58) through (63), the complex number time coordinates in each component system are written as

$$\bar{t}_{a} = t_{a}\exp(j\theta^{a}_{t}) \qquad t_{a} = t^{o}_{a}C^{a}_{1t} \qquad \theta^{a}_{t} = \theta^{a}_{\eta t} - \theta^{a}_{\iota t} \qquad \text{exhaust aperture system} \qquad (172)$$

$$\bar{t}_{f} = t_{f}\exp(j\theta^{f}_{t}) \qquad t_{f} = t^{o}_{f}C^{f}_{1t} \qquad \theta^{f}_{t} = \theta^{f}_{\eta t} - \theta^{f}_{\iota t} \qquad \text{combustion chamber foramen system} \qquad (173)$$

$$\bar{t}_{p} = t_{p}\exp(j\theta^{p}_{t}) \qquad t_{p} = t^{o}_{p}C^{p}_{1t} \qquad \theta^{p}_{t} = \theta^{p}_{\eta t} - \theta^{p}_{\iota t} \qquad \text{personnel compartment walls} \qquad (174)$$

$$\bar{t}_{b} = t_{b}\exp(j\theta^{b}_{t}) \qquad t_{b} = t^{o}_{b}C^{b}_{1t} \qquad \theta^{b}_{t} = \theta^{b}_{\eta t} - \theta^{b}_{\iota t} \qquad \text{rocket body} \qquad (175)$$

$$\bar{t}_{e} = t_{e}\exp(j\theta^{e}_{t}) \qquad t_{e} = t^{o}_{e}C^{e}_{\beta t} \qquad \theta^{e}_{t} = \theta^{e}_{\eta t} - \theta^{e}_{\iota t} \qquad \text{exhaust gases} \qquad (176)$$

$$\bar{t}_{c} = t_{c}\exp(j\theta^{c}_{t}) \qquad t_{c} = t^{o}_{c}C^{c}_{\beta t} \qquad \theta^{c}_{t} = \theta^{c}_{\eta t} - \theta^{c}_{\iota t} \qquad \text{combustion chamber gases} \qquad (177)$$

The six internal phase angles of the time coordinates of the six rocket components depend on the internal phase angles of the complex time counting and segmentation numbers that enter through equations (145) through (148) and equations (172) through (177) and which for ambient wave fields are written for the six component systems of a polyphase rocket as a linear sum[12]

$$\theta^{a}_{\eta t} = \theta^{ga}_{\eta t} + \theta^{\gamma a}_{\eta t} \qquad \theta^{a}_{\iota t} = \theta^{ga}_{\iota t} + \theta^{\gamma a}_{\iota t} \qquad \text{exhaust aperture system} \qquad (178)$$

$$\theta^{f}_{\eta t} = \theta^{gf}_{\eta t} + \theta^{\gamma f}_{\eta t} \qquad \theta^{f}_{\iota t} = \theta^{gf}_{\iota t} + \theta^{\gamma f}_{\iota t} \qquad \text{combustion chamber foramen system} \qquad (179)$$

$$\theta^{p}_{\eta t} = \theta^{gp}_{\eta t} + \theta^{\gamma p}_{\eta t} \qquad \theta^{p}_{\iota t} = \theta^{gp}_{\iota t} + \theta^{\gamma p}_{\iota t} \qquad \text{personnel compartment walls} \qquad (180)$$

$$\theta^{b}_{\eta t} = \theta^{gb}_{\eta t} + \theta^{\gamma b}_{\eta t} \qquad \theta^{b}_{\iota t} = \theta^{gb}_{\iota t} + \theta^{\gamma b}_{\iota t} \qquad \text{rocket body} \qquad (181)$$

$$\theta^{e}_{\eta t} = \theta^{ge}_{\eta t} + \theta^{\gamma e}_{\eta t} \qquad \theta^{e}_{\iota t} = \theta^{ge}_{\iota t} + \theta^{\gamma e}_{\iota t} \qquad \text{exhaust gases} \qquad (182)$$

$$\theta^{c}_{\eta t} = \theta^{gc}_{\eta t} + \theta^{\gamma c}_{\eta t} \qquad \theta^{c}_{\iota t} = \theta^{gc}_{\iota t} + \theta^{\gamma c}_{\iota t} \qquad \text{combustion chamber gases} \qquad (183)$$

These internal phase angles of the complex time coordinate counting and segmentation numbers are calculated for the various components of a polyphase rocket by using the approximate linear equation (149) for ambient wave fields as follows[12]

$$\theta^{ga}_{\eta t} = f^{oa}_{\eta t}\mathfrak{E}^{ga} \qquad \theta^{ga}_{\iota t} = f^{oa}_{\iota t}\mathfrak{E}^{ga} \qquad \theta^{\gamma a}_{\eta t} = f^{oa}_{\eta t}\mathfrak{E}^{\gamma a} \qquad \theta^{\gamma a}_{\iota t} = f^{oa}_{\iota t}\mathfrak{E}^{\gamma a} \qquad (184)$$

$$\theta^{gf}_{\eta t} = f^{of}_{\eta t}\mathfrak{E}^{gf} \qquad \theta^{gf}_{\pi t} = f^{of}_{\pi t}\mathfrak{E}^{gf} \qquad \theta^{\gamma f}_{\eta t} = f^{of}_{\eta t}\mathfrak{E}^{\gamma f} \qquad \theta^{\gamma f}_{\pi t} = f^{of}_{\pi t}\mathfrak{E}^{\gamma f} \qquad (185)$$

$$\theta^{gp}_{\eta t} = f^{op}_{\eta t}\mathfrak{E}^{gp} \qquad \theta^{gp}_{\pi t} = f^{op}_{\pi t}\mathfrak{E}^{gp} \qquad \theta^{\gamma p}_{\eta t} = f^{op}_{\eta t}\mathfrak{E}^{\gamma p} \qquad \theta^{\gamma p}_{\pi t} = f^{op}_{\pi t}\mathfrak{E}^{\gamma p} \qquad (186)$$

$$\theta^{gb}_{\eta t} = f^{ob}_{\eta t}\mathfrak{E}^{gb} \qquad \theta^{gb}_{\pi t} = f^{ob}_{\pi t}\mathfrak{E}^{gb} \qquad \theta^{\gamma b}_{\eta t} = f^{ob}_{\eta t}\mathfrak{E}^{\gamma b} \qquad \theta^{\gamma b}_{\pi t} = f^{ob}_{\pi t}\mathfrak{E}^{\gamma b} \qquad (187)$$

$$\theta^{ge}_{\eta t} = f^{oe}_{\eta t}\mathfrak{E}^{ge} \qquad \theta^{ge}_{\pi t} = f^{oe}_{\pi t}\mathfrak{E}^{ge} \qquad \theta^{\gamma e}_{\eta t} = f^{oe}_{\eta t}\mathfrak{E}^{\gamma e} \qquad \theta^{\gamma e}_{\pi t} = f^{oe}_{\pi t}\mathfrak{E}^{\gamma e} \qquad (188)$$

$$\theta^{gc}_{\eta t} = f^{oc}_{\eta t}\mathfrak{E}^{gc} \qquad \theta^{gc}_{\pi t} = f^{oc}_{\pi t}\mathfrak{E}^{gc} \qquad \theta^{\gamma c}_{\eta t} = f^{oc}_{\eta t}\mathfrak{E}^{\gamma c} \qquad \theta^{\gamma c}_{\pi t} = f^{oc}_{\pi t}\mathfrak{E}^{\gamma c} \qquad (189)$$

where the f^{o} – functions are frequency dependent material parameters. The internal phase angles of the time counting and segmentation numbers are then obtained from equations (178) through (183) after using equations (58) through (63) and are written in the form of the approximate linear equation (149) for ambient wave fields as

$$\theta^{a}_{\eta t} = f^{oa}_{\eta t}\mathfrak{E}^{a} \qquad \theta^{a}_{\pi t} = f^{oa}_{\pi t}\mathfrak{E}^{a} \qquad \text{exhaust aperture system} \qquad (190)$$

$$\theta^{f}_{\eta t} = f^{of}_{\eta t}\mathfrak{E}^{f} \qquad \theta^{f}_{\pi t} = f^{of}_{\pi t}\mathfrak{E}^{f} \qquad \text{combustion chamber foramen system} \qquad (191)$$

$$\theta^{p}_{\eta t} = f^{op}_{\eta t}\mathfrak{E}^{p} \qquad \theta^{p}_{\pi t} = f^{op}_{\pi t}\mathfrak{E}^{p} \qquad \text{personnel compartment walls} \qquad (192)$$

$$\theta^{b}_{\eta t} = f^{ob}_{\eta t}\mathfrak{E}^{b} \qquad \theta^{b}_{\pi t} = f^{ob}_{\pi t}\mathfrak{E}^{b} \qquad \text{rocket body} \qquad (193)$$

$$\theta^{e}_{\eta t} = f^{oe}_{\eta t}\mathfrak{E}^{e} \qquad \theta^{e}_{\pi t} = f^{oe}_{\pi t}\mathfrak{E}^{e} \qquad \text{exhaust gases} \qquad (194)$$

$$\theta^{c}_{\eta t} = f^{oc}_{\eta t}\mathfrak{E}^{c} \qquad \theta^{c}_{\pi t} = f^{oc}_{\pi t}\mathfrak{E}^{c} \qquad \text{combustion chamber gases} \qquad (195)$$

where $\mathfrak{E}^{a}, \mathfrak{E}^{f}, \mathfrak{E}^{p}, \mathfrak{E}^{b}, \mathfrak{E}^{e}$ and $\mathfrak{E}^{c}$ are given by equations (58) through (63). Equation (145) gives the internal phase angle of the time coordinates for ambient wave fields and for each component of a polyphase rocket as

$$\theta^{a}_{t} = f^{oa}_{t}\mathfrak{E}^{a} \qquad \theta^{f}_{t} = f^{of}_{t}\mathfrak{E}^{f} \qquad \theta^{p}_{t} = f^{op}_{t}\mathfrak{E}^{p} \qquad (196)$$

$$\theta^{b}_{t} = f^{ob}_{t}\mathfrak{E}^{b} \qquad \theta^{e}_{t} = f^{oe}_{t}\mathfrak{E}^{e} \qquad \theta^{c}_{t} = f^{oc}_{t}\mathfrak{E}^{c} \qquad (197)$$

where[12]

$$f^{oa}_{t} = f^{oa}_{\eta t} - f^{oa}_{\pi t} \qquad f^{of}_{t} = f^{of}_{\eta t} - f^{of}_{\pi t} \qquad f^{op}_{t} = f^{op}_{\eta t} - f^{op}_{\pi t} \qquad (198)$$

$$f^{ob}_{t} = f^{ob}_{\eta t} - f^{ob}_{\pi t} \qquad f^{oe}_{t} = f^{oe}_{\eta t} - f^{oe}_{\pi t} \qquad f^{oc}_{t} = f^{oc}_{\eta t} - f^{oc}_{\pi t} \qquad (199)$$

The corresponding time coordinate factors for a polyphase rocket are obtained from equations (146) through (148) as

$$C^{a}_{1t} = \cos\theta^{a}_{\eta t}\sec\theta^{a}_{\pi t} \qquad \theta^{a}_{t} = \theta^{a}_{\eta t} - \theta^{a}_{\pi t} \qquad \text{exhaust aperture system} \qquad (200)$$

$$C^{f}_{1t} = \cos\theta^{f}_{\eta t}\sec\theta^{f}_{\pi t} \qquad \theta^{f}_{t} = \theta^{f}_{\eta t} - \theta^{f}_{\pi t} \qquad \text{combustion chamber foramen system} \qquad (201)$$

$$C^{p}_{1t} = \cos\theta^{p}_{\eta t}\sec\theta^{p}_{\pi t} \qquad \theta^{p}_{t} = \theta^{p}_{\eta t} - \theta^{p}_{\pi t} \qquad \text{personnel compartment walls} \qquad (202)$$

$$C_{1t}^{b} = \cos\theta_{\eta t}^{b} \sec\theta_{\pi t}^{b} \qquad \theta_{t}^{b} = \theta_{\eta t}^{b} - \theta_{\pi t}^{b} \qquad \text{rocket body} \tag{203}$$

and for the degenerate low-density systems

$$C_{2t}^{e} = \cos\theta_{\eta t}^{e} \cos\theta_{\pi t}^{e} \qquad \theta_{t}^{e} = \theta_{\eta t}^{e} - \theta_{\pi t}^{e} \qquad \text{exhaust gases} \tag{204}$$

$$C_{3t}^{e} = \sec\theta_{\eta t}^{e} \cos\theta_{\pi t}^{e} \qquad \theta_{t}^{e} = \theta_{\eta t}^{e} - \theta_{\pi t}^{e} \qquad \text{exhaust gases} \tag{205}$$

$$C_{2t}^{c} = \cos\theta_{\eta t}^{c} \cos\theta_{\pi t}^{c} \qquad \theta_{t}^{c} = \theta_{\eta t}^{c} - \theta_{\pi t}^{c} \qquad \text{combustion chamber gases} \tag{206}$$

$$C_{3t}^{c} = \sec\theta_{\eta t}^{c} \cos\theta_{\pi t}^{c} \qquad \theta_{t}^{c} = \theta_{\eta t}^{c} - \theta_{\pi t}^{c} \qquad \text{combustion chamber gases} \tag{207}$$

which can be used in equations (172) through (177).

For the case when only an an ambient gravity field, which is a natural ambient field component of gravity in matter, is acting on a polyphase rocket, the time coordinate factors $C_{\sigma t}$ and the internal phase angles θ_t of the time coordinates are given by

$$C_{1t}^{ag} = \cos\theta_{\eta t}^{ag} \sec\theta_{\pi t}^{ag} \qquad \theta_{t}^{ag} = \theta_{\eta t}^{ag} - \theta_{\pi t}^{ag} \qquad \text{exhaust aperture system} \tag{208}$$

$$C_{1t}^{fg} = \cos\theta_{\eta t}^{fg} \sec\theta_{\pi t}^{fg} \qquad \theta_{t}^{fg} = \theta_{\eta t}^{fg} - \theta_{\pi t}^{fg} \qquad \text{combustion chamber foramen system} \tag{209}$$

$$C_{1t}^{pg} = \cos\theta_{\eta t}^{pg} \sec\theta_{\pi t}^{pg} \qquad \theta_{t}^{pg} = \theta_{\eta t}^{pg} - \theta_{\pi t}^{pg} \qquad \text{personnel compartment walls} \tag{210}$$

$$C_{1t}^{bg} = \cos\theta_{\eta t}^{bg} \sec\theta_{\pi t}^{bg} \qquad \theta_{t}^{bg} = \theta_{\eta t}^{bg} - \theta_{\pi t}^{bg} \qquad \text{rocket body} \tag{211}$$

and for the degenerate low-density systems

$$C_{2t}^{eg} = \cos\theta_{\eta t}^{eg} \cos\theta_{\pi t}^{eg} \qquad \theta_{t}^{eg} = \theta_{\eta t}^{eg} - \theta_{\pi t}^{eg} \qquad \text{exhaust gases} \tag{212}$$

$$C_{3t}^{eg} = \sec\theta_{\eta t}^{eg} \cos\theta_{\pi t}^{eg} \qquad \theta_{t}^{eg} = \theta_{\eta t}^{eg} - \theta_{\pi t}^{eg} \qquad \text{exhaust gases} \tag{213}$$

$$C_{2t}^{cg} = \cos\theta_{\eta t}^{cg} \cos\theta_{\pi t}^{cg} \qquad \theta_{t}^{cg} = \theta_{\eta t}^{cg} - \theta_{\pi t}^{cg} \qquad \text{combustion chamber gases} \tag{214}$$

$$C_{3t}^{cg} = \sec\theta_{\eta t}^{cg} \cos\theta_{\pi t}^{cg} \qquad \theta_{t}^{cg} = \theta_{\eta t}^{cg} - \theta_{\pi t}^{cg} \qquad \text{combustion chamber gases} \tag{215}$$

When the only ambient field present is an ambient gravity field, equation (144) becomes

$$t^{g} = t^{o} C_{\sigma t}^{g} \tag{216}$$

and then combining equations (144) and (216) gives[12]

$$t = t^{g} D_{\sigma t} \tag{217}$$

where

$$D_{\sigma t} = C_{\sigma t} / C^{g}_{\sigma t} \tag{218}$$

where $\sigma = 1,2,3$, and where equation (217) relates a time coordinate in the presence of an arbitrary ambient to the time coordinate for the case when only an ambient gravity field is present. For the case of the component systems of a polyphase rocket equation (216) becomes

$t^{g}_{a} = t^{o}_{a} C^{ag}_{1t}$ exhaust aperture system (219)

$t^{g}_{f} = t^{o}_{f} C^{fg}_{1t}$ combustion chamber foramen system (220)

$t^{g}_{p} = t^{o}_{p} C^{pg}_{1t}$ personnel compartment walls (221)

$t^{g}_{b} = t^{o}_{b} C^{bg}_{1t}$ rocket body (222)

$t^{g}_{e} = t^{o}_{e} C^{eg}_{2t}$ exhaust gases (223)

$t^{g}_{e} = t^{o}_{e} C^{eg}_{3t}$ exhaust gases (224)

$t^{g}_{c} = t^{o}_{c} C^{cg}_{2t}$ combustion chamber gases (225)

$t^{g}_{c} = t^{o}_{c} C^{cg}_{3t}$ combustion chamber gases (226)

while equation (217) becomes

$t_{a} = t^{g}_{a} D^{a}_{1t}$ exhaust aperture system (227)

$t_{f} = t^{g}_{f} D^{f}_{1t}$ combustion chamber foramen system (228)

$t_{p} = t^{g}_{p} D^{p}_{1t}$ personnel compartment walls (229)

$t_{b} = t^{g}_{b} D^{b}_{1t}$ rocket body (230)

$t_{e} = t^{g}_{e} D^{e}_{2t}$ exhaust gases (231)

$t_{e} = t^{g}_{e} D^{e}_{3t}$ exhaust gases (232)

$t_{c} = t^{g}_{c} D^{c}_{2t}$ combustion chamber gases (233)

$t_{c} = t^{g}_{c} D^{c}_{3t}$ combustion chamber gases (234)

where equation (218) becomes

$D^{a}_{1t} = C^{a}_{1t} / C^{ag}_{1t}$ exhaust aperture system (235)

$$D^{f}_{1t} = C^{f}_{1t} / C^{fg}_{1t}$$ combustion chamber foramen system (236)

$$D^{p}_{1t} = C^{p}_{1t} / C^{pg}_{1t}$$ personnel compartment walls (237)

$$D^{b}_{1t} = C^{b}_{1t} / C^{bg}_{1t}$$ rocket body (238)

$$D^{e}_{2t} = C^{e}_{2t} / C^{eg}_{2t}$$ exhaust gases (239)

$$D^{e}_{3t} = C^{e}_{3t} / C^{eg}_{3t}$$ exhaust gases (240)

$$D^{c}_{2t} = C^{c}_{2t} / C^{cg}_{2t}$$ combustion chamber gases (241)

$$D^{c}_{3t} = C^{c}_{3t} / C^{cg}_{3t}$$ combustion chamber gases (242)

so that for a polyphase rocket each component system has a set of temporal coordinate coordinate factors which scales the time coordinates within the system.

For matter in the presence of an ambient field it has been suggested that the particle number density is a complex number in an internal space and can be written for a high-density [hd] system and a low-density [ld] system as[11,12]

$$\bar{n} = n \exp(j\theta_n) \tag{243}$$

$$n = n^{o} \cos\theta_n \qquad [hd] \tag{244}$$

$$n = n^{o} \sec\theta_n \qquad [ld] \tag{245}$$

where n^{o} = zero ambient field value of the particle number density and is given by

$$n^{o} = N / V^{o} \tag{246}$$

where N = integer number of particles and V^{o} = volume of space in which the particles are contained under zero ambient field conditions. The measured particle number density is given by[11,12]

$$n_m = n\cos\theta_n = n^{o}\cos^2\theta_n \qquad [hd] \tag{247}$$

$$n_m = n\cos\theta_n = n^{o} \qquad [ld] \tag{248}$$

In the presence of an ambient field, such as an ambient electromagnetic field or ambient gravity, the number of particles appears as a complex number that is represented as[9-12]

$$\bar{\mathbb{N}} = \mathbb{N} \exp(j\theta_{\mathbb{N}}) \tag{249}$$

where for a large integer number [LN] of particles and a small integer number [SN] of particles[9-12]

$$\mathfrak{N} = \mathrm{N}\sec\theta_{\mathfrak{N}} \qquad [\mathrm{LN}] \qquad (250)$$

$$\mathfrak{N} = \mathrm{N}\cos\theta_{\mathfrak{N}} \qquad [\mathrm{SN}] \qquad (251)$$

and where, for a combined static elastic and viscoelastic system that does not include resonances, $\theta_{\mathfrak{N}}$ is given by[9-12]

$$\tan\theta_{\mathfrak{N}} = (2\mathfrak{E}/\mathrm{G_o})^{1/2} \qquad \mathrm{G_o} = \varsigma_o/\mathrm{t_o} \qquad (252)$$

where $\mathfrak{E}$ = energy density of a shearing ambient field such as is given in equations (33) through (35) for the component systems of a polyphase piston engine or by equations (58) through (63) for the component systems of a polyphase rocket, and where G_o, ζ_o and t_o = zero ambient field values of the shear modulus, viscosity and time factor of the material in a component system. The replacement $G_o \to G_r$, where G_r = an effective dynamical shear modulus, allows a simple treatment of oscillatory systems at a resonance frequency, for example. The measured values of the particle number are given by[9-12]

$$\mathfrak{N}_m = \mathfrak{N}\cos\theta_{\mathfrak{N}} = \mathrm{N} \qquad [\mathrm{LN}] \qquad (253)$$

$$\mathfrak{N}_m = \mathfrak{N}\cos\theta_{\mathfrak{N}} = \mathrm{N}\cos^2\theta_{\mathfrak{N}} \qquad [\mathrm{SN}] \qquad (254)$$

The particle number density can also be written generally using the complex number representations of the volume and particle number given in equations (29) and (249) as

$$\overline{\mathrm{n}} = \mathrm{n}\exp(\mathrm{j}\theta_{\mathrm{n}}) = \overline{\mathfrak{N}}/\overline{\mathrm{V}} \qquad \overline{\mathrm{V}} = \mathrm{V}\exp(\mathrm{j}\theta_{\mathrm{V}}) \qquad (255)$$

so that

$$\mathrm{n} = \mathfrak{N}/\mathrm{V} \qquad \theta_{\mathrm{n}} = \theta_{\mathfrak{N}} - \theta_{\mathrm{V}} \qquad (256)$$

Then combining equations (244), (245), (250) and (251) gives the following four values of the volume of particles in an ambient field[12]

$$\mathrm{V} = \mathrm{V}^{\circ}\cos\theta_{\mathfrak{N}}\sec\theta_{\mathrm{n}} \qquad [\mathrm{hd,SN}] \qquad (257)$$

$$\mathrm{V} = \mathrm{V}^{\circ}\sec\theta_{\mathfrak{N}}\sec\theta_{\mathrm{n}} \qquad [\mathrm{hd,LN}] \qquad (258)$$

$$\mathrm{V} = \mathrm{V}^{\circ}\cos\theta_{\mathfrak{N}}\cos\theta_{\mathrm{n}} \qquad [\mathrm{ld,SN}] \qquad (259)$$

$$\mathrm{V} = \mathrm{V}^{\circ}\sec\theta_{\mathfrak{N}}\cos\theta_{\mathrm{n}} \qquad [\mathrm{ld,LN}] \qquad (260)$$

Combining equations (30), (250), (251) and (256) gives for cylindrical polar coordinates[12]

$$\mathrm{n} = \mathrm{n}^{\circ}\mathrm{C}_{\sigma r}^{-2}\mathrm{C}_{\sigma z}^{-1}\cos\theta_{\mathfrak{N}} \qquad [\mathrm{SN}] \qquad (261)$$

$$\mathrm{n} = \mathrm{n}^{\circ}\mathrm{C}_{\sigma r}^{-2}\mathrm{C}_{\sigma z}^{-1}\sec\theta_{\mathfrak{N}} \qquad [\mathrm{LN}] \qquad (262)$$

where $\sigma = 1$ for [hd] systems and $\sigma = 2,3$ for [ld] systems. A comparison of equations (244) and (245) with (261) and (262) and using equations (10) through (12) gives for cylindrical polar coordinates in homogeneous matter in an ambient field[12]

$$\cos\theta_n = C_{1r}^{-2} C_{1z}^{-1} \cos\theta_{\mathcal{N}} \qquad [\text{hd, SN}] \qquad (263)$$

$$\cos\theta_n = C_{1r}^{-2} C_{1z}^{-1} \sec\theta_{\mathcal{N}} \qquad [\text{hd, LN}] \qquad (264)$$

$$\sec\theta_n = C_{2r}^{-2} C_{2z}^{-1} \cos\theta_{\mathcal{N}} \qquad [\text{ld, SN}] \qquad (265)$$

$$\sec\theta_n = C_{2r}^{-2} C_{2z}^{-1} \sec\theta_{\mathcal{N}} \qquad [\text{ld, LN}] \qquad (266)$$

$$\sec\theta_n = C_{3r}^{-2} C_{3z}^{-1} \cos\theta_{\mathcal{N}} \qquad [\text{ld, SN}] \qquad (267)$$

$$\sec\theta_n = C_{3r}^{-2} C_{3z}^{-1} \sec\theta_{\mathcal{N}} \qquad [\text{ld, LN}] \qquad (268)$$

Equation (264) will be used later to describe the high-density cylindrical rocket body, while equations (266) or (268) will be used to describe the low-density gases in the exhaust chamber and in the combustion chamber of a polyphase rocket. For spherical polar coordinates the equations analogous to equations (261) and (262) are[12]

$$n = n^{\circ} C_{\sigma\rho}^{-3} \cos\theta_{\mathcal{N}} \qquad [\text{SN}] \qquad (269)$$

$$n = n^{\circ} C_{\sigma\rho}^{-3} \sec\theta_{\mathcal{N}} \qquad [\text{LN}] \qquad (270)$$

where $\sigma = 1$ for [hd] systems and $\sigma = 2,3$ for [ld] systems, while the equations analogous to equations (263) through (268) are[12]

$$\cos\theta_n = C_{1\rho}^{-3} \cos\theta_{\mathcal{N}} \qquad [\text{hd, SN}] \qquad (271)$$

$$\cos\theta_n = C_{1\rho}^{-3} \sec\theta_{\mathcal{N}} \qquad [\text{hd, LN}] \qquad (272)$$

$$\sec\theta_n = C_{2\rho}^{-3} \cos\theta_{\mathcal{N}} \qquad [\text{ld, SN}] \qquad (273)$$

$$\sec\theta_n = C_{2\rho}^{-3} \sec\theta_{\mathcal{N}} \qquad [\text{ld, LN}] \qquad (274)$$

$$\sec\theta_n = C_{3\rho}^{-3} \cos\theta_{\mathcal{N}} \qquad [\text{ld, SN}] \qquad (275)$$

$$\sec\theta_n = C_{3\rho}^{-3} \sec\theta_{\mathcal{N}} \qquad [\text{ld, LN}] \qquad (276)$$

where ρ = radial distance in spherical polar coordinates. Equation (272) will be used in the treatment of the high-density spherical Earth.

The six rocket component systems in which ambient fields are located are macroscopic systems of which: (1) the walls of the exhaust aperture, (2) the walls of the combustion chamber foramen, (3) the walls of the personnel compartment and (4) the solid and liquid components of the rockets body are high-density macroscopic systems, while (5) the exhaust gases in the exhaust tube and (6) the combustion product gases in the combustion chamber are low-density macroscopic systems. Thus considering only the large N expressions in equations 262), (264), (266) and (268) combined with equations (244), (245), (258) and (260) give the required expressions for the volume and particle number density for the homogeneous macroscopic [LN] cylindrical systems of a polyphase rocket as[12]

$$V = V^{o} \sec\theta_{\mathbb{N}} \sec\theta_{n} = V^{o} C_{1r}^{2} C_{1z} \qquad [hd, LN] \qquad (277)$$

$$V = V^{o} \sec\theta_{\mathbb{N}} \cos\theta_{n} = V^{o} C_{2r}^{2} C_{1z} \qquad [ld, LN] \qquad (278)$$

$$V = V^{o} \sec\theta_{\mathbb{N}} \cos\theta_{n} = V^{o} C_{3r}^{2} C_{1z} \qquad [ld, LN] \qquad (279)$$

$$n = n^{o} C_{1r}^{-2} C_{1z}^{-1} \sec\theta_{\mathbb{N}} = n^{o} \cos\theta_{n} \qquad [hd] \qquad (280)$$

$$n = n^{o} C_{2r}^{-2} C_{2z}^{-1} \sec\theta_{\mathbb{N}} = n^{o} \sec\theta_{n} \qquad [ld] \qquad (281)$$

$$n = n^{o} C_{3r}^{-2} C_{3z}^{-1} \sec\theta_{\mathbb{N}} = n^{o} \sec\theta_{n} \qquad [ld] \qquad (282)$$

which agree with equations (244), (245), (250) and (256). The mass density of matter in an ambient field can be written as

$$\overline{\rho} = \rho \exp(j\theta_{\rho}) = \overline{\mu n} \qquad (283)$$

where the complex number molecular mass $\overline{\mu}$ is represented by

$$\overline{\mu} = \mu \exp(j\theta_{\mu}) \qquad (284)$$

so that

$$\rho = \mu n \qquad \theta_{\rho} = \theta_{\mu} + \theta_{n} \qquad (285)$$

where n is given by equations (244) and (245) or by equations (280) through (282). In this paper the high-density equations (244), (264) and (280) are applied to the exhaust aperture walls, the combustion chamber foramen walls, the personnel compartment walls and the solid and liquid components of the rocket body, while the low-density equations (245), (266), (268), (281) and (282) are applied to the exhaust gases in the exhaust tube and the combustion product gases in the combustion chamber. The different values of the internal phase angle θ_n for each rocket component, the choice of sec θ_n and cos θ_n for low and high-density systems respectively, the choice of sec $\theta_{\mathbb{N}}$ as the appropriate factor for the large numbers of atoms and molecules in the macroscopic components of a polyphase rocket, the sign of the internal phase angle of the exhaust aperture area $\theta_{A}^{a} \leq 0$ as in equation (27) and the choice of the spatial and temporal coordinate factors $C_{\sigma\alpha}$ and $C_{\sigma t}$ for $\sigma = 1,2,3$, that are associated with the high and low-density component systems of a polyphase rocket , and several other variables are all factors that lead to the possibility of enhancement of thrust and liftoff capability in a polyphase rocket.

This paper introduces the design concepts for a polyphase rocket that has the capability of thrust and liftoff enhancement using ambient fields such as an ambient electromagnetic field. By way of introduction Section 2 of this paper develops the theory of a polyphase piston engine, Section 3 reviews the standard ideas of the geometry, thrust calculation and equation of motion of a conventional rocket, Section 4 introduces the geometrical concepts of a polyphase rocket, Section 5 considers the calculation of particle number density, mass density and mass for the components of a polyphase rocket, Section 6 presents a calculation of the thrust of a polyphase rocket, Section 7 calculates the force of gravity for a polyphase rocket, Section 8 gives presents a theory of the enhancement of liftoff capability for a polyphase rocket, Section 9 gives several solutions to the liftoff equations for a polyphase rocket and finally Section 10 presents a calculation of the magnetic fields required for liftoff enhancement of a polyphase rocket.

2. POLYPHASE PISTON ENGINES. As an introduction to the techniques that will be used for the treatment of polyphase rockets this section develops expressions for the acceleration of a piston of an ambient field enhanced piston engine. First the equations of motion of a piston under zero ambient field conditions is developed and then the case of a piston engine with a piston head, piston body and a working fluid that contain ambient fields is considered and the complex number equation of motion of the piston is developed. The analytical engineering design concepts of a polyphase piston engine, that uses the concepts of space coordinate factors and ambient fields, are developed which may have some use for the development of automobile gasoline engines that have high efficiency and low pollution performance characteristics. This type of engine is a special case of the class of ultrafast coherent heat engines that have already appeared in the literature.[9]

A. Motion Equation for a Piston Under Zero Ambient Field Conditions.

The ambient field free equation of motion of a piston that is driven by gas pressure is written as

$$m_p^o a_p^o = F_{wh}^o = (P_w^o - P_{at}^o)A_h^o \approx P_w^o A_h^o \tag{286}$$

where "o" refers to the case of zero ambient fields, m_p^o = total mass of the piston, a_p^o = acceleration of the piston, F_{wh}^o = force of gas pressure on the piston head, P_w^o = pressure of the working gas, P_{at}^o = atmospheric pressure which for simplicity will be neglected in this paper, and finally where A_h^o = area of the piston head on which the working gas acts. Gravity is neglected in this analysis. The piston mass can be written as

$$m_p^o = m_h^o + m_b^o = m_b^o Q_p^o \tag{287}$$

where

$$Q_p^o = m_p^o / m_b^o = 1 + m_h^o / m_b^o \tag{288}$$

where m_h^o = mass of piston head and m_b^o = mass of piston body. The component masses in equation (287) can be written as

$$m_h^o = \mu_h^o N_h = \rho_h^o V_h^o = \rho_h^o A_h^o L_h^o \tag{289}$$

$$m_b^o = \mu_b^o N_b = \rho_b^o V_b^o = \rho_b^o A_b^o L_b^o \tag{290}$$

where ρ_h^o and ρ_b^o = mass densities of the piston head and piston body which are given by

$$\rho_h^o = \mu_h^o n_h^o \qquad \rho_b^o = \mu_b^o n_b^o \tag{291}$$

where μ_h^o and μ_b^o, N_h and N_b, n_h^o and n_b^o, V_h^o and V_b^o, A_h^o and A_b^o and L_h^o and L_b^o = molecular masses, integer number of molecules, molecular number densities, volumes, cross section areas and lengths of the piston head and piston body respectively. The cross section areas and the volumes of the piston head and body are given by

$$A_h^o = \pi(r_h^o)^2 \quad V_h^o = A_h^o L_h^o = \pi(r_h^o)^2 L_h^o \tag{292}$$

$$A_b^o = \pi(r_b^o)^2 \quad V_b^o = A_b^o L_b^o = \pi(r_b^o)^2 L_b^o \tag{293}$$

where r_h^o and r_b^o = radii of the piston head and body which satisfy the inequality $r_h^o \geq r_b^o$. The pressure of the ideal working gas in the cylinder that acts on the piston head before it is eventually exhausted to the atmosphere is given by

$$P_w^o = n_w^o k_B T_w^o = (N_w / V_w^o) k_B T_w^o \tag{294}$$

where k_B = Boltzmann constant and where for zero ambient fields

$$n_w^o = N_w / V_w^o \tag{295}$$

is the molecular number density of the gas, N_w = integer number of working gas molecules in the cylinder and where V_w^o and T_w^o = volume of the working gas in the cylinder and the absolute temperature of the working gas for zero ambient fields. The volume of the working gas corresponding to N_w gas molecules is given by

$$V_w^o = A_w^o z_w^o = \pi(r_w^o)^2 z_w^o \tag{296}$$

where r_w^o, z_w^o and A_w^o = radius, length and area of a cylinder of working gas that contains N_w molecules. Combining equations (286) through (294) gives the acceleration of the piston as

$$a_p^o = F_{wh}^o / m_p^o = P_w^o A_h^o / m_p^o = n_w^o k_B T_w^o A_h^o (m_h^o + m_b^o)^{-1} \tag{297}$$

$$= n_w^o k_B T_w^o A_h^o (\mu_h^o N_h + \mu_b^o N_b)^{-1}$$

after neglecting friction and the atmospheric pressure, and where

$$F_{wh}^o = P_w^o A_h^o \tag{298}$$

where F_{wh}^o = zero ambient field value of the force on the piston head as shown in equation (286).

B. Motion Equation of a Compound Piston in a Polyphase Piston Engine.

The complex number equation of motion of a piston in a polyphase piston engine, whose components contain the ambient fields shown in equations (33) through (35), is given by

$$\overline{m}_p \overline{a}_p = \overline{F}_{wh} = (\overline{P}_w - \overline{P}_{at})\overline{A}_h \approx \overline{P}_w \overline{A}_h \tag{299}$$

where $\overline{m}_p$ = complex number total mass of piston, $\overline{a}_p$ = complex number acceleration of the piston, $\overline{F}_{wh}$ = force of the gas pressure on the piston head, $\overline{P}_w$ = complex number pressure of the working gas, $\overline{P}_{at}$ = complex number atmospheric pressure which for simplicity is neglected in this paper, and finally where $\overline{A}_h$ = complex number area of the piston head which is adjacent to the working gas. These complex number quantities in equation (299) that are used in this paper are represented as follows

$$\overline{m}_p = m_p \exp(j\theta^p_m) \qquad \overline{a}_p = a_p \exp(j\theta^p_a) \qquad \overline{F}_{wh} = F_{wh} \exp(j\theta^h_{Fw}) \tag{300}$$

$$\overline{P}_w = P_w \exp(j\theta^w_P) \qquad \overline{P}_{at} = P_{at} \exp(j\theta^{at}_P) \qquad \overline{A}_h = A_h \exp(j\theta^h_A) \tag{301}$$

Then the approximate relation in equation (299) can be written as

$$m_p a_p = P_w A_h \qquad \theta^p_m + \theta^p_a = \theta^w_P + \theta^h_A \tag{302}$$

which will be used subsequently to determine the acceleration of the piston.

C. Complex Number Radii and Lengths of the Piston Head and Piston Body.

The complex number radius and length of the high-density piston head in the presence of an ambient field given by equation (33) are represented by

$$\overline{r}_h = r_h \exp(j\theta^h_r) \qquad \overline{L}_h = L_h \exp(j\theta^h_L) \tag{303}$$

where according to equation (19)

$$r_h = r^o_h C^h_{1r} \qquad L_h = L^o_h C^h_{1z} \tag{304}$$

where the coordinate factors C^h_{1r} and C^h_{1z} are given by equation (50) and where r^o_h and L^o_h = zero ambient field values of the radius and length of the piston head as introduced in equations (289) and (292). The measured values of the radius and length of the piston head are given by

$$r_{hm} = r_h \cos\theta^h_r = r^o_h C^h_{1r} \cos\theta^h_r \qquad L_{hm} = L_h \cos\theta^h_L = L^o_h C^h_{1z} \cos\theta^h_L \tag{305}$$

Similarly, the complex radius and length of the high-density piston body in the presence of an ambient field given by equation (34) is represented by

$$\overline{r}_b = r_b \exp(j\theta^b_r) \qquad \overline{L}_b = L_b \exp(j\theta^b_L) \tag{306}$$

where equation (19) gives

$$r_b = r_b^o C_{1r}^b \qquad L_b = L_b^o C_{1z}^b \tag{307}$$

where the coordinate factors C_{1r}^b and C_{1z}^b are given by equation (51) and where r_b^o and L_b^o = zero ambient field values of the radius and length of the piston body as they appear in equations (290) and (293). The measured values of the radius and length of the piston body are

$$r_{bm} = r_b \cos\theta_r^b = r_b^o C_{1r}^b \cos\theta_r^b \qquad L_{bm} = L_b \cos\theta_L^b = L_b^o C_{1z}^b \cos\theta_L^b \tag{308}$$

which are real numbers.

D. Complex Number Areas Of the Piston Head and Piston Body.

The complex number area of the high-density piston head that is located in an ambient field given by equation (33) is represented as in equation (25) by

$$\overline{A}_h = A_h \exp(j\theta_A^h) = \pi\bar{r}_h^2 \tag{309}$$

where the complex number radius $\bar{r}_h$ of the piston head is given by equation (303) and (304) so that equation (309) can be written as in equation (26) and (27) by

$$A_h = \pi r_h^2 = A_h^o (C_{1r}^h)^2 \qquad \theta_A^h = 2\theta_r^h \tag{310}$$

where C_{1r}^h is given by equation (50) and where A_h^o = area of the piston head under zero ambient field conditions and is given by equation (292). The measured cross section area of the piston head is given by

$$A_{hm} = A_h \cos\theta_A^h = A_h^o (C_{1r}^h)^2 \cos(2\theta_r^h) \tag{311}$$

The complex number area of the high-density piston body that is immersed in an ambient field given by equation (34) is given by

$$\overline{A}_b = A_b \exp(j\theta_A^b) = \pi\bar{r}_b^2 \tag{312}$$

where the complex number radius $\bar{r}_b$ of the piston body is given by equations (306) and (307) so that equation (312) can be written as

$$A_b = \pi r_b^2 = A_b^o (C_{1r}^b)^2 \qquad \theta_A^b = 2\theta_r^b \tag{313}$$

where C_{1r}^b is given by equation (51) and where A_b^o = area of piston body under zero ambient field conditions as given by equation (293). The measured cross section area of the piston body is

$$A_{bm} = A_b \cos\theta_A^b = A_b^o (C_{1r}^b)^2 \cos(2\theta_r^b) \tag{314}$$

which is a real number.

E. Complex Number Volumes of the Piston Head and Piston Body.

The complex number volume of the high-density cylindrical piston head that is immersed in an ambient field given by equation (33) is given by

$$\overline{V}_h = V_h \exp(j\theta_V^h) = \overline{A}_h \overline{L}_h \tag{315}$$

where equations (303), (304), (309) and (310) combined with equation (315) give

$$V_h = V_h^o (C_{1r}^h)^2 C_{1z}^h \qquad \theta_V^h = \theta_A^h + \theta_L^h = 2\theta_r^h + \theta_L^h \tag{316}$$

where the space coordinate factors C_{1r}^h and C_{1z}^h are given by equation (50) and where V_h^o = zero ambient field value of the piston head volume given by equation (292). The measured value of the volume of the piston head is given by

$$V_{hm} = V_h \cos\theta_V^h = V_h^o (C_{1r}^h)^2 C_{1z}^h \cos(2\theta_r^h + \theta_L^h) \tag{317}$$

In the same way, the complex number volume of the high-density cylindrical piston body that is located in an ambient field of the form given in equation (34) is written as

$$\overline{V}_b = V_b \exp(j\theta_V^b) = \overline{A}_b \overline{L}_b \tag{318}$$

where equations (306), (307), (312) and (313) combined with equation (318) give

$$V_b = A_b L_b = V_b^o (C_{1r}^b)^2 C_{1z}^b \qquad \theta_V^b = \theta_A^b + \theta_L^b = 2\theta_r^b + \theta_L^b \tag{319}$$

where C_{1r}^b and C_{1z}^b are given by equation (51) and where V_b^o = zero ambient field value of the piston body volume as given by equation (293). The measured value of the volume of the piston body is written as

$$V_{bm} = V_b \cos\theta_V^b = V_b^o (C_{1r}^b)^2 C_{1z}^b \cos(2\theta_r^b + \theta_L^b) \tag{320}$$

which is a real number.

F. Complex Molecular Number Density of the Piston Head and Piston Body.

The complex number $\overline{\mathbb{N}}_h$ of molecules in the high-density piston head that contains an ambient field of the form in equation (33) and the complex number of molecules $\overline{\mathbb{N}}_b$ in the high-density piston body that contains an ambient field of the form given in equation (34) are obtained from equation (249) and the large N equation (250) as

$$\overline{\mathbb{N}}_h = \mathbb{N}_h \exp(j\theta_{\mathbb{N}}^h) \qquad \mathbb{N}_h = N_h \sec\theta_{\mathbb{N}}^h \tag{321}$$

$$\overline{\mathbb{N}}_b = \mathbb{N}_b \exp(j\theta_{\mathbb{N}}^b) \qquad \mathbb{N}_b = N_b \sec\theta_{\mathbb{N}}^b \tag{322}$$

where N_h and N_b = large integer number of molecules in the piston head and piston body respectively, and where the internal phase angles $\theta_{\mathbb{N}}^h$ and $\theta_{\mathbb{N}}^b$ are obtained from equation (252) as

$$\tan\theta^h_{\mathcal{N}} = (2\mathfrak{E}^h / G^o_h)^{1/2} \qquad \tan\theta^b_{\mathcal{N}} = (2\mathfrak{E}^b / G^o_b)^{1/2} \tag{323}$$

where the shear strain energy densities $\mathfrak{E}^h$ and $\mathfrak{E}^b$ are given by equations (33) and (34) and where G^o_h and G^o_b = zero field values of the shear moduli of the materials in the head and body of the piston respectively. The complex number molecular number density $\overline{n}_h$ for the high-density piston head is given by equations (264), (315), (316) and (321) as

$$\overline{n}_h = n_h \exp(j\theta^h_n) = \overline{\mathcal{N}_h} / \overline{V}_h = \mathcal{N}_h / V_h \exp[j(\theta^h_{\mathcal{N}} - \theta^h_V)] \tag{324}$$

where

$$n_h = \mathcal{N}_h / V_h = N_h / V_h \sec\theta^h_{\mathcal{N}} = n^o_h (C^h_{1r})^{-2} (C^h_{1z})^{-1} \sec\theta^h_{\mathcal{N}} = n^o_h \cos\theta^h_n \tag{325}$$

$$\theta^h_n = \theta^h_{\mathcal{N}} - \theta^h_V = \theta^h_{\mathcal{N}} - 2\theta^h_r - \theta^h_L \tag{326}$$

where C^h_{1r} and C^h_{1z} are given by equation (50) and where the [hd,LN] equation (264) was used to obtain equation (325) and where the ambient field free value of the molecular number density n^o_h of the piston head is given by

$$n^o_h = N_h / V^o_h \tag{327}$$

where the zero ambient field volume V^o_h of the high-density piston head is given by equation (292). The measured value of the molecular number density of the high-density piston head is obtained from equations (325) and (326) as

$$n_{hm} = n_h \cos\theta^h_n = n^o_h \cos^2\theta^h_n \tag{328}$$

Similarly the complex molecular number density $\overline{n}_b$ of the high-density piston body is obtained from equations (264), (318), (319) and (322) as

$$\overline{n}_b = n_b \exp(j\theta^b_n) = \overline{\mathcal{N}_b} / \overline{V}_b = \mathcal{N}_b / V_b \exp[j(\theta^b_{\mathcal{N}} - \theta^b_V)] \tag{329}$$

where

$$n_b = \mathcal{N}_b / V_b = N_b / V_b \sec\theta^b_{\mathcal{N}} = n^o_b (C^b_{1r})^{-2} (C^b_{1z})^{-1} \sec\theta^b_{\mathcal{N}} = n^o_b \cos\theta^b_n \tag{330}$$

$$\theta^b_n = \theta^b_{\mathcal{N}} - \theta^b_V = \theta^b_{\mathcal{N}} - 2\theta^b_r - \theta^b_L \tag{331}$$

where C^b_{1r} and C^b_{1z} are given by equation (51) and where the ambient field free value of the molecular number density n^o_b of the piston body is given by

$$n^o_h = N_h / V^o_h \tag{332}$$

where V^o_b = ambient field free value of the volume of the high-density piston body as given by equation (293). The measured value of the molecular number density of the high-density piston body is obtained from equations (330) and (331) as

$$n_{bm} = n_b \cos\theta^b_n = n^o_b \cos^2\theta^b_n \tag{333}$$

which is always a positive real number.

G. Complex Number Mass and Mass Density of the Piston Head and Piston Body.

The complex number molecular mass $\overline{\mu}_h$ of the high-density material in the piston head that contains an ambient field described by equation (33) and the complex number molecular mass $\overline{\mu}_b$ of the high-density piston body that contains an ambient field given by equation (34) are written as

$$\overline{\mu}_h = \mu_h \exp(j\theta_\mu^h) \tag{334}$$

$$\overline{\mu}_b = \mu_b \exp(j\theta_\mu^b) \tag{335}$$

The magnitudes and internal phase angles that appear in equations (334) and (335) can be represented in a general form as

$$\mu_h = \mu_h^o f_{\mu h}(\mathfrak{E}^h) \qquad \theta_\mu^h = \theta_\mu^h(\mathfrak{E}^h) \tag{336}$$

$$\mu_b = \mu_b^o f_{\mu b}(\mathfrak{E}^b) \qquad \theta_\mu^b = \theta_\mu^b(\mathfrak{E}^b) \tag{337}$$

where under zero ambient field conditions

$$f_{\mu h}(0) = 1 \qquad \theta_\mu^h(0) = 0 \tag{338}$$

$$f_{\mu b}(0) = 1 \qquad \theta_\mu^b(0) = 0 \tag{339}$$

where μ_h^o and μ_b^o = zero ambient field values of the molecular masses of the materials in the piston head and piston body respectively as used in equations (289) and (290) and where the parameters $f_{\mu h}$ and $f_{\mu b}$ and the internal phase angles θ_μ^h and θ_μ^b are functions of $\mathfrak{E}^h$ and $\mathfrak{E}^b$ respectively and the atomic structure of the high-density materials. The dependence of the functions $f_{\mu h}$ and $f_{\mu b}$ and θ_μ^h and θ_μ^b on the atomic structure of the materials that constitute the piston head and body respectively can be calculated from the complex number of atoms in a molecule which according to the recipe given in equations (249) and (251) is written as

$$\overline{\mathfrak{N}}_h^{atom} = \mathfrak{N}_h^{atom} \exp(j\theta_{\mathfrak{N}atom}^h) \qquad \mathfrak{N}_h^{atom} = N_h^{atom} \cos\theta_{\mathfrak{N}atom}^h \tag{340}$$

$$\overline{\mathfrak{N}}_b^{atom} = \mathfrak{N}_b^{atom} \exp(j\theta_{\mathfrak{N}atom}^b) \qquad \mathfrak{N}_b^{atom} = N_b^{atom} \cos\theta_{\mathfrak{N}atom}^b \tag{341}$$

where N_h^{atom} and N_b^{atom} = small integer number of atoms in each molecule of the substance constituting the head and body of the polyphase piston. The complex number molecular masses in equation (334) and (335) can also be written as

$$\overline{\mu}_h = m_{hatom}^o \overline{\mathfrak{N}}_h^{atom} - c^{-2}\overline{B}_h^{mol} \tag{342}$$

$$\overline{\mu}_b = m_{batom}^o \overline{\mathfrak{N}}_b^{atom} - c^{-2}\overline{B}_b^{mol} \tag{343}$$

Then after neglecting the complex number molecular binding energies $\overline{B}_h^{mol}$ and $\overline{B}_b^{mol}$ in equations (342) and (343) and after using equations (340) and (341), the magnitudes and the internal phase angles of the molecular masses can be written as

$$\mu_h = m_{hatom}^o N_h^{atom} \cos\theta_{Natom}^h = \mu_h^o \cos\theta_{Natom}^h \qquad \theta_\mu^h = \theta_{Natom}^h \tag{344}$$

$$\mu_b = m_{batom}^o N_b^{atom} \cos\theta_{Natom}^b = \mu_b^o \cos\theta_{Natom}^b \qquad \theta_\mu^b = \theta_{Natom}^b \tag{345}$$

Then a comparison of equations (336) and (337) with (344) and (345) shows that approximately

$$f_{\mu h} = \cos\theta_{Natom}^h \qquad f_{\mu b} = \cos\theta_{Natom}^b \tag{346}$$

and therefore the following inequalities are satisfied

$$f_{\mu h} \le 1 \qquad f_{\mu b} \le 1 \tag{347}$$

which is an interesting conclusion that is used for the calculation of the acceleration of the piston of a polyphase piston engine.

The complex number mass density $\overline{\rho}_h$ of the high-density piston head and the complex number mass density $\overline{\rho}_b$ of the high-density piston body are given by

$$\overline{\rho}_h = \rho_h \exp(j\theta_\rho^h) = \overline{\mu}_h \overline{n}_h \tag{348}$$

$$\overline{\rho}_b = \rho_b \exp(j\theta_\rho^b) = \overline{\mu}_b \overline{n}_b \tag{349}$$

Combining equations (324) through (332) with equations (334) through (349) gives, after using equation (264), the magnitudes and the internal phase angles of the mass densities of the piston head and body as

$$\rho_h = \mu_h n_h = \rho_h^o f_{\mu h} (C_{1r}^h)^{-2} (C_{1z}^h)^{-1} \sec\theta_N^h = \rho_h^o f_{\mu h} \cos\theta_n^h = \rho_h^o \cos\theta_{Natom}^h \cos\theta_n^h \tag{350}$$

$$\rho_b = \mu_b n_b = \rho_b^o f_{\mu b} (C_{1r}^b)^{-2} (C_{1z}^b)^{-1} \sec\theta_N^b = \rho_b^o f_{\mu b} \cos\theta_n^b = \rho_b^o \cos\theta_{Natom}^b \cos\theta_n^b \tag{351}$$

$$\theta_\rho^h = \theta_\mu^h + \theta_n^h \qquad \theta_\rho^b = \theta_\mu^b + \theta_n^b \tag{352}$$

where ρ_h^o and ρ_b^o = zero ambient field values of the mass densities of the piston head and body as given by equation (291). The measured values of the mass densities of the piston head and body are obtained from equations (350) through (352) as

$$\rho_{hm} = \rho_h \cos\theta_\rho^h = \rho_h^o f_{\mu h} \cos_n^h \cos(\theta_\mu^h + \theta_n^h) \tag{353}$$

$$\rho_{bm} = \rho_b \cos\theta_\rho^b = \rho_b^o f_{\mu b} \cos_n^b \cos(\theta_\mu^b + \theta_n^b) \tag{354}$$

which are real numbers.

The complex number masses of the piston head and body in the presence of an ambient field can be written as

$$\overline{m}_h = m_h \exp(j\theta_m^h) = \overline{\mu}_h \overline{\mathcal{N}}_h = \overline{\rho}_h \overline{V}_h = \overline{\rho}_h \overline{A}_h \overline{L}_h \qquad (355)$$

$$\overline{m}_b = m_b \exp(j\theta_m^b) = \overline{\mu}_b \overline{\mathcal{N}}_b = \overline{\rho}_b \overline{V}_b = \overline{\rho}_b \overline{A}_b \overline{L}_b \qquad (356)$$

Combining equations (355) and (356) with either equations (321), (322), (334) and (335) or with equations (315), (316) and (348) through (352) gives

$$m_h = m_h^o f_{\mu h} \sec\theta_{\mathcal{N}}^h \qquad \theta_m^h = \theta_\mu^h + \theta_{\mathcal{N}}^h \qquad (357)$$

$$m_b = m_b^o f_{\mu b} \sec\theta_{\mathcal{N}}^b \qquad \theta_m^b = \theta_\mu^b + \theta_{\mathcal{N}}^b \qquad (358)$$

The measured masses of the piston head and piston body are given by

$$m_{hm} = m_h \cos\theta_m^h = m_h^o f_{\mu h} \sec\theta_{\mathcal{N}}^h \cos(\theta_\mu^h + \theta_{\mathcal{N}}^h) \qquad (359)$$

$$m_{bm} = m_b \cos\theta_m^b = m_b^o f_{\mu b} \sec\theta_{\mathcal{N}}^b \cos(\theta_\mu^b + \theta_{\mathcal{N}}^b) \qquad (360)$$

The complex number total mass of the piston is given by

$$\overline{m}_p = m_p \exp(j\theta_m^p) = \overline{m}_h + \overline{m}_b \qquad (361)$$

from which it follows that

$$m_p \cos\theta_m^p = m_b I_p \qquad (362)$$

$$m_p \sin\theta_m^p = m_b J_p \qquad (363)$$

where

$$I_p = \cos\theta_m^b + m_h / m_b \cos\theta_m^h \qquad (364)$$

$$J_p = \sin\theta_m^b + m_h / m_b \sin\theta_m^h \qquad (365)$$

Equations (362) through (365) give

$$\tan\theta_m^p = J_p / I_p \qquad (366)$$

$$m_p = m_b (I_p^2 + J_p^2)^{1/2} = m_b Q_p \qquad (367)$$

where

$$Q_p = (I_p^2 + J_p^2)^{1/2} = [1 + 2m_h / m_b \cos(\theta_m^h - \theta_m^b) + (m_h / m_b)^2]^{1/2} \qquad (368)$$

where m_h, m_b, θ_m^h and θ_m^b are given by equations (357) and (358). If $m_b >> m_h$ then equations (364) through (368) show that approximately $I_p \approx \cos\theta_m^b$ and $J_p = \sin\theta_m^b$ and

$$\theta_m^p \approx \theta_m^b \qquad m_p \approx m_b \qquad Q_p \approx 1 \tag{369}$$

The measured total mass of the high-density piston is obtained from equations (366) through (369) as

$$m_{pm} = m_p \cos\theta_m^p \tag{370}$$

where m_p and θ_m^p are given by equations (366) and (367).

H. Complex Number Radius, Length, Area and Volume of a Cylinder of the Working Gas in a Piston Engine.

The complex volume associated with a specified large integer number N_w of low-density working gas molecules is obtained from a cylinder that encloses the gas and which has a complex number radius and length given by

$$\bar{r}_w = r_w \exp(j\theta_r^w) \qquad \bar{z}_w = z_w \exp(j\theta_z^w) \tag{371}$$

where equation (9) gives

$$r_w = r_w^o C_{\beta r}^w \qquad z_w = z_w^o C_{\beta z}^w \tag{372}$$

where $\beta = 2{,}3$ and where the spatial coordinate factors $C_{\beta r}^w$ and $C_{\beta z}^w$ are given by the low-density equations (52) and (53) and where r_w^o and z_w^o = zero ambient field values of the radius and length of the cylinder of gas. The measured values of the radius and length of the gas cylinder are given by

$$r_{wm} = r_w \cos\theta_r^w = r_w^o C_{\beta r}^w \cos\theta_r^w \tag{373}$$

$$z_{wm} = z_w \cos\theta_z^w = z_w^o C_{\beta z}^w \cos\theta_z^w \tag{374}$$

The complex number area of the base of the cylinder is given by

$$\bar{A}_w = A_w \exp(j\theta_A^w) = \pi \bar{r}_w^2 \tag{375}$$

so that

$$A_w = \pi r_w^2 \qquad \theta_A^w = 2\theta_r^w \tag{376}$$

Then it follows that

$$A_w = A_w^o (C_{\beta r}^w)^2 \tag{377}$$

where A_w^o = area of the base of the cylinder of gas under zero ambient field conditions and is given by equation (296). The measured value of the area of the base of the cylinder is given by

$$A_{wm} = A_w \cos\theta_A^w = A_w^o (C_{\beta r}^w)^2 \cos(2\theta_r^w) \tag{378}$$

For the working gas of a piston engine the complex number volume of a cylinder of gas is given by

$$\overline{V}_w = V_w \exp(j\theta_V^w) = \overline{A}_w \overline{z}_w \tag{379}$$

with

$$V_w = A_w z_w \qquad \theta_V^w = \theta_A^w + \theta_z^w = 2\theta_r^w + \theta_z^w \tag{380}$$

Then it obviously follows that

$$V_w = V_w^o (C_{\beta r}^w)^2 C_{\beta z}^w \tag{381}$$

where V_w^o = volume of the working gas cylinder under zero ambient field conditions as given by equation (296). Finally, the measured volume of the cylinder of gas is

$$V_{wm} = V_w \cos\theta_V^w = V_w^o (C_{\beta r}^w)^2 C_{\beta z}^w \cos(2\theta_r^w + \theta_z^w) \tag{382}$$

where $\beta = 2,3$ and where in these calculations r_w, z_w, A_w and V_w correspond to an arbitrary volume of a low-density gas containing a large integer number N_w of working gas molecules in the presence of an ambient field as described by equation (35).

I. Complex Number Molecular Number Density and Pressure of the Working Fluid, and the Calculation of the Force Acting on the Piston.

The complex number molecular number density of the low-density working gas of a polyphase piston engine is written as

$$\overline{n}_w = n_w \exp(j\theta_n^w) = \overline{\mathfrak{N}}_w / \overline{V}_w = \mathfrak{N}_w / V_w \exp[j(\theta_{\mathfrak{N}}^w - \theta_V^w)] \tag{383}$$

where the complex number volume $\overline{V}_w$ of working fluid in an ambient field described by equation (35) is given by equations (379) through (382), and where the complex number of gas molecules $\overline{\mathfrak{N}}_w$ associated with the volume $\overline{V}_w$ is obtained from equations (249) and (250) and is written as

$$\overline{\mathfrak{N}}_w = \mathfrak{N}_w \exp(j\theta_{\mathfrak{N}}^w) \qquad \mathfrak{N}_w = N_w \sec\theta_{\mathfrak{N}}^w \tag{384}$$

where N_w = large integer number of molecules contained in V_w and hence in V_w^o and where the internal phase angle $\theta_{\mathfrak{N}}^w$ is given by equation (252) with ς_w^o = viscosity as

$$\tan\theta_{\mathfrak{N}}^w = (2\mathfrak{E}^w / G_w^o)^{1/2} \qquad G_w^o = \varsigma_w^o / t_w^o \qquad t_w^o = \text{time factor} \tag{385}$$

where the shear strain energy density $\mathfrak{E}^w$ is given by equation (35) and where G_w^o = shear modulus of the working fluid. Combining equations (383) and (384) gives

$$n_w = \mathfrak{N}_w / V_w = N_w / V_w \sec\theta_{\mathfrak{N}}^w \qquad \theta_n^w = \theta_{\mathfrak{N}}^w - \theta_V^w = \theta_{\mathfrak{N}}^w - 2\theta_r^w - \theta_z^w \tag{386}$$

The magnitude n_w in equation (386) can be rewritten using equations (266), (268) and (381) as

$$n_w = n_w^o (C_{\beta r}^w)^{-2} (C_{\beta z}^w)^{-1} \sec\theta_{\aleph}^w = n_w^o \sec\theta_n^w \qquad (387)$$

as in equations (245) and (262), where n_w^o = molecular number density of the working gas under zero ambient field conditions and as given in equation (295), and where $C_{\beta r}^w$ and $C_{\beta z}^w$ for $\beta = 2,3$ are given by equations (52) and (53). The measured value of the low-density molecular number density of the working gas or vapor is obtained from equation (387) as

$$n_{wm} = n_w \cos\theta_n^w = n_w^o \qquad (388)$$

so that the measured value of the molecular number density of the working fluid is just equal to the molecular number density under zero ambient field conditions.

The complex number pressure of the working fluid, which is assumed to be an ideal gas in the presence of an ambient field, is given by

$$\overline{P}_w = P_w \exp(j\theta_P^w) = \overline{n}_w k_B \overline{T}_w \qquad P_w = n_w k_B T_w \qquad \theta_P^w = \theta_n^w + \theta_T^w \qquad (389)$$

where the complex number temperature $\overline{T}_w$ of the working gas is given by

$$1/2\overline{\mu}_w \overline{v}_{\mu w}^2 = 3/2k_B\overline{T}_w \qquad 1/2\mu_w v_{\mu w}^2 = 3/2k_B T_w \qquad \theta_\mu^w + 2\theta_{v\mu}^w = \theta_T^w \qquad (390)$$

where $\overline{T}_w = T_w \exp(j\theta_T^w)$, $\overline{\mu}_w = \mu_w \exp(j\theta_\mu^w)$ and $\overline{v}_{\mu w} = v_{\mu w} \exp(j\theta_{v\mu}^w)$ and where k_B = Boltzmann constant, so that with $\mu_w = \mu_w^o f_{\mu w}$ and $v_{\mu w} = v_{\mu w}^o (C_{\beta\rho}^w / C_{\beta t}^w)$ it follows that

$$T_w = f_{Tw} T_w^o \qquad f_{Tw} = f_{\mu w} (C_{\beta\rho}^w)^2 (C_{\beta t}^w)^{-2} \qquad f_{\mu w} = \cos\theta_{\aleph atom}^w \qquad (391)$$

and where the pressure magnitude is given by equations (386), (387), (389) and (391) as

$$P_w = N_w / V_w \, k_B T_w \sec\theta_{\aleph}^w = P_w^o f_{Tw} (C_{\beta r}^w)^{-2} (C_{\beta z}^w)^{-1} \sec\theta_{\aleph}^w = P_w^o f_{Tw} \sec\theta_n^w \qquad (392)$$

where P_w^o = ideal gas pressure in the piston cylinder under zero ambient field conditions and as given by equation (294). The measured value of the low-density working gas pressure is obtained from equations (389) and (392) as

$$P_{wm} = P_w \cos\theta_P^w = P_w^o f_{Tw} \sec\theta_n^w \cos(\theta_n^w + \theta_T^w) \qquad (393)$$

and it is clear that the conclusions in equations (392) and (393) are true only for an ideal gas which is only an approximation to the properties of an actual working gas and is used in this paper only for heuristic purposes to demonstrate the effects of using active space and time coordinates on the calculation of the force acting on a polyphase piston and its resulting acceleration.

The complex number force exerted by the working fluid on the surface of the piston head is given by

$$\overline{F}_{wh} = F_{wh} \exp(j\theta_F^{wh}) = \overline{P}_w \overline{A}_h \qquad (394)$$

whose magnitude and internal phase angle are obtained from equation (394) and which after using equations (266), (268), (310) and (389) through (392) are written as

$$F_{wh} = P_w A_h = F^o_{wh} f_{Tw} (C^h_{1r})^2 (C^w_{\beta r})^{-2} (C^w_{\beta z})^{-1} \sec\theta^w_{\aleph} = F^o_{wh} f_{Tw} (C^h_{1r})^2 \sec\theta^w_n \tag{395}$$

$$\theta^{wh}_F = \theta^w_P + \theta^h_A = \theta^w_n + \theta^w_T + 2\theta^h_r \tag{396}$$

where F^o_{wh} = force of the working fluid on the piston head under zero ambient field conditions and as appears in equation (298) and where $\beta = 2,3$. The measured value of the force exerted on the piston head is obtained from equations (395) and (396) as

$$F_{whm} = F_{wh} \cos\theta^{wh}_F = F^o_{wh} f_{Tw} (C^h_{1r})^2 \sec\theta^w_n \cos(\theta^w_n + \theta^w_T + 2\theta^h_r) \tag{397}$$

where C^h_{1r} is given by equation (50).Because of the conditions $C^h_{1r} \geq 1$ and $\theta^h_r \leq 0$ it follows from equations (395) and (397) that $F_{wh} \geq F^o_{wh}$ and $F_{whm} \geq F^o_{wh}$.

J. Acceleration of a Piston in a Polyphase Piston Engine.

The complex number dynamical equation of motion of a compound piston (consisting of a head and body) moving in a polyphase piston engine with ambient fields described by equations (33) through (35) is given by equation (299) which can be rewritten as the following magnitude and phase angle equations

$$a_p = m_p^{-1} P_w A_h \qquad \theta_{ap} = \theta^w_P + \theta^h_A - \theta^p_m \tag{398}$$

Combining equations (264), (266), (268), (297), (310), (357), (358), (367), (368), (387), (392) and (398) allows the magnitude of the acceleration to be written as

$$\begin{aligned} a_p &= a^o_p \Phi_{p1} f_{Tw} f^{-1}_{\mu b} (C^w_{\beta r})^{-2} (C^w_{\beta z})^{-1} (C^h_{1r})^2 \cos\theta^b_{\aleph} \sec\theta^w_{\aleph} \\ &= a^o_p \Phi_{p1} \Phi_{p2} \Phi_{p3} \Phi_{p4} [(C^b_{1r})^2 C^b_{1z} \cos\theta^b_{\aleph}][(C^w_{\beta r})^{-2} (C^w_{\beta z})^{-1} \sec\theta^w_{\aleph}] \end{aligned} \tag{399}$$

which can be rewritten as

$$a_p = a^o_p S_p \sec\theta^b_n \sec\theta^w_n \qquad a_p \geq a^o_p \tag{400}$$

where a^o_p = acceleration of the piston under zero ambient field conditions and as shown in equation (297), and where

$$S_p = \Phi_{p1} \Phi_{p2} \Phi_{p3} \Phi_{p4} \tag{401}$$

where

$$\Phi_{p1} = Q^o_p / Q_p \approx 1 \qquad \Phi_{p2} = f_{\mu w} (f_{\mu b} C^b_{1z})^{-1} \leq 1 \tag{402}$$

$$\Phi_{p3} = (C^h_{1r})^2 (C^b_{1r})^{-2} \approx 1 \qquad \Phi_{p4} = (C^w_{\beta\rho})^2 (C^w_{\beta t})^{-2} \approx 1 \tag{403}$$

where Q^o_p is given by equation (288), Q_p is given by equation (368), $f_{\mu b}$ is given by equation (346) and where θ^b_n and θ^w_n are given by equations (264) and (266) or (268) as

$$\sec\theta_n^b = (C_{1r}^b)^2 C_{1z}^b \cos\theta_{\aleph}^b \tag{404}$$

$$\sec\theta_n^w = (C_{\beta r}^w)^{-2}(C_{\beta z}^w)^{-1} \cos\theta_{\aleph}^w \tag{405}$$

and where the space coordinate factors $C_{1\alpha}^b$ and $C_{\beta\alpha}^w$ for α = r,z and β = 2,3 are given by equations (51) through (53). Equation (400) shows that for a polyphase piston engine

$$a_p \geq a_p^o \tag{406}$$

so that an the presence of ambient fields enhances the acceleration of a piston and therefore the a polyphase piston engine should be more efficient than a conventional piston engine. Under the assumption of an ideal working gas, the internal phase angle relation in equation (398) can be written as

$$\theta_a^p = \theta_P^w + \theta_A^h - \theta_m^p = \theta_n^w + \theta_T^w + \theta_A^h - \theta_m^p \tag{407}$$

Within the approximation $\theta_m^p \approx \theta_m^b$ of equation (369) the phase angle equation (407) becomes

$$\begin{aligned}\theta_a^p &= \theta_n^w + \theta_T^w + \theta_A^h - \theta_m^b \\ &= \theta_n^w + \theta_T^w + \theta_A^h - \theta_n^b - \theta_\mu^b - \theta_V^b \\ &= \theta_n^w + \theta_T^w - \theta_n^b - |\theta_A^h| + |\theta_V^b| - \theta_\mu^b\end{aligned} \tag{408}$$

Equations (400) and (407) or (408) describe the equation of motion of a polyphase piston engine.

The measured value of the acceleration of a piston in a polyphase piston engine is obtained from equations (400) and (407) or (408) as

$$a_{pm} = a_p \cos\theta_a^p = a_p^o S_p \sec\theta_n^b \sec\theta_n^w \cos\theta_a^p \tag{409}$$

The cosine factor in equation (409) tends to lower the measured acceleration of the piston, but the choice $\theta_a^p = 0$ can always be engineered by adjusting the ambient field strengths of the polyphase piston engine as given in equations (33) through (35). Equation (407) shows that the condition $\theta_a^p = 0$ is equivalent to

$$\theta_m^p = \theta_n^w + \theta_T^w + \theta_A^h = \theta_n^w + \theta_T^w - |\theta_A^h| \tag{410}$$

while equation (408) gives approximately

$$\theta_n^b = \theta_n^w + \theta_T^w + \theta_A^h - \theta_V^b - \theta_\mu^b = \theta_n^w + \theta_T^w - |\theta_A^h| + |\theta_V^b| - \theta_\mu^b \tag{411}$$

For this engineered condition, $a_{pm} = a_p$ because $\theta_a^p = 0$, and equation (400) gives the measured acceleration but equation (411) shows that θ_n^b and θ_n^w are not independent variables, and this makes equation (400) more complicated to solve and and must in general require computer solutions if a particular value of the acceleration gain is specified, in which case the solutions are generally limited to certain limited values of the zero ambient field acceleration. This is described in detail in Section 9 for polyphase rockets.

3. CONVENTIONAL ROCKET THRUST AND ROCKET MOTION EQUATIONS. This section reviews the conventional calculation of the thrust, motion and liftoff condition for a rocket that corresponds to the case of zero ambient fields in a polyphase rocket.

A. General Description of a Rocket Mass and Geometry.

The real number equation of motion of a rocket that is moving vertically from the surface of the Earth is given by

$$dp_R^o / dt^o = T_e^o - F_G^o \tag{412}$$

where the zero superscript indicates that all ambient fields are set to zero, and where p_R^o = rocket momentum, T_e^o = rocket exhaust thrust and F_G^o = force of gravity. The momentum of the rocket is given by

$$p_R^o = m_R^o v_R^o \tag{413}$$

where v_R^o = vertical velocity of the rocket and where m_R^o = total instantaneous mass of a rocket which can be written as

$$m_R^o = m_b^o + \Delta m_a^o + \Delta m_f^o + \Delta m_p^o = m_b^o Q_R^o \tag{414}$$

where

$$Q_R^o = m_R^o / m_b^o = 1 + \Delta m_a^o / m_b^o + \Delta m_f^o / m_b^o + \Delta m_p^o / m_b^o \tag{415}$$

where m_b^o = mass of the rocket body including the liquid or solid fuel, Δm_a^o = mass of the cylindrical shell representing the exhaust aperture system, Δm_f^o = mass of the cylindrical shell representing combustion chamber and combustion chamber foramen system and Δm_p^o = mass of the cylindrical shell of the personnel section. The mass of the gases in the exhaust system and the combustion chamber at any instant are neglected due to their low-density. At the time of initial launch the mass m_b^o of the rocket body, which includes the fuel, is the dominant mass term in equation (414).

The rocket component masses that appear in equations (414) and (415) can be written as

$$\Delta m_a^o = \mu_a^o \Delta N_a = \rho_a^o \Delta V_a^o = \rho_a^o \Delta A_a^o L_a^o \tag{416}$$

$$\Delta m_f^o = \mu_f^o \Delta N_f = \rho_f^o \Delta V_f^o = \rho_f^o \Delta A_f^o L_f^o \tag{417}$$

$$\Delta m_p^o = \mu_p^o \Delta N_p = \rho_p^o \Delta V_p^o = \rho_p^o \Delta A_p^o L_p^o \tag{418}$$

$$m_b^o = \mu_b^o N_b = \rho_b^o V_b^o = \rho_b^o A_b^o L_b^o \tag{419}$$

where the cross-sectional areas are written as

$$\Delta A_a^o = A_{a\delta}^o - A_a^o = \pi[(r_{a\delta}^o)^2 - (r_a^o)^2] = \pi[(\delta_a^o)^2 + 2r_a^o\delta_a^o] \tag{420}$$

$$\Delta A_f^o = A_{f\delta}^o - A_f^o = \pi[(r_{f\delta}^o)^2 - (r_f^o)^2] = \pi[(\delta_f^o)^2 + 2r_f^o\delta_f^o] \tag{421}$$

$$\Delta A_p^o = A_{p\delta}^o - A_p^o = \pi[(r_{p\delta}^o)^2 - (r_p^o)^2] = \pi[(\delta_p^o)^2 + 2r_p^o\delta_p^o] \tag{422}$$

$$A_b^o = \pi(r_b^o)^2 \tag{423}$$

and the volumes of the rocket components are

$$\Delta V_a^o = \Delta A_a^o L_a^o \qquad \Delta V_f^o = \Delta A_f^o L_f^o \qquad \Delta V_p^o = \Delta A_p^o L_p^o \qquad V_b^o = A_b^o L_b^o \tag{424}$$

where ΔA_a^o, ΔA_f^o and ΔA_p^o = annular areas of the exhaust gas aperture tube, combustion chamber foramen tube and the cylindrical shell of the personnel habitat respectively; A_b^o = cross-sectional area of the solid cylinder body part of the rocket; ΔV_a^o, ΔV_f^o and ΔV_p^o = volumes of the cylindrical shells of the exhaust gas aperture tube, combustion chamber foramen tube and personnel compartment cylinder; V_b^o = volume of solid body cylinder of the rocket; L_a^o, L_f^o and L_p^o = lengths of the exhaust gas aperture tube, combustion chamber foramen tube and the personnel quarters tube, and where

$$A_{a\delta}^o = \pi(r_{a\delta}^o)^2 \qquad A_a^o = \pi(r_a^o)^2 \tag{425}$$

$$A_{f\delta}^o = \pi(r_{f\delta}^o)^2 \qquad A_f^o = \pi(r_f^o)^2 \tag{426}$$

$$A_{p\delta}^o = \pi(r_{p\delta}^o)^2 \qquad A_p^o = \pi(r_p^o)^2 \tag{427}$$

where $A_{a\delta}^o$, $A_{f\delta}^o$ and $A_{p\delta}^o$ and A_a^o, A_f^o and A_p^o = respectively the outer and inner cross-sectional areas of the exhaust gas aperture tube, the combustion chamber foramen tube and the personnel quarters cylinder; $r_{a\delta}^o$, $r_{f\delta}^o$ and $r_{p\delta}^o$ and r_a^o, r_f^o and r_p^o = respectively the outer and inner radii of the exhaust aperture tube, combustion chamber foramen system and the personnel quarters tube. The outer and inner radii are related by

$$r_{a\delta}^o = r_a^o + \delta_a^o \qquad r_{f\delta}^o = r_f^o + \delta_f^o \qquad r_{p\delta}^o = r_p^o + \delta_p^o \tag{428}$$

where δ_a^o, δ_f^o and δ_p^o = radial wall thickness of the exhaust aperture tube, combustion chamber foramen tube and the personnel quarters tube. The outer circumferences of the exhaust gas tube, combustion chamber foramen tube and the personnel quarters tube are given by

$$s_{a\delta}^o = 2\pi r_{a\delta}^o \qquad s_{f\delta}^o = 2\pi r_{f\delta}^o \qquad s_{p\delta}^o = 2\pi r_{p\delta}^o \tag{429}$$

while the corresponding inner circumferences are

$$s_a^o = 2\pi r_a^o \qquad s_f^o = 2\pi r_f^o \qquad s_p^o = 2\pi r_p^o \tag{430}$$

and finally the circumference of the solid rocket body cylinder is $s_b^o = 2\pi r_b^o$. In equations (416) through (419) $\rho_a^o, \rho_f^o, \rho_p^o$ and ρ_b^o = average mass densities of the wall of the exhaust aperture tube, combustion chamber foramen wall, cylindrical personnel chamber wall and the solid body portion of the rocket and are calculated as

$$\rho_a^o = \mu_a^o n_a^o \qquad \rho_f^o = \mu_f^o n_f^o \qquad \rho_p^o = \mu_p^o n_p^o \qquad \rho_b^o = \mu_b^o n_b^o \tag{431}$$

where

$$n_a^o = \Delta N_a / \Delta V_a^o = N_a / V_a^o \qquad n_f^o = \Delta N_f / \Delta V_f^o = N_f / V_f^o \tag{432}$$

$$n_p^o = \Delta N_p / \Delta V_p^o = N_p / V_p^o \qquad n_b^o = N_b / V_b^o \tag{433}$$

where $\mu_a^o, \mu_f^o, \mu_p^o$ and μ_b^o and n_a^o, n_f^o, n_p^o and n_b^o= respectively the molecular masses and the average molecular number densities of the material in the exhaust gas aperture wall, combustion chamber foramen wall, personnel chamber wall and the solid body component of the rocket; $\Delta N_a, \Delta N_f, \Delta N_p$ and N_b = integer number of molecules in the exhaust gas aperture cylindrical shell of volume ΔV_a^o, integer number of molecules in the combustion chamber foramen cylindrical shell of volume ΔV_f^o, integer number of molecules in the cylindrical shell wall of the personnel cabin of volume ΔV_p^o, and the integer number of molecules in the solid cylinder of the body of the rocket of volume V_b^o; and where N_a, N_f and N_p = integer number of molecules in the small cylindrical volume elements V_a^o, V_f^o and V_p^o located in the walls of the exhaust gas aperture system, combustion chamber foramen system and personnel cabin respectively. The masses of the volume elements in these four systems are written as

$$m_a^o = \mu_a^o N_a = \rho_a^o V_a^o \qquad m_f^o = \mu_f^o N_f = \rho_f^o V_f^o \tag{434}$$

$$m_p^o = \mu_p^o N_p = \rho_p^o V_p^o \qquad m_b^o = \mu_b^o N_b = \rho_b^o V_b^o \tag{435}$$

Note that in this paper the indices a, f, p, and b refer to the high-density solid or liquid components of a rocket while the indices e (exhaust gases) and c (combustion chamber gases) refer to the low-density gases.

B. Conventional Calculation of Rocket Thrust.

The conventional expression for the rocket total thrust per unit area of the exhaust aperture is given by

$$\tau_e^o = \tau_{ef}^o + P_e^o - P_{at}^o \tag{436}$$

where the thrust per unit area associated with the mass flow is given as

$$\tau_{ef}^o = \rho_e^o (v_e^o)^2 \tag{437}$$

while the rocket thrust that appears in equation (412) is given by

$$T_e^o = \tau_e^o A_e^o = [\rho_e^o (v_e^o)^2 + P_e^o - P_{at}^o] A_a^o \tag{438}$$

where τ_e^o = total rocket thrust per unit aperture area, τ_{ef}^o = rocket thrust per unit area of exhaust aperture due to material flow, P_e^o = exhaust gas pressure, P_{at}^o = atmospheric pressure, T_e^o = total rocket thrust, A_a^o = rocket exhaust aperture area, v_e^o = velocity of the exhaust gases and where ρ_e^o = mass density of the exhaust gases which is given by

$$\rho_e^o = \mu_e^o n_e^o = m_e^o / V_e^o \qquad (439)$$

where m_e^o = mass of the exhaust gases in some volume element V_e^o of the exhaust gases and which is given by

$$m_e^o = \mu_e^o N_e = \mu_e^o n_e^o V_e^o = \rho_e^o V_e^o \qquad (440)$$

where μ_e^o = average molecular mass of the exhaust gases, N_e = integer number of gas molecules in the volume V_e^o, and where n_e^o = molecular number density which is given by

$$n_e^o = N_e / V_e^o \qquad (441)$$

where V_e^o = volume element of gas which contains an integer number N_e of molecules of gas and which for a choice of a cylindrical volume element can be written as

$$V_e^o = A_e^o z_e^o = \pi (r_e^o)^2 z_e^o \qquad A_e^o = \pi (r_e^o)^2 \qquad (442)$$

where r_e^o, z_e^o, A_e^o and V_e^o = radius, length, circular area and volume of a cylindrical volume element of gases in the exhaust tubes. The quantities r_e^o, z_e^o, A_e^o and V_e^o refer to an arbitrarily large integer number N_e of exhaust gas molecules so that statistical fluctuations are not considered. The mass flux of molecules coming out of the exhaust nozzle is given by

$$f_e^o = \rho_e^o v_e^o A_a^o \qquad (443)$$

where A_a^o = exhaust aperture area. The kinetic energy of a volume element of gas in the exhaust tube is calculated as

$$K_e^o = 1/2 m_e^o (v_e^o) = 1/2 \mu_e^o N_e (v_e^o)^2 = 1/2 \mu_e^o n_e^o V_e^o (v_e^o)^2 \qquad (444)$$

so that the average kinetic energy per unit volume of exhaust gas is given by

$$k_e^o = 1/2 \rho_e^o (v_e^o)^2 = 1/2 \mu_e^o n_e^o (v_e^o)^2 = 1/2 \rho_e^o (v_e^o)^2 = 1/2 \tau_{ef}^o \qquad (445)$$

and the average kinetic energy per molecule of the exhaust gases in the exhaust tubes is

$$\varepsilon_{Ke}^o = 1/2 \mu_e^o (v_e^o)^2 \qquad (446)$$

which is associated with the transport of the exhaust gases.

C. Relation Between Exhaust Gas Pressure and Combustion Gas Pressure.

The exhaust gas pressure P_e^o that appears in the rocket thrust equations (436) and

(438) is in general a complicated function of the fuel reaction rate in the combustion chamber, exhaust gas density and the nozzle aperture area and length. The chemical reactions of the burning rocket fuel occur in the combustion chamber of a rocket engine, and in the combustion chamber the gases exist at high pressure, density and temperature but are moving at a relatively low velocity. When these chemical reaction product gases pass from the combustion chamber to the nozzle they attain a state of relatively low pressure, density and temperature but are moving at high velocity. It is the high exhaust velocity that produces large values of thrust as can be seen from the quadratic dependence of the rocket thrust on the velocity of the exhaust gases as shown in equations (436) through (438). This paper chooses a very oversimplified calculation of the exhaust gas pressure P_e^o that is based on Bernoulli's law which can be written as

$$1/2\rho_e^o(v_e^o)^2 + P_e^o = 1/2\rho_c^o(v_c^o)^2 + P_c^o = K^o \qquad (447)$$

where K^o = constant, P_e^o and P_c^o = respectively the gas pressures in the exhaust tube and the combustion chamber, v_e^o and v_c^o = bulk velocity of the gases in the exhaust tube and the combustion chamber respectively and where the mass density ρ_e^o of the exhaust gases in the exhaust tube is given by equation (439) and where the mass density ρ_c^o of the gases in the combustion chamber can be written as

$$\rho_c^o = \mu_c^o n_c^o = m_c^o / V_c^o \qquad (448)$$

where m_c^o = mass of the combustion chamber gases in a volume element V_c^o of the combustion chamber gases and which is given by

$$m_c^o = \mu_c^o N_c = \mu_c^o n_c^o V_c^o = \rho_c^o V_c^o \qquad (449)$$

where μ_c^o = average molecular mass of the combustion chamber gases and where n_c^o = molecular number density of the gases in the combustion chamber and which is given by

$$n_c^o = N_c / V_c^o \qquad V_c^o = A_c^o z_c^o = \pi(r_c^o)^2 z_c^o \qquad A_c^o = \pi(r_c^o)^2 \qquad (450)$$

where N_c = integer number of gas molecules in a small volume element V_c^o of the combustion chamber gas. The relatively high pressure P_c^o of the gases in the combustion chamber can be calculated approximately by assuming the gases behave like an ideal gas and therefore in terms of $v_{\mu c}^o$ = root mean square velocity of combustion gas molecules

$$P_c^o = n_c^o k_B T_c^o = 1/3 n_c^o \mu_c^o (v_{\mu c}^o)^2 \qquad (451)$$

where k_B = Boltzmann constant and T_c^o = absolute temperature of the gases in the combustion chamber under zero ambient field conditions. The relatively low pressure of the exhaust gases in the exhaust tube is obtained from the Bernoulli equation (447) as

$$P_e^o = P_c^o - 1/2\rho_e^o(v_e^o)^2 + 1/2\rho_c^o(v_c^o)^2 \qquad (452)$$

The velocities v_e^o and v_c^o are related by the mass conservation law that can be written in the following form

$$\rho_e^o v_e^o A_a^o = \rho_c^o v_c^o A_f^o \qquad (453)$$

where A_a^o and A_c^o = areas of the exhaust tube aperture and the combustion chamber foramen respectively. The mass conservation equation (453) can be used to eliminate v_c^o from equation (452) with the result that

$$P_e^o = P_c^o - 1/2\rho_e^o(v_e^o)^2[1-(\rho_e^o/\rho_c^o)(A_a^o/A_f^o)^2] \tag{454}$$

Equation (454) shows that the exhaust gas pressure P_e^o is a decreasing function of the exhaust gas velocity and is also a decreasing function of the exhaust gas mass density provided that the ratio ρ_e^o/ρ_c^o remains constant. Thus a high exhaust velocity implies a low exhaust pressure. If h_e^o and t_e^o is the distance and time over which the acceleration of a gas molecules occurs in the rocket exhaust tube, then the velocity of the gas molecule is related to these quantities by

$$v_e^o = a_e^o t_e^o \qquad h_e^o = 1/2a_e^o(t_e^o)^2 \qquad (v_e^o)^2 = 2a_e^o h_e^o \tag{455}$$

where v_e^o and a_e^o = velocity and acceleration of the exhaust gas molecules.

D. Adapted Condition for Rocket Thrust.

The atmospheric pressure that appears in the rocket thrust equations (436) and (438) can be calculated by assuming the atmosphere can be represented as an ideal gas whose density and temperature vary with height. If it is assumed that the space and time coordinates in the atmosphere are not intrinsically affected by gravity and are therefore real numbers it follows that the atmospheric pressure is given by

$$P_{at}^o = n_{at}^o k_B T_{at}^o \tag{456}$$

where k_B = Boltzmann constant, T_{at}^o = absolute temperature of the atmosphere at a specified elevation and where n_{at}^o = molecular number density of the atmosphere at a specified elevation and which is given by

$$n_{at}^o = N_{at}/V_{at}^o \tag{457}$$

where N_{at} = integer number of gas molecules in a volume element V_{at}^o of the atmosphere. Rocket engine designers wish to optimize the thrust given by equation (438). It has been shown that the maximum thrust occurs when the rocket nozzle has been designed to operate in what is called the adapted condition, which is given by

$$P_e^o = P_{at}^o \tag{458}$$

so that equations (436) through (438) give the maximum thrust per unit area of the exhaust aperture and the maximum total thrust as

$$\tau_e^o = \tau_{ef}^o = \rho_e^o(v_e^o)^2 \tag{459}$$

$$T_e^o = \tau_e^o A_a^o = \rho_e^o(v_e^o)^2 A_a^o \tag{460}$$

The value of the exhaust velocity corresponding to the adapted condition in equation (458) is obtained from equation (454) to be given by the following equation

$$(v_e^o)^2 = 2(P_c^o - P_{at}^o)(\rho_e^o)^{-1}[1-(\rho_e^o/\rho_c^o)(A_a^o/A_f^o)^2]^{-1} \qquad P_e^o = P_{at}^o \qquad (461)$$

and gives the exhaust gas velocity for maximum thrust, and the maximum values of the thrust are then obtained by substituting equation (461) into equations (459) and (460) with the results

$$\tau_e^o = 2(P_c^o - P_{at}^o)[1-(\rho_e^o/\rho_c^o)(A_a^o/A_f^o)^2]^{-1} \qquad (462)$$

$$T_e^o = 2(P_c^o - P_{at}^o)A_a^o[1-(\rho_e^o/\rho_c^o)(A_a^o/A_f^o)^2]^{-1} \qquad (463)$$

Calculations of rocket dynamics is much simpler when the nozzle and gas flow are designed for the adapted condition.

E. Conventional Gravity Force on a Rocket at the Earth's Surface.

The conventional force of gravity that appears in the rocket motion equation (412) is given by

$$F_G^o = m_R^o g^o \qquad (464)$$

where g^o = acceleration of gravity of the Earth and is given by

$$g^o = -GM_E^o/\rho_o^2 \qquad (465)$$

where G = Newtonian gravitational constant, ρ_o = radial distance of the rocket from the center of the Earth with $\rho_o \geq R_E^o$ where R_E^o = radius of the Earth, M_E^o = mass of the Earth which can also be written as

$$M_E^o = \mu_E^o N_E = \rho_E^o V_E^o \qquad V_E^o = 4/3\pi(R_E^o)^3 \qquad (466)$$

where V_E^o = volume of the Earth, μ_E^o = zero ambient field value of the average molecular mass of the solid Earth material and where N_E = integer number of molecule in the Earth. At the Earth's surface the force of gravity is given by

$$F_{GE}^o = m_R^o g_E^o = -Gm_R^o M_E^o(R_E^o)^{-2} = -Gm_b^o Q_R^o M_E^o(R_E^o)^{-2} \qquad (467)$$

where Q_R^o = rocket mass factor given by equation (415) and where g_E^o = conventional acceleration of gravity at the Earth's surface and is given by

$$g_E^o = -GM_E^o(R_E^o)^{-2} = -4/3G\rho_E^o R_E^o \qquad (468)$$

where ρ_E^o = conventional constant average density of the Earth as defined by equation (466).

F. Conventional Rocket Liftoff Parameter.

The time derivative of the rocket momentum that appears in the rocket motion equation (412) is written as

$$dp^o_R / dt^o = m^o_R dv^o_R / dt^o + v^o_R dm^o_R / dt^o \tag{469}$$

where the rate of change of the rocket mass due to the exhausting of the reaction products of the burned fuel is given by equation (443) as

$$dm^o_R / dt^o = dm^o_b / dt^o = -\rho^o_e v^o_e A^o_a \tag{470}$$

Then combining equations (412) through (470) gives the rocket motion equation at the Earth's surface as

$$\begin{aligned} m^o_R dv^o_R / dt^o &= \rho^o_e v^o_e v^o_R A^o_a + T^o_e - F^o_{GE} \\ &= (\rho^o_e v^o_e v^o_R + \tau^o_e) A^o_a - m^o_R g^o_E \\ &= [\rho^o_e v^o_e v^o_R + \rho^o_e (v^o_e)^2 + P^o_e - P^o_{at}] A^o_a - m^o_R g^o_E \\ &= \rho^o_e v^o_e v^o_R A^o_a + m^o_R g^o_E (s^o_L - 1) \end{aligned} \tag{471}$$

where s^o_L = liftoff parameter of a conventional rocket and which is defined by

$$s^o_L = T^o_e / F^o_{GE} = (m^o_R g^o_E)^{-1} T^o_e = (m^o_R g^o_E)^{-1} \tau^o_e A^o_a = (m^o_R g^o_E)^{-1} [\rho^o_e (v^o_e)^2 + P^o_e - P^o_{at}] A^o_a \tag{472}$$

For the adapted condition of maximum thrust given by equation (458), the liftoff parameter given in equation (472) becomes

$$s^o_L = (m^o_R g^o_E)^{-1} \rho^o_e (v^o_e)^2 A^o_a = (m^o_b Q^o_R g^o_E)^{-1} \rho^o_e (v^o_e)^2 A^o_a \tag{473}$$

which is a simple parameter which can be used for practical calculations. Rocket liftoff occurs when $s^o_L = 1$, so that at the instant of liftoff $v^o_R = 0$ and $dv^o_R / dt^o = 0$ in equation (471), and at the instant of liftoff for a rocket, the velocity and acceleration of the rocket are zero. After liftoff $s^o_L > 1$ and the acceleration increases rapidly as the rocket slowly rises.

4. GEOMETRY OF A POLYPHASE ROCKET. This section considers the complex number radii, lengths, cross-sectional areas and volumes of the various components of a polyphase rocket.

A. Complex Number Radii and Lengths of the Exhaust Aperture Wall, Combustion Chamber Foramen Wall, Personnel Compartment Wall and the Body of a Polyphase Rocket.

The exhaust aperture system of a polyphase rocket is modeled as a high-density cylindrical shell whose walls contain ambient fields of the form given in equation (58).

The complex number radius of the aperture can be represented according to equation (64) as

$$\bar{r}_a = r_a \exp(j\theta_r^a) \qquad r_a = r_a^o C_{1r}^a \qquad \theta_r^a = \theta_{\eta r}^a - \theta_{\pi r}^a \tag{474}$$

where r_a^o = zero ambient field value of the aperture radius which makes its appearance in equations (420), (425) and (428) and where the high-density space coordinate factor C_{1r}^a is given in equation (92). The circumference of the exhaust aperture for a polyphase rocket is written as

$$\bar{s}_a = s_a \exp(j\theta_s^a) = 2\pi\bar{r}_a \tag{475}$$

which gives

$$s_a = 2\pi r_a = 2\pi r_a^o C_{1r}^a = s_a^o C_{1r}^a \qquad \theta_s^a = \theta_r^a = \theta_{\eta r}^a - \theta_{\pi r}^a \tag{476}$$

where s_a^o is given by equation (430). The complex number outer radius of the exhaust aperture tube is written as

$$\bar{r}_{a\delta} = r_{a\delta} \exp(j\theta_{r\delta}^a) = \bar{r}_a + \bar{\delta}_a = r_a \exp(j\theta_r^a) + \delta_a \exp(j\theta_\delta^a) \tag{477}$$

with

$$r_{a\delta} = r_{a\delta}^o C_{1r\delta}^a \qquad \theta_{r\delta}^a = \theta_{\eta r\delta}^a - \theta_{\pi r\delta}^a \tag{478}$$

where $r_{a\delta}^o$ is given by equation (428) and where $\bar{\delta}_a$ = complex number wall thickness of the rocket exhaust aperture system and can be written as

$$\bar{\delta}_a = \delta_a \exp(j\theta_\delta^a) \qquad \delta_a = \delta_a^o C_{1\delta}^a \qquad \theta_\delta^a = \theta_{\eta\delta}^a - \theta_{\pi\delta}^a \tag{479}$$

where δ_a^o = zero ambient field value of the wall thickness of the exhaust aperture tube and which appears in equations (420) and (428). Because there is approximately a uniform ambient field in the cylindrical shell it follows that

$$r_{a\delta} = r_a + \delta_a \qquad C_{1r\delta}^a = C_{1\delta}^a = C_{1r}^a \qquad \theta_{r\delta}^a = \theta_\delta^a = \theta_r^a \tag{480}$$

$$\theta_{\eta r\delta}^a = \theta_{\eta\delta}^a = \theta_{\eta r}^a \qquad \theta_{\pi r\delta}^a = \theta_{\pi\delta}^a = \theta_{\pi r}^a \tag{481}$$

and then equation (477) agrees with the first part of equation (428), and equation (478) can be written as

$$r_{a\delta} = r_{a\delta}^o C_{1r}^a \qquad \theta_{r\delta}^a = \theta_{\eta r}^a - \theta_{\pi r}^a \tag{482}$$

where $r_{a\delta}^o$ is given by equation (428). The complex number circumference of the outer surface of the cylindrical shell of the exhaust tube is given by

$$\bar{s}_{a\delta} = s_{a\delta} \exp(j\theta_{s\delta}^a) = 2\pi\bar{r}_{a\delta} \tag{483}$$

where

$$s_{a\delta} = 2\pi r_{a\delta} = 2\pi r^{o}_{a\delta} C^{a}_{1r} = s^{o}_{a\delta} C^{a}_{1r} \qquad \theta^{a}_{s\delta} = \theta^{a}_{r\delta} = \theta^{a}_{s} = \theta^{a}_{r} \tag{484}$$

where $s^{o}_{a\delta}$ is given by equation (429). Then the measured values of the radius and circumference of the exhaust aperture is obtained from equations (474) and (476) as

$$r_{am} = r_{a} \cos\theta^{a}_{r} = r^{o}_{a} C^{a}_{1r} \cos(\theta^{a}_{\eta r} - \theta^{a}_{\pi r}) \tag{485}$$

$$s_{am} = s_{a} \cos\theta^{a}_{s} = s^{o}_{a} C^{a}_{1r} \cos(\theta^{a}_{\eta r} - \theta^{a}_{\pi r}) \tag{486}$$

while the measured values of the outer radius and circumference of the exhaust aperture cylinder is obtained from equation (482) and (484) as

$$r_{a\delta m} = r_{a\delta} \cos\theta^{a}_{r\delta} = r^{o}_{a\delta} C^{a}_{1r} \cos(\theta^{a}_{\eta r} - \theta^{a}_{\pi r}) \tag{487}$$

$$s_{a\delta m} = s_{a\delta} \cos\theta^{a}_{s\delta} = s^{o}_{a\delta} C^{a}_{1r} \cos(\theta^{a}_{\eta r} - \theta^{a}_{\pi r}) \tag{488}$$

the measured value of the shell thickness is

$$\delta_{am} = \delta_{a} \cos\theta^{a}_{\delta} = \delta^{o}_{a} C^{a}_{1r} \cos(\theta^{a}_{\eta r} - \theta^{a}_{\pi r}) \tag{489}$$

and therefore equations (485), (487) and (489) give

$$r_{a\delta m} = r_{am} + \delta_{am} \tag{490}$$

The length of the exhaust aperture system cylindrical shell is a complex number, which in conjunction with equation (64), can be written as

$$\overline{L}_{a} = L_{a} \exp(j\theta^{a}_{L}) \qquad L_{a} = L^{o}_{a} C^{a}_{1z} \qquad \theta^{a}_{L} = \theta^{a}_{\eta L} - \theta^{a}_{\pi L} \tag{491}$$

where L^{o}_{a} = zero ambient field value of the length of the cylindrical shell exhaust aperture system and where C^{a}_{1z} = spatial coordinate factor for the high-density wall of the exhaust tube and is given by equation (92). The measured values of the length of a polyphase rocket exhaust tube is obtained from equation (491) as

$$L_{am} = L_{a} \cos\theta^{a}_{L} = L^{o}_{a} C^{a}_{1z} \cos(\theta^{a}_{\eta L} - \theta^{a}_{\pi L}) \tag{492}$$

Finally, note that because $C^{a}_{1r} \geq 1$ and $C^{a}_{1z} \geq 1$ it follows from equations (92), (474), (476), (485) and (486) that

$$r_{a} \geq r^{o}_{a} \qquad r_{am} \geq r^{o}_{a} \qquad s_{a} \geq s^{o}_{a} \qquad s_{am} \geq s^{o}_{a} \tag{493}$$

and equations (92), (482), (484), (487) and (488) give

$$r_{a\delta} \geq r^{o}_{a\delta} \qquad r_{a\delta m} \geq r^{o}_{a\delta} \qquad s_{a\delta} \geq s^{o}_{a\delta} \qquad s_{a\delta m} \geq s^{o}_{a\delta} \tag{494}$$

while equations (92), (479), (489), (491) and (492) yield

$$\delta_{a} \geq \delta^{o}_{a} \qquad \delta_{am} \geq \delta^{o}_{a} \qquad L_{a} \geq L^{o}_{a} \qquad L_{am} \geq L^{o}_{a} \tag{495}$$

Therefore due to the ambient fields there will be small changes of the dimensions of the exhaust aperture system of a polyphase rocket.

In this paper the combustion chamber of a polyphase rocket is described as a high-density cylindrical shell whose wall contain an ambient field of the form given in equation (59). The complex number radius of the combustion chamber foramen can be represented according to equation (65) as

$$\bar{r}_f = r_f \exp(j\theta^f_r) \qquad r_f = r^o_f C^f_{1r} \qquad \theta^f_r = \theta^f_{\eta r} - \theta^f_{\iota r} \tag{496}$$

where r^o_f = zero ambient field value of the combustion chamber foramen radius and which appears in equations (421), (426) and (428) and where the high-density space coordinate factor C^f_{1r} is given by equation (93). The complex number circumference of the combustion chamber for a polyphase rocket is written as

$$\bar{s}_f = s_f \exp(j\theta^f_s) = 2\pi\bar{r}_f \tag{497}$$

which gives

$$s_f = 2\pi r_f = 2\pi r^o_f C^f_{1r} = s^o_f C^f_{1r} \qquad \theta^f_s = \theta^f_r = \theta^f_{\eta r} - \theta^f_{\iota r} \tag{498}$$

where s^o_f is given by equation (430). The complex number outer radius of the combustion chamber cylindrical shell is written as

$$\bar{r}_{f\delta} = r_{f\delta} \exp(j\theta^f_{r\delta}) = \bar{r}_f + \bar{\delta}_f = r_f \exp(j\theta^f_r) + \delta_f \exp(j\theta^f_\delta) \tag{499}$$

with

$$r_{f\delta} = r^o_{f\delta} C^f_{1r\delta} \qquad \theta^f_{r\delta} = \theta^f_{\eta r\delta} - \theta^f_{\iota r\delta} \tag{500}$$

where $r^o_{f\delta}$ is given by equation (428) and where $\bar{\delta}_f$ = complex number wall thickness of the cylindrical shell of the combustion chamber and can be written as

$$\bar{\delta}_f = \delta_f \exp(j\theta^f_\delta) \qquad \delta_f = \delta^o_f C^f_{1\delta} \qquad \theta^f_\delta = \theta^f_{\eta\delta} - \theta^f_{\iota\delta} \tag{501}$$

where δ^o_f = zero ambient field value of the wall thickness of the combustion chamber and which appears in equations (421) and (428). Because there is an approximately uniform ambient field in the cylindrical shell of the combustion chamber it follows that

$$r_{f\delta} = r_f + \delta_f \qquad C^f_{1r\delta} = C^f_{1\delta} = C^f_{1r} \qquad \theta^f_{r\delta} = \theta^f_\delta = \theta^f_r \tag{502}$$

$$\theta^f_{\eta r\delta} = \theta^f_{\eta\delta} = \theta^f_{\eta r} \qquad \theta^f_{\iota r\delta} = \theta^f_{\iota\delta} = \theta^f_{\iota r} \tag{503}$$

and equation (499) agrees with the second part of equation (428), and equation (500) can be written as

$$r_{f\delta} = r^o_{f\delta} C^f_{1r} \qquad \theta^f_{r\delta} = \theta^f_{\eta r} - \theta^f_{\iota r} \tag{504}$$

where $r^{o}_{f\delta}$ is given by equation (428). The complex number circumference of the outer surface of the cylindrical shell of the combustion chamber is given by

$$\bar{s}_{f\delta} = s_{f\delta}\exp(j\theta^{f}_{s\delta}) = 2\pi\bar{r}_{f\delta} \tag{505}$$

where

$$s_{f\delta} = 2\pi r_{f\delta} = 2\pi r^{o}_{f\delta}C^{f}_{1r} = s^{o}_{f\delta}C^{f}_{1r} \qquad \theta^{f}_{s\delta} = \theta^{f}_{r\delta} = \theta^{f}_{s} = \theta^{f}_{r} \tag{506}$$

where $s^{o}_{f\delta}$ is given by equation (429). The measured values of the radius and circumference of the combustion chamber foramen is obtained from equations (496) and (498) as

$$r_{fm} = r_{f}\cos\theta^{f}_{r} = r^{o}_{f}C^{f}_{1r}\cos(\theta^{f}_{\eta r} - \theta^{f}_{\pi r}) \tag{507}$$

$$s_{fm} = s_{f}\cos\theta^{f}_{s} = s^{o}_{f}C^{f}_{1r}\cos(\theta^{f}_{\eta r} - \theta^{f}_{\pi r}) \tag{508}$$

and the measured values of the outer radius and circumference of the cylindrical shell of the combustion chamber is obtained from equation (504) and (506) as

$$r_{f\delta m} = r_{f\delta}\cos\theta^{f}_{r\delta} = r^{o}_{f\delta}C^{f}_{1r}\cos(\theta^{f}_{\eta r} - \theta^{f}_{\pi r}) \tag{509}$$

$$s_{f\delta m} = s_{f\delta}\cos\theta^{f}_{s\delta} = s^{o}_{f\delta}C^{f}_{1r}\cos(\theta^{f}_{\eta r} - \theta^{f}_{\pi r}) \tag{510}$$

the measured value of the shell thickness is

$$\delta_{fm} = \delta_{f}\cos\theta^{f}_{\delta} = \delta^{o}_{f}C^{f}_{1r}\cos(\theta^{f}_{\eta r} - \theta^{f}_{\pi r}) \tag{511}$$

and therefore equations (507), (509) and (511) give

$$r_{f\delta m} = r_{fm} + \delta_{fm} \tag{512}$$

The complex number length of the combustion chamber cylindrical shell can be written, with the help of equation (65), as

$$\bar{L}_{f} = L_{f}\exp(j\theta^{f}_{L}) \qquad L_{f} = L^{o}_{f}C^{f}_{1z} \qquad \theta^{f}_{L} = \theta^{f}_{\eta L} - \theta^{f}_{\pi L} \tag{513}$$

where L^{o}_{f} = zero ambient field value of the length of the combustion chamber and where C^{f}_{1z} = spatial coordinate factor for the high-density wall of the combustion chamber foramen and is given by equation (93). The measured values of the length of a polyphase rocket combustion chamber is obtained from equation (513) as

$$L_{fm} = L_{f}\cos\theta^{f}_{L} = L^{o}_{f}C^{f}_{1z}\cos(\theta^{f}_{\eta L} - \theta^{f}_{\pi L}) \tag{514}$$

Note that because $C^{f}_{1r} \geq 1$ and $C^{f}_{1z} \geq 1$ it follows from equations (93), (496), (498), (507) and (508) that

$$r_{f} \geq r^{o}_{f} \qquad r_{fm} \geq r^{o}_{f} \qquad s_{f} \geq s^{o}_{f} \qquad s_{fm} \geq s^{o}_{f} \tag{515}$$

and equations (93), (504), (506), (509) and (510) give

$$r_{f\delta} \geq r_{f\delta}^{o} \qquad r_{f\delta m} \geq r_{f\delta}^{o} \qquad s_{f\delta} \geq s_{f\delta}^{o} \qquad s_{f\delta m} \geq s_{f\delta}^{o} \tag{516}$$

while equations (93), (501), (511), (513) and (514) yield

$$\delta_{f} \geq \delta_{f}^{o} \qquad \delta_{fm} \geq \delta_{f}^{o} \qquad L_{f} \geq L_{f}^{o} \qquad L_{fm} \geq L_{f}^{o} \tag{517}$$

Therefore as in the case of the exhaust aperture system, there will be small changes of the dimensions of the combustion chamber of a polyphase rocket due to the presence of ambient fields.

In this paper the walls of the personnel compartment of a polyphase rocket is described as a high-density cylindrical shell whose wall contains an ambient field of the form given in equation (60). The complex number radius of the personnel compartment can be represented according to equation (66) as

$$\bar{r}_{p} = r_{p} \exp(j\theta_{r}^{p}) \qquad r_{p} = r_{p}^{o} C_{1r}^{p} \qquad \theta_{r}^{p} = \theta_{\eta r}^{p} - \theta_{\pi r}^{p} \tag{518}$$

where r_{p}^{o} = zero ambient field value of the radius of the personnel compartment and which appears in equations (422), (427) and (428) and where the high-density radial space coordinate factor C_{1r}^{p} is given by equation (94). The complex number circumference of the personnel compartment for a polyphase rocket is written as

$$\bar{s}_{p} = s_{p} \exp(j\theta_{s}^{p}) = 2\pi\bar{r}_{p} \tag{519}$$

which gives

$$s_{p} = 2\pi r_{p} = 2\pi r_{p}^{o} C_{1r}^{p} = s_{p}^{o} C_{1r}^{p} \qquad \theta_{s}^{p} = \theta_{r}^{p} = \theta_{\eta r}^{p} - \theta_{\pi r}^{p} \tag{520}$$

where s_{p}^{o} is given by equation (430). The complex number outer radius of the personnel compartment cylindrical shell is written as

$$\bar{r}_{p\delta} = r_{p\delta} \exp(j\theta_{r\delta}^{p}) = \bar{r}_{p} + \bar{\delta}_{p} = r_{p} \exp(j\theta_{r}^{p}) + \delta_{p} \exp(j\theta_{\delta}^{p}) \tag{521}$$

with

$$r_{p\delta} = r_{p\delta}^{o} C_{1r\delta}^{p} \qquad \theta_{r\delta}^{p} = \theta_{\eta r\delta}^{p} - \theta_{\pi r\delta}^{p} \tag{522}$$

where $r_{p\delta}^{o}$ is given by equation (428) and where $\bar{\delta}_{p}$ = complex number wall thickness of the cylindrical shell of the personnel compartment and can be written as

$$\bar{\delta}_{p} = \delta_{p} \exp(j\theta_{\delta}^{p}) \qquad \delta_{p} = \delta_{p}^{o} C_{1\delta}^{p} \qquad \theta_{\delta}^{p} = \theta_{\eta\delta}^{p} - \theta_{\pi\delta}^{p} \tag{523}$$

where δ_{p}^{o} = zero ambient field value of the wall thickness of the cylindrical shell of the personnel compartment and which appears in equations (422) and (428). Because there is an approximately uniform ambient field in the cylindrical shell of the personnel compartment it follows that

$$r_{p\delta} = r_p + \delta_p \qquad C^p_{1r\delta} = C^p_{1\delta} = C^p_{1r} \qquad \theta^p_{r\delta} = \theta^p_{\delta} = \theta^p_{r} \tag{524}$$

$$\theta^p_{\eta r\delta} = \theta^p_{\eta\delta} = \theta^p_{\eta r} \qquad \theta^p_{\pi r\delta} = \theta^p_{\pi\delta} = \theta^p_{\pi r} \tag{525}$$

and equation (521) agrees with the third part of equation (428), and equation (522) can be written as

$$r_{p\delta} = r^o_{p\delta} C^p_{1r} \qquad \theta^p_{r\delta} = \theta^p_{\eta r} - \theta^p_{\pi r} \tag{526}$$

where $r^o_{p\delta}$ is given by equation (428). The complex number circumference of the outer surface of the cylindrical shell of the personnel compartment is given by

$$\bar{s}_{p\delta} = s_{p\delta} \exp(j\theta^p_{s\delta}) = 2\pi\bar{r}_{p\delta} \tag{527}$$

where

$$s_{p\delta} = 2\pi r_{p\delta} = 2\pi r^o_{p\delta} C^p_{1r} = s^o_{p\delta} C^p_{1r} \qquad \theta^p_{s\delta} = \theta^p_{r\delta} = \theta^p_{s} = \theta^p_{r} \tag{528}$$

where $s^o_{p\delta}$ is given by equation (429). The measured values of the radius and circumference of the personnel compartment are obtained from equations (518) and (520) as

$$r_{pm} = r_p \cos\theta^p_r = r^o_p C^p_{1r} \cos(\theta^p_{\eta r} - \theta^p_{\pi r}) \tag{529}$$

$$s_{pm} = s_p \cos\theta^p_s = s^o_p C^p_{1r} \cos(\theta^p_{\eta r} - \theta^p_{\pi r}) \tag{530}$$

and the measured values of the outer radius and circumference of the cylindrical shell of the personnel compartment are obtained from equations (526) and (528) as

$$r_{p\delta m} = r_{p\delta} \cos\theta^p_{r\delta} = r^o_{p\delta} C^p_{1r} \cos(\theta^p_{\eta r} - \theta^p_{\pi r}) \tag{531}$$

$$s_{p\delta m} = s_{p\delta} \cos\theta^p_{s\delta} = s^o_{p\delta} C^p_{1r} \cos(\theta^p_{\eta r} - \theta^p_{\pi r}) \tag{532}$$

The measured value of the shell thickness is

$$\delta_{pm} = \delta_p \cos\theta^p_{\delta} = \delta^o_p C^p_{1r} \cos(\theta^p_{\eta r} - \theta^p_{\pi r}) \tag{533}$$

and therefore equations (529), (531) and (533) give

$$r_{p\delta m} = r_{pm} + \delta_{pm} \tag{534}$$

The complex number length of the personnel compartment cylindrical shell can be written, with the help of equation (66), as

$$\bar{L}_p = L_p \exp(j\theta^p_L) \qquad L_p = L^o_p C^p_{1z} \qquad \theta^p_L = \theta^p_{\eta L} - \theta^p_{\pi L} \tag{535}$$

where L^o_p = zero ambient field value of the length of the personnel compartment and where C^p_{1z} = spatial coordinate factor for the length of the high-density wall of the personnel compartment and is given by equation (94). The measured values of the length of the personnel compartment of a polyphase rocket is obtained from equation (535) as

$$L_{pm} = L_p \cos\theta_L^p = L_p^o C_{1z}^p \cos(\theta_{\eta L}^p - \theta_{\pi L}^p) \tag{536}$$

Note that because $C_{1r}^p \geq 1$ and $C_{1z}^p \geq 1$ it follows from equations (94), (518), (520), (529) and (530) that

$$r_p \geq r_p^o \qquad r_{pm} \geq r_p^o \qquad s_p \geq s_p^o \qquad s_{pm} \geq s_p^o \tag{537}$$

and equations (94), (526), (528), (531) and (532) give

$$r_{p\delta} \geq r_{p\delta}^o \qquad r_{p\delta m} \geq r_{p\delta}^o \qquad s_{p\delta} \geq s_{p\delta}^o \qquad s_{p\delta m} \geq s_{p\delta}^o \tag{538}$$

while equations (94), (523), (533), (535) and (536) yield

$$\delta_p \geq \delta_p^o \qquad \delta_{pm} \geq \delta_p^o \qquad L_p \geq L_p^o \qquad L_{pm} \geq L_p^o \tag{539}$$

Therefore as in the case of the exhaust aperture system and combustion chamber walls, there will be small increases of the dimensions of the personnel compartment walls of a polyphase rocket due to the presence of ambient fields.

The complex number radius and length of the high-density liquid or solid fuel of the rocket body, which contains an ambient field of the form given in equation (61), are represented as

$$\bar{r}_b = r_b \exp(j\theta_r^b) \qquad \bar{L}_b = L_b \exp(j\theta_L^b) \tag{540}$$

where equation (67) gives

$$r_b = r_b^o C_{1r}^b \qquad L_b = L_b^o C_{1z}^b \tag{541}$$

where the high-density space coordinate factors C_{1r}^b and C_{1z}^b are given by equation (95) and where r_b^o and L_b^o = zero ambient field values of the radius and length of the rocket body as appears in equations (423) and (424). The measured values of the radius and length of the rocket body are given by

$$r_{bm} = r_b \cos\theta_r^b = r_b^o C_{1r}^b \cos\theta_r^b \qquad L_{bm} = L_b \cos\theta_L^b = L_b^o C_{1z}^b \cos\theta_L^b \tag{542}$$

which are real numbers. Equations (541) and (542) show that

$$r_b \geq r_b^o \qquad r_{bm} \geq r_b^o \qquad L_b \geq L_b^o \qquad L_{bm} \geq L_b^o \tag{543}$$

Therefore, as in the previous cases there will be small increases in the radius and the length of the rocket body due to the action of the applied ambient fields. These should be minor effects and not produce any structural weakening of the materials that constitute the exhaust aperture system, combustion chamber system, personnel chamber walls or the rocket body walls.

B. Complex Number Cross-Section Areas of the Rocket Exhaust Aperture Wall, Combustion Chamber Foramen Wall, Personnel Compartment Wall and the Rocket Body.

The complex number cross-section areas of the cylindrical shell of the exhaust aperture system in which there is a uniform ambient field of the form given by equation (58) is given by equations (474) through (482) as

$$\Delta\overline{A}_a = \Delta A_a \exp(j\theta^a_{\Delta A}) = \overline{A}_{a\delta} - \overline{A}_a = \pi(\overline{r}^2_{a\delta} - \overline{r}^2_a) \tag{544}$$

$$\Delta A_a = \Delta A^o_a (C^a_{1r})^2 \qquad \theta^a_{\Delta A} = \theta^a_{A\delta} = \theta^a_A = 2\theta^a_r \tag{545}$$

where

$$\overline{A}_a = A_a \exp(j\theta^a_A) = \pi\overline{r}^2_a \tag{546}$$

$$A_a = \pi r^2_a = A^o_a (C^a_{1r})^2 \qquad \theta^a_A = 2\theta^a_r \tag{547}$$

$$\overline{A}_{a\delta} = A_{a\delta} \exp(j\theta^a_{A\delta}) = \pi\overline{r}^2_{a\delta} \tag{548}$$

$$A_{a\delta} = \pi r^2_{a\delta} = A^o_{a\delta} (C^a_{1r})^2 \qquad \theta^a_{A\delta} = 2\theta^a_{r\delta} = 2\theta^a_r \tag{549}$$

where $\overline{A}_a$ and $\overline{A}_{a\delta}$ = complex number areas of the inner and outer cylinders of the exhaust tube, $\overline{r}_a$ and $\overline{r}_{a\delta}$ = complex number radii of the inner an outer cylinders of the exhaust tube as given in equations (474) through (490), θ^a_r and $\theta^a_{r\delta}$ = internal phase angles of these complex inner and outer radii and where ΔA^o_a, A^o_a and $A^o_{a\delta}$ = zero ambient field values of these areas as given by equations (420) and (425). The measured values of the complex number areas are obtained from equations (544) through (549) as

$$\Delta A_{am} = \Delta A_a \cos\theta^a_{\Delta A} = \Delta A^o_a (C^a_{1r})^2 \cos(2\theta^a_r) \tag{550}$$

$$A_{am} = A_a \cos\theta^a_A = A^o_a (C^a_{1r})^2 \cos(2\theta^a_r) \tag{551}$$

$$A_{a\delta m} = A_{a\delta} \cos\theta^a_{A\delta} = A^o_{a\delta} (C^a_{1r})^2 \cos(2\theta^a_r) \tag{552}$$

which are real numbers.

Similarly, the complex number cross-section areas of the cylindrical shell of the combustion chamber foramen in which there is a uniform ambient field of the form given by equation (59) is given by equations (496) through (504) as

$$\Delta\overline{A}_f = \Delta A_f \exp(j\theta^f_{\Delta A}) = \overline{A}_{f\delta} - \overline{A}_f = \pi(\overline{r}^2_{f\delta} - \overline{r}^2_f) \tag{553}$$

$$\Delta A_f = \Delta A^o_f (C^f_{1r})^2 \qquad \theta^f_{\Delta A} = \theta^f_{A\delta} = \theta^f_A = 2\theta^f_r \tag{554}$$

where

$$\overline{A}_f = A_f \exp(j\theta^f_A) = \pi\bar{r}_f^2 \tag{555}$$

$$A_f = \pi r_f^2 = A_f^o (C^f_{1r})^2 \qquad \theta^f_A = 2\theta^f_r \tag{556}$$

$$\overline{A}_{f\delta} = A_{f\delta} \exp(j\theta^f_{A\delta}) = \pi\bar{r}_{f\delta}^2 \tag{557}$$

$$A_{f\delta} = \pi r_{f\delta}^2 = A_{f\delta}^o (C^f_{1r})^2 \qquad \theta^f_{A\delta} = 2\theta^f_{r\delta} = 2\theta^f_r \tag{558}$$

where $\overline{A}_f$ and $\overline{A}_{f\delta}$ = complex number areas of the inner and outer cylinders of the combustion chamber, $\bar{r}_f$ and $\bar{r}_{f\delta}$ = complex number radii of the inner an outer cylinders of the combustion chamber as given in equations (496) through (512), θ^f_r and $\theta^f_{r\delta}$ = internal phase angles of these complex inner and outer radii and where ΔA^o_f, A^o_f and $A^o_{f\delta}$ = zero ambient field values of these areas as given by equations (421) and (426). The measured values of the complex number areas are obtained from equations (553) through (558) as

$$\Delta A_{fm} = \Delta A_f \cos\theta^f_{\Delta A} = \Delta A^o_f (C^f_{1r})^2 \cos(2\theta^f_r) \tag{559}$$

$$A_{fm} = A_f \cos\theta^f_A = A^o_f (C^f_{1r})^2 \cos(2\theta^f_r) \tag{560}$$

$$A_{f\delta m} = A_{f\delta} \cos\theta^f_{A\delta} = A^o_{f\delta} (C^f_{1r})^2 \cos(2\theta^f_r) \tag{561}$$

which are real numbers.

Similarly, the complex number cross-section areas of the cylindrical shell of the personnel compartment in which there is a uniform ambient field of the form given by equation (60) is given by equations (518) through (526) as

$$\Delta\overline{A}_p = \Delta A_p \exp(j\theta^p_{\Delta A}) = \overline{A}_{p\delta} - \overline{A}_p = \pi(\bar{r}_{p\delta}^2 - \bar{r}_p^2) \tag{562}$$

$$\Delta A_p = \Delta A^o_p (C^p_{1r})^2 \qquad \theta^p_{\Delta A} = \theta^p_{A\delta} = \theta^p_A = 2\theta^p_r \tag{563}$$

where

$$\overline{A}_p = A_p \exp(j\theta^p_A) = \pi\bar{r}_p^2 \tag{564}$$

$$A_p = \pi r_p^2 = A^o_p (C^p_{1r})^2 \qquad \theta^p_A = 2\theta^p_r \tag{565}$$

$$\overline{A}_{p\delta} = A_{p\delta} \exp(j\theta^p_{A\delta}) = \pi\bar{r}_{p\delta}^2 \tag{566}$$

$$A_{p\delta} = \pi r_{p\delta}^2 = A^o_{p\delta} (C^p_{1r})^2 \qquad \theta^p_{A\delta} = 2\theta^p_{r\delta} = 2\theta^p_r \tag{567}$$

where $\overline{A}_p$ and $\overline{A}_{p\delta}$ = complex number areas of the inner and outer cylinders of the personnel compartment, $\bar{r}_p$ and $\bar{r}_{p\delta}$ = complex number radii of the inner an outer cylinders of the personnel compartment as given in equations (518) through (534), θ^p_r and $\theta^p_{r\delta}$ = internal phase angles of these complex inner and outer radii and where ΔA^o_p, A^o_p and $A^o_{p\delta}$ = zero ambient field values of the three complex number areas

$\Delta\overline{A}_p$, $\overline{A}_p$ and $\overline{A}_{p\delta}$ and are as given by equations (422) and (427). The measured values of the complex number areas are obtained from equations (562) through (567) as

$$\Delta A_{pm} = \Delta A_p \cos\theta^p_{\Delta A} = \Delta A^o_p (C^p_{1r})^2 \cos(2\theta^p_r) \tag{568}$$

$$A_{pm} = A_p \cos\theta^p_A = A^o_p (C^p_{1r})^2 \cos(2\theta^p_r) \tag{569}$$

$$A_{p\delta m} = A_{p\delta} \cos\theta^p_{A\delta} = A^o_{p\delta} (C^p_{1r})^2 \cos(2\theta^f_r) \tag{570}$$

which are real numbers.

The complex number cross-sectional area of the high-density liquid or solid fuel of the body of a polyphase rocket in which there is an ambient field of the form described in equation (61) is given by

$$\overline{A}_b = A_b \exp(j\theta^b_A) = \pi\bar{r}_b^2 \qquad A_b = \pi r_b^2 = A^o_b (C^b_{1r})^2 \qquad \theta^b_A = 2\theta^b_r \tag{571}$$

where $\bar{r}_b$ = complex number radius of the rocket body as given by equations (540) and (541), r_b = magnitude of the complex number radius as given by equation (67), C^b_{1r} = radial space coordinate factor for high-density rocket body and which is given by equation (95), and where A^o_b = zero ambient field value of the cross-sectional area of the rocket body as given by equation (423). The measured value of the rocket body cross-sectional area is calculated as

$$A_{bm} = A_b \cos\theta^b_A = A^o_b (C^b_{1r})^2 \cos(2\theta^b_r) \tag{572}$$

which is a real number.

C. Complex Number Volumes of the Rocket Exhaust Aperture Wall, Combustion Chamber Foramen Wall, Personnel Compartment Wall and Rocket Body.

The rocket exhaust aperture system is represented by a cylindrical shell in which there is an ambient field that is described by equation (58). The complex number volume of the exhaust aperture system is calculated from equations (491) and (544) through (549) as

$$\Delta\overline{V}_a = \Delta V_a \exp(j\theta^a_{\Delta V}) = \Delta\overline{A}_a \overline{L}_a \tag{573}$$

where $\overline{L}_a$ = length of the exhaust aperture cylindrical shell as represented by equation (491) and where $\Delta\overline{A}_a$ = cross-sectional area of the exhaust aperture wall as given by equations (544) and (545). Then equation (573) gives

$$\Delta V_a = \Delta A_a L_a = \Delta V^o_a (C^a_{1r})^2 C^a_{1z} \qquad \theta^a_{\Delta V} = \theta^a_{\Delta A} + \theta^a_L = \theta^a_V = 2\theta^a_r + \theta^a_L \tag{574}$$

where ΔV^o_a = zero ambient field value of the volume of the exhaust aperture wall as given by equation (424) and where C^a_{1r} and C^a_{1z} are given by equation (92). The measured value of the volume of the exhaust aperture cylindrical shell is given as

$$\Delta V_{am} = \Delta V_a \cos\theta^a_{\Delta V} = \Delta V^o_a (C^a_{1r})^2 C^a_{1z} \cos(2\theta^a_r + \theta^a_L) \quad (575)$$

which is a real number.

In a similar fashion the cylindrical shell of the combustion chamber foramen system contains an ambient field described by equation (59). The complex number volume of the combustion chamber is calculated from equations (513) and (553) through (558) as

$$\Delta\overline{V}_f = \Delta V_f \exp(j\theta^f_{\Delta V}) = \Delta\overline{A}_f \overline{L}_f \quad (576)$$

where $\overline{L}_f$ = length of the combustion chamber cylindrical shell as represented by equation (513) and where $\Delta\overline{A}_f$ = cross-sectional area of the combustion chamber wall as given by equations (553) and (554). Then equation (576) gives

$$\Delta V_f = \Delta A_f L_f = \Delta V^o_f (C^f_{1r})^2 C^f_{1z} \qquad \theta^f_{\Delta V} = \theta^f_{\Delta A} + \theta^f_L = \theta^f_V = 2\theta^f_r + \theta^f_L \quad (577)$$

where ΔV^o_f = zero ambient field value of the volume of the combustion chamber wall as given by equation (424) and where C^f_{1r} and C^f_{1z} are given by equation (93). The measured value of the volume of the combustion chamber cylindrical shell is given as

$$\Delta V_{fm} = \Delta V_f \cos\theta^f_{\Delta V} = \Delta V^o_f (C^f_{1r})^2 C^f_{1z} \cos(2\theta^f_r + \theta^f_L) \quad (578)$$

which is a real number.

In a similar fashion the cylindrical shell of the personnel compartment contains an ambient field described by equation (60). The complex number volume of the personnel compartment cylindrical shell is calculated from equations (535) and (562) through (567) as

$$\Delta\overline{V}_p = \Delta V_p \exp(j\theta^p_{\Delta V}) = \Delta\overline{A}_p \overline{L}_p \quad (579)$$

where $\overline{L}_p$ = length of the personnel compartment cylindrical shell as represented by equation (535) and where $\Delta\overline{A}_p$ = cross-sectional area of the personnel compartment wall as given by equations (562) and (563). Then equation (579) gives

$$\Delta V_p = \Delta A_p L_p = \Delta V^o_p (C^p_{1r})^2 C^p_{1z} \qquad \theta^p_{\Delta V} = \theta^p_{\Delta A} + \theta^p_L = \theta^p_V = 2\theta^p_r + \theta^p_L \quad (580)$$

where ΔV^o_p = zero ambient field value of the volume of the personnel compartment wall as given by equation (424) and where C^p_{1r} and C^p_{1z} are given by equation (94). The measured value of the volume of the personnel compartment cylindrical shell is given as

$$\Delta V_{pm} = \Delta V_p \cos\theta^p_{\Delta V} = \Delta V^o_p (C^p_{1r})^2 C^p_{1z} \cos(2\theta^p_r + \theta^p_L) \quad (581)$$

which is a real number.

The complex number volume of the solid or liquid fuel of the cylindrical rocket body in which there is an ambient field described by equation (61) is obtained from equations (540) and (571) as

$$\overline{V}_b = V_b \exp(j\theta_V^b) = \overline{A}_b \overline{L}_b \tag{582}$$

where equations (540), (541), (571) and (582) give

$$V_b = A_b L_b = V_b^o (C_{1r}^b)^2 C_{1z}^b \qquad \theta_V^b = \theta_A^b + \theta_L^b = 2\theta_r^b + \theta_L^b \tag{583}$$

where C_{1r}^b and C_{1z}^b are space coordinate factors that are given by equation (95) and where V_b^o = zero ambient field value of the volume of the rocket body and is given by equation (424). The measured value of the volume of the body of a polyphase rocket is given by

$$V_{bm} = V_b \cos\theta_V^b = V_b^o (C_{1r}^b)^2 C_{1z}^b \cos(2\theta_r^b + \theta_L^b) \tag{584}$$

which is a real number.

5. MOLECULAR NUMBER, MOLECULAR NUMBER DENSITY, MASS AND MASS DENSITY OF THE COMPONENT SYSTEMS OF A POLYPHASE ROCKET. This section utilizes the theory of spatial coordinate factors that was reviewed in Section 1 to determine expressions for the molecular number, molecular number density, mass density and mass of the four high-density components of a polyphase rocket. This will be used subsequently for the calculation of the liftoff parameter of a polyphase rocket.

A. Complex Molecular Number and Molecular Number Density of the High Density Components of a Polyphase Rocket.

In the presence of ambient fields, the complex number of molecules $\overline{\mathcal{N}}_a$ in a complex number volume element $\overline{V}_a$ of the exhaust aperture wall, the complex number of molecules $\overline{\mathcal{N}}_f$ in a complex number volume element $\overline{V}_f$ of the combustion chamber foramen wall, the complex number of molecules $\overline{\mathcal{N}}_p$ in a complex number volume element $\overline{V}_p$ of the personnel compartment wall and the complex number of molecules $\overline{\mathcal{N}}_b$ in the volume $\overline{V}_b$ of the rocket body (which consists mainly of solid or liquid fuel) can be obtained from equations (249) and (250) as

$$\overline{\mathcal{N}}_a = \mathcal{N}_a \exp(j\theta_{\mathcal{N}}^a) \qquad \mathcal{N}_a = N_a \sec\theta_{\mathcal{N}}^a \tag{585}$$

$$\overline{\mathcal{N}}_f = \mathcal{N}_f \exp(j\theta_{\mathcal{N}}^f) \qquad \mathcal{N}_f = N_f \sec\theta_{\mathcal{N}}^f \tag{586}$$

$$\overline{\mathcal{N}}_p = \mathcal{N}_p \exp(j\theta_{\mathcal{N}}^p) \qquad \mathcal{N}_p = N_p \sec\theta_{\mathcal{N}}^p \tag{587}$$

$$\overline{\mathcal{N}}_b = \mathcal{N}_b \exp(j\theta_{\mathcal{N}}^b) \qquad \mathcal{N}_b = N_b \sec\theta_{\mathcal{N}}^b \tag{588}$$

where N_a, N_f, N_p and N_b = large integer number of molecules in the volume elements of the high-density exhaust aperture wall, combustion chamber foramen wall, personnel compartment wall and the rocket body respectively, and where the internal phase angles $\theta_{\mathcal{N}}^a, \theta_{\mathcal{N}}^f, \theta_{\mathcal{N}}^p$ and $\theta_{\mathcal{N}}^b$ are obtained from equation (252) as

$$\tan\theta_{\mathcal{N}}^a = (2\mathcal{E}^a / G_a^o)^{1/2} \qquad \tan\theta_{\mathcal{N}}^f = (2\mathcal{E}^f / G_f^o)^{1/2} \tag{589}$$

$$\tan\theta_{\mathcal{N}}^{p} = (2\mathfrak{E}^{p} / G_{p}^{o})^{1/2} \qquad \tan\theta_{\mathcal{N}}^{b} = (2\mathfrak{E}^{b} / G_{b}^{o})^{1/2} \qquad G_{b}^{o} = \varsigma_{b}^{o} / t_{b}^{o} \tag{590}$$

where the shearing energy densities $\mathfrak{E}^{a}, \mathfrak{E}^{f}, \mathfrak{E}^{p}$ and $\mathfrak{E}^{b}$ are given by equations (58) through (61), $G_{a}^{o}, G_{f}^{o}, G_{p}^{o}$ and G_{b}^{o} = shear moduli of the materials in the exhaust aperture walls, combustion chamber walls, personnel compartment walls and the mainly liquid or solid fuel rocket body respectively for zero ambient fields and $\varsigma_{b}^{o}, t_{b}^{o}$ = viscosity and time factor of a liquid fuel rocket body. The corresponding complex number volume elements for these high-density and large molecular number [hd,LN] systems are obtained from equations (258) and (264) or more directly from equation (277) as

$$\overline{V}_{a} = V_{a}\exp(j\theta_{V}^{a}) \qquad V_{a} = V_{a}^{o}\sec\theta_{\mathcal{N}}^{a}\sec\theta_{n}^{a} = V_{a}^{o}(C_{1r}^{a})^{2}C_{1z}^{a} \tag{591}$$

$$\overline{V}_{f} = V_{f}\exp(j\theta_{V}^{f}) \qquad V_{f} = V_{f}^{o}\sec\theta_{\mathcal{N}}^{f}\sec\theta_{n}^{f} = V_{f}^{o}(C_{1r}^{f})^{2}C_{1z}^{f} \tag{592}$$

$$\overline{V}_{p} = V_{p}\exp(j\theta_{V}^{p}) \qquad V_{p} = V_{p}^{o}\sec\theta_{\mathcal{N}}^{p}\sec\theta_{n}^{p} = V_{p}^{o}(C_{1r}^{p})^{2}C_{1z}^{p} \tag{593}$$

$$\overline{V}_{b} = V_{b}\exp(j\theta_{V}^{b}) \qquad V_{b} = V_{b}^{o}\sec\theta_{\mathcal{N}}^{b}\sec\theta_{n}^{b} = V_{b}^{o}(C_{1r}^{b})^{2}C_{1z}^{b} \tag{594}$$

where $V_{a}^{o}, V_{f}^{o}, V_{p}^{o}$ and V_{b}^{o} = zero ambient field values of the volume elements which contain the integer numbers of molecules N_a, N_f, N_p and N_b respectively, and where the internal phase angles $\theta_{n}^{a}, \theta_{n}^{f}, \theta_{n}^{p}$ and θ_{n}^{b} of the molecular number densities will now be determined.

The complex molecular number densities $\overline{n}_{a}, \overline{n}_{f}, \overline{n}_{p}$ and $\overline{n}_{b}$ of the [hd,LN] exhaust aperture wall, combustion chamber foramen wall, personnel compartment wall and rocket body respectively are obtained from equations (585) through (588) and (591) through (594) as

$$\overline{n}_{a} = n_{a}\exp(j\theta_{n}^{a}) = \overline{\mathcal{N}}_{a} / \overline{V}_{a} = \mathcal{N}_{a} / V_{a}\exp[j(\theta_{\mathcal{N}}^{a} - \theta_{V}^{a})] \tag{595}$$

$$n_{a} = \mathcal{N}_{a} / V_{a} = N_{a} / V_{a}\sec\theta_{\mathcal{N}}^{a} = n_{a}^{o}(C_{1r}^{a})^{-2}(C_{1z}^{a})^{-1}\sec\theta_{\mathcal{N}}^{a} = n_{a}^{o}\cos\theta_{n}^{a} \tag{596}$$

$$\theta_{n}^{a} = \theta_{\mathcal{N}}^{a} - \theta_{V}^{a} = \theta_{\mathcal{N}}^{a} - 2\theta_{r}^{a} - \theta_{L}^{a} \tag{597}$$

$$n_{a}^{o} = N_{a} / V_{a}^{o} \tag{598}$$

$$n_{am} = n_{a}\cos\theta_{n}^{a} = n_{a}^{o}\cos^{2}\theta_{n}^{a} = n_{a}^{o}\cos^{2}(\theta_{\mathcal{N}}^{a} - 2\theta_{r}^{a} - \theta_{L}^{a}) \tag{599}$$

$$\overline{n}_{f} = n_{f}\exp(j\theta_{n}^{f}) = \overline{\mathcal{N}}_{f} / \overline{V}_{f} = \mathcal{N}_{f} / V_{f}\exp[j(\theta_{\mathcal{N}}^{f} - \theta_{V}^{f})] \tag{600}$$

$$n_{f} = \mathcal{N}_{f} / V_{f} = N_{f} / V_{f}\sec\theta_{\mathcal{N}}^{f} = n_{f}^{o}(C_{1r}^{f})^{-2}(C_{1z}^{f})^{-1}\sec\theta_{\mathcal{N}}^{f} = n_{f}^{o}\cos\theta_{n}^{f} \tag{601}$$

$$\theta_{n}^{f} = \theta_{\mathcal{N}}^{f} - \theta_{V}^{f} = \theta_{\mathcal{N}}^{f} - 2\theta_{r}^{f} - \theta_{L}^{f} \tag{602}$$

$$n_{f}^{o} = N_{f} / V_{f}^{o} \tag{603}$$

$$n_{fm} = n_{f}\cos\theta_{n}^{f} = n_{f}^{o}\cos^{2}\theta_{n}^{f} = n_{f}^{o}\cos^{2}(\theta_{\mathcal{N}}^{f} - 2\theta_{r}^{f} - \theta_{L}^{f}) \tag{604}$$

$$\bar{n}_p = n_p \exp(j\theta_n^p) = \bar{\mathfrak{N}}_p / \bar{V}_p = \mathfrak{N}_p / V_p \exp[j(\theta_{\mathfrak{N}}^p - \theta_V^p)] \tag{605}$$

$$n_p = \mathfrak{N}_p / V_p = N_p / V_p \sec\theta_{\mathfrak{N}}^p = n_p^o (C_{1r}^p)^{-2} (C_{1z}^p)^{-1} \sec\theta_{\mathfrak{N}}^p = n_p^o \cos\theta_n^p \tag{606}$$

$$\theta_n^p = \theta_{\mathfrak{N}}^p - \theta_V^p = \theta_{\mathfrak{N}}^p - 2\theta_r^p - \theta_L^p \tag{607}$$

$$n_p^o = N_p / V_p^o \tag{608}$$

$$n_{pm} = n_p \cos\theta_n^p = n_p^o \cos^2\theta_n^p = n_p^o \cos^2(\theta_{\mathfrak{N}}^p - 2\theta_r^p - \theta_L^p) \tag{609}$$

$$\bar{n}_b = n_b \exp(j\theta_n^b) = \bar{\mathfrak{N}}_b / \bar{V}_b = \mathfrak{N}_b / V_b \exp[j(\theta_{\mathfrak{N}}^b - \theta_V^b)] \tag{610}$$

$$n_b = \mathfrak{N}_b / V_b = N_b / V_b \sec\theta_{\mathfrak{N}}^b = n_b^o (C_{1r}^b)^{-2} (C_{1z}^b)^{-1} \sec\theta_{\mathfrak{N}}^b = n_b^o \cos\theta_n^b \tag{611}$$

$$\theta_n^b = \theta_{\mathfrak{N}}^b - \theta_V^b = \theta_{\mathfrak{N}}^b - 2\theta_r^b - \theta_L^b \tag{612}$$

$$n_b^o = N_b / V_b^o \tag{613}$$

$$n_{bm} = n_b \cos\theta_n^b = n_b^o \cos^2\theta_n^b = n_b^o \cos^2(\theta_{\mathfrak{N}}^b - 2\theta_r^b - \theta_L^b) \tag{614}$$

where n_a^o, n_f^o, n_p^o and n_b^o = zero ambient field values of the molecular number densities of the materials composing the exhaust aperture system, combustion chamber foramen system, personnel compartment walls and the rocket body, and where $C_{1r}^a, C_{1z}^a, C_{1r}^f, C_{1z}^f, C_{1r}^p, C_{1z}^p, C_{1r}^b$ and C_{1z}^b = high-density spatial coordinate factors for the exhaust aperture system, combustion chamber foramen system, personnel compartment walls and the rocket body.

B. Complex Molecular Mass, Mass Density and Mass of the High-density Components of a Polyphase Rocket.

The complex number molecular masses $\bar{\mu}_a, \bar{\mu}_f, \bar{\mu}_p$ and $\bar{\mu}_b$ of the exhaust aperture Wall, combustion chamber foramen wall, personnel compartment wall and the rocet body corresponding respectively to the ambient field energy densities given in equations (58) through (61) are written as

$$\bar{\mu}_a = \mu_a \exp(j\theta_\mu^a) \qquad \mu_a = \mu_a^o f_{\mu a}(\mathfrak{E}^a) \qquad \theta_\mu^a = \theta_\mu^a(\mathfrak{E}^a) \tag{615}$$

$$\bar{\mu}_f = \mu_f \exp(j\theta_\mu^f) \qquad \mu_f = \mu_f^o f_{\mu f}(\mathfrak{E}^f) \qquad \theta_\mu^f = \theta_\mu^f(\mathfrak{E}^f) \tag{616}$$

$$\bar{\mu}_p = \mu_p \exp(j\theta_\mu^p) \qquad \mu_p = \mu_p^o f_{\mu p}(\mathfrak{E}^p) \qquad \theta_\mu^p = \theta_\mu^p(\mathfrak{E}^p) \tag{617}$$

$$\bar{\mu}_b = \mu_b \exp(j\theta_\mu^b) \qquad \mu_b = \mu_b^o f_{\mu b}(\mathfrak{E}^b) \qquad \theta_\mu^b = \theta_\mu^b(\mathfrak{E}^b) \tag{618}$$

where for zero ambient field conditions

$$f_{\mu a}(0) = 1 \qquad \theta_{\mu a}(0) = 0 \qquad f_{\mu f}(0) = 1 \qquad \theta_{\mu f}(0) = 0 \tag{619}$$

$$f_{\mu p}(0) = 1 \qquad \theta_{\mu p}(0) = 0 \qquad f_{\mu b}(0) = 1 \qquad \theta_{\mu b}(0) = 0 \tag{620}$$

and where $\mu_a^o, \mu_f^o, \mu_p^o$ and μ_b^o = zero ambient field values of the of the molecular masses of the materials in exhaust aperture walls, combustion chamber walls, personnel chamber walls and the rocket body as used in equations (416) through (419), and where the parameters $f_{\mu a}, f_{\mu f}, f_{\mu p}$ and $f_{\mu b}$ and the internal phase angles $\theta_\mu^a, \theta_\mu^f, \theta_\mu^p$ and θ_μ^b are depend on the energy density of the ambient fields and the atomic structure of the materials of the high-density components of a polyphase rocket which contain these fields. Their values can be determined approximately from the complex number of atoms in a molecule of the substances composing the four high-density macroscopic rocket component systems, which according to equations (249) and (251) are written as

$$\overline{\mathbb{N}}_a^{atom} = \mathbb{N}_a^{atom} \exp(j\theta_{\mathbb{N}atom}^a) \qquad \mathbb{N}_a^{atom} = N_a^{atom} \cos\theta_{\mathbb{N}atom}^a \tag{621}$$

$$\overline{\mathbb{N}}_f^{atom} = \mathbb{N}_f^{atom} \exp(j\theta_{\mathbb{N}atom}^f) \qquad \mathbb{N}_f^{atom} = N_f^{atom} \cos\theta_{\mathbb{N}atom}^f \tag{622}$$

$$\overline{\mathbb{N}}_p^{atom} = \mathbb{N}_p^{atom} \exp(j\theta_{\mathbb{N}atom}^p) \qquad \mathbb{N}_p^{atom} = N_p^{atom} \cos\theta_{\mathbb{N}atom}^p \tag{623}$$

$$\overline{\mathbb{N}}_b^{atom} = \mathbb{N}_b^{atom} \exp(j\theta_{\mathbb{N}atom}^b) \qquad \mathbb{N}_b^{atom} = N_b^{atom} \cos\theta_{\mathbb{N}atom}^b \tag{624}$$

where $N_a^{atom}, N_f^{atom}, N_p^{atom}$ and N_b^{atom} = small integer number of atoms in each molcule of the substances constituting the exhaust aperture walls, combustion chamber foramen walls, personnel compartment walls and the rocket body. The complex number molecular masses that appear in equations (615) through (618) can also be written in the following way

$$\overline{\mu}_a = m_{aatom}^o \overline{\mathbb{N}}_a^{atom} - c^{-2}\overline{B}_a^{mol} \qquad \overline{\mu}_f = m_{fatom}^o \overline{\mathbb{N}}_f^{atom} - c^{-2}\overline{B}_f^{mol} \tag{625}$$

$$\overline{\mu}_p = m_{patom}^o \overline{\mathbb{N}}_p^{atom} - c^{-2}\overline{B}_p^{mol} \qquad \overline{\mu}_b = m_{batom}^o \overline{\mathbb{N}}_b^{atom} - c^{-2}\overline{B}_b^{mol} \tag{626}$$

where $m_{aatom}^o, m_{fatom}^o, m_{patom}^o$ and m_{batom}^o = zero ambient field masses of an atom of the substances constituting the exhaust aperture walls, combustion chamber foramen walls, personnel compartment walls and the rocket body. To a first approximation the complex number molecular binding energies $\overline{B}_a^{mol}, \overline{B}_f^{mol}, \overline{B}_p^{mol}$ and $\overline{B}_b^{mol}$ in equations (625) and (626) can be neglected and after using equations (621) through (624) the magnitudes and internal phase angles of the molecular masses can be written as

$$\mu_a = m_{aatom}^o N_a^{atom} \cos\theta_{\mathbb{N}atom}^a = \mu_a^o \cos\theta_{\mathbb{N}atom}^a \qquad \theta_\mu^a = \theta_{\mathbb{N}atom}^a \tag{627}$$

$$\mu_f = m_{fatom}^o N_f^{atom} \cos\theta_{\mathbb{N}atom}^f = \mu_f^o \cos\theta_{\mathbb{N}atom}^f \qquad \theta_\mu^f = \theta_{\mathbb{N}atom}^f \tag{628}$$

$$\mu_p = m_{patom}^o N_p^{atom} \cos\theta_{\mathbb{N}atom}^p = \mu_p^o \cos\theta_{\mathbb{N}atom}^p \qquad \theta_\mu^p = \theta_{\mathbb{N}atom}^p \tag{629}$$

$$\mu_b = m_{batom}^o N_b^{atom} \cos\theta_{\mathbb{N}atom}^b = \mu_b^o \cos\theta_{\mathbb{N}atom}^b \qquad \theta_\mu^b = \theta_{\mathbb{N}atom}^b \tag{630}$$

A comparison of equations (615) through (618) with equations (627) through (630) gives

$$f_{\mu a} = \cos\theta_{\mathbb{N}atom}^a \qquad f_{\mu f} = \cos\theta_{\mathbb{N}atom}^f \qquad f_{\mu p} = \cos\theta_{\mathbb{N}atom}^p \qquad f_{\mu b} = \cos\theta_{\mathbb{N}atom}^b \tag{631}$$

and therefore the following inequalities are satisfied

$$f_{\mu a} \le 1 \qquad f_{\mu f} \le 1 \qquad f_{\mu p} \le 1 \qquad f_{\mu b} \le 1 \tag{632}$$

which in itself is an interesting conclusion which will be used to simplify the expression for the liftoff parameter of a polyphase rocket.

The complex number mass densities $\overline{\rho}_a, \overline{\rho}_f, \overline{\rho}_p$ and $\overline{\rho}_b$ of the exhaust aperture walls, combustion chamber foramen walls, personnel compartment walls and the rocket body can be written as

$$\overline{\rho}_a = \rho_a \exp(j\theta_\rho^a) = \overline{\mu}_a \overline{n}_a \qquad \overline{\rho}_f = \rho_f \exp(j\theta_\rho^f) = \overline{\mu}_f \overline{n}_f \tag{633}$$

$$\overline{\rho}_p = \rho_p \exp(j\theta_\rho^p) = \overline{\mu}_p \overline{n}_p \qquad \overline{\rho}_b = \rho_b \exp(j\theta_\rho^b) = \overline{\mu}_b \overline{n}_b \tag{634}$$

Combining equations (595) through (598), (600) through (603), (605) through (608) and (610) through (613) with equations (627) through (631), (633) and (634) gives the magnitudes and internal phase angles of the mass densities of the [hd,LN] components of a polyphase rocket as

$$\rho_a = \mu_a n_a = \rho_a^o f_{\mu a} (C_{1r}^a)^{-2} (C_{1z}^a)^{-1} \sec\theta_{\mathcal{N}}^a = \rho_a^o f_{\mu a} \cos\theta_n^a = \rho_a^o \cos\theta_{\mathcal{N}atom}^a \cos\theta_n^a \tag{635}$$

$$\rho_f = \mu_f n_f = \rho_f^o f_{\mu f} (C_{1r}^f)^{-2} (C_{1z}^f)^{-1} \sec\theta_{\mathcal{N}}^f = \rho_f^o f_{\mu f} \cos\theta_n^f = \rho_f^o \cos\theta_{\mathcal{N}atom}^f \cos\theta_n^f \tag{636}$$

$$\rho_p = \mu_p n_p = \rho_p^o f_{\mu p} (C_{1r}^p)^{-2} (C_{1z}^p)^{-1} \sec\theta_{\mathcal{N}}^p = \rho_p^o f_{\mu p} \cos\theta_n^p = \rho_p^o \cos\theta_{\mathcal{N}atom}^p \cos\theta_n^p \tag{637}$$

$$\rho_b = \mu_b n_b = \rho_b^o f_{\mu b} (C_{1r}^b)^{-2} (C_{1z}^b)^{-1} \sec\theta_{\mathcal{N}}^b = \rho_b^o f_{\mu b} \cos\theta_n^b = \rho_b^o \cos\theta_{\mathcal{N}atom}^b \cos\theta_n^b \tag{638}$$

$$\theta_\rho^a = \theta_\mu^a + \theta_n^a \qquad \theta_\rho^f = \theta_\mu^f + \theta_n^f \qquad \theta_\rho^p = \theta_\mu^p + \theta_n^p \qquad \theta_\rho^b = \theta_\mu^b + \theta_n^b \tag{639}$$

where $\theta_n^a, \theta_n^f, \theta_n^p$ and θ_n^b and $\theta_\mu^a, \theta_\mu^f, \theta_\mu^p$ and θ_μ^b are given by equations (597), (602), (607), (612) and (627) through (630) respectively. The measured values of the mass densities of the [hd,LN] components of a polyphase rocket are obtained from equations (635) through (639) as

$$\rho_{am} = \rho_a \cos\theta_\rho^a = \rho_a^o f_{\mu a} \cos_n^a \cos(\theta_\mu^a + \theta_n^a) = \rho_a^o \cos\theta_{\mathcal{N}atom}^a \cos_n^a \cos(\theta_\mu^a + \theta_n^a) \tag{640}$$

$$\rho_{fm} = \rho_f \cos\theta_\rho^f = \rho_f^o f_{\mu f} \cos_n^f \cos(\theta_\mu^f + \theta_n^f) = \rho_f^o \cos\theta_{\mathcal{N}atom}^f \cos_n^f \cos(\theta_\mu^f + \theta_n^f) \tag{641}$$

$$\rho_{pm} = \rho_p \cos\theta_\rho^p = \rho_p^o f_{\mu p} \cos_n^p \cos(\theta_\mu^p + \theta_n^p) = \rho_p^o \cos\theta_{\mathcal{N}atom}^p \cos_n^p \cos(\theta_\mu^p + \theta_n^p) \tag{642}$$

$$\rho_{bm} = \rho_b \cos\theta_\rho^b = \rho_b^o f_{\mu b} \cos_n^b \cos(\theta_\mu^b + \theta_n^b) = \rho_b^o \cos\theta_{\mathcal{N}atom}^b \cos_n^b \cos(\theta_\mu^b + \theta_n^b) \tag{643}$$

which are real numbers.

Now first consider the complex masses associated with the solid cylindrical volume elements that contain the numbers of molecules given by equations (585) through (588) which are

$$\overline{m}_a = m_a \exp(j\theta_m^a) = \overline{\mu}_a \overline{\mathcal{N}_a} = \overline{\rho}_a \overline{V}_a = \overline{\rho}_a \overline{A}_a \overline{L}_a \tag{644}$$

$$\overline{m}_f = m_f \exp(j\theta_m^f) = \overline{\mu}_f \overline{\mathcal{N}_f} = \overline{\rho}_f \overline{V}_f = \overline{\rho}_f \overline{A}_f \overline{L}_f \tag{645}$$

$$\overline{m}_p = m_p \exp(j\theta_m^p) = \overline{\mu}_p \overline{\mathcal{N}_p} = \overline{\rho}_p \overline{V}_p = \overline{\rho}_p \overline{A}_p \overline{L}_p \tag{646}$$

$$\overline{m}_b = m_b \exp(j\theta_m^b) = \overline{\mu}_b \overline{\mathcal{N}_b} = \overline{\rho}_b \overline{V}_b = \overline{\rho}_b \overline{A}_b \overline{L}_b \tag{647}$$

where

$$m_a = m_a^o f_{\mu a} \sec\theta_{\mathcal{N}}^a \qquad \theta_m^a = \theta_\mu^a + \theta_{\mathcal{N}}^a = \theta_\mu^a + \theta_n^a + \theta_V^a \tag{648}$$

$$m_f = m_f^o f_{\mu f} \sec\theta_{\mathcal{N}}^f \qquad \theta_m^f = \theta_\mu^f + \theta_{\mathcal{N}}^f = \theta_\mu^f + \theta_n^f + \theta_V^f \tag{649}$$

$$m_p = m_p^o f_{\mu p} \sec\theta_{\mathcal{N}}^p \qquad \theta_m^p = \theta_\mu^p + \theta_{\mathcal{N}}^p = \theta_\mu^p + \theta_n^p + \theta_V^p \tag{650}$$

$$m_b = m_b^o f_{\mu b} \sec\theta_{\mathcal{N}}^b \qquad \theta_m^b = \theta_\mu^b + \theta_{\mathcal{N}}^b = \theta_\mu^b + \theta_n^b + \theta_V^b \tag{651}$$

where the zero ambient field masses m_a^o, m_f^o, m_p^o and m_b^o of the small volume elements are given by equations (434) and (435). In a similar way the complex number masses of the polyphase rocket exhaust aperture walls, combustion chamber foramen walls, personnel compartment walls and the solid cylinder body of the rocket are written as

$$\Delta\overline{m}_a = \Delta m_a \exp(j\theta_m^a) = \overline{\mu}_a \Delta\overline{\mathcal{N}_a} = \overline{\rho}_a \Delta\overline{V}_a = \overline{\rho}_a \Delta\overline{A}_a \overline{L}_a \tag{652}$$

$$\Delta\overline{m}_f = \Delta m_f \exp(j\theta_m^f) = \overline{\mu}_f \Delta\overline{\mathcal{N}_f} = \overline{\rho}_f \Delta\overline{V}_f = \overline{\rho}_f \Delta\overline{A}_f \overline{L}_f \tag{653}$$

$$\Delta\overline{m}_p = \Delta m_p \exp(j\theta_m^p) = \overline{\mu}_p \Delta\overline{\mathcal{N}_p} = \overline{\rho}_p \Delta\overline{V}_p = \overline{\rho}_p \Delta\overline{A}_p \overline{L}_p \tag{654}$$

$$\overline{m}_b = m_b \exp(j\theta_m^b) = \overline{\mu}_b \overline{\mathcal{N}_b} = \overline{\rho}_b \overline{V}_b = \overline{\rho}_b \overline{A}_b \overline{L}_b \tag{655}$$

The magnitudes and internal phase angles in equations (652) through (655) for a polyphase rocket are given as in equations (648) through (651) by

$$\Delta m_a = \Delta m_a^o f_{\mu a} \sec\theta_{\mathcal{N}}^a = \Delta m_a^o \cos\theta_{\mathcal{N}atom}^a \sec\theta_{\mathcal{N}}^a \qquad \theta_m^a = \theta_\mu^a + \theta_{\mathcal{N}}^a \tag{656}$$

$$\Delta m_f = \Delta m_f^o f_{\mu f} \sec\theta_{\mathcal{N}}^f = \Delta m_f^o \cos\theta_{\mathcal{N}atom}^f \sec\theta_{\mathcal{N}}^f \qquad \theta_m^f = \theta_\mu^f + \theta_{\mathcal{N}}^f \tag{657}$$

$$\Delta m_p = \Delta m_p^o f_{\mu p} \sec\theta_{\mathcal{N}}^p = \Delta m_p^o \cos\theta_{\mathcal{N}atom}^p \sec\theta_{\mathcal{N}}^p \qquad \theta_m^p = \theta_\mu^p + \theta_{\mathcal{N}}^p \tag{658}$$

$$m_b = m_b^o f_{\mu b} \sec\theta_{\mathcal{N}}^b = m_b^o \cos\theta_{\mathcal{N}atom}^b \sec\theta_{\mathcal{N}}^b \qquad \theta_m^b = \theta_\mu^b + \theta_{\mathcal{N}}^b \tag{659}$$

where $\Delta m_a^o, \Delta m_f^o, \Delta m_p^o$ and m_b^o are given by equations (416) through (419). The measured masses of the high-density components of a polyphase rocket are obtained from equations (656) through (659) as

$$\Delta m_{am} = \Delta m_a \cos\theta_m^a = \Delta m_a^o f_{\mu a} \sec\theta_{\mathcal{N}}^a \cos\theta_m^a = \Delta m_a^o \cos\theta_{\mathcal{N}atom}^a \sec\theta_{\mathcal{N}}^a \cos\theta_m^a \tag{660}$$

$$\Delta m_{fm} = \Delta m_f \cos\theta_m^f = \Delta m_f^o f_{\mu f} \sec\theta_{\mathcal{N}}^f \cos\theta_m^f = \Delta m_f^o \cos\theta_{\mathcal{N}atom}^f \sec\theta_{\mathcal{N}}^f \cos\theta_m^f \tag{661}$$

$$\Delta m_{pm} = \Delta m_p \cos\theta^p_m = \Delta m^o_p f_{\mu p} \sec\theta^p_{\aleph} \cos\theta^p_m = \Delta m^o_p \cos\theta^p_{\aleph atom} \sec\theta^p_{\aleph} \cos\theta^p_m \qquad (662)$$

$$m_{bm} = m_b \cos\theta^b_m = m^o_b f_{\mu b} \sec\theta^b_{\aleph} \cos\theta^b_m = m^o_b \cos\theta^b_{\aleph atom} \sec\theta^b_{\aleph} \cos\theta^b_m \qquad (663)$$

The complex number total mass of a polyphase rocket is given by

$$\overline{m}_R = m_R \exp(j\theta^R_m) = \overline{m}_b + \Delta\overline{m}_a + \Delta\overline{m}_f + \Delta\overline{m}_p = \overline{Q}_R \overline{m}_b = Q_R m_b \exp[j(\theta^R_Q + \theta^b_m)] \qquad (664)$$

The real and imaginary parts of equation (664) are written as

$$m_R \cos\theta^R_m = m_b I \qquad m_R \sin\theta^R_m = m_b J \qquad (665)$$

where

$$I = \cos\theta^b_m + \Delta m_a / m_b \cos\theta^a_m + \Delta m_f / m_b \cos\theta^f_m + \Delta m_p / m_b \cos\theta^p_m \qquad (666)$$

$$J = \sin\theta^b_m + \Delta m_a / m_b \sin\theta^a_m + \Delta m_f / m_b \sin\theta^f_m + \Delta m_p / m_b \sin\theta^p_m \qquad (667)$$

Then equations (659), (666) and (667) give the magnitude and internal phase angle of the total mass of a polyphase rocket as

$$m_R = m_b (I^2 + J^2)^{1/2} = m_b Q_R = m^o_b f_{\mu b} Q_R \sec\theta^b_{\aleph} \qquad (668)$$

$$\tan\theta^R_m = J / I \qquad \theta^R_Q = \theta^R_m - \theta^b_m \qquad (669)$$

where

$$Q_R = (I^2 + J^2)^{1/2} = [1 + Q_{R1} + Q_{R2} + Q_{R3} + Q_{R4} + Q_{R5} + Q_{R6} + Q_{R7}]^{1/2} \qquad (670)$$

where

$$Q_{R1} = 2\Delta m_a / m_b \cos(\theta^b_m - \theta^a_m) \qquad (671)$$

$$Q_{R2} = 2\Delta m_f / m_b \cos(\theta^b_m - \theta^f_m) \qquad (672)$$

$$Q_{R3} = 2\Delta m_p / m_b \cos(\theta^b_m - \theta^p_m) \qquad (673)$$

$$Q_{R4} = 2(\Delta m_a / m_b)(\Delta m_f / m_b)\cos(\theta^a_m - \theta^f_m) \qquad (674)$$

$$Q_{R5} = 2(\Delta m_f / m_b)(\Delta m_p / m_b)\cos(\theta^f_m - \theta^p_m) \qquad (675)$$

$$Q_{R6} = 2(\Delta m_a / m_b)(\Delta m_p / m_b)\cos(\theta^a_m - \theta^p_m) \qquad (676)$$

$$Q_{R7} = (\Delta m_a / m_b)^2 + (\Delta m_f / m_b)^2 + (\Delta m_p / m_b)^2 \qquad (677)$$

where $\Delta m_a, \theta^a_m; \Delta m_f, \theta^f_m; \Delta m_p, \theta^p_m$ and m_b, θ^b_m are given by equations (656) through (659). The measured value of the total mass of a polyphase rocket is obtained from equations (631) and (668) through (670) as

$$m_{Rm} = m_R \cos\theta^R_m = m_b Q_R \cos\theta^R_m = m^o_b f_{\mu b} Q_R \sec\theta^b_{\aleph} \cos\theta^R_m \qquad (678)$$

In general because the fuel is part of the rocket body the following conditions are valid at liftoff

$$m_b >> \Delta m_a \qquad m_b >> \Delta m_f \qquad m_b >> \Delta m_p \qquad (679)$$

and therefore equations (659), (663) and (665) through (679) give

$$m_R \approx m_b \qquad m_{Rm} \approx m_{bm} \qquad \theta^R_m \approx \theta^b_m \qquad Q_R \approx 1 \qquad \theta^R_Q \approx 0 \qquad (680)$$

where m_{bm} = measured mass of rocket body as given by equation (663).

6. CALCULATION OF THE THRUST OF A POLYPHASE ROCKET. This section calculates the complex number thrust of a polyphase rocket and suggests that it is larger than the thrust of a comparable conventional chemical rocket.

A. Exhaust Gas Velocity and Acceleration.

For a polyphase rocket, the complex number exhaust gas velocity $\bar{v}_e$ and acceleration $\bar{a}_e$ are related to the complex number length $\bar{h}_e$ of the flight path of a gas molecule and to the corresponding complex number time of flight $\bar{t}_e$ by a simple generalization of the conventional real number equations. The complex number length of the flightpath and the time of flight and of the exhaust gases in the exhaust tube that contains an ambient field of the form given in equation (62) are represented as

$$\bar{h}_e = h_e \exp(j\theta^e_h) \qquad \bar{t}_e = t_e \exp(j\theta^e_t) \qquad (681)$$

whose magnitudes are related to the zero ambient field values h^o_e and t^o_e of the flight path distance and the time of flight by equations (8), (11), (12) and (144), (147), (148) respectively as

$$h_e = h^o_e C^e_{\beta z} \qquad t_e = t^o_e C^e_{\beta t} \qquad (682)$$

with $\beta = 2,3$. The complex number constant acceleration of the exhaust gases in the exhaust tube is represented by[13]

$$\bar{a}_e = a_e \exp(j\theta^e_a) = 2\bar{h}_e / \bar{t}^2_e \qquad (683)$$

whose magnitude and internal phase angle are given by[13]

$$a_e = a^o_e C^e_{\beta z} (C^e_{\beta t})^{-2} \qquad \theta^e_a = \theta^e_h - 2\theta^e_t \qquad a^o_e = 2h^o_e (t^o_e)^{-2} \qquad (684)$$

where a^o_e = zero ambient field value of the constant acceleration as described in equation (455). The complex number exhaust velocity for constant acceleration is given by[13]

$$\bar{v}_e = v_e \exp(j\theta^e_v) = \bar{a}_e \bar{t}_e \qquad (685)$$

whose magnitude and internal phase angle are

$$v_e = a_e t_e = v_e^o C_{\beta z}^e (C_{\beta t}^e)^{-1} \qquad \theta_v^e = \theta_a^e + \theta_t^e = \theta_h^e - \theta_t^e \qquad v_e^o = a_e^o t_e^o \tag{686}$$

The complex number flightpath length given in equations (681) through (683) can be written as

$$\bar{h}_e = h_e \exp(j\theta_h^e) = 1/2\bar{a}_e \bar{t}_e^2 \tag{687}$$

$$h_e = 1/2 a_e t_e^2 = h_e^o C_{\beta z}^e \qquad \theta_h^e = \theta_a^e + 2\theta_t^e \tag{688}$$

which also satisfies the relation

$$\bar{v}_e^2 = 2\bar{a}_e \bar{h}_e \qquad v_e^2 = 2a_e h_e \qquad 2\theta_v^e = \theta_a^e + \theta_h^e \qquad (v_e^o)^2 = 2a_e^o h_e^o \tag{689}$$

The condition $\beta = 2,3$ means that the exhaust gases are treated as a low-density system.

B. Complex Number Radius, Cross-Sectional Area and Volume of a Cylindrical Volume Element of Exhaust Gas in the Rocket Exhaust Tube.

The complex number radius and height of a cylinder of exhaust gas that contains a large integer number N_e of exhaust gas molecules are represented as

$$\bar{r}_e = r_e \exp(j\theta_r^e) \qquad \bar{z}_e = z_e \exp(j\theta_z^e) \tag{690}$$

where equation (19) gives for the low-density exhaust gases

$$r_e = r_e^o C_{\beta r}^e \qquad z_e = z_e^o C_{\beta z}^e \tag{691}$$

where $C_{\beta r}^e$ and $C_{\beta z}^e$ with $\beta = 2,3$ and θ_r^e and θ_z^e are given by equations (96) and (97), and where r_e^o and z_e^o are the zero ambient field values of the radius and length of the cylinder of gas. The measured values of the radius and length of the cylinder are given by

$$r_{em} = r_e \cos\theta_r^e = r_e^o C_{\beta r}^e \cos\theta_r^e \tag{692}$$

$$z_{em} = z_e \cos\theta_z^e = z_e^o C_{\beta z}^e \cos\theta_z^e \tag{693}$$

The complex number area of the base of the cylinder of gas is given by

$$\bar{A}_e = A_e \exp(j\theta_A^e) = \pi\bar{r}_e^2 \qquad A_e = \pi r_e^2 \qquad \theta_A^e = 2\theta_r^e \tag{694}$$

and then it follows from equations (691) and (694) that

$$A_e = A_e^o (C_{\beta r}^e)^2 \qquad A_e^o = \pi(r_e^o)^2 \tag{695}$$

where A_e^o = zero ambient field value of the base of the gas cylinder. The measured value of the area of the base of the cylinder is given by

$$A_{em} = A_e \cos\theta_A^e = A_e^o (C_{\beta r}^e)^2 \cos(2\theta_r^e) \tag{696}$$

The complex number volume of the cylinder of gas is given by

$$\overline{V}_e = V_e \exp(j\theta_V^e) = \overline{A}_e \overline{z}_e \tag{697}$$

whose magnitude and internal phase angle is given by

$$V_e = A_e z_e = \pi r_e^2 z_e = V_e^o (C_{\beta r}^e)^2 C_{\beta z}^e = V_e^o \sec\theta_{\mathbb{N}}^e \cos\theta_n^e \tag{698}$$

$$\theta_V^e = \theta_A^e + \theta_z^e = 2\theta_r^e + \theta_z^e \tag{699}$$

where V_e^o = zero ambient field value of the volume of the exhaust gas cylinder as given by equation (442). The measured value of the volume of the cylinder of exhaust gas is

$$V_{em} = V_e \cos\theta_V^e = V_e^o (C_{\beta r}^e)^2 C_{\beta z}^e \cos(2\theta_r^e + \theta_z^e) \tag{700}$$

where r_e, z_e, A_e and V_e refer to an arbitrary cylindrical volume of exhaust gas that contains a large integer number N_e of molecules in an ambient field described by equation (62).

C. Molecular Number, Molecular Number Density, Mass and Mass Density of the Exhaust Gases in the Polyphase Rocket Exhaust Tube.

For a polyphase rocket, the large complex number of molecules in a volume element of exhaust gas in the exhaust tube is written in the form of equations (249), (250) and (252) and in terms of the shear modulus G_e^o, viscosity ς_e^o and time factor t_e^o as

$$\overline{\mathbb{N}}_e = \mathbb{N}_e \exp(j\theta_{\mathbb{N}}^e) \qquad \mathbb{N}_e = N_e \sec\theta_{\mathbb{N}}^e \qquad \tan\theta_{\mathbb{N}}^e = (2\mathbb{E}^e / G_e^o)^{1/2} \qquad G_e^o = \varsigma_e^o / t_e^o \tag{701}$$

where N_e = large integer number of exhaust gas molecules that are located in a complex number volume element given by equation (697). The complex molecular number density of the [ld,LN] exhaust gases in the rocket exhaust tube is given by

$$\overline{n}_e = n_e \exp(j\theta_n^e) = \overline{\mathbb{N}}_e / \overline{V}_e = \mathbb{N}_e / V_e \exp[j(\theta_{\mathbb{N}}^e - \theta_V^e)] \tag{702}$$

so that after using equation (701) the magnitude and internal phase angles of the complex molecular number density can be written as

$$n_e = \mathbb{N}_e / V_e = N_e / V_e \sec\theta_{\mathbb{N}}^e \qquad \theta_n^e = \theta_{\mathbb{N}}^e - \theta_V^e = \theta_{\mathbb{N}}^e - 2\theta_r^e - \theta_z^e \tag{703}$$

The magnitude n_e of the complex molecular number density given in equation (703) can be rewritten by using equation (698) to represent V_e with the result that

$$n_e = n_e^o (C_{\beta r}^e)^{-2} (C_{\beta z}^e)^{-1} \sec\theta_{\mathbb{N}}^e = n_e^o \sec\theta_n^e \qquad n_e^o = N_e / V_e^o \tag{704}$$

and where equations (266) or (268) are used written in the form

$$\sec\theta_n^e = (C_{\beta r}^e)^{-2} (C_{\beta z}^e)^{-1} \sec\theta_{\mathbb{N}}^e \tag{705}$$

for an [ld,LN] system of exhaust gases has also been used to obtain the result in equation (705) and where n_e^o = zero ambient field value of the exhaust gas molecular number density as given by equation (441). The measured value of the exhaust gas molecular number density in the rocket exhaust tube

$$n_{em} = n_e \cos\theta_n^e = n_e^o \tag{706}$$

which is just the zero ambient field value.

The complex number mass associated with a complex number volume element $\overline{V}_e$ of exhaust gas in the exhaust tube of a polyphase rocket is given by

$$\overline{m}_e = m_e \exp(j\theta_m^e) = \overline{\mu}_e \overline{\mathfrak{N}}_e = \overline{\mu}_e \overline{n}_e \overline{V}_e \tag{707}$$

where the complex number molecular mass $\overline{\mu}_e$ depends on the energy density given in equation (62), and is written in analogy to equations (615) through (618) as

$$\overline{\mu}_e = \mu_e \exp(j\theta_\mu^e) \qquad \mu_e = \mu_e^o f_{\mu e}(\mathfrak{E}^e) \qquad \theta_\mu^e = \theta_\mu^e(\mathfrak{E}^e) \tag{708}$$

subject to the zero ambient field conditions

$$f_{\mu e}(0) = 1 \qquad \theta_{\mu e}(0) = 0 \tag{709}$$

and where μ_e^o = zero ambient field value of the average molecular mass of the exhaust gas. It will be assumed that the integer number of atoms in a gas molecule is small so that in analogy to equations (621) through (624) the complex number of atoms in an exhaust gas molecule of a polyphase rocket is written as

$$\overline{\mathfrak{N}}_e^{atom} = \mathfrak{N}_e^{atom} \exp(j\theta_{\mathfrak{N}atom}^e) \qquad \mathfrak{N}_e^{atom} = N_e^{atom} \cos\theta_{\mathfrak{N}atom}^e \tag{710}$$

where N_e^{atom} = small integer number of atoms per exhaust gas molecule. The complex number molecular mass of an exhaust gas molecule can be written in analogy to equations (625) and (626) as

$$\overline{\mu}_e = m_{eatom}^o \overline{\mathfrak{N}}_e^{atom} - c^{-2}\overline{B}_e^{mol} \tag{711}$$

where m_{eatom}^o = zero ambient field mass of an atom of exhaust gas. After neglecting the molecular binding energy $\overline{B}_e^{mol}$, the combination of equations (710) and (711) gives the magnitude and internal phase angle of the molecular mass of an exhaust gas molecule as

$$\mu_e = m_{eatom}^o N_e^{atom} \cos\theta_{\mathfrak{N}atom}^e = \mu_e^o \cos\theta_{\mathfrak{N}atom}^e \qquad \theta_\mu^e = \theta_{\mathfrak{N}atom}^e \tag{712}$$

And therefore a comparison of equation (708) and (712) shows that

$$f_{\mu e} = \cos\theta_{\mathfrak{N}atom}^e \qquad f_{\mu e} \le 1 \tag{713}$$

Then equation (707) gives the magnitude and the internal phase angle of the mass of a volume element of exhaust gas after using equations (701) and (713) as

$$m_e = \mu_e \mathcal{N}_e = \mu_e^o f_{\mu e} N_e \sec\theta_{\mathcal{N}}^e = m_e^o \cos\theta_{\mathcal{N}atom}^e \sec\theta_{\mathcal{N}}^e \qquad \theta_m^e = \theta_\mu^e + \theta_{\mathcal{N}}^e \quad (714)$$

where m_e^o = is given by equation (440). The measured mass of the exhaust gases in the exhaust tube is given by

$$m_{em} = m_e \cos\theta_m^e = m_e^o \cos\theta_{\mathcal{N}atom}^e \sec\theta_{\mathcal{N}}^e \cos(\theta_\mu^e + \theta_{\mathcal{N}}^e) \quad (715)$$

so that in an ambient field $m_{em} \le m_e^o$.

The complex number mass density of the exhaust gases in the rocket exhaust tube of a polyphase rocket is given by

$$\bar{\rho}_e = \rho_e \exp(j\theta_\rho^e) = \bar{m}_e / \bar{V}_e = \bar{\mu}_e \bar{n}_e \quad (716)$$

so that after using equations (266), (268), (698), (704), (708), (713) and (714) the magnitude of the exhaust gas mass density is given by

$$\rho_e = m_e/V_e = \mu_e n_e = \rho_e^o f_{\mu e} (C_{\beta r}^e)^{-2} (C_{\beta z}^e)^{-1} \sec\theta_{\mathcal{N}}^e = \rho_e^o f_{\mu e} \sec\theta_n^e = \rho_e^o \cos\theta_{\mathcal{N}atom}^e \sec\theta_n^e \quad (717)$$

where $C_{\beta r}^e$ and $C_{\beta z}^e$ are given by either equations (96) or (97), and where the internal phase angle of the exhaust gas mass density is given by

$$\theta_\rho^e = \theta_m^e - \theta_V^e = \theta_\mu^e + \theta_{\mathcal{N}}^e - \theta_V^e = \theta_\mu^e + \theta_n^e \quad (718)$$

The measured value of the mass density of the low-density rocket exhaust gas is obtained from equations (717) and (718) as

$$\rho_{em} = \rho_e \cos\theta_\rho^e = \rho_e^o f_{\mu e} \sec\theta_n^e \cos(\theta_\mu^e + \theta_n^e) = \rho_e^o \cos\theta_{\mathcal{N}atom}^e \sec\theta_n^e \cos(\theta_\mu^e + \theta_n^e) \quad (719)$$

Assuming that $\theta_\mu^e << \theta_n^e$, and $\theta_{\mathcal{N}atom}^e << \theta_n^e$ allows equations (717) and (719) to be written as the following approximations

$$\rho_e \approx \rho_e^o \sec\theta_n^e \qquad \rho_{em} \approx \rho_e^o \quad (720)$$

which shows that $\rho_e \ge \rho_e^o$ for a polyphase rocket.

D. Velocity and Acceleration of the Flow of Combustion Chamber Gas.

For a polyphase rocket, the complex number combustion chamber gas velocity $\bar{v}_c$ and acceleration $\bar{a}_c$ are related to the complex number length $\bar{h}_c$ of of the flight path of a gas molecule and to the corresponding complex number time of flight $\bar{t}_c$ by a simple generalization of the conventional real number equations. The complex number length of the flight path and the time of flight and of an element of combustion chamber gases that contains an ambient field of the form given in equation (63) are represented as

$$\bar{h}_c = h_c \exp(j\theta_h^c) \qquad \bar{t}_c = t_c \exp(j\theta_t^c) \quad (721)$$

whose magnitudes are related to the zero ambient field values h_c^o and t_c^o of the flight path distance and the time of flight by equations (8), (11), (12) and (144), (147), (148) respectively as

$$h_c = h_c^o C_{\beta z}^c \qquad t_c = t_c^o C_{\beta t}^c \tag{722}$$

with $\beta = 2,3$. The complex number constant acceleration of the combustion chamber gases is represented by[13]

$$\bar{a}_c = a_c \exp(j\theta_a^c) = 2\bar{h}_c / \bar{t}_c^2 \tag{723}$$

whose magnitude and internal phase angle are given by[13]

$$a_c = a_c^o C_{\beta z}^c (C_{\beta t}^c)^{-2} \qquad \theta_a^c = \theta_h^c - 2\theta_t^c \qquad a_c^o = 2h_c^o (t_c^o)^{-2} \tag{724}$$

where a_c^o = zero ambient field value of the constant acceleration as described in equation (455). For constant acceleration the complex number combustion gas velocity is given by[13]

$$\bar{v}_c = v_c \exp(j\theta_v^c) = \bar{a}_c \bar{t}_c \tag{725}$$

whose magnitude and internal phase angle are

$$v_c = a_c t_c = v_c^o C_{\beta z}^c (C_{\beta t}^c)^{-1} \qquad \theta_v^c = \theta_a^c + \theta_t^c = \theta_h^c - \theta_t^c \qquad v_c^o = a_c^o t_c^o \tag{726}$$

The complex number flight path length given in equations (721) through (723) can be written as

$$\bar{h}_c = h_c \exp(j\theta_h^c) = 1/2 \bar{a}_c \bar{t}_c^2 \tag{727}$$

$$h_c = 1/2 a_c t_c^2 = h_c^o C_{\beta z}^c \qquad \theta_h^c = \theta_a^c + 2\theta_t^c \tag{728}$$

which also satisfies the relation

$$\bar{v}_c^2 = 2\bar{a}_c \bar{h}_c \qquad v_c^2 = 2a_c h_c \qquad 2\theta_v^c = \theta_a^c + \theta_h^c \qquad (v_e^o)^2 = 2a_e^o h_e^o \tag{729}$$

The condition $\beta = 2,3$ means that the combustion chamber gases are treated as a low-density system.

E. Complex Number Pressure, Molecular Number Density and Mass Density of the Gases in the Combustion Chamber of a Polyphase Rocket.

In order to calculate the complex number gas pressure in the exhaust tube of a polyphase rocket it is first necessary to calculate the complex number gas pressure and mass density of the gases in the combustion chamber. First the combustion chamber gas pressure is calculated by considering a complex number volume element of combustion chamber gas given by

$$\bar{V}_c = V_c \exp(j\theta_V^c) = \bar{A}_c \bar{z}_c \tag{730}$$

where

$$\overline{A}_c = A_c \exp(j\theta_A^c) = \pi\bar{r}_c^2 \qquad A_c = \pi r_c^2 \qquad \theta_A^c = 2\theta_r^c \tag{731}$$

where

$$\bar{r}_c = r_c \exp(j\theta_r^c) \qquad \bar{z}_c = z_c \exp(j\theta_z^c) \tag{732}$$

where from equation (19) for low-density systems

$$r_c = r_c^o C_{\beta r}^c \qquad z_c = z_c^o C_{\beta z}^c \tag{733}$$

and where $C_{\beta r}^c$ and $C_{\beta z}^c$ with $\beta = 2,3$ and θ_r^c and θ_z^c are given by equations (98) and (99), and where r_c^o and z_c^o are the zero ambient field values of the radius and length of the cylinder of gas. The complex number area of the base of the cylinder of gas in the combustion chamber is given by

$$\overline{A}_c = A_c \exp(j\theta_A^c) = \pi\bar{r}_c^2 \qquad A_c = \pi r_c^2 \qquad \theta_A^c = 2\theta_r^c \tag{734}$$

and then equations (733) and (734) give

$$A_c = A_c^o (C_{\beta r}^c)^2 \qquad A_c^o = \pi(r_c^o)^2 \tag{735}$$

where A_c^o = zero ambient field value of the base of the gas cylinder. The complex number volume of the cylindrical volume element of gas in the combustion chamber is given by

$$\overline{V}_c = V_c \exp(j\theta_V^c) = \overline{A}_c \bar{z}_c \tag{736}$$

whose magnitude and internal phase angle is given by

$$V_c = A_c z_c = \pi r_c^2 z_c = V_c^o (C_{\beta r}^c)^2 C_{\beta z}^c = V_c^o \sec\theta_{\mathcal{N}}^c \cos\theta_n^c \tag{737}$$

$$\theta_V^c = \theta_A^c + \theta_z^c = 2\theta_r^c + \theta_z^c \tag{738}$$

where V_c^o = zero ambient field value of the volume of the combustion chamber gas cylinder as given by equation (450). As a first approximation, because the bulk matter velocity is low in the combustion chamber, it is assumed that the complex number pressure in the combustion chamber for a hot gas under the influence of an ambient field of the form given in equation (63) can be represented by an ideal gas equation of the form

$$\overline{P}_c = P_c \exp(j\theta_P^c) = \bar{n}_c k_B \overline{T}_c = (\overline{\mathcal{N}}_c / \overline{V}_c) k_B \overline{T}_c \tag{739}$$

where k_B = Boltzmann constant and where from equations (249), (250) and (252) with shear modulus G_c^o and time factor t_c^o

$$\overline{\mathcal{N}}_c = \mathcal{N}_c \exp(j\theta_{\mathcal{N}}^c) \qquad \mathcal{N}_c = N_c \sec\theta_{\mathcal{N}}^c \qquad \tan\theta_{\mathcal{N}}^c = (2\mathcal{E}^c / G_c^o)^{1/2} \qquad G_c^o = \varsigma_c^o / t_c^o \tag{740}$$

where ς_c^o = viscosity and N_c = large integer number of combustion chamber molecules located in a complex number volume element given by equation (736). The complex molecular number density of the [ld,LN] gases in the combustion chamber is given by

$$\bar{n}_c = n_c \exp(j\theta_n^c) = \bar{\mathbb{N}}_c / \bar{V}_c = \mathbb{N}_c / V_c \exp[j(\theta_{\mathbb{N}}^c - \theta_V^c)] \quad (741)$$

so that after using equation (740) the magnitude and internal phase angles of the complex molecular number density can be written as

$$n_c = \mathbb{N}_c / V_c = N_c / V_c \sec\theta_{\mathbb{N}}^c \qquad \theta_n^c = \theta_{\mathbb{N}}^c - \theta_V^c = \theta_{\mathbb{N}}^c - 2\theta_r^c - \theta_z^c \quad (742)$$

The magnitude n_c of the complex molecular number density in the combustion chamber given in equation (742) can be rewritten by using equation (737) to represent V_c with the result that

$$n_c = n_c^o (C_{\beta r}^c)^{-2} (C_{\beta z}^c)^{-1} \sec\theta_{\mathbb{N}}^c = n_c^o \sec\theta_n^c \qquad n_c^o = N_c / V_c^o \quad (743)$$

and where equations (266) or (268) are used written in the form

$$\sec\theta_n^c = (C_{\beta r}^c)^{-2} (C_{\beta z}^c)^{-1} \sec\theta_{\mathbb{N}}^c \quad (744)$$

for an [ld,LN] system of combustion chamber gases where $C_{\beta r}^c$ and $C_{\beta z}^c$ with $\beta = 2,3$ are given by equations (98) and (99), and where n_c^o = zero ambient field value of the molecular number density of the combustion chamber gases as given by equation (450). The complex number temperature $\bar{T}_c$ that appears in equation (739) is related to the complex number molecular mass $\bar{\mu}_c$ and the complex number root mean squared molecular velocity $\bar{v}_{\mu c}$ by

$$1/2\bar{\mu}_c \bar{v}_{\mu c}^2 = 3/2 k_B \bar{T}_c \quad 1/2\mu_c v_{\mu c}^2 = 3/2 k_B T_c \quad \theta_T^c = \theta_\mu^c + 2\theta_{v\mu}^c \quad (745)$$

where k_B = Boltzmann constant and where

$$\bar{\mu}_c = \mu_c \exp(j\theta_\mu^c) \qquad \bar{v}_{\mu c} = v_{\mu c} \exp(j\theta_{v\mu}^c) \qquad \bar{T}_c = T_c \exp(j\theta_T^c) \quad (746)$$

Equation (745) shows that

$$T_c = f_{Tc} T_c^o \qquad f_{Tc} = f_{\mu c} (C_{\beta\rho}^c)^2 (C_{\beta t}^c)^{-2} \qquad f_{\mu c} = \cos\theta_{\mathbb{N}atom}^c \quad (747)$$

where $f_{\mu c}$ is given by equations (752) through (758) and where for $\beta = 2,3$, equations (98) and (99) give

$$C_{2\rho}^c = \cos\theta_{\eta\rho}^c \cos\theta_{\mathrm{n}\rho}^c \quad (748)$$

$$C_{3\rho}^c = \sec\theta_{\eta\rho}^c \cos\theta_{\mathrm{n}\rho}^c \quad (749)$$

where ρ = radial coordinate in a spherical polar coordinate system ρ,φ,ψ and where $C_{\beta t}^c$ is given by equations (206) and (207). The magnitude and internal phase angle of the complex number combustion gas pressure can then be written as

$$P_c = n_c k_B T_c = P_c^o f_{Tc} \sec\theta_n^c \qquad \theta_P^c = \theta_n^c + \theta_T^c \tag{750}$$

where P_c^o = combustion gas pressure under zero ambient field condition and as given by equation (451). The measured pressure in the combustion chamber is given by

$$P_{cm} = P_c \cos\theta_P^c = P_c^o f_{Tc} \sec\theta_n^c \cos(\theta_n^c + \theta_T^c) \approx P_c^o \tag{751}$$

which is approximately equal to the zero ambient field pressure if $\theta_T^c << \theta_n^c$ and $f_{Tc} \approx 1$ as has generally been assumed in previous work.[8-11]

The calculation of the complex number mass density of the gases in the combustion chamber will now be accomplished. The complex number mass associated with a complex number volume element $\overline{V}_c$ of gas in the combustion chamber of a polyphase rocket is given by

$$\overline{m}_c = m_c \exp(j\theta_m^c) = \overline{\mu}_c \overline{\mathfrak{N}}_c = \overline{\mu}_c \overline{n}_c \overline{V}_c \tag{752}$$

where the complex number molecular mass $\overline{\mu}_c$ depends on the energy density given in equation (63), and is written in analogy to equation (708) as

$$\overline{\mu}_c = \mu_c \exp(j\theta_\mu^c) \qquad \mu_c = \mu_c^o f_{\mu c}(\mathfrak{E}^c) \qquad \theta_\mu^c = \theta_\mu^c(\mathfrak{E}^c) \tag{753}$$

subject to the zero ambient field conditions

$$f_{\mu c}(0) = 1 \qquad \theta_{\mu c}(0) = 0 \tag{754}$$

and where μ_c^o = zero ambient field value of the average molecular mass of the combustion chamber gas. It will be assumed that the integer number of atoms in a gas molecule is small so that in analogy to equation (710) the complex number of atoms in a combustion chamber gas molecule of a polyphase rocket is written as

$$\overline{\mathfrak{N}}_c^{\,atom} = \mathfrak{N}_c^{atom} \exp(j\theta_{\mathfrak{N}atom}^c) \qquad \mathfrak{N}_c^{atom} = N_c^{atom} \cos\theta_{\mathfrak{N}atom}^c \tag{755}$$

where N_c^{atom} = small integer number of atoms per combustion chamber gas molecule. The complex number molecular mass of a combustion chamber gas molecule can be written in analogy to equation (711) as

$$\overline{\mu}_c = m_{catom}^o \overline{\mathfrak{N}}_c^{\,atom} - c^{-2} \overline{B}_c^{mol} \tag{756}$$

where m_{catom}^o = zero ambient field mass of an atom of combustion chamber gas. After neglecting the complex number molecular binding energy $\overline{B}_c^{mol}$, the combination of equations (755) and (756) gives the magnitude and internal phase angle of the molecular mass of a combustion chamber gas molecule as

$$\mu_c = m_{catom}^o N_c^{atom} \cos\theta_{\mathfrak{N}atom}^c = \mu_c^o \cos\theta_{\mathfrak{N}atom}^c \qquad \theta_\mu^c = \theta_{\mathfrak{N}atom}^c \tag{757}$$

And therefore a comparison of equation (753) and (757) shows that

$$f_{\mu c} = \cos\theta_{\mathfrak{N}atom}^c \qquad f_{\mu c} \le 1 \tag{758}$$

and $f_{\mu c} \approx 1$ if $\theta^c_{Natom} \approx 0$. Then equation (752) gives the magnitude and the internal phase angle of the mass of a volume element of combustion chamber gas after using equations (740) and (758) as

$$m_c = \mu_c \mathcal{N}_c = \mu^o_c f_{\mu c} N_c \sec\theta^c_{\mathcal{N}} = m^o_c \cos\theta^c_{Natom} \sec\theta^c_{\mathcal{N}} \qquad \theta^c_m = \theta^c_\mu + \theta^c_{\mathcal{N}} \qquad (759)$$

where m^o_c = is given by equation (449). The measured mass of a volume element of the combustion chamber gases of a polyphase rocket is given by

$$m_{cm} = m_c \cos\theta^c_m = m^o_c \cos\theta^c_{Natom} \sec\theta^c_{\mathcal{N}} \cos(\theta^c_\mu + \theta^c_{\mathcal{N}}) \qquad (760)$$

so that in an ambient field $m_{cm} \le m^o_c$.

The complex number mass density of the gases in the combustion chamber of a polyphase rocket is given by

$$\overline{\rho}_c = \rho_c \exp(j\theta^c_\rho) = \overline{m}_c / \overline{V}_c = \overline{\mu}_c \overline{n}_c \qquad (761)$$

so that after using equations (266), (268), (737), (743), (753), (758) and (759) the magnitude of the combustion chamber gas mass density is given by

$$\rho_c = m_c/V_c = \mu_c n_c = \rho^o_c f_{\mu c} (C^c_{\beta r})^{-2} (C^c_{\beta z})^{-1} \sec\theta^c_{\mathcal{N}} = \rho^o_c f_{\mu c} \sec\theta^c_n = \rho^o_c \cos\theta^c_{Natom} \sec\theta^c_n \qquad (762)$$

where $C^c_{\beta r}$ and $C^c_{\beta z}$ are given by either equations (98) or (99), and where the internal phase angle of the combustion chamber gas mass density is given by

$$\theta^c_\rho = \theta^c_m - \theta^c_V = \theta^c_\mu + \theta^c_{\mathcal{N}} - \theta^c_V = \theta^c_\mu + \theta^c_n \qquad (763)$$

The measured value of the mass density of the low-density combustion chamber gas is obtained from quations (762) and (763) as

$$\rho_{cm} = \rho_c \cos\theta^c_\rho = \rho^o_c f_{\mu c} \sec\theta^c_n \cos(\theta^c_\mu + \theta^c_n) = \rho^o_c \cos\theta^c_{\mathcal{N}} \sec\theta^c_n \cos(\theta^c_\mu + \theta^c_n) \qquad (764)$$

which shows that $\rho_{cm} \le \rho^o_c$ if $\theta^c_\mu \ge 0$.

F. Calculation of the Complex Number Exhaust Gas Pressure.

The complex number exhaust gas pressure is required for the calculation of the thrust of a polyphase rocket and is written as

$$\overline{P}_e = P_e \exp(j\theta^e_P) \qquad (765)$$

The magnitude P_e and internal phase angle θ^e_P are complicated functions of the chemical burn rate in the combustion chamber, nozzle length and area and the strength of the applied ambient fields. The rocket nozzle is designed to convert the low velocity, low-density, high temperature and high-pressure gases in the combustion chamber into the high-velocity, low-density, low-temperature and low-pressure gases of the rocket exhaust. High exhaust velocity means large thrust. Thus the exhaust gas velocity and pressure depend on the thermodynamic properties of the gases. In this paper, for heuristic purposes to show the effects of ambient fields on the thrust and liftoff conditions of a polyphase rocket, a very simple relation between the exhaust gas velocity and pressure and the combustion chamber gas velocity and pressure will be assumed that is based on Bernoulli's law. Therefore in analogy to equation (447) that is valid for the case of zero ambient fields, the Bernoulli equation for a polyphase rocket is written as the following complex number equation

$$1/2\overline{\rho}_e\overline{v}_e^2 + \overline{P}_e = 1/2\overline{\rho}_c\overline{v}_c^2 + \overline{P}_c \qquad (766)$$

where $\overline{\rho}_e, \overline{v}_e, \overline{P}_e$ and $\overline{\rho}_c, \overline{v}_c, \overline{P}_c$ = complex number mass density, flow velocity and pressure of the exhaust gases and combustion chamber gases respectively as given by equations (685), (716), (765) and (761), (725), (739). In addition the flow of gas in a polyphase rocket must satisfy a mass conservation equation that is given, in analogy to equation (453) by

$$\overline{\rho}_e\overline{v}_e\overline{A}_a = \overline{\rho}_c\overline{v}_c\overline{A}_f \qquad (767)$$

where $\overline{A}_a$ and $\overline{A}_f$ = complex number areas of the exhaust gas aperture and the combustion chamber foramen respectively as given by equations (546), (547) and (555), (556). The mass conservation equation (767) can be used to to eliminate the combustion chamber velocity from Bernoulli's equation (766) for a polyphase rocket with the result that

$$\overline{P}_e = \overline{P}_c - 1/2\overline{\rho}_e\overline{v}_e^2[1 - (\overline{\rho}_e/\overline{\rho}_c)(\overline{A}_a/\overline{A}_f)^2] \qquad (768)$$

which is the analog of equation (454). The real an imaginary parts of equation (768) are easily determined to be

$$P_e \cos\theta_P^e = I_x \qquad P_e \sin\theta_P^e = I_y \qquad (769)$$

where

$$I_x = P_c \cos\theta_P^c - 1/2\rho_e v_e^2[\cos\Psi_1 - (\rho_e/\rho_c)(A_a/A_f)^2 \cos\Psi_2] \qquad (770)$$

$$I_y = P_c \sin\theta_P^c - 1/2\rho_e v_e^2[\sin\Psi_1 - (\rho_e/\rho_c)(A_a/A_f)^2 \sin\Psi_2] \qquad (771)$$

where P_c and θ_P^c are given by equation (750) and where the angles Ψ_1 and Ψ_2 are given by

$$\Psi_1 = \theta_\rho^e + 2\theta_v^e \qquad \Psi_2 = 2\theta_\rho^e - \theta_\rho^c + 2\theta_v^e + 2(\theta_A^a - \theta_A^f) \qquad (772)$$

Then the magnitude and the internal phase angle of the complex number exhaust gas pressure is given as

$$P_e = (I_x^2 + I_y^2)^{1/2} \qquad \tan\theta_P^e = I_y/I_x \qquad P_e \le P_c \qquad (773)$$

Combining equations (769) through (773) allows the magnitude P_e to be rewritten simply as

$$P_e = (P_c^2 + S_1 + S_2 + S_3 + S_4 + S_5)^{1/2} \tag{774}$$

where

$$S_1 = -\rho_e v_e^2 P_c \cos(\Psi_1 - \theta_P^c) \tag{775}$$

$$S_2 = \rho_e v_e^2 P_c (\rho_e / \rho_c)(A_a / A_f)^2 \cos(\Psi_2 - \theta_P^c) \tag{776}$$

$$S_3 = -1/2(\rho_e v_e^2)^2 (\rho_e / \rho_c)(A_a / A_f)^2 \cos(\Psi_2 - \Psi_1) \tag{777}$$

$$S_4 = 1/4(\rho_e v_e^2)^2 \tag{778}$$

$$S_5 = 1/4(\rho_e v_e^2)^2 (\rho_e / \rho_c)^2 (A_a / A_f)^4 \tag{779}$$

The measured value of the exhaust gas pressure is given by

$$P_{em} = P_e \cos\theta_P^e = I_x \tag{780}$$

where I_x is given by equation (770).

G. Calculation of the Complex Number Atmospheric Pressure.

The atmospheric pressure is a complex number due to the gravitational field of the Earth which is assumed to be an ambient field in matter. The atmospheric gas pressure is calculated by considering a complex number volume element of gas in the atmosphere which given by

$$\overline{V}_{at} = V_{at} \exp(j\theta_V^{at}) = \overline{A}_{at}\overline{z}_{at} \tag{781}$$

where

$$\overline{A}_{at} = A_{at} \exp(j\theta_A^{at}) = \pi\overline{r}_{at}^2 \qquad A_{at} = \pi r_{at}^2 \qquad \theta_A^{at} = 2\theta_r^{at} \tag{782}$$

where

$$\overline{r}_{at} = r_{at} \exp(j\theta_r^{at}) \qquad \overline{z}_{at} = z_{at} \exp(j\theta_z^{at}) \tag{783}$$

where from equation (19) for low-density [ld] systems

$$r_{at} = r_{at}^o C_{\beta r}^{at} \qquad z_{at} = z_{at}^o C_{\beta z}^{at} \tag{784}$$

and where $C_{\beta r}^{at}$ and $C_{\beta z}^{at}$ with $\beta = 2,3$ and θ_r^{at} and θ_z^{at} are obtained from equations (11) and (12) as

$$C_{2\alpha}^{at} = \cos\theta_{\eta\alpha}^{at} \cos\theta_{\pi\alpha}^{at} \qquad \theta_\alpha^{at} = \theta_{\eta\alpha}^{at} - \theta_{\pi\alpha}^{at} \qquad \text{atmospheric gases} \tag{785}$$

$$C_{3\alpha}^{at} = \sec\theta_{\eta\alpha}^{at} \cos\theta_{\pi\alpha}^{at} \qquad \theta_\alpha^{at} = \theta_{\eta\alpha}^{at} - \theta_{\pi\alpha}^{at} \qquad \text{atmospheric gases} \tag{786}$$

with $\alpha = r,z$, and where r^{o}_{at} and z^{o}_{at} are the zero ambient field values of the radius and length of a cylinder of atmospheric gases. The complex number area of the base of the cylinder of [ld] atmospheric gases is given by

$$\overline{A}_{at} = A_{at} \exp(j\theta^{at}_{A}) = \pi \overline{r}^{2}_{at} \qquad A_{at} = \pi r^{2}_{at} \qquad \theta^{at}_{A} = 2\theta^{at}_{r} \tag{787}$$

and then equations (784) and (787) give for [ld] atmospheric gases

$$A_{at} = A^{o}_{at} (C^{at}_{\beta r})^{2} \qquad A^{o}_{at} = \pi (r^{o}_{at})^{2} \tag{788}$$

where A^{o}_{at} = zero ambient field value of the base of the gas cylinder. The complex number volume of the cylindrical volume element of atmospheric gases is given by the following equation

$$\overline{V}_{at} = V_{at} \exp(j\theta^{at}_{V}) = \overline{A}_{at} \overline{z}_{at} \tag{789}$$

whose magnitude and internal phase angle is given by

$$V_{at} = A_{at} z_{at} = \pi r^{2}_{at} z_{at} = V^{o}_{at} (C^{at}_{\beta r})^{2} C^{at}_{\beta z} = V^{o}_{at} \sec\theta^{at}_{\mathfrak{N}} \cos\theta^{at}_{n} \tag{790}$$

$$\theta^{at}_{V} = \theta^{at}_{A} + \theta^{at}_{z} = 2\theta^{at}_{r} + \theta^{at}_{z} \tag{791}$$

where V^{o}_{at} = zero ambient field value of the volume of the cylinder of atmospheric gases as given by

$$V^{o}_{at} = A^{o}_{at} z^{o}_{at} = \pi (r^{o}_{at})^{2} z^{o}_{at} \qquad A^{o}_{at} = \pi (r^{o}_{at})^{2} \tag{792}$$

The ambient field acting on the atmosphere can be written in analogy to equations (62) and (63) as

$$\mathfrak{E}^{at} = \mathfrak{E}^{gat} + \mathfrak{E}^{\gamma at} \qquad \text{atmospheric gases} \tag{793}$$

where the dominant ambient field in the atmosphere is due to gravity but thermal diffusion and fluid flow fields are also important. As a first approximation it is assumed that the complex number pressure of the [ld] atmosphere under the influence of an ambient field of the form given in equation (793) can be represented by an ideal gas equation of the form

$$\overline{P}_{at} = P_{at} \exp(j\theta^{at}_{P}) = \overline{n}_{at} k_{B} \overline{T}_{at} = (\overline{\mathfrak{N}}_{at} / \overline{V}_{at}) k_{B} \overline{T}_{at} \tag{794}$$

where from equations (249), (250) and (252)

$$\overline{\mathfrak{N}}_{at} = \mathfrak{N}_{at} \exp(j\theta^{at}_{\mathfrak{N}}) \qquad \mathfrak{N}_{at} = N_{at} \sec\theta^{at}_{\mathfrak{N}} \qquad \tan\theta^{at}_{\mathfrak{N}} = (2\mathfrak{E}^{at} / G^{o}_{at})^{1/2} \qquad G^{o}_{at} = \varsigma^{o}_{at} / t^{o}_{at} \tag{795}$$

where N_{at} = large integer [LN] number of atmospheric gas molecules that is located in a complex number volume element given by equation (789) and where ς^{o}_{at} and t^{o}_{at} = viscosity and time factor of the atmospheric gases. The complex molecular number density of the [ld,LN] gases in the atmosphere at ground level is given by

$$\overline{n}_{at} = n_{at}\exp(j\theta_n^{at}) = \overline{\mathbb{N}}_{at} / \overline{V}_{at} = \mathbb{N}_{at} / V_{at}\exp[j(\theta_{\mathbb{N}}^{at} - \theta_V^{at})] \tag{796}$$

so that after using equation (795) the magnitude and internal phase angles of the complex molecular number density of the atmospheric gases can be written as

$$n_{at} = \mathbb{N}_{at} / V_{at} = N_{at} / V_{at}\sec\theta_{\mathbb{N}}^{at} \qquad \theta_n^{at} = \theta_{\mathbb{N}}^{at} - \theta_V^{at} = \theta_{\mathbb{N}}^{at} - 2\theta_r^{at} - \theta_z^{at} \tag{797}$$

The magnitude n_{at} of the complex molecular number density of the atmospheric gases given in equation (797) can be rewritten by using equation (790) to represent V_{at} with the result that

$$n_{at} = n_{at}^o (C_{\beta r}^{at})^{-2} (C_{\beta z}^{at})^{-1} \sec\theta_{\mathbb{N}}^{at} = n_{at}^o \sec\theta_n^{at} \qquad n_{at}^o = N_{at} / V_{at}^o \tag{798}$$

and where equations (266) or (268) have been used written in the form

$$\sec\theta_n^{at} = (C_{\beta r}^{at})^{-2} (C_{\beta z}^{at})^{-1} \sec\theta_{\mathbb{N}}^{at} \tag{799}$$

for an [ld,LN] cylindrical volume element of atmospheric gases where $C_{\beta r}^{at}$ and $C_{\beta z}^{at}$ with $\beta = 2,3$ are given by equations (785) and (786), and where n_{at}^o = zero ambient field value of the molecular number density of the atmospheric gases as given by

$$n_{at}^o = N_{at} / V_{at}^o \tag{800}$$

The complex number mass associated with a complex number volume element $\overline{V}_{cat}$ of atmospheric gases is given by

$$\overline{m}_c = m_c \exp(j\theta_m^c) = \overline{\mu}_c \overline{\mathbb{N}}_c = \overline{\mu}_c \overline{n}_c \overline{V}_c \tag{801}$$

where the complex number molecular mass $\overline{\mu}_{at}$ depends on the energy density given in equation (793), and is written in analogy to equation (708) as

$$\overline{\mu}_{at} = \mu_{at}\exp(j\theta_\mu^{at}) \qquad \mu_{at} = \mu_{at}^o f_{\mu at}(\mathfrak{E}^{at}) \qquad \theta_\mu^{at} = \theta_\mu^{at}(\mathfrak{E}^{at}) \tag{802}$$

subject to the zero ambient field conditions

$$f_{\mu at}(0) = 1 \qquad \theta_{\mu at}(0) = 0 \tag{803}$$

and where μ_{at}^o = zero ambient field value of the average molecular mass of the atmospheric gases. It will be assumed that the integer number of atoms in a gas molecule is small so that in analogy to equation (710) the complex number of atoms in a gas molecule of the atmosphere is written as

$$\overline{\mathbb{N}}_{at}^{atom} = \mathbb{N}_{at}^{atom}\exp(j\theta_{\mathbb{N}atom}^{at}) \qquad \mathbb{N}_{at}^{atom} = N_{at}^{atom}\cos\theta_{\mathbb{N}atom}^{at} \tag{804}$$

where N_{at}^{atom} = small integer number of atoms per atmospheric gas molecule. The complex number molecular mass of an atmospheric gas molecule can be written in analogy to equation (711) as

$$\overline{\mu}_{at} = m_{atatom}^o \overline{\mathbb{N}}_{at}^{atom} - c^{-2}\overline{B}_{at}^{mol} \tag{805}$$

where m^o_{catom} = zero ambient field mass of an atom of combustion chamber gas. After neglecting the complex number molecular binding energy $\overline{B}^{mol}_{at}$, the combination of equations (804) and (805) gives the magnitude and internal phase angle of the molecular mass of an atmospheric gas molecule as

$$\mu_{at} = m^o_{atatom} N^{atom}_{at} \cos\theta^{at}_{Natom} = \mu^o_{at} \cos\theta^{at}_{Natom} \qquad \theta^{at}_{\mu} = \theta^{at}_{Natom} \qquad (806)$$

And therefore a comparison of equation (802) and (806) gives the following expression for $f_{\mu at}$

$$f_{\mu at} = \cos\theta^{at}_{Natom} \qquad f_{\mu at} \leq 1 \qquad (807)$$

and $f_{\mu at} \approx 1$ if $\theta^{at}_{Natom} \approx 0$.The complex number temperature $\overline{T}_{at}$ that appears in equation (794) is related to the complex number molecular mass $\overline{\mu}_{at}$ and the complex number root mean squared molecular velocity $\overline{v}_{\mu at}$ by

$$1/2\overline{\mu}_{at}\overline{v}^2_{\mu at} = 3/2k_B\overline{T}_{at} \qquad 1/2\mu_{at}v^2_{\mu at} = 3/2k_B T_{at} \qquad \theta^{at}_T = \theta^{at}_{\mu} + 2\theta^{at}_{v\mu} \qquad (808)$$

where

$$\overline{\mu}_{at} = \mu_{at}\exp(j\theta^{at}_{\mu}) \qquad \overline{v}_{\mu at} = v_{\mu at}\exp(j\theta^{at}_{v\mu}) \qquad \overline{T}_{at} = T_{at}\exp(j\theta^{at}_T) \qquad (809)$$

Equation (808) shows that

$$T_{at} = f_{Tat}T^o_{at} \qquad f_{Tat} = f_{\mu at}(C^{at}_{\beta\rho})^2(C^{at}_{\beta t})^{=2} \qquad (810)$$

where $f_{\mu at}$ is given by equations (802) and (807) and where for $\beta = 2,3$ equations (11), (12), (147) and (148) give

$$C^{at}_{2\rho} = \cos\theta^{at}_{\eta\rho}\cos\theta^{at}_{\pi\rho} \qquad C^{at}_{2t} = \cos\theta^{at}_{\eta t}\cos\theta^{at}_{\pi t} \qquad (811)$$

$$C^{at}_{3\rho} = \sec\theta^{at}_{\eta\rho}\cos\theta^{at}_{\pi\rho} \qquad C^{at}_{3t} = \sec\theta^{at}_{\eta t}\cos\theta^{at}_{\pi t} \qquad (812)$$

where ρ = radial coordinate in a spherical polar coordinate system ρ,φ,ψ. The magnitude and internal phase angle of the complex number atmospheric pressure can then be written as

$$P_{at} = n_{at}k_B T_{at} = P^o_{at}f_{Tat}\sec\theta^{at}_n \qquad \theta^{at}_P = \theta^{at}_n + \theta^{at}_T \qquad (813)$$

where P^o_{at} = atmospheric pressure under zero ambient field condition and as given by equation (456). The measured atmospheric pressure at any specified height above the Earth's surface is given by

$$P_{atm} = P_{at}\cos\theta^{at}_P = P^o_{at}f_{Tat}\sec\theta^{at}_n\cos(\theta^{at}_n + \theta^{at}_T) \approx P^o_{at} \qquad (814)$$

which is approximately the conventional value P^o_{at} of the atmospheric pressure if $\theta^{at}_T << \theta^{at}_n$ and if $f_{Tat} \approx 1$ and $C^{at}_{\beta\rho} \approx C^{at}_{\beta t}$ as may possibly be the case.

H. Calculation of the Thrust of a Polyphase Rocket.

The complex number total thrust per unit area for a polyphase rocket is obtained as a complex number generalization of the conventional-real number-zero ambient field-expression of the total thrust per unit area that appears in equation (436), and is written as

$$\overline{\tau}_e = \tau_e \exp(j\theta_\tau^e) = \overline{\tau}_{ef} + \overline{P}_e - \overline{P}_{at} = \overline{\rho}_e \overline{v}_e^2 + \overline{P}_e - \overline{P}_{at} \tag{815}$$

where $\overline{\tau}_e$ = total complex number thrust per unit area of the exhaust aperture, $\overline{\tau}_{ef}$ = complex number thrust per unit area due to material flow and is given by

$$\overline{\tau}_{ef} = \tau_{ef} \exp(j\theta_\tau^{ef}) = \overline{\rho}_e \overline{v}_e^2 \tag{816}$$

where $\overline{\rho}_e$ is given by equation (716) and $\overline{v}_e$ is given by equation (685), and where $\overline{P}_e$ and $\overline{P}_{at}$ = complex number exhaust gas pressure nand complex number atmospheric pressure given by equations (768) and (794) respectively. The thrust of a polyphase rocket is then given by

$$\overline{T}_e = T_e \exp(j\theta_T^e) = \overline{\tau}_e \overline{A}_a = (\overline{\rho}_e \overline{v}_e^2 + \overline{P}_e - \overline{P}_{at})\overline{A}_a \tag{817}$$

It will be assumed that for maximum efficiency the rocket nozzle has been designed to operate at the adapted condition, which is

$$\overline{P}_e = \overline{P}_{at} \qquad P_e = P_{at} \qquad \theta_P^e = \theta_P^{at} \tag{818}$$

The adapted condition of equation (818) can be combined with equation (768) to give the following expression for the complex number exhaust velocity for maximum thrust

$$\overline{v}_e^2 = 2(\overline{P}_c - \overline{P}_{at})(\overline{\rho}_e)^{-1}[1 - (\overline{\rho}_e / \overline{\rho}_c)(\overline{A}_a / \overline{A}_f)^2]^{-1} \tag{819}$$

The maximum values of the complex number thrust per unit aperture area and the complex number total thrust are then obtained from equations (815) through (819) as

$$\overline{\tau}_e = 2(\overline{P}_c - \overline{P}_{at})[1 - (\overline{\rho}_e / \overline{\rho}_c)(\overline{A}_a / \overline{A}_f)^2]^{-1} \tag{820}$$

$$\overline{T}_e = 2(\overline{P}_c - \overline{P}_{at})\overline{A}_a[1 - (\overline{\rho}_e / \overline{\rho}_c)(\overline{A}_a / \overline{A}_f)^2]^{-1} \tag{821}$$

For the adapted condition in equation (818) the total thrust per unit area of the exhaust aperture and the total thrust given in equations (815) and (817) can be written as

$$\overline{\tau}_e = \tau_e \exp(j\theta_\tau^e) = \overline{\rho}_e \overline{v}_e^2 \tag{822}$$

$$\overline{T}_e = T_e \exp(j\theta_T^e) = \overline{\tau}_e \overline{A}_a = \overline{\rho}_e \overline{v}_e^2 \overline{A}_a \tag{823}$$

which are actually the conditions for maximum thrust in a polyphase rocket. The magnitude and internal phase angle of the the thrust per unit area in equation (822) can be written as

$$\tau_e = \rho_e v_e^2 \qquad \theta_\tau^e = \theta_\rho^e + 2\theta_v^e = \theta_\mu^e + \theta_n^e + 2\theta_v^e \tag{824}$$

and the magnitude can be rewritten using equations (459), (686) and (717) as

$$\tau_e = \tau_e^o f_{\mu e} (C_{\beta z}^e)^2 (C_{\beta t}^e)^{-2} \sec\theta_n^e = \tau_e^o (C_{\beta z}^e)^2 (C_{\beta t}^e)^{-2} \cos\theta_{\mathbb{N}atom}^e \sec\theta_n^e \tag{825}$$

with $\beta = 2,3$ for the low-density exhaust gases. The measured thrust per unit area is obtained from equations (824) and (825) as

$$\tau_{em} = \tau_e^o (C_{\beta z}^e)^2 (C_{\beta t}^e)^{-2} \cos\theta_{\mathbb{N}atom}^e \sec\theta_n^e \cos(\theta_\mu^e + \theta_n^e + 2\theta_v^e) \tag{826}$$

The magnitude and inernal phase angle of the total thrust is obtained from equation (823) as

$$T_e = \tau_e A_a = \rho_e v_e^2 A_a \tag{827}$$

$$\theta_T^e = \theta_\tau^e + \theta_A^e = \theta_\rho^e + 2\theta_v^e + \theta_A^e = \theta_\mu^e + \theta_n^e + 2\theta_v^e - |\theta_A^e| \tag{828}$$

and the magnitude of the total thrust can be rewritten as

$$T_e = T_e^o f_{\mu e} (C_{\beta z}^e)^2 (C_{\beta t}^e)^{-2} (C_{1r}^a)^2 \sec\theta_n^e = T_e^o (C_{\beta z}^e)^2 (C_{\beta t}^e)^{-2} (C_{1r}^a)^2 \cos\theta_{\mathbb{N}atom}^e \sec\theta_n^e \tag{829}$$

after using equations (460), (547), (686) and (717). The measured value of the rocket exhaust thrust is given by

$$T_{em} = T_e \cos\theta_T^e = T_e^o (C_{\beta z}^e)^2 (C_{\beta t}^e)^{-2} (C_{1r}^a)^2 \cos\theta_{\mathbb{N}atom}^e \sec\theta_n^e \cos(\theta_\mu^e + \theta_n^e + 2\theta_v^e - |\theta_A^e|) \tag{830}$$

Note that the rocket exhaust thrust T_e given in equation (829) can easily be expressed in terms of $\sec\theta_{\mathbb{N}}^e$ instead of $\sec\theta_n^e$ by using the equations (266) or (268) in the form

$$\sec\theta_n^e = (C_{\beta r}^e)^{-2} (C_{\beta z}^e)^{-1} \sec\theta_{\mathbb{N}}^e \qquad [\mathrm{ld, LN}] \tag{831}$$

where $\beta = 2,3$. Assuming the inequalities $\theta_\mu^e << \theta_n^e, \theta_v^e << \theta_n^e$ and $\theta_{\mathbb{N}atom}^e << \theta_n^e$ allows equations (825), (826), (829) and (830) to be written approximately as

$$\tau_e \approx \tau_e^o (C_{\beta z}^e)^2 (C_{\beta t}^e)^{-2} \sec\theta_n^e \tag{832}$$

$$\tau_{em} \approx \tau_e^o (C_{\beta z}^e)^2 (C_{\beta t}^e)^{-2} \tag{833}$$

$$T_e \approx T_e^o (C_{\beta z}^e)^2 (C_{\beta t}^e)^{-2} (C_{1r}^a)^2 \sec\theta_n^e \tag{834}$$

$$T_{em} \approx T_e^o (C_{\beta z}^e)^2 (C_{\beta t}^e)^{-2} (C_{1r}^a)^2 \sec\theta_n^e \cos(\theta_n^e - |\theta_A^e|) \tag{835}$$

The further assumption that $C_{\beta\alpha}^e \approx C_{\beta t}^e$ allows these four equations to be written as the additional approximations

$$\tau_e \approx \tau_e^o \sec\theta_n^e \tag{836}$$

$$\tau_{em} \approx \tau_e^o \tag{837}$$

$$T_e \approx T_e^o (C_{1r}^a)^2 \sec\theta_n^e \tag{838}$$

$$T_{em} \approx T_e^o (C_{1r}^a)^2 \sec\theta_n^e \cos(\theta_n^e - |\theta_A^e|) \tag{839}$$

equations (832) through (839) show that the following important inequalities are valid

$$\tau_e \geq \tau_e^o \qquad \tau_{em} \approx \tau_e^o \qquad T_e \geq T_e^o \qquad T_{em} \geq T_e^o \tag{840}$$

and therefore thrust enhancement is possible in a polyphase rocket.

7. FORCE OF GRAVITY ON A POLYPHASE ROCKET AT THE EARTH'S SURFACE. This section considers the gravitational force on a polyphase rocket at the Earth's surface using complex number active space coordinates and their associated space coordinate factors and assuming that an ambient gravitational field exists within matter so that the gravitational force on a polyphase rocket must be represented as a complex number in an internal space. The liftoff equations of a polyphase rocket that are treated in Section 8 include a gravitational force $\overline{F}_{GE}$ at the surface of the Earth that acts on the complex number mass $\overline{m}_R$ of a polyphase rocket under the action of a complex number acceleration of gravity $\overline{g}_E$ at the Earth's surface, and this gravitational force is written as

$$\overline{F}_{GE} = F_{GE}\exp(j\theta_{FG}^E) = \overline{m}_R \overline{g}_E \tag{841}$$

whose magnitude and internal phase angle are

$$F_{GE} = m_R g_E \qquad \theta_{FG}^E = \theta_m^R + \theta_g^E = \theta_m^R + \theta_M^E - 2\theta_R^E = \theta_m^R + \theta_\rho^E + \theta_R^E \tag{842}$$

where the complex number mass of the rocket is given in equation (664) and where

$$\overline{g}_E = g_E \exp(j\theta_g^E) = -G\overline{M}_E / \overline{R}_E^2 = -4/3G\overline{\rho}_E\overline{R}_E \tag{843}$$

whose magnitude and internal phase angle are

$$g_E = -GM_E R_E^{-2} = -4/3\pi G\rho_E R_E \qquad \theta_g^E = \theta_M^E - 2\theta_R^E = \theta_\rho^E + \theta_R^E \tag{844}$$

and where the complex number mass of the Earth is written as

$$\overline{M}_E = M_E \exp(j\theta_M^E) = \overline{\mu}_E \overline{\aleph}_E = \overline{\rho}_E \overline{V}_E \tag{845}$$

where $\overline{\rho}_E = \rho_E \exp(j\theta_\rho^E)$ and $\overline{\mu}_E = \mu_E \exp(j\theta_\mu^E)$ = complex number average mass density and molecular mass respectively of the materials in the solid Earth, and where the complex number volume of the Earth is given by[12]

$$\overline{V}_E = V_E \exp(j\theta_V^E) = 4/3\pi\overline{R}_E^3 \qquad V_E = V_E^o \sec\theta_\aleph^E \sec\theta_n^E = V_E^o (C_{1R}^E)^3 \tag{846}$$

where V_E^o = conventional zero ambient field value of the Earth's volume as given in equation (466). The magnitude and internal phase angle of the Earth's mass are

$$M_E = \mu_E \mathfrak{N}_E = 4/3\pi R_E^3 \rho_E \qquad \theta_M^E = \theta_\mu^E + \theta_{\mathfrak{N}}^E = \theta_\rho^E + 3\theta_R^E \tag{847}$$

where $\overline{R}_E$ complex number radius of the Earth and is represented as

$$\overline{R}_E = R_E \exp(j\theta_R^E) \qquad R_E = R_E^o C_{1R}^E \qquad \theta_R^E = \theta_{\eta R}^E - \theta_{\pi R}^E \leq 0 \tag{848}$$

where the high-density space coordinate factor C_{1R}^R is defined by the equations (8) and (9)and is obtained from equation (10) for the [hd] Earth materials as

$$C_{1R}^E = \cos\theta_{\eta R}^E \sec\theta_{\pi R}^E \qquad C_{1R}^E \geq 1 \tag{849}$$

where $\theta_{\eta R}^E$ and $\theta_{\pi R}^E$ are the internal phase angles of the complex number spatial counting and segmentation numbers respectively of the complex number Earth's radius.[12]

Equation (845) shows that the complex number average mass density of the Earth $\overline{\rho}_E$ can also be represented as

$$\overline{\rho}_E = \rho_E \exp(j\theta_\rho^E) = \overline{\mu}_E \overline{n}_E = \overline{M}_E / \overline{V}_E \tag{850}$$

and where $\overline{n}_E$ = complex number molecular number density is given by

$$\overline{n}_E = n_E \exp(j\theta_n^E) = \overline{\mathfrak{N}_E} / \overline{V}_E = \mathfrak{N}_E / V_E \exp[j(\theta_{\mathfrak{N}}^E - \theta_V^E)] \tag{851}$$

$$n_E = \mathfrak{N}_E / V_E = N_E / V_E \sec\theta_{\mathfrak{N}}^E = n_E^o (C_{1R}^E)^{-3} \sec\theta_{\mathfrak{N}}^E = n_E^o \cos\theta_n^E \tag{852}$$

$$\theta_n^E = \theta_{\mathfrak{N}}^E - \theta_V^E = \theta_{\mathfrak{N}}^E - 3\theta_R^E \tag{853}$$

$$n_E^o = N_E / V_E^o \tag{854}$$

$$n_{Em} = n_E \cos\theta_n^E = n_E^o \cos^2\theta_n^E = n_E^o \cos^2(\theta_{\mathfrak{N}}^E - 3\theta_R^E) \tag{855}$$

and where $\overline{\mathfrak{N}_E}$ = complex number of molecules in the Earth, N_E = integer number of molecules in the Earth, and $\overline{\mu}_E$ = complex number average molecular mass of the Earth's costituent materials. In analogy to equations (585) through (588) and because the Earth is a large system of molecules, $\overline{\mathfrak{N}_E}$ = complex number of molecules in the Earth and is given by

$$\overline{\mathfrak{N}_E} = \mathfrak{N}_E \exp(j\theta_{\mathfrak{N}}^E) \qquad \mathfrak{N}_E = N_E \sec\theta_{\mathfrak{N}}^E \tag{856}$$

where N_E = integer number of molecules in the Earth, and where the internal phase angle $\theta_{\mathfrak{N}}^E$ is obtained from equations (252) or (589) and (590) as

$$\tan\theta_{\mathfrak{N}}^E = (2\mathfrak{E}^E / G_E^o)^{1/2} \tag{857}$$

where G_E^o = zero field shear modulus of the Earth and where the energy density of the ambient field in the Earth can be written in analogy to equations (58) through (61) as

$$\mathfrak{E}^E = \mathfrak{E}^{gE} + \mathfrak{E}^{\gamma E} \qquad \text{solid earth} \tag{858}$$

where gravity is the dominant ambient field but the energy density $\mathfrak{E}^{\gamma E}$ that describes rotational effects, thermal convection and magnetic fields is also important. The complex number average molecular mass $\overline{\mu}_E$ of the Earth's constituent matter can be written in a form analogous to that given in equations (615) through (632) as follows

$$\overline{\mu}_E = \mu_E \exp(j\theta_\mu^E) \qquad \mu_E = \mu_E^o f_{\mu E}(\mathfrak{E}^E) \qquad \theta_\mu^E = \theta_\mu^E(\mathfrak{E}^E) \tag{859}$$

subject to the zero ambient field conditions

$$f_{\mu E}(0) = 1 \qquad \theta_{\mu E}(0) = 0 \tag{860}$$

and where μ_E^o = zero ambient field value of the average molecular mass of the solid Earth. It will be assumed that the integer number of atoms in a molecule of Earth material is small so that in analogy to equations (621) through (624) the complex number of atoms in a molecule of solid Earth material is written as

$$\overline{\mathfrak{N}}_E^{atom} = \mathfrak{N}_E^{atom} \exp(j\theta_{\mathfrak{N}atom}^E) \qquad \mathfrak{N}_E^{atom} = N_E^{atom} \cos\theta_{\mathfrak{N}atom}^E \tag{861}$$

where N_E^{atom} = average small integer number of atoms per molecule of solid Earth material. The complex number molecular mass of a molecule of solid Earth material can be written in analogy to equations (625) and (626) as

$$\overline{\mu}_E = m_{Eatom}^o \overline{\mathfrak{N}}_E^{atom} - c^{-2}\overline{B}_E^{mol} \tag{862}$$

where m_{Eatom}^o = zero ambient field mass of an atom of solid Earth material. After neglecting the complex number molecular binding energy $\overline{B}_E^{mol}$, the combination of equations (861) and (862) gives the magnitude and internal phase angle of the molecular mass of a molecule of solid Earth material as

$$\mu_E = m_{Eatom}^o N_E^{atom} \cos\theta_{\mathfrak{N}atom}^E = \mu_E^o \cos\theta_{\mathfrak{N}atom}^E \qquad \theta_\mu^E = \theta_{\mathfrak{N}atom}^E \tag{863}$$

And therefore a comparison of equation (859) and (863) shows that

$$f_{\mu E} = \cos\theta_{\mathfrak{N}atom}^E \qquad f_{\mu E} \le 1 \tag{864}$$

which is valid for small molecules with only a few atoms. For large molecules a secant function replaces the cosine term in equation (864).

Equations (842) and (844) show that both m_R and M_E are required to calculate the gravitational force on a polyphase rocket. The value of the magnitude m_R of the complex number mass of a polyphase rocket is given by equation (668) as

$$m_R = m_b^o f_{\mu b} Q_R \sec\theta_{\mathfrak{N}}^b = m_b^o Q_R \cos\theta_{\mathfrak{N}atom}^b \sec\theta_{\mathfrak{N}}^b \tag{865}$$

where m_b^o = zero ambient field value of the mass of the rocket body, while the value of the magnitude M_E of the complex number mass of the Earth follows from equations (466), (847), (856) and (863) as

$$M_E = M_E^o f_{\mu E} \sec\theta_{\mathfrak{N}}^E = M_E^o \cos\theta_{\mathfrak{N}atom}^E \sec\theta_{\mathfrak{N}}^E \tag{866}$$

where M_E^o = the zero ambient field value of the Earth's mass and is given by equation (466). Combining equations (844) and (866) and using equation (870) gives the magnitude g_E of the acceleration of gravity at the Earth's surface as

$$g_E = g_E^o f_{\mu E} (C_{1R}^E)^{-2} \sec\theta_N^E = g_E^o f_{\mu E} C_{1R}^E \cos\theta_n^E \qquad g_E \le g_E^o \tag{867}$$

where the zero ambient field value of the acceleration of gravity at the Earth's surface g_E^o is given by equation (468). Combining equations (659), (842), (865) and (867) gives the magnitude F_{GE} of the complex number force of gravity on a polyphase rocket at the Earth' surface as

$$F_{GE} = m_R g_E = F_{GE}^o f_{\mu b} f_{\mu E} (Q_R / Q_R^o)(C_{1R}^E)^{-2} \sec\theta_N^b \sec\theta_N^E \qquad F_{GE} \le F_{GE}^o \tag{868}$$

where F_{GE}^o = zero ambient field value of the gravitational force acting on the rocket at the Earth's suface and is given by equation (467) and where the factors $f_{\mu b}$ and $f_{\mu E}$ are given by equations (631) and (864), and where Q_R and Q_R^o are given by equations (670) and (415) and where the internal phase angles θ_N^b and θ_N^E are given by equations (590) and (857). Finally, it is possible to express the gravitational force F_{GE} acting on a polyphase rocket at the Earth' surface in terms of the angles θ_n^b and θ_n^E instead of the angles θ_N^b and θ_N^E by using equation (264) for the [hd,LN] cylindrical rocket body and equation (272) for the [hd,LN] spherical Earth, and which for the case at hand are written as

$$\cos\theta_n^b = (C_{1r}^b)^{-2} (C_{1z}^b)^{-1} \sec\theta_N^b \qquad [\text{hd,LN}] \tag{869}$$

$$\cos\theta_n^E = (C_{1R}^E)^{-3} \sec\theta_N^E \qquad [\text{hd,LN}] \tag{870}$$

Combining equations (868) through (870) gives

$$F_{GE} = F_{GE}^o f_{\mu b} f_{\mu E} (Q_R / Q_R^o) C_{1R}^E C_{1z}^b (C_{1r}^b)^2 \cos\theta_n^b \cos\theta_n^E \qquad F_{GE} \le F_{GE}^o \tag{871}$$

which is an alternative expression for the the force of gravity on a polyphase rocket.

<u>8. DETERMINATION OF THE LIFTOFF CAPABILITY OF A POLYPHASE ROCKET.</u> This section uses complex number active space and time coordinates and complex number molecular number densities to determine the dynamical equations of motion of a polyphase rocket in a gravitational field and develops the concept of a complex number liftoff parameter that is used to predict the liftoff enhancement of a polyphase rocket compared to a conventional rocket.

A. Complex Number Dynamical Equations of Motion of a Polyphase Rocket in Earth's Gravitational Field and the Complex Number Liftoff Parameter.

Consider now the equations of motion of a polyphase rocket that are the complex number generalizations of the real number equations for a conventional rocket that appear in equations (412), (413) and (469) through (473). Six ambient fields are applied to the basic components of a polyphase rocket as described by equations (58) through (63), which require the space and time coordinates and the kinematical and dynamical quantities to be represented as complex numbers in an internal space. For a polyphase rocket the complex number generalizations of the real number equations of motion of a conventional rocket are

$$d\overline{p}_R / d\overline{t} = \overline{T}_e - \overline{F}_{GE} \tag{872}$$

$$d\overline{p}_R / d\overline{t} = \overline{m}_R d\overline{v}_R / d\overline{t} + \overline{v}_R d\overline{m}_R / d\overline{t} \tag{873}$$

$$d\overline{m}_R / d\overline{t} = -\overline{\rho}_e \overline{v}_e \overline{A}_a \tag{874}$$

where $\overline{T}_e$ = complex number rocket thrust as given by equation (817), $\overline{F}_{GE}$ = complex number force of gravity acting on a polyphase rocket and as given by equations (841) through (871), $\overline{t}$ = complex number time as given by equation (1), $\overline{\rho}_e$ = complex number mass density of the exhaust gases in the exhaust tube as given by equation (716), $\overline{v}_e$ = complex number exhaust gas velocity as given by equation (685), $\overline{A}_a$ = complex number area of the exhaust aperture as given by equation (546), and where the complex number velocity of the rocket $\overline{v}_R$ is represented as

$$\overline{v}_R = v_R \exp(j\theta_v^R) \tag{875}$$

and the complex number rocket momentum is given by

$$\overline{p}_R = p_R \exp(j\theta_p^R) = \overline{m}_R \overline{v}_R \qquad p_R = m_R v_R \qquad \theta_p^R = \theta_m^R + \theta_v^R \tag{876}$$

and where the complex number polyphase rocket mass $\overline{m}_R = m_R \exp(j\theta_m^R)$ has been treated in equations (664) through (680). Then the polyphase rocket motion equation (872) can be written as

$$\begin{aligned}\overline{m}_R d\overline{v}_R / d\overline{t} &= \overline{\rho}_e \overline{v}_e \overline{v}_R \overline{A}_a + \overline{T}_e - \overline{F}_{GE} \\ &= (\overline{\rho}_e \overline{v}_e \overline{v}_R + \overline{\tau}_e)\overline{A}_a - \overline{m}_R \overline{g}_E \\ &= (\overline{\rho}_e \overline{v}_e \overline{v}_R + \overline{\rho}_e \overline{v}_e^2 + \overline{P}_e - \overline{P}_{at})\overline{A}_a - \overline{m}_R \overline{g}_E \\ &= \overline{\rho}_e \overline{v}_e \overline{v}_R \overline{A}_a + \overline{m}_R \overline{g}_E (\overline{s}_L - 1)\end{aligned} \tag{877}$$

where the complex number liftoff factor is given by

$$\overline{s}_L = \overline{T}_e / \overline{F}_{Ge} = (\overline{m}_R \overline{g}_E)^{-1} \overline{T}_e = (\overline{m}_R \overline{g}_E)^{-1} \overline{\tau}_e \overline{A}_a = (\overline{m}_R \overline{g}_E)^{-1} (\overline{\rho}_e \overline{v}_e^2 + \overline{P}_e - \overline{P}_{at})\overline{A}_a \tag{878}$$

where $\overline{g}_E$ = complex number acceleration of gravity at the Earth's surface as given by equation (843). In this paper is is assumed that the polyphase rocket is operated in the adapted mode of equation (818) so that the complex number rocket thrust per unit area of the exhaust aperture $\overline{\tau}_e$ and the rocket thrust $\overline{T}_e$ are given by equations (822) and (823) and the complex number liftoff parameter given for the general case in equation (878) becomes

$$\overline{s}_L = s_L \exp(j\theta_{sL}) = \overline{T}_e / \overline{F}_{Ge} = (\overline{m}_R \overline{g}_E)^{-1} \overline{T}_e = (\overline{m}_R \overline{g}_E)^{-1} \overline{\tau}_e \overline{A}_a = (\overline{m}_R \overline{g}_E)^{-1} \overline{\rho}_e \overline{v}_e^2 \overline{A}_a \tag{879}$$

Note that equation (877) shows that the condition $\bar{s}_L = 1$, or equivalently $s_L = 1$ and $\theta_{sL} = 0$, corresponds to zero velocity and acceleration of the rocket, so that $\bar{v}_R = 0$ and $d\bar{v}_R / d\bar{t} = 0$. Equations (872) through (879) are complex number generalizations of the conventional real number (zero ambient field) rocket launch equations (412) through (473).

B. Magnitude and Internal Phase Angle of the Complex Number Rocket Liftoff Parameter.

The magnitude s_L of the complex number liftoff parameter for a polyphase rocket that is operating at the adapted condition is obtained from equation (879) as

$$s_L = T_e / F_{GE} = (m_R g_E)^{-1} T_e = (m_R g_E)^{-1} \tau_e A_a = (m_R g_E)^{-1} \rho_e v_e^2 A_a \tag{880}$$

Equivalent expressions for s_L can be obtained by combining equations (829), (871) and (880) with the result that

$$s_L = s_L^o S_R \sec\theta_n^b \sec\theta_n^e \qquad s_L \geq s_L^o \tag{881}$$

where s_L^o = liftoff parameter for a conventional rocket as given by equation (473) and where S_R is given by

$$S_R = \Phi_{R1}\Phi_{R2}\Phi_{R3}\Phi_{R4} \approx 1 \tag{882}$$

where

$$\Phi_{R1} = Q_R^o / Q_R \approx 1 \tag{883}$$

$$\Phi_{R2} = f_{\mu e}(f_{\mu b} f_{\mu E})^{-1}(C_{1z}^b)^{-1}(C_{1R}^E)^{-1} \sec\theta_n^E \approx 1 \tag{884}$$

$$\Phi_{R3} = (C_{1r}^a)^2 (C_{1r}^b)^{-2} \approx 1 \tag{885}$$

$$\Phi_{R4} = (C_{\beta z}^e)^2 (C_{\beta t}^e)^{-2} \approx 1 \tag{886}$$

where θ_n^b, θ_n^e and θ_n^E = internal phase angles of the complex molecular number densities of the materials in the [hd] rocket body, [ld] exhaust gases and [hd] Earth material as given by equations (612), (703) and (853), Q_R^o and Q_R = rocket mass factors as given in equations (415) and (670), $f_{\mu b}, f_{\mu e}$ and $f_{\mu E}$ = molecular mass factors given by equations (631), (713) and (864) and where the space and time coordinate factors $C_{1r}^a, C_{1r}^b, C_{1z}^b, C_{\beta z}^e, C_{\beta t}^e$ and C_{1R}^E are given by equations (92), (95), (96), (97), (204), (205) and (849). Because $S_R \approx 1$, equation (881) can be written approximately as the following simple equation

$$s_L = s_L^o \sec\theta_n^b \sec\theta_n^e \qquad s_L \geq s_L^o \tag{887}$$

Using equations (831), (869) and (870) it is possible to express equations (881) through (887) in terms of $\theta_{\mathcal{N}}^b, \theta_{\mathcal{N}}^e, \theta_{\mathcal{N}}^E$ instead of $\theta_n^b, \theta_n^e, \theta_n^E$, but does not lead to such simple and interpretable results. The particular forms of equations (881) and (887) arise from the [hd] nature of the rocket body and the [ld] nature of the exhaust gases. Both equations

(881) and (887) show that $s_L \geq s_L^o$ and therefore liftoff enhancement is possible for a polyphase rocket.

The corresponding internal phase angle of the complex liftoff parameter given in equation (879) is written as

$$\begin{aligned}\theta_{sL} &= \theta_\rho^e + 2\theta_v^e + \theta_A^a - \theta_m^R - \theta_g^E \qquad (888)\\ &= \theta_n^e + \theta_\mu^e + 2\theta_v^e + \theta_A^a - \theta_m^R - \theta_g^E\end{aligned}$$

Using the approximation $\theta_m^R \approx \theta_m^b$ of equation (680) allows the phase angle equation (888) to be written as

$$\begin{aligned}\theta_{sL} &= \theta_\rho^e + 2\theta_v^e + \theta_A^a - \theta_m^b - \theta_g^E \qquad (889)\\ &= \theta_n^e - \theta_n^b + \theta_\mu^e - \theta_\mu^b + \theta_A^a - \theta_V^b + 2\theta_v^e - \theta_g^E\\ &= (\theta_n^e - \theta_n^b) + (\theta_\mu^e - \theta_\mu^b) + (\theta_A^a + \theta_h^e - \theta_V^b) + (\theta_a^e - \theta_g^E)\\ &= (\theta_n^e - \theta_n^b) + (\theta_\mu^e - \theta_\mu^b) + (|\theta_V^b| - |\theta_A^a| - |\theta_h^e|) + (|\theta_g^E| - |\theta_a^e|)\end{aligned}$$

where equations (612), (651) and (689) were used to obtain this result. Equations (880) through (889) give the complex number liftoff parameter for a polyphase rocket whose ambient fields are described in equations (58) through (63). The measured liftoff parameter is obtained from equations (881) and (889) as

$$s_{Lm} = s_L \cos\theta_{sL} = s_L^o S_R \sec\theta_n^b \sec\theta_n^e \cos\theta_{sL} \qquad (890)$$

The measured value of the liftoff parameter contains the factor $\cos\theta_{sL}$ so that its value is lower than the value of the magnitude of the complex number liftoff parameter as given by equation (881). However, by a proper adjustment of the ambient fields (except the gravity field) acting on the exhaust gases and on the rocket body it is possible to make $\theta_{sL} = 0$ and $\cos\theta_{sL} = 1$ so that the measured liftoff parameter will have a relatively large value. For the condition $\theta_{sL} = 0$ the measured value of the liftoff parameter becomes equal to the value of the magnitude of the liftoff parameter that is given by equations (881) or (887), and the complex number liftoff parameter becomes a real number. But from equation (877) it is clear that at the liftoff boundary $\bar{s}_L = 1$, the complex number liftoff parameter is just a real number and can be written as $s_L = 1$ and $\theta_{sL} = 0$. Thus at the liftoff point the magnitude and measured values of the liftoff parameter are equal and in fact they are equal to unity. The liftoff conditions for a polyphase rocket will be studied in the next section where the liftoff equation (881) with $s_L > 1$ is rewritten and solutions of this equation are derived for an underpowered ($s_L^o < 1$) and an overpowered ($s_L^o > 1$) conventional rocket.

9. LIFTOFF EQUATIONS FOR A POLYPHASE ROCKET AND THEIR SOLUTIONS. This section develops the liftoff equations for a polyphase rocket and solves these equations for the internal phase angle θ_n^e of the molecular number density of the exhaust gases and the internal phase angle θ_n^b of the molecular number density of the material constituting the rocket body that are required for the liftoff of a rocket which is

conventionally underpowered and unable to attain liftoff for the case of zero ambient field conditions. Then the strengths of the ambient electromagnetic fields that are applied to the polyphase rocket exhaust gases in the exhaust tube and to the rocket body are determined that are required for the enhanced liftoff of an underpowered conventional rocket. The same analysis applies to the case where a conventional rocket is overpowered and able to attain liftoff, but the rocket engineer wishes to improve the liftoff efficiency by applying ambient fields to the various rocket components.

A. Liftoff Equations for a Polyphase Rocket.

Two initial conditions of a conventional rocket are possible when the ambient fields are turned on so as to produce a polyphase rocket. A conventional rocket can be underpowered so that the conventional liftoff parameter satisfies the condition $s_L^o < 1$ where s_L^o is given by equation (473) and therefore liftoff is not possible without ambient field enhancement, or a conventional rocket that is overpowered with $s_L^o > 1$ and liftoff is possible and the ambient fields only increase the efficiency of the liftoff. For either case the lift off condition for a polyphase rocket is given by $s_L > 1$ while equation (881) always requires that $s_L \geq s_L^o$. For a polyphase rocket with ambient field enhancement, the complex number liftoff condition is obtained from equation (877) and can be written as

$$\bar{s}_L = s_L^c \qquad s_L = s_L^c \qquad \theta_{sL} = 0 \qquad s_L^c > 1 \tag{891}$$

and corresponds to two real number liftoff equations, where $s_L^c > 1$ is the liftoff parameter magnitude specified by a rocket design engineer. For a polyphase rocket with ambient field enhancement the complex number liftoff boundary, corresponding zero velocity and acceleration of the rocket, is obtained from equation (877) as

$$\bar{s}_L = 1 \qquad s_L = 1 \qquad \theta_{sL} = 0 \tag{892}$$

The first real number liftoff equation that appears in equation (891) is the magnitude condition $s_L = s_L^c$, which combined with equation (881) gives

$$\cos\theta_n^e \cos\theta_n^b = s \tag{893}$$

where the augmented liftoff parameter s is defined by

$$s = s_L^o S_R / s_L = s_L^o S_R / s_L^c \approx s_L^o / s_L^c \tag{894}$$

where $s_L^c \geq s_L^o$ and $s_L^c > 1$ with s_L^o given by equation (473), and where $S_R \approx 1$ is given by equation (882). Because $s_L^c \geq s_L^o$ and $S_R \approx 1$ it follows that $s \leq 1$ and equation (893) is mathematically possible, and can be used for either an initially underpowered or overpowered conventional rocket. The second real number liftoff equation in equation (891) is the phase angle equation $\theta_{sL} = 0$, which combined with equation (888) gives

$$\theta_m^R = \theta_n^e + \theta_\mu^e + 2\theta_v^e + \theta_A^a - \theta_g^E \tag{895}$$

or approximately from equation (889)

$$\theta_n^b = \theta_n^e + \theta_\mu^e - \theta_\mu^b + \theta_A^a - \theta_V^b + 2\theta_v^e - \theta_g^E \tag{896}$$

$$= \theta_n^e + (\theta_\mu^e - \theta_\mu^b) + (\theta_A^a + \theta_h^e - \theta_V^b) + (\theta_a^e - \theta_g^E)$$

$$= \theta_n^e + (\theta_\mu^e - \theta_\mu^b) + (|\theta_V^b| - |\theta_A^a| - |\theta_h^e|) + (|\theta_g^E| - |\theta_a^e|)$$

which is rather complicated if all angles (except those associated with the ambient gravity field) can be varied.

With all of the variables involved in the design of a polyphase rocket it is clear that equations (893) and (896) need to be solved numerically using a computer. However for heuristic purposes to find out the general conditions under which equations (893) and (896) have simultaneous solutions it is necessary to have some analytical solutions. The liftoff equation (896) needs to be simplified if analytical solutions to the coupled equations (893) and (896) are to be found. The simplification used in this paper is to write the $\theta_{sL} = 0$ condition given by equation (896) in the form

$$\theta_n^b = \theta_n^e + \theta_c \tag{897}$$

where θ_c = constant given by

$$\theta_c = \theta_\mu^e - \theta_\mu^b + \theta_A^a - \theta_V^b + 2\theta_v^e - \theta_g^E \tag{898}$$

$$= (\theta_\mu^e - \theta_\mu^b) + (\theta_A^a + \theta_h^e - \theta_V^b) + (\theta_a^e - \theta_g^E)$$

$$= (\theta_\mu^e - \theta_\mu^b) + (|\theta_V^b| - |\theta_A^a| - |\theta_h^e|) + (|\theta_g^E| - |\theta_a^e|)$$

Combining equations (893) and (897) gives the liftoff equations for a polyphase rocket as

$$\cos\theta_n^e \cos(\theta_n^e + \theta_c) = s \qquad 0 \le s \le 1 \tag{899}$$

$$\theta_n^b = \theta_n^e + \theta_c \tag{900}$$

with s defined in equation (894), and are approximate equations that correspond to the liftoff condition in equation (891). The liftoff equations (899) and (900) are valid for $\theta_c \ge 0$ or $\theta_c \le 0$. It will be shown that a solution to the liftoff equations (899) and (900) for the case of arbitrary values of θ_c is not in general possible for the entire range $0 \le s \le 1$, and this is indicated in Figure 1a for the special case of $\theta_c = \pi/4$. On the other hand for the choice $\theta_c = 0$, a solution to the liftoff equations (899) and (900) is possible for the entire range $0 \le s \le 1$ as seen from Figure 1b, and for this special case equations (899) and (900) become

$$\cos\theta_n^e = s^{1/2} \qquad \theta_n^b = \theta_n^e \tag{901}$$

Equation (898) shows that the value of the internal phase angle constant θ_c can be selected by a proper choice of the applied ambient fields represented by equations (58) through (63) because these fields determine the values of the internal phase angles that appear on the right hand side of equation (898), with the exception of the acceleration of gravity internal phase angle θ_g^E which is beyond the control of rocket engineering. For instance, equation (898) shows that the condition $\theta_c = 0$ can be obtained by adjusting the fields within the various systems of a polyphase rocket.

B. Analytical Solutions of the Polyphase Rocket Liftoff Equations.

The solutions to equations (899) and (900) fall into two cases according to whether $\theta_c \geq 0$ or $\theta_c \leq 0$, but with $|\theta_c| < \pi/2$.

Case a. $\theta_c \geq 0$.

The general relation between the internal phase angles corresponding to the condition $\theta_c \geq 0$ is obtained from equation (898) as

$$\theta^e_\mu - \theta^b_V - \theta^E_g \geq \theta^b_\mu - \theta^a_A - \theta^e_h - \theta^e_a \qquad (902)$$

or equivalently

$$\theta^e_\mu + |\theta^b_V| + |\theta^E_g| \geq \theta^b_\mu + |\theta^a_A| + |\theta^e_h| + |\theta^e_a| \qquad (903)$$

For this case the solution to the polyphase rocket liftoff equations (899) and (900) is given by

$$\cos\theta^{e+}_n = [1/2(A+B)]^{1/2} \qquad \theta^{e+}_n \geq 0 \qquad (904)$$

$$\theta^{b+}_n = \theta^{e+}_n + \theta_c \qquad (905)$$

where

$$A = \sin^2\theta_c + 2s\cos\theta_c \qquad A \geq 0 \qquad (906)$$

$$B = \sin|\theta_c|[\sin^2\theta_c + 4s\cos\theta_c - 4s^2]^{1/2} \qquad B \geq 0 \qquad (907)$$

where A and B are functions of s and θ_c, and where s is defined in equation (894) and the angle θ_c is defined by equation (898).

Equations (906) and (907) show that the point $s = 0$ corresponds to values of A and B given by

$$A = \sin^2\theta_c \qquad B = \sin^2\theta_c \qquad A + B = 2\sin^2\theta_c \qquad s = 0 \qquad (908)$$

from which equations (904) and (905) give

$$\theta^{e+}_n = \pi/2 - \theta_c \qquad \theta^{b+}_n = \pi/2 \qquad s = 0 \qquad (909)$$

as shown in Figure 1a for $\theta_c = \pi/4$. It is also shown in Figure 1a that the angle θ^{e+}_n has a zero at the point $s = s_{o1}$ that will now be determined. At the point $s = s_{o1}$ the following conditions are valid

$$\theta^{e+}_n = 0 \qquad \theta^{b+}_n = \theta_c \qquad s = s_{o1} \qquad (910)$$

where s_{o1} is obtained from equation (904) by the relation

$$A + B = 2 \qquad s = s_{o1} \tag{911}$$

where the general expressions for A and B are given by equations (906) and (907). The solution to equation (911) is

$$s = s_{o1} = \cos\theta_c \tag{912}$$

at which point A and B have the specific values

$$A = 1 + \cos^2\theta_c \qquad B = \sin^2\theta_c \qquad s = s_{o1} \tag{913}$$

as determined from equations (906) and (907).

The range of the validity of the solutions in equations (904) through (907) depends on the size of the angle $\theta_c < \pi/2$. For large values of θ_c the range of validity of equations (904) and (905) is determined by the condition $\theta_n^{b+} < \pi/2$, so that the upper bound on the angle θ_n^{b+} is given by

$$\theta_n^{b+} = \theta_n^{e+} + \theta_c = \pi/2 \tag{914}$$

For $\pi/3 \le \theta_c < \pi/2$ there are two points that satisfy equation (914) and they are $s = 0$ and $s = s_{\pi/2}^{(1)}$, where $s = 0$ corresponds to $\theta_n^{e+} = \pi/2 - \theta_c$ and $\theta_n^{b+} = \pi/2$, and where the point $s_{\pi/2}^{(1)}$ is determined by the general solution of equation (914) for $\theta_c \ge \pi/3$. In the range $0 < s < s_{\pi/2}^{(1)}$, the angle θ_n^{b+} satisfies the condition $\theta_n^{b+} < \pi/2$ and is a proper solution. On the other hand for the case when $\theta_c \le \pi/3$, the limiting value of s is determined by the requirement that the function B must be a real number, and so for $\theta_c \le \pi/3$ the solution in equation (907) is valid only in the following range of values of the parameter s for which liftoff is possible

$$0 \le s \le s_1 \tag{915}$$

where

$$s_1 = 1/2(1 + \cos\theta_c) \tag{916}$$

as shown in Figure 1a for the particular case of $\theta_c = \pi/4$. At the point $s = s_1$ the values of A and B are obtained from equations (906) and (907) to be

$$A = 1 + \cos\theta_c = 2s_1 \qquad s = s_1 \tag{917}$$

$$B = 0 \qquad s = s_1 \tag{918}$$

so that the value of θ_n^{e+} at the limiting point $s = s_1$ is designated by $\theta_n^{e+} = \theta_{n1}^{e+}$ and is obtained from equation (904) as

$$\cos\theta_{n1}^{e+} = s_1^{1/2} = [1/2(1 + \cos\theta_c)]^{1/2} \qquad s = s_1 \tag{919}$$

Thus there is a region of the parameter s for which there are no real number solutions and a gap exists that is given by

$$s_1 \leq s \leq 1 \tag{920}$$

with a gap width given by

$$\Delta s_1 = 1 - s_1 = 1/2(1 - \cos\theta_c) \tag{921}$$

in which it is not possible to have liftoff parameter enhancement and in which liftoff of an underpowered rocket is not possible. However, the gap width given by equation (921) can always be reduced in size by adjusting the ambient fields in a polyphase rocket so as to lower the value of the angle θ_c given by equation (898).

The critical angle $\theta_c = \pi/3$ is the dividing point between the two types of ranges of validity of the solutions given in equations (904) and (905). This is discerned by noting that the condition for $s^{(1)}_{\pi/2} = s_1$ is obtained from equations (914) and (919) as

$$\cos^{-1}(s_1^{1/2}) + \theta_c = \pi/2 \tag{922}$$

where s_1 is given by equation (916). Simple inspection of equation (922) shows that its solution is $\theta_c = \pi/3 = 1.0472$. Thus for the case $\pi/3 \leq \theta_c < \pi/2$ the solutions in equations (904) and (905) are valid in the range of s given by

$$0 \leq s \leq s^{(1)}_{\pi/2} \qquad 0 \leq s^{(1)}_{\pi/2} \leq 3/4 \tag{923}$$

while for the case $0 \leq \theta_c \leq \pi/3$ the solutions in equations (904) and (905) are valid in the range

$$0 \leq s \leq s_1 \qquad 3/4 \leq s_1 \leq 1 \tag{924}$$

Note that the angles θ_n^{e+} and θ_n^{b+} that appear in Figure 1b for $\theta_c = \pi/2$ are not solutions of equations (899) and (900) because $\theta_n^{b+} \geq \pi/2$, but are rather the limiting boundaries of the angle solutions θ_n^{e+} and θ_n^{b+}. The value $s_1 = 1/2$ that appears in Figure 1b is of mathematical interest only because it is not associated with an actual solution, whereas the ranges in equation (924) correspond to actual solutions.

For the special case of $\theta_c = \pi/4$, equations (906) and (907) give the following results

$$A = 1/2 + \sqrt{2}s \qquad B = \sqrt{2}/2(1/2 + 2\sqrt{2}s - 4s^2)^{1/2} \qquad 0 \leq s \leq s_1 \tag{925}$$

and then equations (904) and (905) give θ_n^{e+} and θ_n^{b+}. For the point s = 0, equations (904), (905), (908), (909) and (925) give

$$A = 1/2 \qquad B = 1/2 \qquad \theta_n^{e+} = \pi/4 = 0.7854 \qquad \theta_n^{b+} = \pi/2 = 1.5708 \tag{926}$$

Equation (912) gives $s_{o1} = \sqrt{2}/2 = 0.7071$, and at this point equations (904), (905), (910) through (913) and (925) give

$$A = 3/2 \qquad B = 1/2 \qquad \theta_n^{e+} = 0 \qquad \theta_n^{b+} = \pi/4 = 0.7854 \qquad (927)$$

and for the point $s_1 = 1/2(1+\sqrt{2}/2) = 0.8536$ given by equation (916), equations (904), (905), (917) through (919) and (925) give

$$A = 2s_1 = 1.7071 \qquad B = 0 \qquad \theta_n^{e+} = \pi/8 = 0.3927 \qquad \theta_n^{b+} = 3\pi/8 = 1.1781 \qquad (928)$$

Equations (920), (921) and Figure 1a show that for $\theta_c = \pi/4$ there is a gap in the region $0.8536 \le s \le 1$ of width $\Delta s_1 = 1/2(1-\sqrt{2}/2) = 0.1464$ in which there are no real number solutions.

For the special case of $\theta_c = \pi/3$, equations (906) and (907) give the following equations

$$A = 3/4 + s \qquad B = \sqrt{3}/2(3/4 + 2s - 4s^2)^{1/2} \qquad 0 \le s \le s_1 \qquad (929)$$

Equations (904) and (905) give θ_{N1}^{+} and θ_{N2}^{+}, and for the point $s = 0$ equations (904), (905), (908), (909) and (929) give

$$A = 3/4 \qquad B = 3/4 \qquad \theta_n^{e+} = \pi/6 = 0.5236 \qquad \theta_n^{b+} = \pi/2 = 1.5708 \qquad (930)$$

Equation (912) gives $s_{o1} = 1/2$, and at this point equations (904), (905), (910) through (913) and (929) give

$$A = 5/4 \qquad B = 3/4 \qquad \theta_n^{e+} = 0 \qquad \theta_n^{b+} = \pi/3 = 1.0472 \qquad (931)$$

Equation (916) gives $s_1 = 1/2(1+1/2) = 3/4$, and at this point equations (904), (905), (917) through (919) and (929) give

$$A = 2s_1 = 3/2 \qquad B = 0 \qquad \theta_n^{e+} = \pi/6 = 0.5236 \qquad \theta_n^{b+} = \pi/2 = 1.5708 \qquad (932)$$

Equations (920) and (921) show that for $\theta_c = \pi/3$ there is a gap in the region $3/4 \le s \le 1$ of width $\Delta s_1 = 1/2(1-1/2) = 1/4$ in which no real number solutions exist.

For the specific value $\theta_c = \pi/2$, equations (906) and (907) give the following results

$$A = 1 \qquad B = (1-4s^2)^{1/2} \qquad (933)$$

and equation (904) gives

$$\cos\theta_n^{e+} = \{1/2[1+(1-4s^2)^{1/2}]\}^{1/2} \qquad (934)$$

while equation (912) gives $s_{o1} = 0$ which can also be deduced from equation (934), and at this point $s = s_{o1} = 0$ and equations (904), (905), (910) through (913), (933) and (934) give

$$A = 1 \qquad B = 1 \qquad \theta_n^{e+} = 0 \qquad \theta_n^{b+} = \pi/2 \tag{935}$$

Equation (916) gives $s_1 = 1/2$, and at this point equations (904), (905), (917) through (919), (933) and (934) give

$$A = 1 \qquad B = 0 \qquad \theta_n^{e+} = \pi/4 \qquad \theta_n^{b+} = 3\pi/4 \tag{936}$$

Equations (933) and (934) are valid in the range described by equations (915), (916), (919) and (921) as

$$0 \le s \le 1/2 \qquad s_1 = 1/2 \qquad \Delta s_1 = 1/2 \tag{937}$$

where for this special case there is a gap region as can be seen from the graphs of θ_n^{e+} and θ_n^{b+} which is shown in Figure 1b, but it has no physical meaning because the case of $\theta_c = \pi/2$ is not a solution but is rather a limiting case. Note that the internal phase angles θ_n^{e+} and θ_n^{b+} that appear in Figure 1b for $\theta_c = \pi/2$ are not solutions of equations (899) and (900) because $\theta_n^{b+} \ge \pi/2$, but are rather the limiting boundaries of the angles θ_n^{e+} and θ_n^{b+}. The value $s_1 = 1/2$ that appears in equation (935) and Figure 1b is of mathematical interest only because it is not associated with an actual solution, whereas the ranges in equation (924) correspond to actual solutions.

The particular value $\theta_c = 0$ placed in equations (906) and (907) gives the following results

$$A = 2s \qquad B = 0 \tag{938}$$

and equation (904) gives

$$\cos\theta_n^{e+} = s^{1/2} \tag{939}$$

for which case equations (912), (915), (916), (919) and (921) become respectively

$$s_{o1} = 1 \qquad 0 \le s \le 1 \qquad s_1 = 1 \qquad \theta_{n1}^{e+} = 0 \qquad \Delta s_1 = 0 \tag{940}$$

and there is no gap for this special case as shown in Figure 1c. For this case a liftoff solution is possible over the entire range $0 \le s \le 1$ of the parameter s. For s = 0

$$A = 0 \qquad B = 0 \qquad \theta_n^{e+} = \pi/2 \qquad \theta_n^{b+} = \pi/2 \tag{941}$$

while for s = 1

$$A = 2 \qquad B = 0 \qquad \theta_n^{e+} = 0 \qquad \theta_n^{b+} = 0 \tag{942}$$

The condition $\theta_c = 0$ can be written using equation (898) as

$$\theta_\mu^e - \theta_V^b - \theta_g^E = \theta_\mu^b - \theta_A^a - \theta_h^e - \theta_a^e \tag{943}$$

or equivalently as

$$\theta^{e}_{\mu} + |\theta^{b}_{V}| + |\theta^{E}_{g}| = \theta^{b}_{\mu} + |\theta^{a}_{A}| + |\theta^{e}_{h}| + |\theta^{e}_{a}| \tag{944}$$

This condition can be satisfied by adjusting the relative strengths of the ambient fields in the six component systems of a polyphase rocket that are listed in equations (58) through (63).

Equations (904) through (907) give the solutions of the liftoff equation for a polyphase rocket and specifically give the internal phase angle solutions θ^{e+}_{n} and θ^{b+}_{n} in terms of the parameter $s \leq 1$ given by equation (894) and in terms of the angle $\theta_c \geq 0$ given by equation (898). The value of the phase angle θ_c is determined by the relative strengths of the ambient fields that are applied to the six component sytems of an idealized polyphase rocket. The special case of $\theta_c = 0$ gives the full range $0 \leq s \leq 1$ for liftoff enhancement, and is perhaps the optimum condition and safest condition in which to operate a polyphase rocket.

Case b. $\theta_c \leq 0$.

The general relation between the internal phase angles corresponding to the condition $\theta_c \leq 0$ is obtained from equation (898) as

$$\theta^{e}_{\mu} - \theta^{b}_{V} - \theta^{E}_{g} \leq \theta^{b}_{\mu} - \theta^{a}_{A} - \theta^{e}_{h} - \theta^{e}_{a} \tag{945}$$

or equivalently

$$\theta^{e}_{\mu} + |\theta^{b}_{V}| + |\theta^{E}_{g}| \leq \theta^{b}_{\mu} + |\theta^{a}_{A}| + |\theta^{e}_{h}| + |\theta^{e}_{a}| \tag{946}$$

In this case the solution to equation (899) and (900) is given by

$$\cos\theta^{e-}_{n} = [1/2(A - B)]^{1/2} \qquad \theta^{e-}_{n} \geq 0 \tag{947}$$

$$\theta^{b-}_{n} = \theta^{e-}_{n} - |\theta_{c}| \tag{948}$$

where A and B are given by equations (906) and (907) and for the same range of values of s as given by equations (915) and (916). It is clear from equation (899) that in general for the range of values of s given by equations (915) and (916)

$$\theta^{e-}_{n} = \theta^{e+}_{n} + |\theta_{c}| \qquad \theta^{b+}_{n} = \theta^{b-}_{n} + |\theta_{c}| \tag{949}$$

so that the solutions for Case a and Case b differ only by an additive constant. Combining equations (905), (948) and (949) gives the following general results

$$\theta^{b+}_{n} = \theta^{e-}_{n} \qquad \theta^{b-}_{n} = \theta^{e+}_{n} \tag{950}$$

in the range $0 \leq s \leq s_1$.

For the specific values $s = 0$ and $s = s_{o1} = \cos\theta_c$, equations (947) and (948) give the following results

$$\theta_n^{e-} = \pi/2 \qquad \theta_n^{b-} = \pi/2 - |\theta_c| \qquad s = 0 \tag{951}$$

$$\theta_n^{e-} = |\theta_c| \qquad \theta_n^{b-} = 0 \qquad s = s_{01} \tag{952}$$

where at these points A and B are given by equations (908) and (913) and therefore

$$A - B = 0 \qquad s = 0 \tag{953}$$

$$A - B = 2\cos^2\theta_c \qquad s = s_{o1} \tag{954}$$

which, when placed in equations (947) and (948), give the results that appear in equations (951) and (952). Note that θ_n^{e-} does not have any zero values for $\theta_c \neq 0$, however θ_n^{b-} has a zero at the point $s = s_{o1}$ as expected if the second part of equation (950) is combined with the first part of equation (910).

C. Calculation of the Internal Phase Angle θ_n^c of the Gases in the Combustion Chamber in terms of the Internal Phase Angle θ_n^e of the Exhaust Gases in the Rocket Exhaust Tube.

Because the exhaust gases have a high velocity and a low pressure while the combustion gases in the combustion chamber have a low velocity and a high pressure, Bernoulli's equation (447) for the case of zero ambient fields can be written approximately as

$$1/2\rho_e^o(v_e^o)^2 \approx 1/2\mu_e^o n_e^o(v_e^o)^2 \approx P_c^o \approx n_c^o k_B T_c^o \approx 1/3 n_c^o \mu_c^o (v_{\mu c}^o)^2 \tag{955}$$

where μ_e^o and v_e^o = molecular mass and exhaust gas velocity of the gases in the exhaust tube and μ_c^o and $v_{\mu c}^o$ = molecular mass and the root mean square thermal velocity of the molecules in the combustion chamber. Bernoulli's equation (766) for the case when an ambient field is present can be written approximately as

$$1/2\overline{\rho}_e \overline{v}_e^2 \approx \overline{P}_c \approx \overline{n}_c k_B \overline{T}_c \tag{956}$$

Combining equations (716) and (956) gives

$$1/2\overline{\mu}_e \overline{n}_e \overline{v}_e^2 \approx \overline{n}_c k_B \overline{T}_c \tag{957}$$

Combining equations (745) and (957) also gives

$$1/2\overline{\mu}_e \overline{n}_e \overline{v}_e^2 \approx 1/3\overline{\mu}_c \overline{n}_c \overline{v}_{\mu c}^2 \tag{958}$$

Equation (957) can be written as the following magnitude and phase angle equations

$$1/2\mu_e n_e v_e^2 \approx n_c k_B T_c \qquad \theta_\mu^e + \theta_n^e + 2\theta_v^e \approx \theta_n^c + \theta_T^c \tag{959}$$

and similarly equation (958) gives

$$1/2\mu_e n_e v_e^2 \approx 1/3\mu_c n_c v_{\mu c}^2 \qquad \theta_\mu^e + \theta_n^e + 2\theta_v^e \approx \theta_\mu^c + \theta_n^c + 2\theta_{v\mu}^c \tag{960}$$

either of which may possibly allow a calculation of the required value of θ_n^c in terms of θ_n^e which is given by the solution of the liftoff equations (899) and (900) as in equations (904) and (947). But these are useful only if the remaining phase angles in equations (959) and (960) are known. The mass conservation equation (453) for the case of zero ambient fields is written as

$$\rho_e^o v_e^o A_a^o = \rho_c^o v_c^o A_f^o \tag{961}$$

and for the case when an ambient field is present the mass conservation law in equation (767) is written as

$$\overline{\rho}_e \overline{v}_e \overline{A}_a = \overline{\rho}_c \overline{v}_c \overline{A}_f \tag{962}$$

which can be written as the following magnitude and phase angle equations

$$\rho_e v_e A_a = \rho_c v_c A_f \qquad \theta_n^e + \theta_\mu^e + \theta_v^e + \theta_A^a = \theta_n^c + \theta_\mu^c + \theta_v^c + \theta_A^f \tag{963}$$

and which can be used to relate θ_n^c and θ_n^e provided that the other phase angles in equation (963) are known. The internal phase angles θ_n^e and θ_n^c are useful to calculate the actual magnitudes of the intrinsic pressure and bulk moduli of the exhaust gases and the combustion gases in an ambient field as for example

$$K_e = n_e \partial P_e / \partial n_e \tag{964}$$

$$K_c = n_c \partial P_c / \partial n_c = P_c = n_c k_B T_c = P_c^o f_{Tc} \sec\theta_n^c = K_c^o f_{Tc} \sec\theta_n^c \tag{965}$$

where P_e and P_c are given by equations (773) and (750) respectively and where f_{Tc} is given by equation (747), and also to calculate the electromagnetic fields that are required for liftoff enhancement of a polyphase rocket.

<u>10. AMBIENT ELECTROMAGNETIC FIELDS REQUIRED FOR THE LIFTOFF OF A POLYPHASE ROCKET.</u> This section presents a calculation of the electromagnetic fields in the exhaust gas tube, the combustion chamber and the body of a polyphase rocket that are required for the liftoff enhancement. These electromagnetic fields are expressed in terms of the critical internal phase angles θ_n^e, θ_n^c and θ_n^b that have been determined in Section 9 from the solutions of the liftoff equations of a polyphase rocket. However, first the static elastic energy density that is induced in matter by an ambient field must be calculated. Then the elastic energies that are induced in these three polyhase rocket components will be set equal to the energy densities of the corresponding ambient electromagnetic fields that are applied to the three rocket systems.

A. Static Elastic and Viscoelastic Energy Density, Pressure and Bulk Modulus Induced in Matter by the Presence of an Ambient Field.

For a mechanical system or a thermodynamic system where temperature effects on the induced pressure can be ignored, the general relations for calculating the the induced pressure and the induced bulk modulus in terms of the induced energy density (excitation energy density) are given for both [ld] and [hd] matter by

$$P = n\partial\mathfrak{E}/\partial n - \mathfrak{E} \tag{966}$$

$$K = \partial P/\partial n = n^2\partial^2\mathfrak{E}/\partial n^2 \tag{967}$$

The application of an ambient field to matter is associated with changes in the molecular number densities as described by equation (245) for [ld] matter and by equation (244) for [hd] matter. For [ld] matter equation (245) shows that an ambient field increases the molecular number density and therefore represents a compression field, while for [hd] matter equation (244) indicates that an ambient field decreases the molecular number density and represents a tension field. The specific expression for the excitation energy density that is associated with a change of the density of matter is (after ignoring an additive constant and higher order terms)[12]

$$\mathfrak{E} = 1/2K_o n_o^{-2}(n - n_o)^2 = 1/2K_o(n/n_o - 1)^2 \tag{968}$$

Combining equations (966) through (968) gives the excitation values of the pressure and the bulk modulus as

$$P = 1/2K_o n_o^{-2}(n^2 - n_o^2) = 1/2K_o[(n/n_o)^2 - 1] \tag{969}$$

$$K = K_o(n/n_o)^2 \tag{970}$$

which are general (mechanical) expressions for calculating the pressure and bulk modulus that are associated with a change in the molecular number density of matter.

Consider now the application of equations (968) through (970) to the case of an ambient field that is applied to [ld] and [hd] matter and is associated with the induction of an internal phase angle θ_n of the molecular number density. Combining equations (245) and (968) through (970) gives the compression (C) fields in [ld] matter as

$$\mathfrak{E}_C = 1/2K_o(\sec\theta_n - 1)^2 \qquad [\text{ld}] \tag{971}$$

$$P_C = 1/K_o(\sec^2\theta_n - 1) = 1/2K_o\tan^2\theta_n \qquad [\text{ld}] \tag{972}$$

$$K_C = K_o\sec^2\theta_n \qquad [\text{ld}] \tag{973}$$

Combining equations (244) and (968) through (970) gives the following tension (T) field in [hd] matter

$$\mathfrak{E}_T = 1/2K_o(1 - \cos\theta_n)^2 \qquad [\text{hd}] \tag{974}$$

$$P_T = 1/K_o(\cos^2\theta_n - 1) = -1/2K_o\sin^2\theta_n \qquad [\text{hd}] \tag{975}$$

$$K_T = K_o\cos^2\theta_n \qquad [\text{hd}] \tag{976}$$

In the limit $\theta_n \to 0$ it follows from equations (971) through (976) that

$$\mathfrak{E}_C \to 0 \quad P_C \to 0 \quad K_C \to K_o \quad \mathfrak{E}_T \to 0 \quad P_T \to 0 \quad K_T \to K_o \tag{977}$$

as expected.

The shearing energy density that is associated with the induction of an internal phase angle $\theta_{\mathfrak{N}}$ of the number of molecules by the application of an ambient field to a combined elastic and viscoelastic piece of matter with viscosity ς^{o} is given by[9-12]

$$\mathfrak{E}_{S} = 1/2 G_{o} \tan^{2}\theta_{\mathfrak{N}} \qquad G_{o} = \varsigma_{o}/t^{o} \tag{978}$$

which is valid for [ld] or [hd] matter, and where t^{o} = time factor. Then the induced pressure and the bulk modulus can be calculated from equations (966) and (967), but this has to be done for [hd] and [ld] matter by using equations (244) and (245) because the molecular number density is changing according to these equations. Combining equations (245), (966), (967) and (978) gives the following results for [ld] matter

$$\mathfrak{E}_{SC} = 1/2 G_{o} \tan^{2}\theta_{\mathfrak{N}} \qquad \text{[ld]} \tag{979}$$

$$P_{SC} = G_{o} B \tan\theta_{\mathfrak{N}} \qquad \text{[ld]} \tag{980}$$

$$K_{SC} = G_{o} H \sec^{2}\theta_{\mathfrak{N}} \qquad \text{[ld]} \tag{981}$$

and combining equations (244), (966), (967) and (978) gives for the case of [hd] matter

$$\mathfrak{E}_{ST} = 1/2 G_{o} \tan^{2}\theta_{\mathfrak{N}} \qquad \text{[hd]} \tag{982}$$

$$P_{ST} = -G_{o} E \tan\theta_{\mathfrak{N}} \qquad \text{[hd]} \tag{983}$$

$$K_{ST} = G_{o} I \sec^{2}\theta_{\mathfrak{N}} \qquad \text{[hd]} \tag{984}$$

where

$$B = \cot\theta_{n} \sec^{2}\theta_{\mathfrak{N}}\, d\theta_{\mathfrak{N}}/d\theta_{n} - 1/2\tan\theta_{\mathfrak{N}} \qquad \text{[ld]} \tag{985}$$

$$C = (D - 1/2\, d\theta_{\mathfrak{N}}/d\theta_{n})\tan\theta_{\mathfrak{N}} \qquad \text{[ld]} \tag{986}$$

$$H = (C + B\, d\theta_{\mathfrak{N}}/d\theta_{n})\cot\theta_{n} \qquad \text{[ld]} \tag{987}$$

$$D = -\csc^{2}\theta_{n}\, d\theta_{\mathfrak{N}}/d\theta_{n} + 2\cot\theta_{n}\tan\theta_{\mathfrak{N}}(d\theta_{\mathfrak{N}}/d\theta_{n})^{2} + \cot\theta_{n}\, d^{2}\theta_{\mathfrak{N}}/d\theta_{n}^{2} \tag{988}$$

$$E = \cot\theta_{n} \sec^{2}\theta_{\mathfrak{N}}\, d\theta_{\mathfrak{N}}/d\theta_{n} + 1/2\tan\theta_{\mathfrak{N}} \qquad \text{[hd]} \tag{989}$$

$$F = (D + 1/2\, d\theta_{\mathfrak{N}}/d\theta_{n})\tan\theta_{\mathfrak{N}} \qquad \text{[hd]} \tag{990}$$

$$I = (F + E\, d\theta_{\mathfrak{N}}/d\theta_{n})\cot\theta_{n} \qquad \text{[hd]} \tag{991}$$

which are exact equations. Note that the shear field in equations (979) through (984) are indexed to the compression and tension modes of matter because the pressure and bulk modulus are different functions of $\theta_{\mathfrak{N}}$ and θ_{n} for [ld] and [hd] matter and also because the internal phase angles $\theta_{\mathfrak{N}}$ and θ_{n} are related to eachother differently for the [ld] and [hd] cases of matter.

A particularly interesting (and perhaps universal) connection between $\theta_{\mathfrak{N}}$ and θ_n is given by the following equations[12]

$$\tan\theta_{\mathfrak{N}} = \sec\theta_n - 1 \qquad \theta_n \ge \theta_{\mathfrak{N}} \qquad 0 \le \theta_n \le \pi/2 \qquad 0 \le \theta_{\mathfrak{N}} \le \pi/2 \qquad \text{[ld]} \qquad (992)$$

$$\tan\theta_{\mathfrak{N}} = 1 - \cos\theta_n \qquad \theta_n \ge \theta_{\mathfrak{N}} \qquad 0 \le \theta_n \le \pi/2 \qquad 0 \le \theta_{\mathfrak{N}} \le \pi/4 \qquad \text{[hd]} \qquad (993)$$

which follows from the assumption that for static elasticity[12]

$$\mathfrak{E}_C = K_o(K_o + G_o)^{-1}\mathfrak{E} \qquad \mathfrak{E}_{SC} = G_o(K_o + G_o)^{-1}\mathfrak{E} \qquad \text{[ld]} \qquad (994)$$

$$\mathfrak{E}_T = K_o(K_o + G_o)^{-1}\mathfrak{E} \qquad \mathfrak{E}_{ST} = G_o(K_o + G_o)^{-1}\mathfrak{E} \qquad \text{[hd]} \qquad (995)$$

where the total energy density for the for the [ld] compression case and the [hd] tension case of static elasticity are given as

$$\mathfrak{E} = \mathfrak{E}_C + \mathfrak{E}_{SC} \qquad \text{[ld]} \qquad (996)$$

$$\mathfrak{E} = \mathfrak{E}_T + \mathfrak{E}_{ST} \qquad \text{[hd]} \qquad (997)$$

The small angle approximation of equation (992) or equation (993) is given by

$$\theta_{\mathfrak{N}} \approx \theta_n^2/2 \qquad \theta_n \ge \theta_{\mathfrak{N}} \qquad \text{[ld] or [hd]} \qquad (998)$$

and then it is easy to see from equations (985) through (991) that the limits $\theta_n \to 0$ and $\theta_{\mathfrak{N}} \to 0$ in the form of equation (998) yield the results

$$B \to 1 \qquad C \to 0 \qquad H \to 1 \qquad D \to 0 \qquad E \to 1 \qquad F \to 0 \qquad I \to 1 \qquad (999)$$

and in this limit equations (979) through (984) give

$$\mathfrak{E}_{SC} \to 0 \qquad P_{SC} \to 0 \qquad K_{SC} \to G_o \qquad \mathfrak{E}_{ST} \to 0 \qquad P_{ST} \to 0 \qquad K_{ST} \to G_o \qquad (1000)$$

The limiting values in equations (999) and (1000) depend critically on the form of the limiting relation between $\theta_{\mathfrak{N}}$ and θ_n given in equation (998). The limits given in equation (1000) can be obtained more generally by placing equations (992) and (993) into equation (978) to express the induced shearing energy density in terms of θ_n and then using equations (966) and (967) to calculate the pressure and bulk modulus that are induced by a shearing ambient field, with the results that for [ld] matter

$$\mathfrak{E}_{SC} = 1/2G_o(\sec\theta_n - 1)^2 \qquad \text{[ld]} \qquad (1001)$$

$$P_{SC} = 1/2G_o(\sec^2\theta_n - 1) = 1/2G_o\tan^2\theta_n \qquad \text{[ld]} \qquad (1002)$$

$$K_{SC} = G_o\sec^2\theta_n \qquad \text{[ld]} \qquad (1003)$$

while for [hd] matter

$$\mathfrak{E}_{ST} = 1/2G_o(1-\cos\theta_n)^2 \qquad [hd] \qquad (1004)$$

$$P_{ST} = 1/2G_o(\cos^2\theta_n - 1) = -1/2G_o\sin^2\theta_n \qquad [hd] \qquad (1005)$$

$$K_{ST} = G_o\cos^2\theta_n \qquad [hd] \qquad (1006)$$

from which the limiting values of the induced energy density, pressure and bulk modulus given in equation (1000) are easily obtained when $\theta_n \to 0$.

The total energy density, pressure and bulk modulus that are induced in [ld] matter by an ambient field are obtained from equations (971) through (973) and (979) through (981) as

$$\mathfrak{E} = \mathfrak{E}_C + \mathfrak{E}_{SC} = 1/2K_o(\sec\theta_n - 1)^2 + 1/2G_o\tan^2\theta_{\mathcal{N}} \qquad [ld] \qquad (1007)$$

$$P = P_C + P_{SC} = 1/2K_o\tan^2\theta_n + G_oB\tan\theta_{\mathcal{N}} \qquad [ld] \qquad (1008)$$

$$K = K_C + K_{SC} = K_o\sec^2\theta_n + G_oH\sec^2\theta_{\mathcal{N}} \qquad [ld] \qquad (1009)$$

while for [hd] matter the total energy density, pressure and bulk modulus that are induced by an ambient field are obtained from equations (974) through (976) and (982) through (984) as

$$\mathfrak{E} = \mathfrak{E}_T + \mathfrak{E}_{ST} = 1/2K_o(1-\cos\theta_n)^2 + 1/2G_o\tan^2\theta_{\mathcal{N}} \qquad [hd] \qquad (1010)$$

$$P = P_T + P_{ST} = -1/2K_o\sin^2\theta_n - G_oE\tan\theta_{\mathcal{N}} \qquad [hd] \qquad (1011)$$

$$K = K_T + K_{ST} = K_o\cos^2\theta_n + G_oI\sec^2\theta_{\mathcal{N}} \qquad [hd] \qquad (1012)$$

However, equation (992) established the validity of equations (1001) through (1003) and when these equations are used in equations (1007) through (1009) the result for [ld] matter is

$$\mathfrak{E} = \mathfrak{E}_C + \mathfrak{E}_{SC} = 1/2(K_o + G_o)(\sec\theta_n - 1)^2 \qquad [ld] \qquad (1013)$$

$$P = P_C + P_{SC} = 1/2(K_o + G_o)\tan^2\theta_n \qquad [ld] \qquad (1014)$$

$$K = K_C + K_{SC} = (K_o + G_o)\sec^2\theta_n \qquad [ld] \qquad (1015)$$

and placing equations (993) and (1004) through (1006) into equations (1010) through (1012) gives for [hd] matter

$$\mathfrak{E} = \mathfrak{E}_T + \mathfrak{E}_{ST} = 1/2(K_o + G_o)(1-\cos\theta_n)^2 \qquad [hd] \qquad (1016)$$

$$P = P_T + P_{ST} = -1/2(K_o + G_o)\sin^2\theta_n \qquad [hd] \qquad (1017)$$

$$K = K_T + K_{ST} = (K_o + G_o)\cos^2\theta_n \qquad [hd] \qquad (1018)$$

Equations (1013) through (1015) and (1016) through (1018) can also be expressed in terms of the shearing angle $\theta_{\mathbb{N}}$ alone by using equations (992) and (993) respectively. Equations (1013) through (1018) give the total energy density, pressure and bulk modulus that are induced in matter by the application of an ambient field. Note that $\mathfrak{E} \rightarrow 0, P \rightarrow 0$ and $K \rightarrow K_o + G_o$ when the internal phase angles approach zero when the ambient field is turned off. Note also that equations (975), (983), (1005), (1011) and (1017) show that the pressure induced in [hd] matter by an ambient field has a negative value and represents a uniform state of tension. Also, for the application of an ambient field, the bulk modulus is increased for [ld] matter and is decreased for [hd] matter.

B. Static Elastic and Viscoelastic Calculation of the Magnitudes of the Ambient Electromagnetic Wave Fields That are Applied to the Exhaust Gases in the Exhaust Tube, to the Combustion Chamber Gases and to the Rocket Body, and that are Required for Liftoff Enhancement of a Polyphase Rocket.

The use of ambient electromagnetic wave fields is perhaps the only practical way to facilitate the operation of an internal phase rocket engine. The preceding calculations of this section give, among other things, the calculated values of the internal phase angle θ^e_n of the molecular number density of the exhaust gases, the internal phase angle θ^c_n of the molecular number density of the combustion chamber gases and the internal phase angle θ^b_n of the molecular number density of the rocket body that are required for the liftoff enhancement of an underpowered (for zero ambient field conditions) polyphase rocket. This section determines the magnitudes of the effective magnetic fields and magnetic inductions of the ambient electromagnetic wave fields that induce the internal phase angle θ^e_n in the exhaust gases, the internal phase angle θ^c_n in the combustion chamber gases and the internal phase angle θ^b_n in the rocket body.

Using a real number approximate formalism, the changes in the compression (C) energy densities of the [ld] exhaust gases in the exhaust tube and the combustion gases in the combustion chamber resulting from the application of ambient electromagnetic wave fields are written approximately for small changes of volume as[12]

$$\mathfrak{E}^e_C = 1/2K^o_e(\Delta n_e/n^o_e)^2 = 1/2K^o_e[(n_e - n^o_e)/n^o_e]^2 = 1/2K^o_e(\sec\theta^e_n - 1)^2 \quad \text{[ld]} \quad (1019)$$

$$\mathfrak{E}^c_C = 1/2K^o_c(\Delta n_c/n^o_c)^2 = 1/2K^o_c[(n_c - n^o_c)/n^o_c]^2 = 1/2K^o_c(\sec\theta^c_n - 1)^2 \quad \text{[ld]} \quad (1020)$$

where the relative changes in the [ld] molecular number densities of the exhaust gases and the combustion gases due to the application of ambient electromagnetic wave fields is obtained from equations (704) and (743) as

$$\Delta n_e/n^o_e = (n_e - n^o_e)/n^o_e = (n^o_e \sec\theta^e_n - n^o_e)/n^o_e = \sec\theta^e_n - 1 \quad \text{[ld]} \quad (1021)$$

$$\Delta n_c/n^o_c = (n_c - n^o_c)/n^o_c = (n^o_c \sec\theta^c_n - n^o_c)/n^o_c = \sec\theta^c_n - 1 \quad \text{[ld]} \quad (1022)$$

and where K^o_e and K^o_c = constant magnitudes of the bulk moduli of the rocket exhaust gases in the exhaust tube and the combustion gases in the combustion chamber for zero ambient electromagnetic wave fields and are given by

$$K^o_e = n^o_e \partial P^o_e/\partial n^o_e \qquad K^o_c = n^o_c \partial P^o_c/\partial n^o_c = P^o_c = n^o_c k T^o_c \quad (1023)$$

where P_e^o and P_c^o are given by equations (454) and (451) respectively. Therefore in this approximation the ambient electromagnetic wave field affects the compression energy density of the gases by way of the relative increase in the molecular number density of the gases.

The energy density of the ambient electromagnetic wave field that is applied to the exhaust gases and the combustion chamber gases can be written approximately in the following forms

$$\mathfrak{E}_e^\gamma = 1/2\mu_e H_{\gamma e}'^2 + 1/2\varepsilon_e E_{\gamma e}'^2 = 1/2\mu_e H_{\gamma e}^2 = 1/(2\mu_e) B_{\gamma e}^2 \tag{1024}$$

$$\mathfrak{E}_c^\gamma = 1/2\mu_c H_{\gamma c}'^2 + 1/2\varepsilon_c E_{\gamma c}'^2 = 1/2\mu_c H_{\gamma c}^2 = 1/(2\mu_c) B_{\gamma c}^2 \tag{1025}$$

where $H'_{\gamma e}, E'_{\gamma e}$ and $H'_{\gamma c}, E'_{\gamma c}$ = magnetic and electric field components of the ambient electromagnetic wave field applied to the exhaust gases and the combustion chamber gases respectively and $H_{\gamma e}$, $B_{\gamma e}$ and $H_{\gamma c}$, $B_{\gamma c}$ = effective magnetic fields and magnetic inductions of the applied ambient electromagnetic wave field and which are given by

$$H_{\gamma e} = \sqrt{2} H'_{\gamma e} \qquad B_{\gamma e} = \sqrt{2} B'_{\gamma e} \tag{1026}$$

$$H_{\gamma c} = \sqrt{2} H'_{\gamma c} \qquad B_{\gamma c} = \sqrt{2} B'_{\gamma c} \tag{1027}$$

and where μ_e, ε_e and μ_c, ε_c = magnetic permeability and electric permittivity of the exhaust gases and the combustion chamber gases that are located inambient fields. Equating the energy densities in equations (1019) and (1024) and equating the energy densities in equations (1020) and (1025) gives

$$H_{\gamma e}^C = (K_e^o/\mu_e)^{1/2}(\sec\theta_n^e - 1) \qquad B_{\gamma e}^C = (\mu_e K_e^o)^{1/2}(\sec\theta_n^e - 1) \quad [\text{ld}] \tag{1028}$$

$$H_{\gamma c}^C = (K_c^o/\mu_c)^{1/2}(\sec\theta_n^c - 1) \qquad B_{\gamma c}^C = (\mu_c K_c^o)^{1/2}(\sec\theta_n^c - 1) \quad [\text{ld}] \tag{1029}$$

Assuming $\mu_e = \mu_e^o$, where μ_e^o = magnetic permeability of the exhaust gases for zero ambient electromagnetic wave fields, gives the following approximate equations for the effective magnetic field and magnetic induction of the ambient electromagnetic waves that must be applied to the exhaust gases in the exhaust tube of a polyphase rocket to obtain a liftoff enhancement associated with the internal phase angle θ_n^e

$$H_{\gamma e}^C = (K_e^o/\mu_e^o)^{1/2}(\sec\theta_n^e - 1) \qquad B_{\gamma e}^C = (\mu_e^o K_e^o)^{1/2}(\sec\theta_n^e - 1) \quad [\text{ld}] \tag{1030}$$

$$H_{\gamma c}^C = (K_c^o/\mu_c^o)^{1/2}(\sec\theta_n^c - 1) \qquad B_{\gamma c}^C = (\mu_c^o K_c^o)^{1/2}(\sec\theta_n^c - 1) \quad [\text{ld}] \tag{1031}$$

Values of θ_n^e for use in equations (1030) can be obtained as solutions to the liftoff equations (899) and (900) as given by equations (904) and (947) in terms of the parameters s and θ_c that are associated with an underpowered polyphase rocket and given by equations (894) and (898) respectively, and then the values of θ_n^c for use in equation (1031) can be obtained from equations (959), (960) or (963).

The isotropic tension (T) energy density of the [hd] liquid or solid fuel rocket body resulting from the application of an ambient electromagnetic wave field is written approximately for small changes of volume as[12]

$$\mathfrak{E}_T^b = 1/2K_b^o(\Delta n_b / n_b^o)^2 = 1/2K_b^o[(n_b - n_b^o)/n_b^o]^2 = 1/2K_b^o(\cos\theta_n^b - 1)^2 \quad [hd] \quad (1032)$$

where the relative change in the [hd] molecular number density of the rocket body due to the application of an ambient electromagnetic wave field is obtained from equation (611) as

$$\Delta n_b / n_b^o = (n_b - n_b^o)/n_b^o = (n_b^o \cos\theta_n^b - n_b^o)/n_b^o = \cos\theta_n^b - 1 \quad [hd] \quad (1033)$$

and where K_b^o = constant magnitude of the bulk (isotropic tension) modulus of the rocket body material for a zero ambient electromagnetic wave field. Therefore in this approximation the ambient electromagnetic wave field affects the isotropic tension energy density of the solid or liquid material of the rocket body by way of the relative decrease in the molecular number density of the solids or liquids.

The energy density of the ambient electromagnetic wave field that is applied to the rocket body can be written approximately in the following forms

$$\mathfrak{E}_b^\gamma = 1/2\mu_b H_{\gamma b}'^2 + 1/2\varepsilon_b E_{\gamma b}'^2 = 1/2\mu_b H_{\gamma b}^2 = 1/(2\mu_b)B_{\gamma b}^2 \quad (1034)$$

where $H'_{\gamma b}$ and $E'_{\gamma b}$ = magnetic and electric field components of the ambient electromagnetic wave field applied to the rocket body, $H_{\gamma b}$ and $B_{\gamma b}$ = effective magnetic field and magnetic induction of the applied ambient electromagnetic wave field and which are given by

$$H_{\gamma b} = \sqrt{2}H'_{\gamma b} \qquad B_{\gamma b} = \sqrt{2}B'_{\gamma b} \quad (1035)$$

and where μ_b and ε_b = constant magnetic permeability and electric permittivity of the rocket body. Equating the energy densities in equations (1032) and (1034) gives the magnetic field and the magnetic induction as

$$H_{\gamma b}^T = (K_b^o / \mu_b)^{1/2}(1 - \cos\theta_n^b) \qquad B_{\gamma b}^T = (\mu_b K_b^o)^{1/2}(1 - \cos\theta_n^b) \quad [hd] \quad (1036)$$

Assuming $\mu_b = \mu_b^o$, where μ_b^o = magnetic permeability of the rocket body material for zero ambient electromagnetic wave fields, gives the following approximate equations for the effective magnetic field and magnetic induction of the ambient electromagnetic waves that must be applied to the solid or liquid material that constitutes the rocket body of a polyphase rocket to obtain a liftoff enhancement associated with the internal phase angle θ_n^b

$$H_{\gamma b}^T = (K_b^o / \mu_b^o)^{1/2}(1 - \cos\theta_n^b) \qquad B_{\gamma b}^T = (\mu_b^o K_b^o)^{1/2}(1 - \cos\theta_n^b) \quad [hd] \quad (1037)$$

Values of the angle θ_n^b for use in equation (1037) can be obtained from equations from solutions of the liftoff equations (899) and (900) as given by equations (905) and (948) in terms of the parameters s and θ_c that are associated with a polyphase rocket and given by equations (894) and (898) respectively.

From equations (979), (982), (1001) and (1004) it follows that the shear energy densities that are induced by the shearing component of an ambient field are written for the [ld] gases of the exhaust tube and the combustion chamber and for the [hd] material of the rocket body (where $G^{o}_{b} = \varsigma^{o}_{b} / t^{o}_{b}$ applies only to liquid fuel) as

$$\mathfrak{E}^{e}_{SC} = 1/2 G^{o}_{e} \tan^{2} \theta^{e}_{\aleph} \qquad G^{o}_{e} = \varsigma^{o}_{e} / t^{o}_{e} \qquad [ld] \qquad (1038)$$

$$\mathfrak{E}^{c}_{SC} = 1/2 G^{o}_{c} \tan^{2} \theta^{c}_{\aleph} \qquad G^{o}_{c} = \varsigma^{o}_{c} / t^{o}_{c} \qquad [ld] \qquad (1039)$$

$$\mathfrak{E}^{b}_{ST} = 1/2 G^{o}_{b} \tan^{2} \theta^{b}_{\aleph} \qquad G^{o}_{b} = \varsigma^{o}_{b} / t^{o}_{b} \qquad [hd] \qquad (1040)$$

which can be rewritten in the following forms if equations (992) and (993) are assumed

$$\mathfrak{E}^{e}_{SC} = 1/2 G^{o}_{e} (\sec \theta^{e}_{n} - 1)^{2} \qquad G^{o}_{e} = \varsigma^{o}_{e} / t^{o}_{e} \qquad [ld] \qquad (1041)$$

$$\mathfrak{E}^{c}_{SC} = 1/2 G^{o}_{c} (\sec \theta^{c}_{n} - 1)^{2} \qquad G^{o}_{c} = \varsigma^{o}_{c} / t^{o}_{c} \qquad [ld] \qquad (1042)$$

$$\mathfrak{E}^{b}_{ST} = 1/2 G^{o}_{b} (1 - \cos \theta^{b}_{n})^{2} \qquad G^{o}_{b} = \varsigma^{o}_{b} / t^{o}_{b} \qquad [hd] \qquad (1043)$$

Equating the electromagnetic energy densities in equations (1024), (1025) and (1034) respectively with the shearing energy densities in equations (1038) through (1040) gives the following effective magnetic fields and magnetic inductions

$$H^{SC}_{\gamma e} = (G^{o}_{e} / \mu^{o}_{e})^{1/2} \tan \theta^{e}_{\aleph} \qquad [ld] \qquad (1044)$$

$$B^{SC}_{\gamma e} = (\mu^{o}_{e} G^{o}_{e})^{1/2} \tan \theta^{e}_{\aleph} \qquad [ld] \qquad (1045)$$

$$H^{SC}_{\gamma c} = (G^{o}_{c} / \mu^{o}_{c})^{1/2} \tan \theta^{c}_{\aleph} \qquad [ld] \qquad (1046)$$

$$B^{SC}_{\gamma c} = (\mu^{o}_{c} G^{o}_{c})^{1/2} \tan \theta^{c}_{\aleph} \qquad [ld] \qquad (1047)$$

$$H^{ST}_{\gamma b} = (G^{o}_{b} / \mu^{o}_{b})^{1/2} \tan \theta^{b}_{\aleph} \qquad [hd] \qquad (1048)$$

$$B^{ST}_{\gamma b} = (\mu^{o}_{b} G^{o}_{b})^{1/2} \tan \theta^{b}_{\aleph} \qquad [hd] \qquad (1049)$$

or if equations (992) and (993) are assumed

$$H^{SC}_{\gamma e} = (G^{o}_{e} / \mu^{o}_{e})^{1/2} (\sec \theta^{e}_{n} - 1) \qquad [ld] \qquad (1050)$$

$$B^{SC}_{\gamma e} = (\mu^{o}_{e} G^{o}_{e})^{1/2} (\sec \theta^{e}_{n} - 1) \qquad [ld] \qquad (1051)$$

$$H^{SC}_{\gamma c} = (G^{o}_{c} / \mu^{o}_{c})^{1/2} (\sec \theta^{c}_{n} - 1) \qquad [ld] \qquad (1052)$$

$$B^{SC}_{\gamma c} = (\mu^{o}_{c} G^{o}_{c})^{1/2} (\sec \theta^{c}_{n} - 1) \qquad [ld] \qquad (1053)$$

$$H_{\gamma b}^{ST} = (G_b^o / \mu_b^o)^{1/2}(1 - \cos\theta_n^b) \quad [hd] \quad (1054)$$

$$B_{\gamma b}^{ST} = (\mu_b^o G_b^o)^{1/2}(1 - \cos\theta_n^b) \quad [hd] \quad (1055)$$

which are the effective magnetic fields and inductions associated with the shearing component of the electromagnetic field and which are required for the enhanced liftoff capability of a polyphase rocket.

For the case when an ambient shearing field is present in addition to an ambient compression field for the [ld] gases in the exhaust tube and in the combustion chamber and in addition to an ambient tension field for the [hd] material of the rocket body, the total induced energy densities for these systems are obtained from equations (1007) and (1010) as

$$\mathfrak{E}^e = \mathfrak{E}_C^e + \mathfrak{E}_{SC}^e = 1/2K_e^o(\sec\theta_n^e - 1)^2 + 1/2G_e^o \tan^2\theta_{\mathcal{N}}^e \quad [ld] \quad (1056)$$

$$\mathfrak{E}^c = \mathfrak{E}_C^c + \mathfrak{E}_{SC}^c = 1/2K_c^o(\sec\theta_n^c - 1)^2 + 1/2G_c^o \tan^2\theta_{\mathcal{N}}^c \quad [ld] \quad (1057)$$

$$\mathfrak{E}^b = \mathfrak{E}_T^b + \mathfrak{E}_{ST}^b = 1/2K_b^o(1 - \cos\theta_n^b)^2 + 1/2G_b^o \tan^2\theta_{\mathcal{N}}^b \quad [hd] \quad (1058)$$

The effective magnetic fields and magnetic inductions of the ambient electromagnetic field that correspond to the energy densities in equations (1056) through (1058) are given without assuming the validity of equations (992) and (993) by

$$H_{\gamma e} = (2\mathfrak{E}^e / \mu_e^o)^{1/2} = [K_e^o / \mu_e^o(\sec\theta_n^e - 1)^2 + G_e^o / \mu_e^o \tan^2\theta_{\mathcal{N}}^e]^{1/2} \quad [ld] \quad (1059)$$

$$= [(H_{\gamma e}^C)^2 + (H_{\gamma e}^{SC})^2]^{1/2}$$

$$B_{\gamma e} = (2\mu_e^o\mathfrak{E}^e)^{1/2} = [\mu_e^o K_e^o(\sec\theta_n^e - 1)^2 + \mu_e^o G_e^o \tan^2\theta_{\mathcal{N}}^e]^{1/2} \quad [ld] \quad (1060)$$

$$= [(B_{\gamma e}^C)^2 + (B_{\gamma e}^{SC})^2]^{1/2}$$

$$H_{\gamma c} = (2\mathfrak{E}^c / \mu_c^o)^{1/2} = [K_c^o / \mu_c^o(\sec\theta_n^c - 1)^2 + G_c^o / \mu_c^o \tan^2\theta_{\mathcal{N}}^c]^{1/2} \quad [ld] \quad (1061)$$

$$= [(H_{\gamma c}^C)^2 + (H_{\gamma c}^{SC})^2]^{1/2}$$

$$B_{\gamma c} = (2\mu_c^o\mathfrak{E}^c)^{1/2} = [\mu_c^o K_c^o(\sec\theta_n^c - 1)^2 + \mu_c^o G_c^o \tan^2\theta_{\mathcal{N}}^c]^{1/2} \quad [ld] \quad (1062)$$

$$= [(B_{\gamma c}^C)^2 + (B_{\gamma c}^{SC})^2]^{1/2}$$

$$H_{\gamma b} = (2\mathfrak{E}^b / \mu_b^o)^{1/2} = [K_b^o / \mu_b^o(1 - \cos\theta_n^b)^2 + G_b^o / \mu_b^o \tan^2\theta_{\mathcal{N}}^b]^{1/2} \quad [hd] \quad (1063)$$

$$= [(H_{\gamma b}^T)^2 + (H_{\gamma b}^{ST})^2]^{1/2}$$

$$B_{\gamma b} = (2\mu_b^o \mathfrak{E}^b)^{1/2} = [\mu_b^o K_b^o (1-\cos\theta_n^b)^2 + \mu_b^o G_b^o \tan^2\theta_{\mathfrak{N}}^b]^{1/2} \qquad [hd] \qquad (1064)$$

$$= [(B_{\gamma b}^T)^2 + (B_{\gamma b}^{ST})^2]^{1/2}$$

where in equations (1059) through (1064) the ambient magnetic fields and magnetic inductions $H_{\gamma e}^C, B_{\gamma e}^C, H_{\gamma e}^{SC}, B_{\gamma e}^{SC}; H_{\gamma c}^C, B_{\gamma c}^C, H_{\gamma c}^{SC}, B_{\gamma c}^{SC}$ and $H_{\gamma b}^T, B_{\gamma b}^T, H_{\gamma b}^{ST}, B_{\gamma b}^{ST}$ are given by equations (1030), (1031), (1037) and (1044) through (1049). Equations (1056) through (1058) can be rewritten within the limits of the validity of equations (992) and (993) in the following manner

$$\mathfrak{E}^e = \mathfrak{E}_C^e + \mathfrak{E}_{SC}^e = 1/2(K_e^o + G_e^o)(\sec\theta_n^e - 1)^2 \qquad [ld] \qquad (1065)$$

$$\mathfrak{E}^c = \mathfrak{E}_C^c + \mathfrak{E}_{SC}^c = 1/2(K_c^o + G_c^o)(\sec\theta_n^c - 1)^2 \qquad [ld] \qquad (1066)$$

$$\mathfrak{E}^b = \mathfrak{E}_T^b + \mathfrak{E}_{ST}^b = 1/2(K_b^o + G_b^o)(1-\cos\theta_n^b)^2 \qquad [hd] \qquad (1067)$$

while under these conditions equations (1059) through (1064) can be written as

$$H_{\gamma e} = (2\mathfrak{E}^e/\mu_e^o)^{1/2} = [(K_e^o + G_e^o)/\mu_e^o]^{1/2}(\sec\theta_n^e - 1) = [(H_{\gamma e}^C)^2 + (H_{\gamma e}^{SC})^2]^{1/2} \quad [ld] \quad (1068)$$

$$B_{\gamma e} = (2\mu_e^o \mathfrak{E}^e)^{1/2} = [\mu_e^o (K_e^o + G_e^o)]^{1/2}(\sec\theta_n^e - 1) = [(B_{\gamma e}^C)^2 + (B_{\gamma e}^{SC})^2]^{1/2} \quad [ld] \quad (1069)$$

$$H_{\gamma c} = (2\mathfrak{E}^c/\mu_c^o)^{1/2} = [(K_c^o + G_c^o)/\mu_c^o]^{1/2}(\sec\theta_n^c - 1) = [(H_{\gamma c}^C)^2 + (H_{\gamma c}^{SC})^2]^{1/2} \quad [ld] \quad (1070)$$

$$B_{\gamma c} = (2\mu_c^o \mathfrak{E}^c)^{1/2} = [\mu_c^o (K_c^o + G_c^o)]^{1/2}(\sec\theta_n^c - 1) = [(B_{\gamma c}^C)^2 + (B_{\gamma c}^{SC})^2]^{1/2} \quad [ld] \quad (1071)$$

$$H_{\gamma b} = (2\mathfrak{E}^b/\mu_b^o)^{1/2} = [(K_b^o + G_b^o)/\mu_b^o]^{1/2}(1-\cos\theta_n^b) = [(H_{\gamma b}^T)^2 + (H_{\gamma b}^{ST})^2]^{1/2} \quad [hd] \quad (1072)$$

$$B_{\gamma b} = (2\mu_b^o \mathfrak{E}^b)^{1/2} = [\mu_b^o (K_b^o + G_b^o)]^{1/2}(1-\cos\theta_n^b) = [(B_{\gamma b}^T)^2 + (B_{\gamma b}^{ST})^2]^{1/2} \quad [hd] \quad (1073)$$

where in equations (1068) through (1073) the ambient magnetic fields and magnetic inductions $H_{\gamma e}^C, B_{\gamma e}^C, H_{\gamma e}^{SC}, B_{\gamma e}^{SC}; H_{\gamma c}^C, B_{\gamma c}^C, H_{\gamma c}^{SC}, B_{\gamma c}^{SC}$ and $H_{\gamma b}^T, B_{\gamma b}^T, H_{\gamma b}^{ST}, B_{\gamma b}^{ST}$ are given by equations (1030), (1031), (1037) and (1050) through (1055). In this way the ambient magnetic fields and ambient magnetic inductions that are required for liftoff enhancement are calculated for three important components of a polyphase rocket.

A static elastic and approximate viscoelastic calculation of the electromagnetic fields that are required to produce a specified enhancement of a polyphase rocket has been presented. A more realistic dynamical and viscoelastic calculation of the required fields is more complicated and has not been accomplished. A simple engineering approach to this calculation may possibly be the substitution of effective, dynamical, frequency-dependent elastic moduli for the static elastic moduli in the manner

$$K_e^o \to K_e^r \quad G_e^o \to G_e^r \qquad K_c^o \to K_c^r \quad G_c^o \to G_c^r \qquad K_b^o \to K_b^r \quad G_b^o \to G_b^r \qquad (1074)$$

where $K_e^r, G_e^r, K_c^r, G_c^r, K_b^r$ and G_b^r correspond to characteristic resonance frequencies of interest. In this way equations (589), (590), (701), (740), (795) and (857) can be altered to account for the resonances associated with wave or oscillatory motions.

11. CONCLUSION. The engineering design concepts for a polyphase rocket are presented which are based on a better, but far from complete, understanding of the nature of space and time and the active or physical coordinates that are used to describe these physical quantities. A polyphase rocket is one in which the basic components such as the exhaust aperture system, exhaust gases and the rocket body are immersed in ambient fields such as for instance - ambient electromagnetic wave fields. An ambient field is a field in bulk matter that induces internal phase angles into all physical quantities (such as space and time coordinates for example which are then called active or physical coordinates) but with the exception of the fundamental constants of nature (including h, c and G), which remain real numbers. An integral part of the design concepts for a polyphase rocket is the use of active space and time coordinates, that are complex numbers in an internal space as written in equation (1), to describe the geometry, kinematics and dynamics of rocket engine operation and rocket motion at liftoff. This introduces the space and time coordinate factors $C_{\sigma\alpha}$ and $C_{\sigma t}$, that are defined in equations (8) through (12), (243) through (282) and (144) through (148), into essentially every calculation of rocket engineering. The use of ambient fields leads to the concept of a polyphase rocket whose liftoff parameter can be increased to the point where even underpowered conventional rockets (corresponding to zero ambient fields) can attain liftoff. Many scientific concepts and engineering methods have been proposed as ways to go in space.[14-19] The material in the present paper may possibly increase the efficiency of a variety of these concepts and methods.

REFERENCES

1. Baker, D., The History of Manned Space Flight, Crown Publishers, New York, 1981.

2. Clarke, A.C., Interplanetary Flight, Harper, New York, 1960.

3. Gatland, K., Space Technology, Harmony, New York, 1981.

4. Ley, W., Rockets, Missiles and Space Travel, Viking, New York, 1957.

5. Ley, W., Rockets, Missiles and Men in Space, Viking, New York, 1968.

6. Miller, R., The Dream Machines, Krieger, Malabar, 1993.

7. Simpson, T., Pioneering the Space Frontier, Bantam, New York, 1986.

8. Weiss, R.A., Gauge Theory of Thermodynamics, K&W Publications, Vicksburg, 1989.

9. Weiss, R.A., Clean Fission, K&W Publications, Vicksburg, 1992.

10. Weiss, R.A., Internal Fission, K&W Publications, Vicksburg, 1995.

11. Weiss, R.A., Vacuum Fission, K&W Publications, Vicksburg, 1998.

12. Weiss, R.A., "Physical Nature of Space and Time Coordinates," article in Polyphase Fission and Polyphase Rockets, p. 1, K&W Publications, Tucson, 2003.

13. Weiss, R.A., "Kinematics and Dynamics of Bodies in Ambient Fields," article in Polyphase Fission and Polyphase Rockets, p. 95, K&W Publications, Tucson, 2003.

14. Beardsley, T., et al., "The Way to Go in Space", p. 81-97, Scientific American, February 1999.

15. Mark, H., Editor, Encyclopedia of Space Science and Technology, John Wiley, New York, 2003.

16. Sutton, G.P. and Biblarz, O., Rocket Propulsion Elements, John Wiley, New York, 2001.

17. Roy, G.D., Advances in Chemical Propulsion, CRC Press, Boca Raton, 2001.

18. Turner, M.J., Rocket and Spacecraft Propulsion, Springer-Verlag, New York, 2000.

19. Archer, R.D. and Saarlas, M., An Introduction to Aerospace Propulsion, Prentice-Hall, New York, 1995.

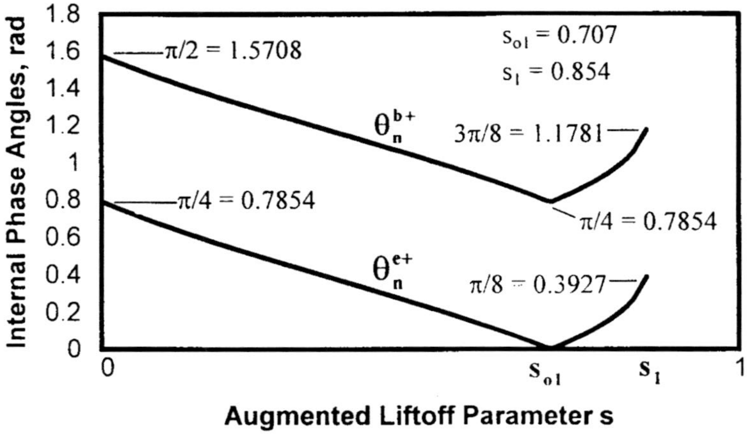

1a. $\theta_c = \pi/4$

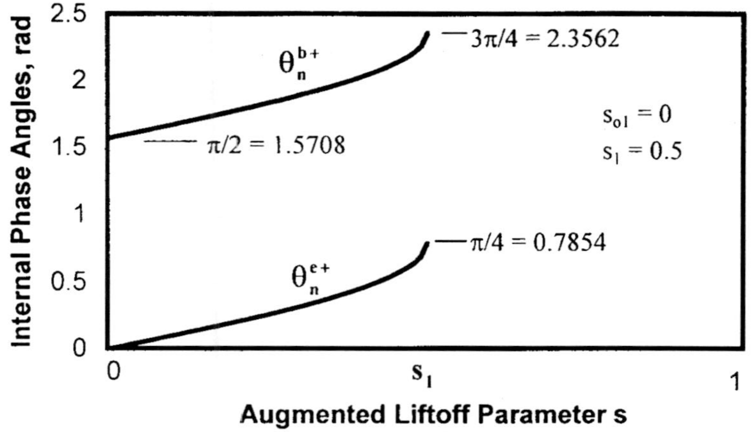

1b. $\theta_c = \pi/2$

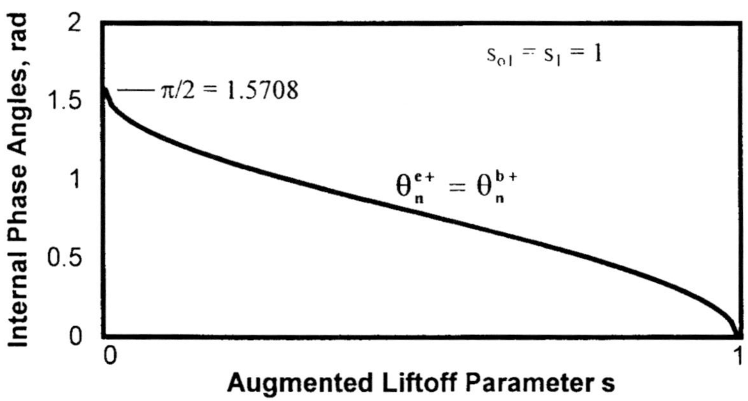

1c. $\theta_c = 0$

Figure 1. Phase angles θ_n^{e+} and θ_n^{b+} versus the augmented liftoff parameters s.

C

MAXWELL'S EQUATIONS WITH BROKEN INTERNAL SYMMETRIES

Richard A. Weiss
K&W Research Company
Tucson, Arizona

ABSTRACT. On account of the broken symmetries of the thermodynamic ground state and excited states of bulk matter and the vacuum, the electric and magnetic fields in bulk matter and the vacuum exhibit broken internal symmetries. Maxwell's equations are formulated for an electromagnetic field with broken internal symmetry. Lorentz covariance is expressed in terms of space and time coordinates that have broken symmetries represented by internal phase angles. Special relativity mechanics in bulk matter and the vacuum with broken symmetries is formulated for particles whose kinematic and dynamic variables exhibit internal phases. Electromagnetic wave equations for broken symmetry matter and vacuum are developed and the gauge conditions for the electromagnetic potential are developed. The vacuum state is shown to have properties that are essentially similar to those of a bulk matter system, and in particular both exhibit broken internal symmetry. The description of electromagnetic effects in matter and the vacuum must properly account for the broken symmetry of the fields and space and time coordinates, and the internal phase angles of the electromagnetic field vectors must be determined jointly with the internal phase angles of the space and time coordinates. A better knowledge of electromagnetic interactions in bulk matter will be useful for understanding electromagnetic wave propagation in the atmosphere and for comprehending the complex processes that occur when high energy microwave beams interact with matter.

1. INTRODUCTION. Electrodynamics is a theory that is based on the Lorentz covariant set of Maxwell's equations and on the symmetry of the gauge group U(1).[1-3] This theory has charges and currents as the sources of the electromagnetic field. Maxwell's equations are a set of partial differential equations that determine the space and time variation of the electric and magnetic fields that are associated with the distribution of charges and currents. Classically, the charges and currents are situated in a passive space and time background (the vacuum) which is assumed to be inert and plays no active part in the determination of the fields. In quantum electrodynamics, the vacuum is taken to be a polarizable medium which can affect the energy levels of charged particle configurations. The active vacuum is one of the great discoveries of twentieth century physics, and has been experimentally verified in a number of ways including a measurement of the Lamb shift of energy levels.[4-6]

In this paper an additional vacuum effect on the electromagnetic field is suggested to manifest itself through the fact that space and time coordinates within asymmetric bulk matter or vacuum acquire internal phase angles (broken symmetries). The electric and magnetic field vectors also acquire broken symmetries. The internal phase angles of the space and time coordinates and of the electromagnetic field vectors are due to the interaction of Minkowski spacetime with bulk matter, the electromagnetic field, and the vacuum. The internal phase angles of the electromagnetic field vectors must be determined jointly with the internal phase angles of the space and time coordinates, and it is the joint solution of Maxwell's equations and the equations of motion of charged particles for a system with broken internal symmetries that accomplishes this task.

The relativistic values of the electromagnetic field vectors in asymmetric bulk matter or vacuum must satisfy the relativistic trace equation for radiation.[7-8] This radiation trace equation relates the renormalized radiation pressure to the corresponding nonrelativistic radiation pressure. The trace equation for radiation is derived from the relativistic trace equation for the ground state of bulk matter which is written as[7,8]

$$\bar{U} + T\left(\frac{d\bar{U}}{dT}\right)_{\bar{P}V} - 3V\frac{d}{dV}(\bar{P}V)_{\bar{U}} = U^a + T\left(\frac{dU^a}{dT}\right)_{p^aV} \quad (1)$$

or equivalently as

$$(1 - \bar{b} + T\frac{\partial}{\partial T} - \bar{b}V\frac{\partial}{\partial V})\bar{E} - 3(1 + \bar{\gamma} + V\frac{\partial}{\partial V} - \bar{\gamma}T\frac{\partial}{\partial T})\bar{P} = \psi^a \quad (2)$$

where

$$\psi^a = (T\frac{\partial}{\partial T} - b^aV\frac{\partial}{\partial V} + 1 - b^a)E^a \quad (3)$$

and where $\bar{U}$, $\bar{E}$, $\bar{P}$, $\bar{\gamma}$, and $\bar{b}$ are complex number representations of the internal energy, energy density, pressure, and the gauge parameters, T = absolute temperature, and V = volume of specified number of particles. The complex number Grüneisen parameter is defined as

$$\bar{\gamma} = \frac{V}{\bar{C}_V}\frac{\partial\bar{P}}{\partial T} = \frac{\partial\bar{P}/\partial T}{\partial\bar{E}/\partial T} = \gamma e^{\theta_\gamma} \quad (4)$$

where γ and θ_γ = magnitude and phase of the Grüneisen parameter respectively. The corresponding equation for radiation in matter with internal phases is derived from equation (2) to be[8]

$$(1 - \bar{b} + T\frac{\partial}{\partial T} - \bar{b}V\frac{\partial}{\partial V})\bar{E}_r - \bar{\beta}_r(T\frac{\partial\bar{P}}{\partial T} - \bar{P}) \quad (5)$$

$$- 3[(1 + \bar{\gamma} + V\frac{\partial}{\partial V} - \bar{\gamma}T\frac{\partial}{\partial T})\bar{P}_r - \bar{\delta}_r(T\frac{\partial\bar{P}}{\partial T} - \bar{P})] = \psi_r^a$$

where

$$\psi_r^a = (T \frac{\partial}{\partial T} - b^a V \frac{\partial}{\partial V} + 1 - b^a)E_r^a - \beta_r^a(T \frac{\partial P^a}{\partial T} - P^a) \quad (6)$$

and where $\bar{E}_r$, $\bar{P}_r$, $\bar{\beta}_r$, and $\bar{\delta}_r$ are the complex number radiation energy density, radiation pressure, and two radiation gauge functions respectively.[8] The radiation Grüneisen parameter is defined by

$$\gamma_r = \frac{\partial P_r}{\partial T} / \frac{\partial E_r}{\partial T} \quad (6A)$$

Throughout this paper the index "a" will refer to nonrelativistic (unrenormalized) calculations. Equation (1) with its right hand side set equal to zero represents the asymmetric ground state of the vacuum, while equation (5) with its right hand side equal to zero represents the excited (radiation) states of the asymmetric vacuum.

The relativistic trace equations for the ground and excited states of bulk matter and the vacuum imply that the ground state and excited state pressure fields have broken symmetries.[8] In turn, this implies that all of the descriptive variables of particles and fields located in asymmetric bulk matter or vacuum also exhibit broken symmetries. Therefore the space and time coordinates as well as the electric and magnetic field vectors will exhibit broken symmetries as manifested by internal phase angles. The space and time coordinates are written as

$$\bar{x} = xe^{j\theta_x} \quad (7)$$

$$\bar{y} = ye^{j\theta_y} \quad (8)$$

$$\bar{z} = ze^{j\theta_z} \quad (9)$$

$$\bar{t} = te^{j\theta_t} \quad (10)$$

and the derivatives with respect to the space and time coordinates are written as

$$\partial/\partial\bar{x} = e^{-j\theta dx} \cos \beta_{x,x} \, \partial/\partial x \quad (11)$$

$$\partial/\partial\bar{t} = e^{-j\theta dt} \cos \beta_{t,t} \, \partial/\partial t \quad (12)$$

where

$$\tan \beta_{x,x} = x \, \partial\theta_x/\partial x \quad (13)$$

$$\tan \beta_{t,t} = t \, \partial\theta_t/\partial t \quad (14)$$

$$\cos \beta_{x,x} = [1 + (x\, \partial\theta_x/\partial x)^2]^{-1/2} \tag{15}$$

$$\cos \beta_{t,t} = [1 + (t\, \partial\theta_t/\partial t)^2]^{-1/2} \tag{16}$$

$$\theta_{dx} = \theta_x + \beta_{x,x} \tag{17}$$

$$\theta_{dt} = \theta_t + \beta_{t,t} \tag{17A}$$

In a condensed notation the coordinates and derivatives are written with $\eta = t, x, y, z$ as follows

$$\bar{\eta} = \eta e^{j\theta_\eta} \tag{18}$$

$$\partial/\partial\bar{\eta} = e^{-j\theta_{d\eta}} \cos \beta_{\eta,\eta}\, \partial/\partial\eta \tag{19}$$

$$\theta_{d\eta} = \theta_\eta + \beta_{\eta,\eta} \tag{20}$$

$$\tan \beta_{\eta,\eta} = \eta\, \partial\theta_\eta/\partial\eta \tag{21}$$

$$\cos \beta_{\eta,\eta} = [1 + (\eta\, \partial\theta_\eta/\partial\eta)^2]^{-1/2} \tag{22}$$

Note that it is the real parts of the complex number quantities such as space and time coordinates, electric and magnetic field vectors, pressure and energy that are the measured quantities.

This paper develops Maxwell's equations for electromagnetic fields that have broken internal symmetries and for space and time coordinates that also have broken internal symmetries. Section 2 considers the fields and coordinates with broken symmetry, while Section 3 develops Maxwell's equations and the equations of motion of charged particles in an electromagnetic field in a broken symmetry vacuum or bulk matter system. Section 4 develops the consequences of assuming the validity of Lorentz covariance for coordinate systems with broken internal symmetry. In Section 5 the electromagnetic wave equations and their gauge conditions are written for systems with internal phase angles. Finally, Section 6 develops the equations of the relativistic vacuum from the corresponding bulk matter equations, and a broken symmetry condition for the vacuum state is suggested. Therefore all of the conclusions for the bulk matter state with broken internal symmetries are also valid for the broken symmetry vacuum state. The conventional coordinates t_a , x_a , y_a , z_a are related to the gauge rotated coordinates t , x , y and z by $t = t_a$, $x = x_a$, $y = y_a$ and $z = z_a$.

2. THE BROKEN SYMMETRY OF ELECTROMAGNETIC FIELDS. For electromagnetic waves within asymmetric bulk matter or vacuum, the electric and magnetic fields

are expected to acquire internal phase angles. This is due to the fact that the spacetime coordinates and the kinematical and dynamical variables of particles in asymmetric bulk matter or vacuum exhibit broken internal symmetries. In particular the particle velocity, and therefore the electric current for charged particles, has an internal phase angle. Therefore the cartesian components of the electric and magnetic field vectors in asymmetric bulk matter or vacuum can be written as

$$\bar{E}_\alpha = E_\alpha e^{j\theta_{E\alpha}} = E_{\alpha R} + jE_{\alpha I} \tag{23}$$

$$\bar{D}_\alpha = D_\alpha e^{j\theta_{D\alpha}} = D_{\alpha R} + jD_{\alpha I} \tag{24}$$

$$\bar{H}_\alpha = H_\alpha e^{j\theta_{H\alpha}} = H_{\alpha R} + jH_{\alpha I} \tag{25}$$

$$\bar{B}_\alpha = B_\alpha e^{j\theta_{B\alpha}} = B_{\alpha R} + jB_{\alpha I} \tag{26}$$

where α = x, y, and z. The phase angles $\theta_{E\alpha}$, $\theta_{D\alpha}$, $\theta_{H\alpha}$, and $\theta_{B\alpha}$ are in general functions of space and time of the general form

$$\theta_{E\alpha} = \theta_{E\alpha}(x,y,z,t,\theta_x,\theta_y,\theta_z,\theta_t) \tag{27}$$

The field vector amplitudes are also functions of space and time, as for example

$$E_\alpha = E_\alpha(x,y,z,t,\theta_x,\theta_y,\theta_z,\theta_t) \tag{28}$$

The imaginary number j will be used to refer to internal phase angles that are associated with broken symmetries, while the imaginary number i will refer to external phase angles. For plane waves the magnitudes of the field vectors in equations (23) through (26) may be written as

$$E_\alpha = A_{E\alpha} e^{i\xi} \tag{29}$$

$$D_\alpha = A_{D\alpha} e^{i\xi} \tag{30}$$

$$H_\alpha = A_{H\alpha} e^{i\xi} \tag{31}$$

$$B_\alpha = A_{B\alpha} e^{i\xi} \tag{32}$$

where

$$\xi = k_x x + k_y y + k_z z - \omega t \tag{32A}$$

where $A_{E\alpha}$, $A_{D\alpha}$, $A_{H\alpha}$, and $A_{B\alpha}$ are constants; k_x, k_y, k_z = wavenumber components, and ω = magnitude of the frequency.

In general the internal phase angles $\theta_{E\alpha}$, $\theta_{D\alpha}$, $\theta_{H\alpha}$, and $\theta_{B\alpha}$ are functions of space and time, and it follows from equations (18) through (22) and (23) through (26) that

$$\partial \bar{E}_\alpha / \partial \bar{\eta} = e^{j\Phi_{E\alpha,\eta}} \cos \beta_{\eta,\eta} \, W_{E\alpha,\eta} \tag{33}$$

$$\partial \bar{D}_\alpha / \partial \bar{\eta} = e^{j\Phi_{D\alpha,\eta}} \cos \beta_{\eta,\eta} \, W_{D\alpha,\eta} \tag{34}$$

$$\partial \bar{H}_\alpha / \partial \bar{\eta} = e^{j\Phi_{H\alpha,\eta}} \cos \beta_{\eta,\eta} \, W_{H\alpha,\eta} \tag{35}$$

$$\partial \bar{B}_\alpha / \partial \bar{\eta} = e^{j\Phi_{B\alpha,\eta}} \cos \beta_{\eta,\eta} \, W_{B\alpha,\eta} \tag{36}$$

where η = t, x, y, z and where

$$W_{E\alpha,\eta} = \sqrt{(\partial E_\alpha/\partial \eta)^2 + (E_\alpha \, \partial \theta_{E\alpha}/\partial \eta)^2} \tag{37}$$

$$W_{D\alpha,\eta} = \sqrt{(\partial D_\alpha/\partial \eta)^2 + (D_\alpha \, \partial \theta_{D\alpha}/\partial \eta)^2} \tag{38}$$

$$W_{H\alpha,\eta} = \sqrt{(\partial H_\alpha/\partial \eta)^2 + (H_\alpha \, \partial \theta_{H\alpha}/\partial \eta)^2} \tag{39}$$

$$W_{B\alpha,\eta} = \sqrt{(\partial B_\alpha/\partial \eta)^2 + (B_\alpha \, \partial \theta_{B\alpha}/\partial \eta)^2} \tag{40}$$

$$\Phi_{E\alpha,\eta} = \theta_{E\alpha} + \beta_{E\alpha,\eta} - \theta_{d\eta} \tag{41}$$

$$\Phi_{D\alpha,\eta} = \theta_{D\alpha} + \beta_{D\alpha,\eta} - \theta_{d\eta} \tag{42}$$

$$\Phi_{H\alpha,\eta} = \theta_{H\alpha} + \beta_{H\alpha,\eta} - \theta_{d\eta} \tag{43}$$

$$\Phi_{B\alpha,\eta} = \theta_{B\alpha} + \beta_{B\alpha,\eta} - \theta_{d\eta} \tag{44}$$

$$\tan \beta_{E\alpha,\eta} = E_\alpha \frac{\partial\theta_{E\alpha}/\partial\eta}{\partial E_\alpha/\partial\eta} \tag{45}$$

$$\tan \beta_{D\alpha,\eta} = D_\alpha \frac{\partial\theta_{D\alpha}/\partial\eta}{\partial D_\alpha/\partial\eta} \tag{46}$$

$$\tan \beta_{H\alpha,\eta} = H_\alpha \frac{\partial\theta_{H\alpha}/\partial\eta}{\partial H_\alpha/\partial\eta} \tag{47}$$

$$\tan \beta_{B\alpha,\eta} = B_\alpha \frac{\partial\theta_{B\alpha}/\partial\eta}{\partial B_\alpha/\partial\eta} \tag{48}$$

and where

$$\theta_{d\eta} = \theta_\eta + \beta_{\eta,\eta} \tag{49}$$

If the electric and magnetic fields have an external time dependence given by equations (29) through (32) it follows that

$$\frac{\partial \bar{E}_\alpha}{\partial \bar{t}} = e^{j\phi_{E\alpha,t}} \cos \beta_{t,t} \, W_{E\alpha,t} \tag{50}$$

and

$$W_{E\alpha,t} = A_{E\alpha} e^{i\xi} \sqrt{-\omega^2 + \left(\frac{\partial\theta_{E\alpha}}{\partial t}\right)^2} = -iA_{E\alpha} e^{i\xi} \sqrt{\omega^2 - \left(\frac{\partial\theta_{E\alpha}}{\partial t}\right)^2} \tag{51}$$

where ξ is given by equation (32A), with similar expressions for the magnetic vector components. Similarly, the derivative with respect to the coordinate $\bar{x}$ is given by

$$\frac{\partial \bar{E}_\alpha}{\partial \bar{x}} = e^{j\phi_{E\alpha,x}} \cos \beta_{x,x} \, W_{E\alpha,x} \tag{52}$$

where

$$W_{E\alpha,x} = A_{E\alpha} e^{i\xi} \sqrt{-k_x^2 + \left(\frac{\partial\theta_{E\alpha}}{\partial x}\right)^2} = iA_{E\alpha} e^{i\xi} \sqrt{k_x^2 - \left(\frac{\partial\theta_{E\alpha}}{\partial x}\right)^2} \tag{53}$$

with similar expressions for the magnetic field vector components and for the derivatives with respect to $\bar{y}$ and $\bar{z}$.

It is sometimes conventient to write the derivatives in equations (33) and (36) in the following alternative forms

$$\frac{\partial \bar{E}_\alpha}{\partial \bar{\eta}} = \cos \beta_{\eta,\eta} \; e^{-j\theta_{d\eta}} \frac{\partial \bar{E}_\alpha}{\partial \eta} = R_{1\eta}(E_{\alpha R}, E_{\alpha I}) + jI_{1\eta}(E_{\alpha R}, E_{\alpha I}) \tag{54}$$

$$\frac{\partial \bar{B}_\alpha}{\partial \bar{\eta}} = \cos \beta_{\eta,\eta} \; e^{-j\theta_{d\eta}} \frac{\partial \bar{B}_\alpha}{\partial \eta} = R_{1\eta}(B_{\alpha R}, B_{\alpha I}) + jI_{1\eta}(B_{\alpha R}, B_{\alpha I}) \tag{55}$$

where

$$R_{1\eta}(E_{\alpha R}, E_{\alpha I}) = \cos \beta_{\eta,\eta} \; (\cos \theta_{d\eta} \; \partial E_{\alpha R}/\partial \eta + \sin \theta_{d\eta} \; \partial E_{\alpha I}/\partial \eta) \tag{56}$$

$$= \cos \beta_{\eta,\eta} \; \cos \Phi_{E\alpha,\eta} \; W_{E\alpha,\eta}$$

$$\sim \cos \beta_{\eta,\eta} \; \cos (\theta_{E\alpha} - \theta_{d\eta}) \; \partial E_\alpha/\partial \eta$$

$$I_{1\eta}(E_{\alpha R}, E_{\alpha I}) = \cos \beta_{\eta,\eta} \; (- \sin \theta_{d\eta} \; \partial E_{\alpha R}/\partial \eta + \cos \theta_{d\eta} \; \partial E_{\alpha I}/\partial \eta) \tag{57}$$

$$= \cos \beta_{\eta,\eta} \; \sin \Phi_{E\alpha,\eta} \; W_{E\alpha,\eta}$$

$$\sim \cos \beta_{\eta,\eta} \; \sin (\theta_{E\alpha} - \theta_{d\eta}) \; \partial E_\alpha/\partial \eta$$

$$R_{1\eta}(B_{\alpha R}, B_{\alpha I}) = \cos \beta_{\eta,\eta} \; (\cos \theta_{d\eta} \; \partial B_{\alpha R}/\partial \eta + \sin \theta_{d\eta} \; \partial B_{\alpha I}/\partial \eta) \tag{58}$$

$$= \cos \beta_{\eta,\eta} \; \cos \Phi_{B\alpha,\eta} \; W_{B\alpha,\eta}$$

$$\sim \cos \beta_{\eta,\eta} \; \cos (\theta_{B\alpha} - \theta_{d\eta}) \; \partial B_\alpha/\partial \eta$$

$$I_{1\eta}(B_{\alpha R},B_{\alpha I}) = \cos \beta_{\eta,\eta}\ (-\sin \theta_{d\eta}\ \partial B_{\alpha R}/\partial\eta + \cos \theta_{d\eta}\ \partial B_{\alpha I}/\partial\eta) \tag{59}$$

$$= \cos \beta_{\eta,\eta}\ \sin \Phi_{B\alpha,\eta}\ W_{B\alpha,\eta}$$

$$\sim \cos \beta_{\eta,\eta}\ \sin (\theta_{B\alpha} - \theta_{d\eta})\ \partial B_{\alpha}/\partial\eta$$

and where

$$E_{\alpha R} = E_{\alpha} \cos \theta_{E\alpha} \tag{60}$$

$$E_{\alpha I} = E_{\alpha} \sin \theta_{E\alpha} \tag{61}$$

$$B_{\alpha R} = B_{\alpha} \cos \theta_{B\alpha} \tag{62}$$

$$B_{\alpha I} = B_{\alpha} \sin \theta_{B\alpha} \tag{63}$$

The second derivatives of the field vectors are obtained by consecutively applying the first derivative operators that appear in equation (54) as follows

$$\frac{\partial^2 \bar{E}_{\alpha}}{\partial \bar{\eta}^2} = \cos \beta_{\eta,\eta}\ e^{-j\theta_{d\eta}} \frac{\partial}{\partial\eta} \left(\cos \beta_{\eta,\eta}\ e^{-j\theta_{d\eta}} \frac{\partial \bar{E}_{\alpha}}{\partial\eta}\right) \tag{64}$$

where $\alpha = x, y, z$ and $\eta = x, y, z, t$. For simplicity it will be assumed that $\beta_{\eta,\eta}$ and $\theta_{d\eta}$ are slowly varying functions of η . Within the limits of this approximation the second derivatives of the field vectors are written as

$$\frac{\partial^2 \bar{E}_{\alpha}}{\partial \bar{\eta}^2} \sim \cos^2 \beta_{\eta,\eta}\ e^{-2j\theta_{d\eta}} \frac{\partial^2 \bar{E}_{\alpha}}{\partial\eta^2} = R_{2\eta}(E_{\alpha R},E_{\alpha I}) + jI_{2\eta}(E_{\alpha R},E_{\alpha I}) \tag{65}$$

$$\sim \cos^2 \beta_{\eta,\eta}\ e^{j\Psi_{E\alpha,\eta}}\ T_{E\alpha,\eta}$$

$$\sim \cos^2 \beta_{\eta,\eta}\ e^{j(\theta_{E\alpha} - 2\theta_{d\eta})} \frac{\partial^2 E_{\alpha}}{\partial\eta^2}$$

$$\frac{\partial^2 \bar{H}_\alpha}{\partial \bar{\eta}^2} \sim \cos^2 \beta_{\eta,\eta}\ e^{-2j\theta_{d\eta}} \frac{\partial^2 \bar{H}_\alpha}{\partial \eta^2} = R_{2\eta}(H_{\alpha R}, H_{\alpha I}) + jI_{2\eta}(H_{\alpha R}, H_{\alpha I}) \tag{66}$$

$$\sim \cos^2 \beta_{\eta,\eta}\ e^{j\Psi_{H\alpha,\eta}}\ T_{H\alpha,\eta}$$

$$\sim \cos^2 \beta_{\eta,\eta}\ e^{j(\theta_{H\alpha} - 2\theta_{d\eta})} \frac{\partial^2 H_\alpha}{\partial \eta^2}$$

where

$$T_{E\alpha,\eta} = \sqrt{\left(\frac{\partial^2 E_\alpha}{\partial \eta^2}\right)^2 + \left(E_\alpha \frac{\partial^2 \theta_{E\alpha}}{\partial \eta^2}\right)^2} \tag{66A}$$

$$T_{H\alpha,\eta} = \sqrt{\left(\frac{\partial^2 H_\alpha}{\partial \eta^2}\right)^2 + \left(H_\alpha \frac{\partial^2 \theta_{H\alpha}}{\partial \eta^2}\right)^2} \tag{66B}$$

$$\Psi_{E\alpha,\eta} = \theta_{E\alpha} + \delta_{E\alpha,\eta} - 2\theta_{d\eta} \tag{66C}$$

$$\Psi_{H\alpha,\eta} = \theta_{H\alpha} + \delta_{H\alpha,\eta} - 2\theta_{d\eta} \tag{66D}$$

and where

$$R_{2\eta}(E_{\alpha R}, E_{\alpha I}) = \cos^2 \beta_{\eta,\eta} \left[\cos(2\theta_{d\eta}) \frac{\partial^2 E_{\alpha R}}{\partial \eta^2} + \sin(2\theta_{d\eta}) \frac{\partial^2 E_{\alpha I}}{\partial \eta^2}\right] \tag{67}$$

$$\sim \cos^2 \beta_{\eta,\eta} \cos \Psi_{E\alpha,\eta}\ T_{E\alpha,\eta}$$

$$\sim \cos^2 \beta_{\eta,\eta} \cos(\theta_{E\alpha} - 2\theta_{d\eta}) \frac{\partial^2 E_\alpha}{\partial \eta^2}$$

$$I_{2\eta}(E_{\alpha R},E_{\alpha I}) = \cos^2 \beta_{\eta,\eta} \left[- \sin(2\theta_{d\eta}) \frac{\partial^2 E_{\alpha R}}{\partial \eta^2} + \cos(2\theta_{d\eta}) \frac{\partial^2 E_{\alpha I}}{\partial \eta^2}\right] \tag{68}$$

$$\sim \cos^2 \beta_{\eta,\eta} \sin \Psi_{E\alpha,\eta} T_{E\alpha,\eta}$$

$$\sim \cos^2 \beta_{\eta,\eta} \sin(\theta_{E\alpha} - 2\theta_{d\eta}) \frac{\partial^2 E_\alpha}{\partial \eta^2}$$

$$R_{2\eta}(H_{\alpha R},H_{\alpha I}) = \cos^2 \beta_{\eta,\eta} \left[\cos(2\theta_{d\eta}) \frac{\partial^2 H_{\alpha R}}{\partial \eta^2} + \sin(2\theta_{d\eta}) \frac{\partial^2 H_{\alpha I}}{\partial \eta^2}\right] \tag{69}$$

$$\sim \cos^2 \beta_{\eta,\eta} \cos \Psi_{H\alpha,\eta} T_{H\alpha,\eta}$$

$$\sim \cos^2 \beta_{\eta,\eta} \cos(\theta_{H\alpha} - 2\theta_{d\eta}) \frac{\partial^2 H_\alpha}{\partial \eta^2}$$

$$I_{2\eta}(H_{\alpha R},H_{\alpha I}) = \cos^2 \beta_{\eta,\eta} \left[- \sin(2\theta_{d\eta}) \frac{\partial^2 H_{\alpha R}}{\partial \eta^2} + \cos(2\theta_{d\eta}) \frac{\partial^2 H_{\alpha I}}{\partial \eta^2}\right] \tag{70}$$

$$\sim \cos^2 \beta_{\eta,\eta} \sin \Psi_{H\alpha,\eta} T_{H\alpha,\eta}$$

$$\sim \cos^2 \beta_{\eta,\eta} \sin(\theta_{H\alpha} - 2\theta_{d\eta}) \frac{\partial^2 H_\alpha}{\partial \eta^2}$$

and

$$\tan \delta_{E\alpha,\eta} = E_\alpha \frac{\partial^2 \theta_{E\alpha}}{\partial \eta^2} \Big/ \frac{\partial^2 E_\alpha}{\partial \eta^2} \tag{70A}$$

$$\tan \delta_{H\alpha,\eta} = H_\alpha \frac{\partial^2 \theta_{H\alpha}}{\partial \eta^2} \Big/ \frac{\partial^2 H_\alpha}{\partial \eta^2} \tag{70B}$$

Similar equations can be written for the D_α and B_α components. In this way the derivatives necessary for evaluating Maxwell's equations in asymmetric bulk matter or vacuum can be evaluated.

3. MAXWELL'S EQUATIONS WITH BROKEN INTERNAL SYMMETRIES. This section develops Maxwell's equations for electromagnetic fields whose field vectors have internal phase angles. The asymmetric bulk matter or vacuum in which the electromagnetic fields exist also have broken internal symmetries in the static pressure field, Grüneisen parameter, and in the space and time coordinates of each point in the system. The internal phase angles of the ambient medium such as θ_P, θ_γ, θ_x, θ_y, θ_z and θ_t must be calculated in conjunction with the internal phase angles of the electromagnetic field vectors. The quantities of θ_P and θ_γ are obtained from the ground state relativistic trace equation (1) and equation (4) that defines the relativistic Grüneisen function.

The unrenormalized Maxwell equations for charges and currents are written as[9-17]

$$\vec{\nabla}_a \cdot \vec{B}^a = 0 \tag{71}$$

$$\vec{\nabla}_a \cdot \vec{D}^a = \rho_q^a \tag{72}$$

$$\vec{\nabla}_a \times \vec{H}^a = \partial\vec{D}^a/\partial t_a + \vec{j}^a \tag{73}$$

$$\vec{\nabla}_a \times \vec{E}^a = -\partial\vec{B}^a/\partial t_a \tag{74}$$

where $\vec{B}^a$ = unrenormalized magnetic induction vector, $\vec{D}^a$ = unrenormalized electric displacement vector, ρ_q^a = unrenormalized charge density, $\vec{H}^a$ = unrenormalized magnetic field vector, $\vec{j}^a$ = unrenormalized current density vector, and $\vec{E}^a$ = unrenormalized electric field vector. Equations (71) through (74) represent eight equations, six of which are independent. The simplest constitutive equations are the following

$$\vec{B}^a = \mu^a\vec{H}^a \tag{75}$$

$$\vec{D}^a = \varepsilon^a\vec{E}^a \tag{76}$$

where μ^a = unrenormalized magnetic permeability, and ε^a = unrenormalized dielectric constant (permittivity). More complicated constitutive equations are often used.[18] In general

$$\mu^a = \mu^a(P^a,\gamma^a) \qquad \varepsilon^a = \varepsilon^a(P^a,\gamma^a) \tag{77}$$

where P^a and γ^a are functions of density and temperature.

Within asymmetric bulk matter or vacuum a similar set of Maxwell's equations must be valid except now the renormalized electric and magnetic field vectors must have internal phases, and the space and time coordinates must also have internal phase angles. Therefore equations (71) through (74) can be written for the electromagnetic field in bulk matter or vacuum with broken internal symmetries as follows

$$\bar{\vec{\nabla}} \cdot \bar{\vec{B}} = 0 \tag{78}$$

$$\bar{\vec{\nabla}} \cdot \bar{\vec{D}} = \bar{\rho}_q \tag{79}$$

$$\bar{\vec{\nabla}} \times \bar{\vec{H}} = \partial\bar{\vec{D}}/\partial\bar{t} + \bar{\vec{j}} \tag{80}$$

$$\bar{\vec{\nabla}} \times \bar{\vec{E}} = -\partial\bar{\vec{B}}/\partial\bar{t} \tag{81}$$

where $\bar{\vec{B}}$ = renormalized complex number magnetic induction vector, $\bar{\vec{D}}$ = renormalized complex number electric displacement vector, $\bar{\rho}_q$ = renormalized charge density, $\bar{\vec{H}}$ = renormalized complex number magnetic field vector, $\bar{\vec{j}}$ = renormalized complex number current density vector, and $\bar{\vec{E}}$ = renormalized complex number electric field vector. Equations (78) through (81) are complex number vector equations and represent a total of sixteen equations, twelve of which are indepenent. Note that $\bar{\rho}_q \neq \rho^a_q$ as can be seen from equations (72) and (79) on account of $\bar{\vec{D}} \neq \vec{D}^a$. Since $\bar{\rho}_q = \bar{n}q$ and $\rho^a_q = n^a q^a$ it follows that $\bar{n} = ne^{j\theta_n}$ and $n = n^a \cos\beta_{x,x} \cos\beta_{y,y} \cos\beta_{z,z}$ and $\theta_n = -\theta_x - \theta_y - \theta_z - \beta_{x,x} - \beta_{y,y} - \beta_{z,z}$ where it was assumed that $q = q^a$. The renormalized current density is given by $\bar{\vec{j}} = \bar{n}q\bar{\vec{v}}$, where $\bar{\vec{v}}$ = vector particle velocity with internal phase. The internal phase angle of the particle velocity is a function of θ_x, θ_y, θ_z, θ_t, and clearly $\bar{\vec{j}} \neq \vec{j}^a$.

The simplest renormalized constitutive equations are written as

$$\bar{\vec{B}} = \mu\bar{\vec{H}} \tag{82}$$

$$\bar{\vec{D}} = \varepsilon\bar{\vec{E}} \tag{83}$$

where μ = renormalized magnetic permeability, and ε = renormalized permittivity. Taking account of the fact that equations (82) and (83) are vector equations with real and imaginary parts, it is clear that they represent twelve equations.

The state equations for μ and ε will be assumed to be given by

$$\mu = \mu^a(P,\gamma) \tag{84}$$

$$\varepsilon = \varepsilon^a(P,\gamma) \tag{85}$$

where the functions are evaluated at the values of the renormalized pressure and Grüneisen function as determined from a solution of equation (1).

The component form of the renormalized Maxwell equations (78) through (81) are written as

$$\partial\bar{B}_x/\partial\bar{x} + \partial\bar{B}_y/\partial\bar{y} + \partial\bar{B}_z/\partial\bar{z} = 0 \tag{86}$$

$$\partial\bar{D}_x/\partial\bar{x} + \partial\bar{D}_y/\partial\bar{y} + \partial\bar{D}_z/\partial\bar{z} = \bar{\rho}_q \tag{87}$$

$$\partial\bar{H}_y/\partial\bar{x} - \partial\bar{H}_x/\partial\bar{y} = \partial\bar{D}_z/\partial\bar{t} + \bar{j}_z \tag{88}$$

$$\partial\bar{H}_z/\partial\bar{y} - \partial\bar{H}_y/\partial\bar{z} = \partial\bar{D}_x/\partial\bar{t} + \bar{j}_x \tag{89}$$

$$\partial\bar{H}_x/\partial\bar{z} - \partial\bar{H}_z/\partial\bar{x} = \partial\bar{D}_y/\partial\bar{t} + \bar{j}_y \tag{90}$$

$$\partial\bar{E}_y/\partial\bar{x} - \partial\bar{E}_x/\partial\bar{y} = -\,\partial\bar{B}_z/\partial\bar{t} \tag{91}$$

$$\partial\bar{E}_z/\partial\bar{y} - \partial\bar{E}_y/\partial\bar{z} = -\,\partial\bar{B}_x/\partial\bar{t} \tag{92}$$

$$\partial\bar{E}_x/\partial\bar{z} - \partial\bar{E}_z/\partial\bar{x} = -\,\partial\bar{B}_y/\partial\bar{t} \tag{93}$$

Using equations (33) through (49) to evaluate derivatives allows Maxwell's equations to be rewritten as

$$W_{Bx,x} \cos \Phi_{Bx,x} \cos \beta_{x,x} + W_{By,y} \cos \Phi_{By,y} \cos \beta_{y,y} + W_{Bz,z} \cos \Phi_{Bz,z} \cos \beta_{z,z} = 0 \tag{94}$$

$$W_{Bx,x} \sin \Phi_{Bx,x} \cos \beta_{x,x} + W_{By,y} \sin \Phi_{By,y} \cos \beta_{y,y} \tag{95}$$
$$+ W_{Bz,z} \sin \Phi_{Bz,z} \cos \beta_{z,z} = 0$$

$$W_{Dx,x} \cos \Phi_{Dx,x} \cos \beta_{x,x} + W_{Dy,y} \cos \Phi_{Dy,y} \cos \beta_{y,y} \tag{96}$$
$$+ W_{Dz,z} \cos \Phi_{Dz,z} \cos \beta_{z,z} = \rho_q \cos \theta_{\rho q}$$

$$W_{Dx,x} \sin \Phi_{Dx,x} \cos \beta_{x,x} + W_{Dy,y} \sin \Phi_{Dy,y} \cos \beta_{y,y} \tag{97}$$
$$+ W_{Dz,z} \sin \Phi_{Dz,z} \cos \beta_{z,z} = \rho_q \sin \theta_{\rho q}$$

$$W_{Hy,x} \cos \Phi_{Hy,x} \cos \beta_{x,x} - W_{Hx,y} \cos \Phi_{Hx,y} \cos \beta_{y,y} \tag{98}$$
$$= W_{Dz,t} \cos \Phi_{Dz,t} \cos \beta_{t,t} + j_z \cos \theta_{jz}$$

$$W_{Hy,x} \sin \Phi_{Hy,x} \cos \beta_{x,x} - W_{Hx,y} \sin \Phi_{Hx,y} \cos \beta_{y,y} \tag{99}$$
$$= W_{Dz,t} \sin \Phi_{Dz,t} \cos \beta_{t,t} + j_z \sin \theta_{jz}$$

$$W_{Hz,y} \cos \Phi_{Hz,y} \cos \beta_{y,y} - W_{Hy,z} \cos \Phi_{Hy,z} \cos \beta_{z,z} \tag{100}$$
$$= W_{Dx,t} \cos \Phi_{Dx,t} \cos \beta_{t,t} + j_x \cos \theta_{jx}$$

$$W_{Hz,y} \sin \Phi_{Hz,y} \cos \beta_{y,y} - W_{Hy,z} \sin \Phi_{Hy,z} \cos \beta_{z,z} \tag{101}$$
$$= W_{Dx,t} \sin \Phi_{Dx,t} \cos \beta_{t,t} + j_x \sin \theta_{jx}$$

$$W_{Hx,z} \cos \Phi_{Hx,z} \cos \beta_{z,z} - W_{Hz,x} \cos \Phi_{Hz,x} \cos \beta_{x,x} \quad (102)$$

$$= W_{Dy,t} \cos \Phi_{Dy,t} \cos \beta_{t,t} + j_y \cos \theta_{jy}$$

$$W_{Hx,z} \sin \Phi_{Hx,z} \cos \beta_{z,z} - W_{Hz,x} \sin \Phi_{Hz,x} \cos \beta_{x,x} \quad (103)$$

$$= W_{Dy,t} \sin \Phi_{Dy,t} \cos \beta_{t,t} + j_y \sin \theta_{jy}$$

$$W_{Ey,x} \cos \Phi_{Ey,x} \cos \beta_{x,x} - W_{Ex,y} \cos \Phi_{Ex,y} \cos \beta_{y,y} \quad (104)$$

$$= - W_{Bz,t} \cos \Phi_{Bz,t} \cos \beta_{t,t}$$

$$W_{Ey,x} \sin \Phi_{Ey,x} \cos \beta_{x,x} - W_{Ex,y} \sin \Phi_{Ex,y} \cos \beta_{y,y} \quad (105)$$

$$= - W_{Bz,t} \sin \Phi_{Bz,t} \cos \beta_{t,t}$$

$$W_{Ez,y} \cos \Phi_{Ez,y} \cos \beta_{y,y} - W_{Ey,z} \cos \Phi_{Ey,z} \cos \beta_{z,z} \quad (106)$$

$$= - W_{Bx,t} \cos \Phi_{Bx,t} \cos \beta_{t,t}$$

$$W_{Ez,y} \sin \Phi_{Ez,y} \cos \beta_{y,y} - W_{Ey,z} \sin \Phi_{Ey,z} \cos \beta_{z,z} \quad (107)$$

$$= - W_{Bx,t} \sin \Phi_{Bx,t} \cos \beta_{t,t}$$

$$W_{Ex,z} \cos \Phi_{Ex,z} \cos \beta_{z,z} - W_{Ez,x} \cos \Phi_{Ez,x} \cos \beta_{x,x} \quad (108)$$

$$= - W_{By,t} \cos \Phi_{By,t} \cos \beta_{t,t}$$

$$W_{Ex,z} \sin \Phi_{Ex,z} \cos \beta_{z,z} - W_{Ez,x} \sin \Phi_{Ez,x} \cos \beta_{x,x} \quad (109)$$

$$= - W_{By,t} \sin \Phi_{By,t} \cos \beta_{t,t}$$

Maxwell's equations can be written in an alternative but equivalent form by using equations (54) through (59) as follows

$$R_{1x}(B_{xR},B_{xI}) + R_{1y}(B_{yR},B_{yI}) + R_{1z}(B_{zR},B_{zI}) = 0 \quad (110)$$

$$I_{1x}(B_{xR},B_{xI}) + I_{1y}(B_{yR},B_{yI}) + I_{1z}(B_{zR},B_{zI}) = 0 \quad (111)$$

$$R_{1x}(D_{xR},D_{xI}) + R_{1y}(D_{yR},D_{yI}) + R_{1z}(D_{zR},D_{zI}) = \rho_q \cos \theta_{\rho q} \quad (112)$$

$$I_{1x}(D_{xR},D_{xI}) + I_{1y}(D_{yR},D_{yI}) + I_{1z}(D_{zR},D_{zI}) = \rho_q \sin \theta_{\rho q} \quad (113)$$

$$R_{1x}(H_{yR},H_{yI}) - R_{1y}(H_{xR},H_{xI}) = R_{1t}(D_{zR},D_{zI}) + j_{zR} \quad (114)$$

$$I_{1x}(H_{yR},H_{yI}) - I_{1y}(H_{xR},H_{xI}) = I_{1t}(D_{zR},D_{zI}) + j_{zI} \quad (115)$$

$$R_{1y}(H_{zR},H_{zI}) - R_{1z}(H_{yR},H_{yI}) = R_{1t}(D_{xR},D_{xI}) + j_{xR} \quad (116)$$

$$I_{1y}(H_{zR},H_{zI}) - I_{1z}(H_{yR},H_{yI}) = I_{1t}(D_{xR},D_{xI}) + j_{xI} \quad (117)$$

$$R_{1z}(H_{xR},H_{xI}) - R_{1x}(H_{zR},H_{zI}) = R_{1t}(D_{yR},D_{yI}) + j_{yR} \quad (118)$$

$$I_{1z}(H_{xR},H_{xI}) - I_{1x}(H_{zR},H_{zI}) = I_{1t}(D_{yR},D_{yI}) + j_{yI} \quad (119)$$

$$R_{1x}(E_{yR},E_{yI}) - R_{1y}(E_{xR},E_{xI}) = - R_{1t}(B_{zR},B_{zI}) \quad (120)$$

$$I_{1x}(E_{yR},E_{yI}) - I_{1y}(E_{xR},E_{xI}) = - I_{1t}(B_{zR},B_{zI}) \quad (121)$$

$$R_{1y}(E_{zR},E_{zI}) - R_{1z}(E_{yR},E_{yI}) = - R_{1t}(B_{xR},B_{xI}) \tag{122}$$

$$I_{1y}(E_{zR},E_{zI}) - I_{1z}(E_{yR},E_{yI}) = - I_{1t}(B_{xR},B_{xI}) \tag{123}$$

$$R_{1z}(E_{xR},E_{xI}) - R_{1x}(E_{zR},E_{zI}) = - R_{1t}(B_{yR},B_{yI}) \tag{124}$$

$$I_{1z}(E_{xR},E_{xI}) - I_{1x}(E_{zR},E_{zI}) = - I_{1t}(B_{yR},B_{yI}) \tag{125}$$

where

$$j_{zR} = j_z \cos \theta_{jz} \qquad j_{zI} = j_z \sin \theta_{jz} \tag{126}$$

$$j_{xR} = j_x \cos \theta_{jx} \qquad j_{xI} = j_x \sin \theta_{jx} \tag{127}$$

$$j_{yR} = j_y \cos \theta_{jy} \qquad j_{yI} = j_y \sin \theta_{jy} \tag{128}$$

The radiation pressure is related to the radiation energy density for isotropic radiation with broken internal symmetry by the following approximate formula

$$\bar{P}_r = P_r e^{j\theta_{Pr}} = \frac{1}{3} \bar{\mathcal{E}}_r = \frac{1}{3} \mathcal{E}_r e^{j\theta_{\mathcal{E}r}} \tag{129}$$

Equation (129) is exact only for symmetrical isotropic radiation. Equation (129) gives the following approximate equations

$$P_r = \frac{1}{3} \mathcal{E}_r \tag{130}$$

$$\theta_{Pr} = \theta_{\mathcal{E}r} \tag{131}$$

where $\bar{\mathcal{E}}_r$ = radiation energy density with broken internal symmetry, that is related to the electromagnetic field vectors as follows

$$\bar{\mathcal{E}}_r = \frac{\varepsilon}{2} (\bar{E}_x^2 + \bar{E}_y^2 + \bar{E}_z^2) + \frac{\mu}{2} (\bar{H}_x^2 + \bar{H}_y^2 + \bar{H}_z^2) \tag{132}$$

Equation (132) is equivalent to the following two equations

$$\mathcal{E}_r \cos \theta_{\mathcal{E}r} = \frac{\varepsilon}{2}[\, E_x^2 \cos(2\theta_{Ex}) + E_y^2 \cos(2\theta_{Ey}) + E_z^2 \cos(2\theta_{Ez})] \tag{133}$$

$$+ \frac{\mu}{2}[\, H_x^2 \cos(2\theta_{Hx}) + H_y^2 \cos(2\theta_{Hy}) + H_z^2 \cos(2\theta_{Hz})]$$

$$\mathcal{E}_r \sin \theta_{\mathcal{E}r} = \frac{\varepsilon}{2}[\, E_x^2 \sin(2\theta_{Ex}) + E_y^2 \sin(2\theta_{Ey}) + E_z^2 \sin(2\theta_{Ez})] \tag{134}$$

$$+ \frac{\mu}{2}[\, H_x^2 \sin(2\theta_{Hx}) + H_y^2 \sin(2\theta_{Hy}) + H_z^2 \sin(2\theta_{Hz})]$$

from which $\mathcal{E}_r$ and $\theta_{\mathcal{E}r}$ can be immediately obtained. The unrenormalized radiation density is given by equation (132) with the bars removed and with the superscript "a" inserted on all quantities.

In addition to Maxwell's equations several other equations are required to form a complete set of equations to determine the phase angles of the spacetime coordinates as well as the phase angles of the electromagnetic field vectors. Six of the additional equations required are the equations of motion for charged bulk matter (plasma). These six equations are given by the complex number vector nonrelativistic Euler equations combined with the Lorentz force as follows[19-22]

$$\rho d\bar{v}_\alpha/d\bar{t} = -\cos \beta_{\alpha,\alpha} \;(\partial\bar{P}/\partial\alpha + \partial\bar{P}_r/\partial\alpha) - \rho\partial\bar{W}/\partial\bar{\alpha} + \rho_q(\vec{\bar{E}} + \vec{\bar{v}} \times \vec{\bar{B}})_\alpha \tag{135}$$

where $\alpha = x, y, z$, ρ = mass density, $\bar{v}_\alpha$ = spatial components of particle velocity with internal phase, $\bar{P}$ = static pressure with internal phase, $\bar{P}_r$ = radiation pressure with internal phase, $\bar{W}$ = external potential (such as gravity) with internal phase, and $\vec{\bar{v}}$ = particle velocity vector with internal phase = $(\bar{v}_x, \bar{v}_y, \bar{v}_z)$. The static pressure and external potential are complex numbers with internal phase angles and are written as

$$\bar{P} = Pe^{j\theta_P} \tag{136}$$

$$\bar{W} = We^{j\theta_W} \tag{137}$$

The time derivatives of the velocity components in equation (135) are given by the following six equations

$$\bar{a}_\alpha = d\bar{v}_\alpha/d\bar{t} = \partial\bar{v}_\alpha/\partial\bar{t} + \sum_\sigma \bar{v}_\sigma \,\partial\bar{v}_\alpha/\partial\bar{\sigma} \tag{138}$$

where the following six equations define the complex number velocity

$$\bar{v}_\alpha = d\bar{\alpha}/d\bar{t} = v_\alpha e^{j\theta_{v\alpha}} \tag{139}$$

$$v_\alpha = \sqrt{\frac{(d\alpha/dt)^2 + (\alpha\, d\theta_\alpha/dt)^2}{1 + (t\, d\theta_t/dt)^2}} \tag{140}$$

$$\theta_{v\alpha} = \theta_\alpha + \beta_{\alpha,t} - \theta_t - \beta_{t,t} \tag{141}$$

where

$$\tan \beta_{\alpha,t} = \alpha \frac{d\theta_\alpha/dt}{d\alpha/dt} \tag{142}$$

$$\frac{d\theta_\alpha}{dt} = \frac{\partial\theta_\alpha}{\partial t} + \frac{\partial\theta_\alpha}{\partial x}\frac{dx}{dt} + \frac{\partial\theta_\alpha}{\partial y}\frac{dy}{dt} + \frac{\partial\theta_\alpha}{\partial z}\frac{dz}{dt} \tag{143}$$

In addition, the continuity equation

$$\partial\bar{\rho}_q/\partial\bar{t} + \vec{\bar{\nabla}} \cdot (\bar{\rho}_q \vec{\bar{v}}) = 0 \tag{144A}$$

is necessary to determine θ_t and $\rho_q(x,y,z,t)$. Equation (144A) has two components because it is a complex number equation.

The Maxwell equations for broken symmetry matter, equations (94) through (109) or equivalently equations (110) through (125), are not sufficient by themselves to determine the internal phase angles of the space and time coordinates. This is because the twelve independent Maxwell equations (88) through (93), the twelve constitutive equations (82) and (83), the two components of the ground state trace equation (1), the two components of the ground state Grüneisen parameter equation (4), the two components of the excited state trace equation (5), the two components of the radiation Grüneisen parameter equation (6A), the two state equations (84) and (85) for the renormalized magnetic permeability and electric permittivity, and the two components of the continuity equation, represent thirty-six equations. However, thus far only thirty-five field and matter variables have been enumerated and these are: E_α , $\theta_{E\alpha}$; H_α , $\theta_{H\alpha}$; B_α , $\theta_{B\alpha}$; $\mathcal{E}$, $\theta_\mathcal{E}$; γ , θ_γ ; $\mathcal{E}_r$, $\theta_{\mathcal{E}r}$; γ_r , $\theta_{\gamma r}$; ε , μ ; and ρ_q . But these thirty-five quantities are related to nineteen kinematic and dynamic variables because of the space and time derivatives of the field vectors in Maxwell's equations (33) through (48) and because of the appearance of the current density (velocity) in Maxwell's equations. The nineteen kinematic and dynamic variables are: x , y , z , v_x , v_y , v_z , a_x , a_y , a_z and the corresponding phase angles θ_x , θ_y , θ_z , θ_{vx} , θ_{vy} , θ_{vz} ,

θ_{ax}, θ_{ay}, θ_{az} and by itself θ_t. In these calculations the magnitude of the time t is taken to be a free and independent variable. Therefore a total of fifty-four unknown quantities need to be calculated, and thus far only thirty-six equations have been enumerated. The additional necessary eighteen equations are: the six equations of motion (135), the six kinematic acceleration equations (138), and the six kinematic velocity equations (139). Thus there are fifty-four equations and fifty-four unknown variables to be determined. Note that the two components of the complex number scalar continuity equation introduces only one new unknown quantity ρ_q, and this leaves the second component equation to determine θ_t which stands by itself because t is taken to be an independent variable.

The relativistic trace equations (1) and (5) play an important part in the calculation of the renormalized electromagnetic fields in asymmetric bulk matter or vacuum. Starting with the unrenormalized ground state energy density and Grüneisen parameter, $\mathcal{E}^a$ and γ^a respectively, equation (1) is used to calculate the renormalized values of the ground state energy density $\mathcal{E}$, $\theta_{\mathcal{E}}$ and the ground state Grüneisen parameter γ, θ_γ. The renormalized values of magnetic permeability μ and dielectric constant ε are expressed in terms of P and γ through equations (84) and (85). In addition to Maxwell's equations, the radiation trace equation (5), in conjunction with equations (133) and (134) that relate the radiation energy density to the electromagnetic field vectors, determines the renormalized field vectors in terms of the corresponding unrenormalized values. The solution of the unrenormalized Maxwell equations (71) through (74) gives the unrenormalized field vectors in terms of the unrenormalized charge density ρ_q^a and current density j^a. The unrenormalized energy density is then calculated in terms of the unrenormalized field vectors using equations (133) and (134). Then equation (5) is applied again in conjuction with equations (133) and (134) and Maxwell's equations to obtain the renormalized field vectors. Finally the renormalized charge and current density are obtained from the renormalized field vectors by using equations (79) and (80).

For electromagnetic waves in the vacuum, the total density ρ and the charge density ρ_q that appear in equations (135) refer to test charges placed within the vacuum to measure the electromagnetic field strengths. Therefore the case of electromagnetic waves in the vacuum is formally equivalent to the case of electromagnetic waves in bulk matter. For the vacuum

$$
\begin{aligned}
&\theta_x = \theta_x^{(v)} &\quad& \theta_y = \theta_y^{(v)} &\quad& \theta_z = \theta_z^{(v)} \\
&\theta_{vx} = \theta_{vx}^{(v)} && \theta_{vy} = \theta_{vy}^{(v)} && \theta_{vz} = \theta_{vz}^{(v)} \\
&\theta_{ax} = \theta_{ax}^{(v)} && \theta_{ay} = \theta_{ay}^{(v)} && \theta_{az} = \theta_{az}^{(v)} \\
&\theta_t = \theta_t^{(v)} && \theta_P = \theta_P^{(v)} && \theta_U = \theta_U^{(v)}
\end{aligned}
\qquad (145)
$$

where (v) refers to the vacuum state (see Section 6).

The energy conservation equation is the first integral of equation (135) and in its simplest form is written as[19]

$$\frac{1}{2}\rho\bar{v}^2 + \bar{P} + \bar{P}_r + \bar{W} - \rho_q \int \vec{\bar{E}} \cdot \vec{\bar{v}}\, d\bar{t} = \text{constant} \qquad (146)$$

where $\vec{\bar{v}}$ = complex number vector velocity whose complex number magnitude is given by

$$\bar{v}^2 = \bar{v}_x^2 + \bar{v}_y^2 + \bar{v}_z^2 \qquad (147)$$

where

$$\bar{v} = v e^{j\theta_v} \qquad (147A)$$

$$\bar{v}_\alpha = v_\alpha e^{j\theta_{v\alpha}} \qquad (147B)$$

Note that

$$\vec{\bar{E}} \cdot \vec{\bar{v}} = \sum_{\alpha=1}^{3} \bar{E}_\alpha \bar{v}_\alpha \qquad (148)$$

Were it possible to neglect the charge density term by making ρ_q vanishingly small, equation (146) becomes

$$\frac{1}{2}\rho\bar{v}^2 + \bar{P}_r + \bar{P} + \bar{W} = \text{constant} \qquad (149)$$

where the mass density ρ refers to a test probe. Finally, the dynamical equations for relativistic bulk matter with broken internal symmetry are given by the following generalization of equation (135) for low pressures[23]

$$\rho\bar{\gamma}^2 \frac{d\bar{v}_\alpha}{d\bar{t}} = -\cos\beta_{\alpha,\alpha}\left[\frac{\partial\bar{P}}{\partial\alpha} + \frac{\partial\bar{P}_r}{\partial\alpha} + \frac{dt}{d\alpha}\left(\frac{\partial\bar{P}}{\partial t} + \frac{\partial\bar{P}_r}{\partial t}\right)\right] \qquad (150)$$

$$- \rho\frac{\partial\bar{W}}{\partial\bar{\alpha}} + \rho_q\bar{\gamma}(\vec{\bar{E}} + \vec{\bar{v}} \times \vec{\bar{B}})_\alpha$$

where $\bar{\gamma}$ = complex velocity factor that is defined in Section 4, and where c = light speed in the vacuum.

4. LORENTZ INVARIANCE IN ASYMMETRIC BULK MATTER AND VACUUM. This section considers the Lorentz invariance of Maxwell's equations in bulk matter and vacuum with broken internal symmetries. Maxwell's equations for symmetric systems, equations (71) through (74), are invariant under the Lorentz transformation of coordinate systems[9,10,16,23-34]

$$x'_a = \gamma_a(x_a - v_a t_a) \tag{151}$$

$$t'_a = \gamma_a(t_a - v_a x_a/c^2) \tag{152}$$

where v_a = relative speed of coordinate systems, and the standard velocity factor is given by

$$\gamma_a = (1 - \beta_a^2)^{-1/2} \tag{153}$$

where $\beta_a = v_a/c$, where c = light speed in vacuum. The Lorentz transformation can be obtained by requiring the form invariance of the Minkowski metric as follows

$$x_a^2 - c^2 t_a^2 = x'^2_a - c^2 t'^2_a \tag{154}$$

General relativity, which is not considered in this paper, uses a Riemann metric.[23-25]

The form of Maxwell's equations for charges and currents in asymmetric bulk matter or vacuum, equations (78) through (81), is the same as that for symmetric bulk matter or vacuum, equations (71) through (74). The only difference is that in asymmetric systems the field vectors, current density, and spacetime coordinates are complex numbers. Therefore by the same analysis that shows the symmetric Maxwell equations (71) through (74) to be form covariant under the real number Lorentz transformation equations (151) through (153), it follows that the asymmetric Maxwell equations (78) through (81) are form covariant under the following complex number Lorentz transformations

$$\bar{x}' = \bar{\gamma}(\bar{x} - \bar{v}\bar{t}) \tag{155}$$

$$\bar{t}' = \bar{\gamma}(\bar{t} - \bar{v}\bar{x}/c^2) \tag{156}$$

where $\bar{v}$ = complex number relative speed of the two coordinate systems, and the complex number velocity factor for an asymmetric system is given by

$$\bar{\gamma} = (1 - \bar{\beta}^2)^{-1/2} \tag{157}$$

where $\bar{\beta} = \bar{v}/c$. Also, simple algebra shows that equations (155) through (157) satisfy

$$\bar{x}'^2 - c^2\bar{t}'^2 = \bar{x}^2 - c^2\bar{t}^2 \tag{158}$$

where $\bar{x}$, $\bar{t}$, $\bar{x}'$ and $\bar{t}'$ are space and time coordinates that exhibit broken internal symmetry, and in general

$$\bar{x} = xe^{j\theta_x} \qquad \bar{x}' = x'e^{j\theta'_x} \tag{159}$$

$$\bar{t} = te^{j\theta_t} \qquad \bar{t}' = t'e^{j\theta'_t} \tag{160}$$

and the relative speed of the coordinate systems is written as

$$\bar{v} = ve^{j\theta_v} \qquad \bar{\beta} = \beta e^{j\theta_v} \tag{161}$$

where θ_x , θ'_x , θ_t , θ'_t and θ_v are functions of P and θ_P of the ambient asymmetric bulk matter or vacuum.

Combining equations (157) and (161) gives

$$\bar{\gamma} = (f - jb)^{-1/2} = \sqrt{\frac{f + jb}{f^2 + b^2}} = \gamma e^{j\theta_\gamma} \tag{162}$$

where

$$f = 1 - \beta^2 \cos(2\theta_v) \tag{163}$$

$$b = \beta^2 \sin(2\theta_v) \tag{164}$$

From equation (162) it follows that for an asymmetric system the magnitude and internal phase angle of the velocity factor are given by

$$\gamma = (f^2 + b^2)^{-1/4} = [1 - 2\beta^2 \cos(2\theta_v) + \beta^4]^{-1/4} \tag{165}$$

$$\tan(2\theta_\gamma) = \frac{\beta^2 \sin(2\theta_v)}{1 - \beta^2 \cos(2\theta_v)} \tag{166}$$

Note that $\beta = v/c$, where now v = magnitude of the complex number velocity that appears in equation (161). Also, if $\theta_v = 0$ then equation (165) reduces to equation (153).

The Lorentz transformations in equations (155) and (156) can be written in the form of real and imaginary components as follows

$$x' \cos \theta'_x = \gamma[x \cos(\theta_x + \theta_\gamma) - vt \cos(\theta_v + \theta_t + \theta_\gamma)] \tag{167}$$

$$x' \sin \theta'_x = \gamma[x \sin(\theta_x + \theta_\gamma) - vt \sin(\theta_v + \theta_t + \theta_\gamma)] \tag{168}$$

$$t' \cos \theta'_t = \gamma[t \cos(\theta_t + \theta_\gamma) - vx/c^2 \cos(\theta_v + \theta_x + \theta_\gamma)] \tag{169}$$

$$t' \sin \theta'_t = \gamma[t \sin(\theta_t + \theta_\gamma) - vx/c^2 \sin(\theta_v + \theta_x + \theta_\gamma)] \tag{170}$$

where γ is given by equation (165). From equations (167) and (168) x' and θ'_x can be calculated as follows

$$x'^2 = \gamma^2[x^2 + v^2t^2 - 2vtx \cos(\theta_t + \theta_v - \theta_x)] \tag{171}$$

$$\tan \theta'_x = \frac{x \sin(\theta_x + \theta_\gamma) - vt \sin(\theta_v + \theta_t + \theta_\gamma)}{x \cos(\theta_x + \theta_\gamma) - vt \cos(\theta_v + \theta_t + \theta_\gamma)} \tag{172}$$

while from equations (169) and (170) t' and θ'_t can be calculated in the following manner

$$t'^2 = \gamma^2[t^2 + v^2x^2/c^4 - 2vxt/c^2 \cos(\theta_x + \theta_v - \theta_t)] \tag{173}$$

$$\tan \theta'_t = \frac{t \sin(\theta_t + \theta_\gamma) - vx/c^2 \sin(\theta_v + \theta_x + \theta_\gamma)}{t \cos(\theta_t + \theta_\gamma) - vx/c^2 \cos(\theta_v + \theta_x + \theta_\gamma)} \tag{174}$$

From equations (171) and (173) the Minkowski interval can be written as

$$x'^2 - c^2t'^2 = \gamma^2[(1 - \beta^2)(x^2 - c^2t^2) - 4vtx \sin \theta_v \sin(\theta_x - \theta_t)] \tag{175}$$

where γ is given by equation (165). If $\theta_v = 0$ equation (175) reduces to equation (154).

Consider now some properties of the velocity factor γ given by equation (165). The first thing to see is that γ is not singular for real values of β. In fact it is easy to show that the roots of the denominator in equation (165) are given by

$$\beta^4 - 2\beta^2 \cos(2\theta_v) + 1 = 0 \tag{176}$$

or

$$\beta = e^{\pm j\theta_V} \tag{177}$$

Only when $\theta_V = 0$ does equation (176) have a real root $\beta = 1$ which agrees with equation (153). By taking the derivative of γ given by equation (165) and setting the result equal to zero it is easy to show that γ has a maximum value (assuming θ_V to be independent of velocity) given by

$$[\gamma]_{max} = [\sin (2\theta_V)]^{-1/2} \tag{178}$$

and this maximum value of γ occurs at a value of β given by

$$[\beta]_{max\ \gamma} = [\cos (2\theta_V)]^{1/2} < 1 \tag{179}$$

Combining equations (166) and (179) gives the following value of θ_γ at the maximum point of γ

$$[\theta_\gamma]_{max\ \gamma} = \pi/4 - \theta_V \tag{180}$$

The values of γ and θ_γ for $\beta = 1$ are obtained from equations (165) and (166) respectively as

$$[\gamma]_{\beta=1} = [2 \sin \theta_V]^{-1/2} \tag{181}$$

$$[\theta_\gamma]_{\beta=1} = \pi/4 - \theta_V/2 \tag{182}$$

The functions γ and θ_γ appear in Figures 1 and 2 respectively. As shown by equations (179) the maximum value of γ occurs for $\beta < 1$, and if θ_V is small the maximum value of γ occurs close to $\beta = 1$. Within asymmetric bulk matter or vacuum γ is nonsingular. For $\beta \sim 0$ it follows from equations (165) and (166) that

$$\gamma \sim 1 + \frac{1}{2} \beta^2 \cos (2\theta_V) \tag{183}$$

$$\theta_\gamma \sim \frac{1}{2} \beta^2 \sin (2\theta_V) \tag{184}$$

For $\beta \to \infty$ it follows from equations (165) and (166) that

$$\gamma \to 1/\beta \to 0 \tag{185}$$

$$\theta_\gamma \to \pi/2 - \theta_V \tag{186}$$

where θ_v is assumed to be independent of particle velocity. It is assumed that θ_v depends on P and θ_P of the ambient medium.

The complex number de Broglie wavelength for a relativistic particle moving at velocity $\bar{v}$ is[35]

$$\bar{\lambda} = h/\bar{p} = h/(m\bar{\gamma}\bar{v}) = \lambda e^{j\theta_\lambda} \tag{187}$$

where $\bar{p}$ = complex number momentum whose magnitude is given by $p = m\gamma v$, and h = Planck's constant. From equation (187) it follows that

$$\lambda = h/(m\gamma v) = \lambda_c/(\gamma\beta) \tag{188}$$

$$\theta_\lambda = -\theta_\gamma - \theta_v \tag{189}$$

where λ_c = Compton wavelength given by[30]

$$\lambda_c = h/(mc) \tag{190}$$

Three special cases can be considered.

<u>Case 1.</u> $\beta = [\beta]_{max\ \gamma} = [\cos(2\theta_v)]^{1/2}$ (191)

It follows from equations (178), (179), (180), (188), and (189) that

$$\lambda = \lambda_c[\tan(2\theta_v)]^{1/2} \tag{192}$$

$$\theta_\lambda = -\pi/4 \tag{193}$$

<u>Case 2.</u> $\beta = 1$ (194)

In this case it follows from equations (181), (182) and (188) that

$$\lambda = \lambda_c\ (2\sin\theta_v)^{1/2} \tag{195}$$

$$\theta_\lambda = -\pi/4 - \theta_v/2 \tag{196}$$

<u>Case 3.</u> $\beta \to \infty$ (197)

In the limit $\beta \to \infty$ equations (185), (186), (188) and (189) give

$$\lambda \to \lambda_c \tag{198}$$

$$\theta_\lambda \to -\pi/2 \tag{199}$$

Therefore as β increases without limit the de Broglie wavelength increases to a limiting value of λ_c. It is assumed that θ_v is independent of velocity.

The total energy of a particle located in asymmetric bulk matter or vacuum is given by the following generalization of the standard results of special relativity[24]

$$\bar{\varepsilon} = \varepsilon e^{j\theta_\varepsilon} = \bar{\gamma} mc^2 \tag{200}$$

where the complex number velocity factor is given by equation (157), and m = proper mass. Therefore the total energy of a particle has the same properties as $\bar{\gamma}$, so that

$$\varepsilon = \gamma mc^2 \tag{201}$$

$$\theta_\varepsilon = \theta_\gamma \tag{202}$$

where the magnitude and internal phase angle of the velocity factor is given by equations (165) and (166) respectively. The kinetic energy of a particle in asymmetric bulk matter or vacuum is given by[24]

$$\bar{\varepsilon}_K = \varepsilon_K e^{j\theta_K} = (\bar{\gamma} - 1)mc^2 \tag{203}$$

The component form of equation (203) is written as

$$\varepsilon_K \cos\theta_K = (\gamma \cos\theta_\gamma - 1)mc^2 \tag{204}$$

$$\varepsilon_K \sin\theta_K = \gamma \sin\theta_\gamma \, mc^2 \tag{205}$$

and therefore for a broken symmetry system

$$\tan\theta_K = \frac{\gamma \sin\theta_\gamma}{\gamma \cos\theta_\gamma - 1} \tag{206}$$

$$\varepsilon_K = mc^2(\gamma^2 - 2\gamma\cos\theta_\gamma + 1)^{1/2} \tag{207}$$

Placing equations (183) and (184) into equations (206) and (207) shows that for $\beta \sim 0$

$$\theta_K \sim 2\theta_v \tag{208}$$

$$\varepsilon_K \sim \frac{1}{2} mv^2 \tag{209}$$

which agrees with the nonrelativistic limit obtained directly from equations (157) and (203) namely

$$\bar{\varepsilon}_K \sim \frac{1}{2} m\bar{v}^2 \tag{210}$$

Figures 3 and 4 show ε_K and θ_K in terms of β .

Specific values of the total energy, kinetic energy, and momentum will now be considered for some characteristic values of β .

<u>Case 1.</u> $\beta \sim 0$

$$\gamma \sim 1 + \beta^2/2 \cos (2\theta_v) \tag{211}$$

$$\theta_\gamma \sim \beta^2/2 \sin (2\theta_v) \tag{212}$$

$$\varepsilon \sim mc^2 + \frac{1}{2} mv^2 \tag{212A}$$

$$\theta_\varepsilon \sim \beta^2/2 \sin (2\theta_v) \tag{212B}$$

$$\varepsilon_K \sim \frac{1}{2} mv^2 \tag{212C}$$

$$\theta_K \sim 2\theta_v \tag{212D}$$

$$p \sim mv \tag{212E}$$

$$\theta_p \sim \theta_v \tag{212F}$$

<u>Case 2.</u> $\beta = [\beta]_{max\ \gamma} = [\cos (2\theta_v)]^{1/2}$

$$\gamma = [\sin (2\theta_v)]^{-1/2} \tag{213}$$

$$\theta_\gamma = \pi/4 - \theta_v \tag{214}$$

$$\varepsilon = mc^2[\sin(2\theta_v)]^{-1/2} \quad (215)$$

$$\theta_\varepsilon = \pi/4 - \theta_v \quad (216)$$

$$\varepsilon_K = mc^2\left[\frac{1}{\sin(2\theta_v)} - \frac{2\cos(\pi/4 - \theta_v)}{\sqrt{\sin(2\theta_v)}} + 1\right]^{1/2} \quad (217)$$

$$\varepsilon_K \sim \varepsilon \text{ for small } \theta_v \quad (218A)$$

$$\tan\theta_K = \frac{\sin(\pi/4 - \theta_v)}{\cos(\pi/4 - \theta_v) - \sqrt{\sin(2\theta_v)}} \quad (218B)$$

$$\theta_K \sim \theta_\gamma \text{ for small } \theta_v \quad (218C)$$

$$p/mc = \gamma\beta = [\cot(2\theta_v)]^{1/2} \quad (218D)$$

$$\theta_p = \pi/4 \quad (218E)$$

Case 3. $\beta = 1$

$$\gamma = (2\sin\theta_v)^{-1/2} \quad (219)$$

$$\theta_\gamma = \pi/4 - \theta_v/2 \quad (220)$$

$$\varepsilon = mc^2(2\sin\theta_v)^{-1/2} = cp \quad (221)$$

$$\theta_\varepsilon = \pi/4 - \theta_v/2 \quad (222)$$

$$\varepsilon_K = mc^2\left[\frac{1}{2\sin\theta_v} - \frac{2\cos(\pi/4 - \theta_v/2)}{\sqrt{2\sin\theta_v}} + 1\right]^{1/2} \quad (223)$$

$$\varepsilon_K \sim \varepsilon \text{ for small } \theta_v \quad (224)$$

$$\tan \theta_K = \frac{\sin (\pi/4 - \theta_v/2)}{\cos (\pi/4 - \theta_v/2) - \sqrt{2 \sin \theta_v}} \tag{224A}$$

$$\theta_K \sim \theta_\gamma \text{ for small } \theta_v \tag{224B}$$

$$p/mc = \gamma\beta = (2 \sin \theta_v)^{-1/2} \tag{224C}$$

$$\theta_p = \pi/4 + \theta_v/2 \tag{224D}$$

<u>Case 4.</u> $\beta \to \infty$

$$\gamma \to 1/\beta \to 0 \tag{225}$$

$$\theta_\gamma \to \pi/2 - \theta_v \tag{226}$$

$$\varepsilon \to mc^2/\beta \to 0 \tag{227}$$

$$\theta_\varepsilon \to \pi/2 - \theta_v \tag{228}$$

$$\varepsilon_K \to mc^2[1 - \frac{1}{\beta} \cos (\pi/2 - \theta_v)] \to mc^2 \tag{229}$$

$$\theta_K \to \pi \tag{230A}$$

$$p/mc = \gamma\beta \to 1 \tag{230B}$$

$$\theta_p \to \pi/2 \tag{230C}$$

In direct analogy to the standard expression for relativistic momentum, the momentum of a particle located in asymmetric bulk matter or vacuum is written as[24]

$$\bar{p} = pe^{j\theta_p} = m\bar{\gamma}\bar{v} \tag{231}$$

so that

$$p = m\gamma v \tag{232}$$

$$\theta_p = \theta_\gamma + \theta_v \tag{233}$$

where γ and θ_γ are given by equations (165) and (166) respectively. From equations (157), (204) and (231) it follows that the single particle energy is

$$\bar{\varepsilon}^2 = c^2\bar{p}^2 + m^2c^4 \tag{234}$$

which shows the four-vector status of $\bar{\varepsilon}$ and $\bar{p}$. Equation (234) has two component equations

$$\varepsilon^2 \cos(2\theta_\varepsilon) = c^2p^2 \cos(2\theta_p) + m^2c^4 \tag{235}$$

$$\varepsilon^2 \sin(2\theta_\varepsilon) = c^2p^2 \sin(2\theta_p) \tag{236}$$

Equations (231) through (236) are equivalent to equations (165), (166), (205) and (206). From equations (235) and (236) it follows that

$$\tan(2\theta_\varepsilon) = \frac{p^2 \sin(2\theta_p)}{p^2 \cos(2\theta_p) + m^2c^2} \tag{237}$$

$$\gamma = \frac{\varepsilon}{mc^2} = [(\frac{p}{mc})^4 + 2(\frac{p}{mc})^2 \cos(2\theta_p) + 1]^{1/4} \tag{238}$$

It should be remembered that for asymmetric matter or vacuum, an interaction potential and a gauge potential needs to be added to obtain the total single particle energy

$$\bar{\varepsilon}_i = \bar{\varepsilon} + \bar{V}_e + \bar{V}_g \tag{239}$$

Equation (232) can be written as

$$p/mc = \gamma\beta \tag{240A}$$

Combining equations (165) and (240A) and setting the derivative of the momentum equal to zero gives the following value of β for maximum momentum

$$[\beta]_{max\ p} = [\cos(2\theta_v)]^{-1/2} = 1/[\beta]_{max\ \gamma} > 1 \tag{240B}$$

for which

$$[\gamma]_{max\ p} = [\cot(2\theta_v)]^{1/2} = [p/mc]_{max\ \gamma} \tag{240C}$$

so that the maximum momentum is

$$[p/mc]_{max} = [\sin(2\theta_v)]^{-1/2} = [\gamma]_{max}$$

where equation (178) has been used. Figures 5 and 6 give p and θ_p in terms of β .

The following arguments show why numerical values of θ_v for the asymmetric vacuum cannot be obtained from the experimental results of the Michelson-Morley experiment.[10] The generalization of the standard relativistic velocity addition formula to a system with broken internal symmetry is[24]

$$\bar{w} = \frac{\bar{u} + \bar{v}}{1 + \bar{u}\bar{v}/c^2} \tag{241}$$

where $\bar{u}$ = particle velocity relative to a reference frame that itself is moving at a velocity $\bar{v}$, and $\bar{w}$ = particle velocity relative to a frame of reference from which the moving frame has a velocity $\bar{v}$. Writing the velocities as

$$\bar{w} = we^{j\theta_w} \qquad \bar{u} = ue^{j\theta_u} \qquad \bar{v} = ve^{j\theta_v} \tag{241A}$$

gives the following velocity addition formula for asymmetric bulk matter or vacuum

$$we^{j\theta_w} = \frac{A + jB}{C + jD} \tag{242}$$

$$w = \left[\frac{A^2 + B^2}{C^2 + D^2}\right]^{1/2} \tag{243}$$

$$\theta_w = \theta_N - \theta_D \tag{244}$$

$$\tan\theta_N = B/A \tag{245}$$

$$\tan\theta_D = D/C \tag{246}$$

$$A = u \cos \theta_u + v \cos \theta_v \tag{247}$$

$$B = u \sin \theta_u + v \sin \theta_v \tag{248}$$

$$C = 1 + uv/c^2 \cos (\theta_u + \theta_v) \tag{249}$$

$$D = uv/c^2 \sin (\theta_u + \theta_v) \tag{250}$$

It is easy to show that

$$A^2 + B^2 = u^2 + v^2 + 2uv \cos (\theta_u - \theta_v) \tag{251}$$

$$C^2 + D^2 = 1 + u^2v^2/c^4 + 2uv/c^2 \cos (\theta_u + \theta_v) \tag{252}$$

Equations (251) and (252) are completely general.

It will be assumed that the internal phase of the massive particle velocity is independent of the magnitude of the velocity so that $\theta_u = \theta_v = \theta$ for massive particles and

$$A^2 + B^2 = (u + v)^2 \tag{253}$$

$$C^2 + D^2 = 1 + u^2v^2/c^4 + 2uv/c^2 \cos(2\theta) \tag{254}$$

Consider now the case of two massive particles each travelling at the light speed so that $u = v = c$ and $\theta = \theta_u = \theta_v = \theta_c \neq 0$, then equations (243), (244) and (247) through (254) yield for two massive particles

$$A = 2c \cos \theta_c \qquad B = 2c \sin \theta_c$$

$$C = 1 + \cos(2\theta_c) \qquad D = \sin(2\theta_c) \tag{255}$$

$$w = c/\cos \theta_c \qquad \theta_N = \theta_D = \theta_c \qquad \theta_w = 0$$

Because $\theta_w = 0$, it follows that the measured relative velocity is given by $w_m = w$. If $\theta_c = 0$ in equation (255), the standard special relativistic result $w = c$ is regained.

For two photons the situation is $u = v = c$ and $\theta_u = \theta_v = \theta_c = 0$ and equations (243), (244) and (247) through (252) give

$$A = 2c \qquad B = 0 \qquad C = 2$$

$$A^2 + B^2 = 4c^2 \qquad C^2 + D^2 = 4 \qquad D = 0 \tag{256}$$

$$\theta_N = \theta_D = 0 \qquad w = c \qquad \theta_w = 0$$

Because $\theta_w = 0$ it follows that the measured relative speed is $w_m = w = c$.

Consider now the case of the Michelson and Morley experiment.[10,24-34] For this case the photons with $u = c$ and $\theta_c = 0$ move relative to the earth which has speed v and internal phase angle of speed θ_v . From equations (247) through (250) it follows that

$$A = c + v \cos \theta_v \tag{257}$$

$$B = v \sin \theta_v \tag{258}$$

$$C = 1 + v/c \cos \theta_v = A/c \tag{259}$$

$$D = v/c \sin \theta_v = B/c \tag{260}$$

Then equations (257) through (260) and equations (245) and (246) give

$$A^2 + B^2 = (c + v)^2 - 4cv \sin^2(\theta_v/2) \tag{261}$$

$$C^2 + D^2 = (A^2 + B^2)/c^2 \tag{262}$$

$$\theta_N = \theta_D = \tan^{-1}[(v \sin \theta_v)/(c + v \cos \theta_v)] \tag{263}$$

Equations (261) and (262) agree with equations (251) and (252) with $u = c$ and $\theta_u = \theta_c = 0$. Combining equations (243) and (244) with equations (261) through (263) yields

$$w = c \qquad \theta_w = 0 \tag{264}$$

for the relative velocity and its internal phase angle for photons moving relative to the earth. Because $\theta_w = 0$ it follows that the measured relative speed is $w_m = w = c$. Thus the Michelson and Morley experiment cannot be used to determine the magnitude v or the internal phase angle θ_v of the earth's velocity because the analysis of this experiment using the broken symmetry form of the velocity addition formula nevertheless yields the standard special relativity result given in equation (264). This is true even though the v and θ_v appear in equations (257) through (262) for the earth moving in spacetime with broken coordinate symmetries. The speed of the photons relative to the earth does not depend on either v or θ_v .

Alternatively, measurements of the velocity factor γ from particle accelerator experiments may eventually produce a maximum value of γ which will immediately determine the value of θ_v. For this, particles with $\beta > 1$ would have to be observed. If no such particles are ever found, it would show that $\theta_v = 0$ for the vacuum, and that the vacuum is symmetric. Experiment can only resolve this issue. Experiments to determine θ_v for asymmetric bulk matter may be easier because θ_v for bulk matter is expected to be larger than $\theta_v^{(v)}$ for the vacuum. Note that astronomical objects with $\beta > 1$ have apparently already been observed, and their explanation in terms of conventional effects can be given only with much difficulty.[36]

Finally, the laws of motion of a relativistic particle in asymmetric bulk matter or vacuum are considered. Newton's law of motion is modified by special relativity to give the following dynamical equation of motion for a force in the direction of motion[24]

$$F^a = \frac{d}{dt}(m\gamma_a v_a) = m\gamma_a^3\, dv_a/dt_a = m\gamma_a^3 a_a \tag{265}$$

where a_a = conventionally calculated acceleration, and γ_a is given by equation (153). The generalization of this equation to the case of particle motion in asymmetric bulk matter or vacuum is

$$\bar{F} = \frac{d}{d\bar{t}}(m\bar{\gamma}\bar{v}) = m\bar{\gamma}^3\, d\bar{v}/d\bar{t} = m\bar{\gamma}^3\, \bar{a} \tag{266}$$

where $\bar{\gamma}$ is given by equation (157) and where $\bar{t}$, $\bar{v}$, $\bar{a}$, and $\bar{F}$ are the gauge rotated time, velocity, acceleration and force respectively. Therefore

$$\bar{a} = ae^{j\theta_a} \tag{267}$$

$$\bar{F} = Fe^{j\theta_F} \tag{268}$$

Combining equation (162) with equations (266) through (268) gives the force in the direction of motion as

$$F = m\gamma^3 a \tag{269}$$

$$\theta_F = 3\theta_\gamma + \theta_a \tag{270}$$

where γ and θ_γ are given by equations (165) and (166) respectively.

5. ELECTROMAGNETIC WAVE EQUATIONS. A direct result of Maxwell's equations is a set of wave equations that describe the time and space dependence of the electric and magnetic field vectors in a material body or vacuum.[9,10] This section considers the construction of electromagnetic wave equations for matter and radiation with broken internal symmetries. The standard equation of telegraphy that determines the electric (or magnetic) field in a conducting medium is[9,10]

$$\nabla_a^2 E_\alpha^a = \varepsilon^a \mu^a \frac{\partial^2 E_\alpha^a}{\partial t_a^2} + \mu^a \sigma^a \frac{\partial E_\alpha^a}{\partial t_a} \tag{271}$$

where $\alpha = x$, y and z , and σ^a = unrenormalized conductivity. The Laplacian operator is defined as

$$\nabla_a^2 = \frac{\partial^2}{\partial x_a^2} + \frac{\partial^2}{\partial y_a^2} + \frac{\partial^2}{\partial z_a^2} \tag{272}$$

The prescription introduced in this paper to handle electromagnetism in matter or vacuum with broken internal symmetries is to use gauge rotated field vectors and gauge rotated space and time coordinates. Applying this prescription to equation (271) yields

$$\bar{\nabla}^2 \bar{E}_\alpha = \varepsilon\mu \frac{\partial^2 \bar{E}_\alpha}{\partial \bar{t}^2} + \mu\sigma \frac{\partial \bar{E}_\alpha}{\partial \bar{t}} \tag{273}$$

The complex number Laplacian is given by

$$\bar{\nabla}^2 = \frac{\partial^2}{\partial \bar{x}^2} + \frac{\partial^2}{\partial \bar{y}^2} + \frac{\partial^2}{\partial \bar{z}^2} \tag{274}$$

The first and second derivative terms in equation (273) have already been evaluated in Section 2. Using the notation developed in equations (54) through (59) and (65) through (70) allows equation (273) to be written as six real number relations as follows

$$\sum_\beta R_{2\beta}(E_{xR}, E_{xI}) \sim \varepsilon\mu R_{2t}(E_{xR}, E_{xI}) + \mu\sigma R_{1t}(E_{xR}, E_{xI}) \tag{275}$$

$$\sum_\beta R_{2\beta}(E_{yR}, E_{yI}) \sim \varepsilon\mu R_{2t}(E_{yR}, E_{yI}) + \mu\sigma R_{1t}(E_{yR}, E_{yI}) \tag{276}$$

$$\sum_\beta R_{2\beta}(E_{zR}, E_{zI}) \sim \varepsilon\mu R_{2t}(E_{zR}, E_{zI}) + \mu\sigma R_{1t}(E_{zR}, E_{zI}) \tag{277}$$

$$\sum_{\beta} I_{2\beta}(E_{xR},E_{yR}) \sim \varepsilon\mu I_{2t}(E_{xR},E_{xI}) + \mu\sigma I_{1t}(E_{xR},E_{xI}) \tag{278}$$

$$\sum_{\beta} I_{2\beta}(E_{yR},E_{yI}) \sim \varepsilon\mu I_{2t}(E_{yR},E_{yI}) + \mu\sigma I_{1t}(E_{yR},E_{yI}) \tag{279}$$

$$\sum_{\beta} I_{2\beta}(E_{zR},E_{zI}) \sim \varepsilon\mu I_{2t}(E_{zR},E_{zI}) + \mu\sigma I_{1t}(E_{zR},E_{zI}) \tag{280}$$

where the sum is over $\beta = x$, y and z .

The standard equations that determine the electromagnetic potentials are written as[9,10]

$$\nabla_a^2 \phi^a - \varepsilon^a \mu^a \frac{\partial^2 \phi^a}{\partial t_a^2} = - \rho_q^a/\varepsilon^a \tag{281}$$

$$\nabla_a^2 A_\alpha^a - \varepsilon^a \mu^a \frac{\partial^2 A_\alpha^a}{\partial t_a^2} = - \mu^a j_\alpha^a \tag{282}$$

The generalization of these equations to electromagnetic fields in asymmetric bulk matter or vacuum is as follows

$$\bar{\nabla}^2 \bar{\phi} - \varepsilon\mu \frac{\partial^2 \bar{\phi}}{\partial \bar{t}^2} = - \bar{\rho}_q/\varepsilon \tag{283}$$

$$\bar{\nabla}^2 \bar{A}_\alpha - \varepsilon\mu \frac{\partial^2 \bar{A}_\alpha}{\partial \bar{t}^2} = - \mu \bar{j}_\alpha \tag{284}$$

where the complex number electromagnetic potentials are written as

$$\bar{\phi} = \phi e^{j\theta_\phi} = \phi_R + j\phi_I \tag{285}$$

$$\bar{A}_\alpha = A_\alpha e^{j\theta_{A\alpha}} = A_{\alpha R} + jA_{\alpha I} \tag{286}$$

Using the notation of equations (65) and (66) allows equation (283) to be written as the following two relations

$$\sum_{\beta} R_{2\beta}(\phi_R,\phi_I) - \varepsilon\mu R_{2t}(\phi_R,\phi_I) \sim - \rho_q/\varepsilon \cos\theta_{\rho q} \tag{287}$$

$$\sum_{\beta} I_{2\beta}(\phi_R,\phi_I) - \epsilon\mu I_{2t}(\phi_R,\phi_I) \sim - \rho_q/\epsilon \sin\theta_{\rho q} \tag{288}$$

while equation (284) can be written as the following six approximations

$$\sum_{\beta} R_{2\beta}(A_{xR},A_{xI}) - \epsilon\mu R_{2t}(A_{xR},A_{xI}) \sim - \mu j_x \cos\theta_{jx} \tag{289}$$

$$\sum_{\beta} R_{2\beta}(A_{yR},A_{yI}) - \epsilon\mu R_{2t}(A_{yR},A_{yI}) \sim - \mu j_y \cos\theta_{jy} \tag{290}$$

$$\sum_{\beta} R_{2\beta}(A_{zR},A_{zI}) - \epsilon\mu R_{2t}(A_{zR},A_{zI}) \sim - \mu j_z \cos\theta_{jz} \tag{291}$$

$$\sum_{\beta} I_{2\beta}(A_{xR},A_{xI}) - \epsilon\mu I_{2t}(A_{xR},A_{xI}) \sim - \mu j_x \sin\theta_{jx} \tag{292}$$

$$\sum_{\beta} I_{2\beta}(A_{yR},A_{yI}) - \epsilon\mu I_{2t}(A_{yR},A_{yI}) \sim - \mu j_y \sin\theta_{jy} \tag{293}$$

$$\sum_{\beta} I_{2\beta}(A_{zR},A_{zI}) - \epsilon\mu I_{2t}(A_{zR},A_{zI}) \sim - \mu j_z \sin\theta_{jz} \tag{294}$$

Finally the gauge conditions for an electromagnetic field with broken internal symmetry is written as[9,10]

$$\vec{\bar{\nabla}} \cdot \vec{\bar{A}} + \epsilon\mu \frac{\partial\bar{\phi}}{\partial\bar{t}} = 0 \tag{295}$$

which can be written in terms of real and imaginary components as

$$\sum_{\beta} R_{1\beta}(A_{\beta R},A_{\beta I}) + \epsilon\mu R_{1t}(\phi_R,\phi_I) = 0 \tag{296}$$

$$\sum_{\beta} I_{1\beta}(A_{\beta R},A_{\beta I}) + \epsilon\mu I_{1t}(\phi_R,\phi_I) = 0 \tag{297}$$

where the sum is over $\beta = x$, y and z .

6. VACUUM WITH BROKEN INTERNAL SYMMETRIES. Of special importance to the propagation of electromagnetic waves are the properties of the vacuum state. The vacuum state may exhibit the same broken internal symmetries as does bulk

matter. Consider the vacuum state to have a zero temperature state coupled to a thermal state in such a way that the vacuum energy density and pressure for low temperatures are given by

$$\bar{E}^{(v)} = \bar{E}_o^{(v)} + \bar{E}_j^{(v)} T^j + \cdots = E^{(v)} e^{j\theta_E^{(v)}} \tag{298}$$

$$\bar{P}^{(v)} = \bar{P}_o^{(v)} + \bar{P}_j^{(v)} T^j + \cdots = P^{(v)} e^{j\theta_P^{(v)}} \tag{299}$$

where $\bar{E}^{(v)}$ and $\bar{P}^{(v)}$ = vacuum energy density and pressure respectively, $\bar{E}_o^{(v)}$ and $\bar{P}_o^{(v)}$ = zero temperature vacuum energy density and pressure respectively, and $\bar{E}_j^{(v)}$ and $\bar{P}_j^{(v)}$ = thermal coefficients for the vacuum energy density and pressure respectively. The vacuum Grüneisen parameter is defined by

$$\bar{\gamma}_o^{(v)} = \gamma_o^{(v)} e^{j\theta_{\gamma o}^{(v)}} \tag{300}$$

$$= \left[\frac{\partial \bar{P}^{(v)}/\partial T}{\partial \bar{E}^{(v)}/\partial T}\right]_{T=o} = \frac{\bar{P}_j^{(v)}}{\bar{E}_j^{(v)}} = \frac{1}{(j-1)} \frac{V}{\bar{U}_j^{(v)}} \frac{\partial \bar{U}_j^{(v)}}{\partial V}$$

where $\bar{U}_j^{(v)} = V\bar{E}_j^{(v)}$, and where j = index that describes the thermal properties of the vacuum.

The energy density $\bar{E}_o^{(v)}$ and Grüneisen parameter $\bar{\gamma}_o^{(v)}$ for the zero temperature vacuum are calculated from the simultaneous solution of two differential equations

$$\bar{E}_o^{(v)} - 3\{[1 + \bar{\gamma}_o^{(v)}]\bar{P}_o^{(v)} - \bar{K}_o^{(v)}\} = 0 \tag{301}$$

$$1 + j + \frac{j\bar{\gamma}_o^{(v)}\bar{P}_o^{(v)}}{\bar{P}_o^{(v)} - \bar{K}_o^{(v)}} + 3n \frac{d\bar{\gamma}_o^{(v)}}{dn} = 0 \tag{302}$$

which are just equations (252) and (253) of Reference 8 with their right hand sides set equal to zero. A trivial solution of equations (252) and (253) of Reference 8 with their right hand sides equal to zero is just $\bar{E}_o^{(v)} = 0$ and $\bar{E}_j^{(v)} = 0$ which is equivalent to the unrenormalized vacuum $E_o^a = 0$ and $E_j^a = 0$. A non-trivial solution is obtained by simultaneously solving equations (301) and (302). It is easy to show that equation (301) can be written as

$$3n^2 \frac{d^2\bar{E}_o^{(v)}}{dn^2} - 3[1 + \bar{\gamma}_o^{(v)}]n \frac{d\bar{E}_o^{(v)}}{dn} + [3\bar{\gamma}_o^{(v)} + 4]\bar{E}_o^{(v)} = 0 \qquad (303)$$

The vacuum radiation energy density and pressure are written as[8]

$$\bar{E}_r^{(v)} = \bar{E}_{or}^{(v)} + \bar{E}_{jr}^{(v)}T^j + \cdots \qquad (304)$$

$$\bar{P}_r^{(v)} = \bar{P}_{or}^{(v)} + \bar{P}_{jr}^{(v)}T^j + \cdots \qquad (305)$$

while the zero temperature radiation Gruneisen parameter for the vacuum is given by

$$\bar{\gamma}_{or}^{(v)} = \gamma_{or}^{(v)} e^{j\theta_{\gamma or}^{(v)}} \qquad (306)$$

$$= \left[\frac{\partial \bar{P}_r^{(v)}/\partial T}{\partial \bar{E}_r^{(v)}/\partial T}\right]_{T=o} = \frac{\bar{P}_{jr}^{(v)}}{\bar{E}_{jr}^{(v)}} = \frac{1}{(j-1)} \frac{V}{\bar{U}_{jr}^{(v)}} \frac{\partial \bar{U}_{jr}^{(v)}}{\partial V}$$

The vacuum radiation equations are then written as[8]

$$\bar{E}_{or}^{(v)} - 3\left\{[1 + \bar{\gamma}_o^{(v)}]\bar{P}_{or}^{(v)} - \bar{K}_{or}^{(v)}\right\} - 3 \frac{\bar{E}_{jr}^{(v)}}{\bar{E}_j^{(v)}} \bar{P}_o^{(v)}[\bar{\gamma}_{or}^{(v)} - \bar{\gamma}_o^{(v)}] = 0 \qquad (307)$$

$$j\bar{E}_j^{(v)}[\bar{\alpha}^{(v)}\bar{K}_{or}^{(v)} - \bar{\beta}^{(v)}\bar{P}_{or}^{(v)}] + \bar{E}_{jr}^{(v)}\bar{S}_{jr}^{(v)} = 0 \qquad (308)$$

which are just equations (287) and (288) of Reference 8 with their right hand sides set equal to zero and a superscript (v) added to indicate a vacuum solution.

Therefore in principle asymmetric vacuum state is formally identical to the asymmetric bulk matter state. In fact, the vacuum is simpler than the bulk matter state as can be seen by comparing equations (301), (302), (307) and (308) with equations (252), (253), (287) and (288) respectively of Reference 8. The vacuum is expected to exhibit a broken internal symmetry state that is described by $\theta_P^{(v)}$ and $\theta_\gamma^{(v)}$. The broken symmetry of the vacuum will impress broken symmetries on the kinematic and dynamic variable of particles moving in the vacuum. Similarly, electromagnetic waves in the vacuum are expected to possess electric and magnetic fields and a spacetime coordinate description that exhibit internal phase angles.

7. CONCLUSION. The effects of the broken symmetry of space and time on electromagnetism in matter and the vacuum is considered, and Maxwell's equations with broken internal symmetries are developed. The Lorentz covariance of these equations is assumed to be valid but must now be represented in the form of complex number Lorentz transformations. The results of the Michelson-Morley experiment cannot be used to place a limit on the magnitude or the internal phase angle of the velocity of a particle moving in the vacuum. Experiments conducted in asymmetric bulk matter may be fruitful because the internal symmetries of spacetime are larger in this case than for the vacuum. The wave equations and gauge conditions for electromagnetic waves with broken internal symmetries are easily developed. Finally, the broken symmetry properties of the vacuum are obtained by solving a set of coupled differential equations which are similar in form to the corresponding equations for asymmetric bulk matter.

ACKNOWLEDGEMENT

The author wishes to thank Elizabeth K. Klein for typing this paper.

REFERENCES

1. O'Raifeartaigh, L., Group Structure of Gauge Theories, Cambridge University Press, Cambridge, 1986.

2. Taylor, J., Gauge Theories of Weak Interactions, Cambridge University Press, Cambridge, 1976.

3. Ramond, P., Field Theory: A Modern Primer, Benjamin, New York, 1981.

4. Mandl, F. and Shaw, G., Quantum Field Theory, John Wiley, New York, 1984.

5. Itzykson, C. and Zuber, J., Quantum Field Theory, McGraw-Hill, New York, 1980.

6. de Wit, B. and Smith, J., Field Theory in Particle Physics, North-Holland, New York, 1986.

7. Weiss, R. A., Relativistic Thermodynamics, Vols. 1 and 2, Exposition Press, New York, 1976.

8. Weiss, R. A., "Thermodynamic Gauge Theory of Solids and Quantum Liquids with Internal Phase", Fifth Army Conference on Applied Mathematics and Computing, West Point, New York, June 15-18, 1987, p. 649.

9. Jackson, J., Classical Electrodynamics, John Wiley, New York, 1975.

10. Panofsky, W. and Phillips, M., Classical Electricity and Magnetism, Addison-Wesley, Reading, MA, 1955.

11. Stratton, J., Electromagnetic Theory, McGraw-Hill, New York, 1941.

12. Smythe, W., Static and Dynamic Electricity, McGraw-Hill, New York, 1950.

13. Becker, R., Electromagnetic Fields and Interactions, Dover, New York, 1964.

14. Jeans, J., The Mathematical Theory of Electricity and Magnetism, Cambridge University Press, Cambridge, 1951.

15. Menzel, D., Mathematical Physics, Prentice-Hall, New York, 1953.

16. Whittaker, E., A History of the Theories of Aether and Electricity, Philosophical Library, New York, 1951.

17. Sommerfeld, A., Electrodynamics, Academic, New York, 1952.

18. Kong, J., Electromagnetic Wave Theory, John Wiley, New York, 1986.

19. Chandrasekhar, S., Plasma Physics, University of Chicago Press, Chicago, 1960.

20. Artsimovich, L. A., A Physicists ABC on Plasma, MIR Publishers, Moscow, 1978.

21. Spitzer, L., Physics of Fully Ionized Gases, John Wiley-Interscience, New York, 1964.

22. Thompson, W. B., An Introduction to Plasma Physics, Addison-Wesley, Reading, Massachusetts, 1962.

23. Weinberg, S., Gravitation and Cosmology: Principles and Applications of the General Theory of Relativity, John Wiley, New York, 1972.

24. Pauli, W., Theory of Relativity, Pergamon, New York, 1958.

25. Møller, C., The Theory of Relativity, Oxford, New York, 1952.

26. Synge, J. L., Relativity: The Special Theory, North-Holland, Amsterdam, 1965.

27. Bergmann, P., Introduction to the Theory of Relativity, Prentice-Hall, New York, 1953.

28. Eddington, A., The Mathematical Theory of Relativity, Cambridge University Press, Cambridge, 1952.

29. Weyl, H., Space-Time-Matter, Dover, New York, 1922.

30. Robertson, H. and Noonan, T., Relativity and Cosmology, Saunders, Philadelphia, 1968.

31. Tolman, R., Relativity Thermodynamics and Cosmology, Oxford, New York, 1934.

32. Aharoni, J., The Special Theory of Relativity, Oxford, New York, 1959.

33. Misner, C., Thorne, K. and Wheeler, J., Gravitation, Freeman, San Francisco, 1973.

34. Zeldovich, Ya. and Novikov, I., Relativistic Astrophysics Volume 1 Stars and Relativity, University of Chicago Press, Chicago, 1971.

35. Born, M., Atomic Physics, Hafner, New York, 1953.

36. Zensus, J. A. and Pearson, T. J., Eds., Superluminal Radio Sources, Cambridge University Press, New York, 1987.

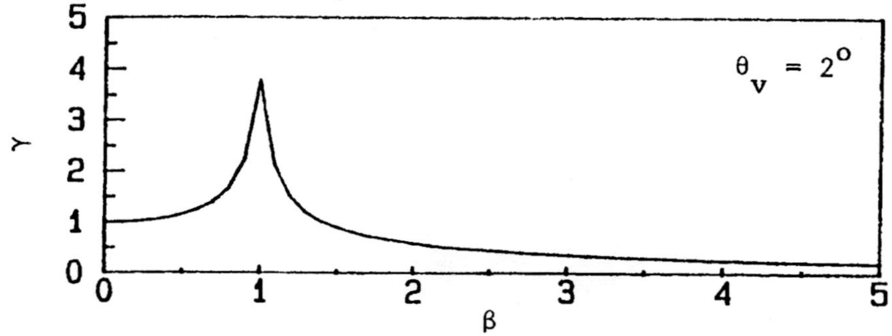

Figure 1. Magnitude of velocity factor versus magnitude of velocity ratio.

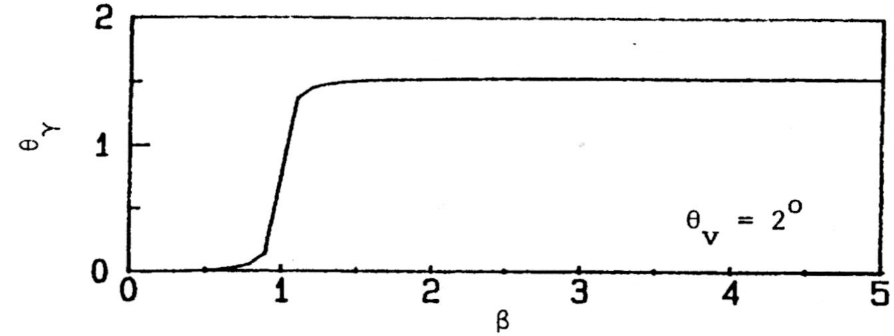

Figure 2. Phase angle of velocity factor versus magnitude of velocity ratio.

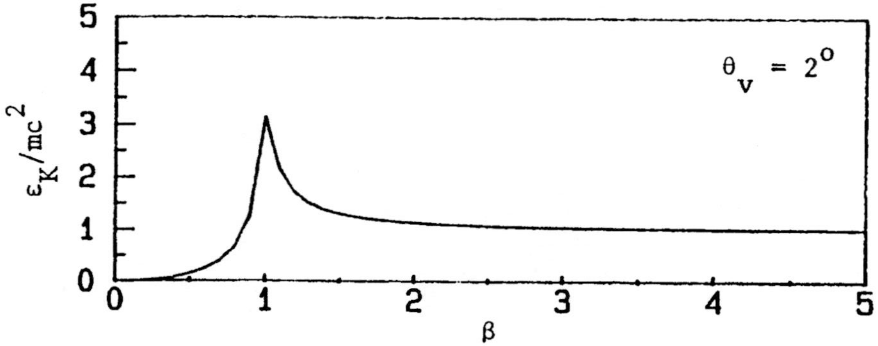

Figure 3. Normalized magnitude of kinetic energy versus magnitude of velocity ratio.

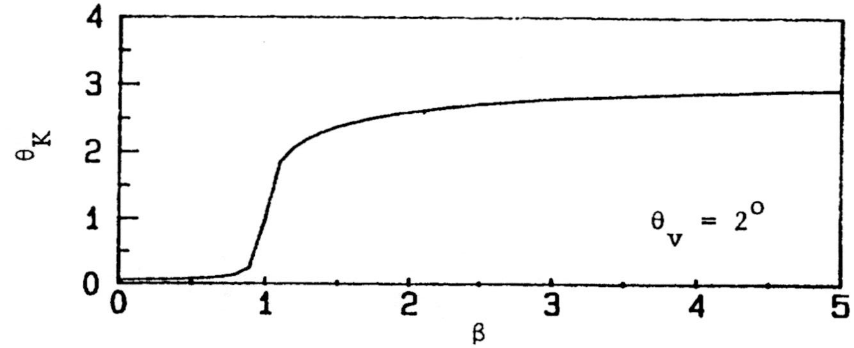

Figure 4. Phase angle of kinetic energy versus magnitude of velocity ratio.

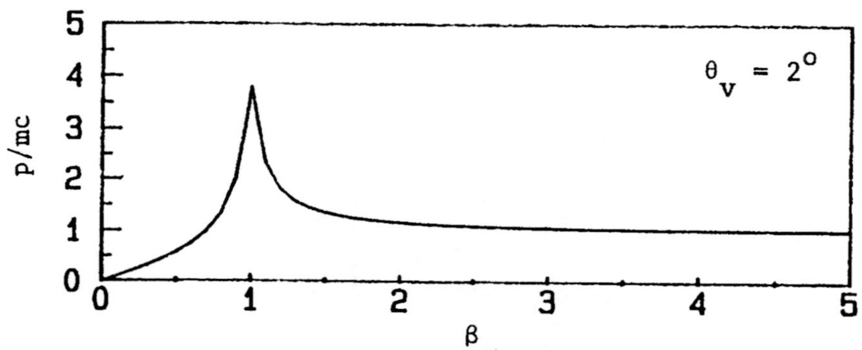

Figure 5. Normalized magnitude of momentum versus magnitude of velocity ratio.

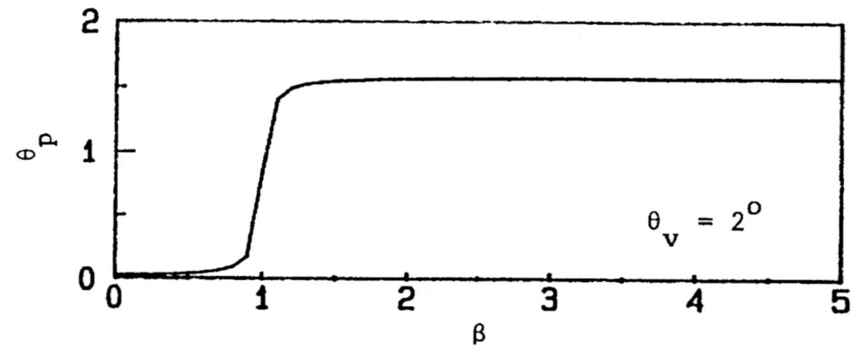

Figure 6. Phase angle of momentum versus magnitude of velocity ratio.

D

ULTRAFAST FLIGHT OF POLYPHASE SPACE VEHICLES

Richard A. Weiss
K&W Research Company
Tucson, Arizona

ABSTRACT. This paper studies the possibility of ultrafast flight of a polyphase space vehicle by assuming the validity of the Lorentz transformation for space and time coordinates in which an ambient phase field is present which induces internal phase angles into these coordinates and makes them complex numbers in an internal space. An ambient phase field is a field that induces internal phase angles into the space and time coordinates and all other physical quantities with the exception of the fundamental constants of nature $\hbar$, c, G ···. Space and time coordinate factors, which give the ratio of the magnitude of a complex number coordinate to its zero ambient phase field value, are delineated for high and low density systems of matter and the vacuum. Lorentz transformations of space and time coordinates that are complex numbers in an internal space are used to develop the corresponding transformations of the space and time coordinate factors as seen by observers on stationary and moving coordinate systems when the observers and the moving body are immersed in a cosmic ambient phase field and when the polyphase space vehicle has the capability of generating an adjustable artificial internal ambient phase field. Applications to the flight of polyphase space vehicles are considered and, based on the forms taken by the space coordinate factors, it is demonstrated that only a vacuum of low-dimensionality and low-dimensional density that contains a cosmic ambient phase field can support the ultrafast flight of a small high-density polyphase space vehicle that contains a properly tuned manmade mechanical, acoustic or electromagnetic ambient phase field. Natural objects cannot undergo ultrafast flight.

1. INTRODUCTION. Mankind will never reach the solar systems of stars in the Galaxy and in other galaxies without developing the understanding and technology of ultrafast subluminal flight and superluminal flight. Within the limits of present technology and understanding it would take thousands of years to reach nearby stars and millions, billions and trillions of years to reach very distant stars and galaxies.[1,2] And therefore it makes sense to investigate whether ultrafast subluminal flight and superluminal flight of polyphase space vehicles is possible within the limits of Einstein's theory of special relativity and its founding principles of the covariance of the laws of physics and the constancy of the speed of light in vacuo. Certainly the Lorentz transformations, as deduced from the Einstein formulation of the relativity of space and time and as developed from the form covariance of Maxwell's equations of the electromagnetic field, show that it is not possible for a body to travel faster than the speed of light in vacuo.[3,4] Any argument about this is a waste of time. Instead, the concept of the Einstein invariance of the laws of physics and the constancy of the speed of light in vacuum should be examined, not so much from the point of view of its violation, but rather to its possible generalization and applicability to space and time coordinates in the presence of a cosmic field such as an ambient phase field, and in fact the Lorentz transformations have already been generalized to the complex number space and time coordinates that are associated with an ambient phase field.[5]

An ambient phase field (APF) is defined as a field whose presence in matter or vacuum forces all space and time coordinates and all physical quantities, with the exception of the constants of nature such as $\hbar$, c, G ⋯, to be complex numbers in an internal space.[5-10] It has been suggested that a manmade APF applied to atomic nuclei leads to the possibility complex number nucleon numbers and complex number dimensions of space and time and ultimately to the concepts of polyphase fission and the polyphase fission map.[6-10] And it has been suggested that a cosmic APF may pervade all space and time, and Lorentz transformations are generalized to the complex number space and time coordinates that are associated with an APF.[5] It is assumed that special relativity expressed in terms of complex number space and time coordinates is valid in the presence of an APF. Finally, it has been suggested that the presence of several manmade APFs in a rocket can be used to enhance the liftoff of a heavy-lift rocket, and this type of rocket has been called a polyphase rocket.[9] As with the polyphase rocket, each component of a polyphase space vehicle contains an APF that determines the internal phase angles of the particle number and the particle number density whose average values for the space vehicle enter the calculation of the real number space and time coordinate factors.[9] The equations for the transformations of the space and time coordinate factors are determined and allow or disallow the ultrafast flight of a polyphase space vehicle.

This Introduction will first delineate the general forms of the space and time coordinate factors for the vacuum and a polyphase space vehicle and introduce the basic physics associated with ultrafast flight. Part A of this Introduction gives the general expressions for the space and time coordinate factors for the case of an observer on Earth when the polyphase space vehicle is at rest in a laboratory on Earth, Part B describes the apparent space and time coordinate factors for a polyphase space vehicle moving relative to an observer on the stationary Earth, Part C shows the space and time coordinate factors of the vacuum, Part D gives the space and time coordinate factors of a moving polyphase space vehicle from the perspective of an observer on a system of coordinates moving relative to the Earth, Part E delineates the space and time coordinate factors of a polyphase space vehicle from the perspective of an observer on a co-moving system of coordinates that is attached to the rear of a polyphase space vehicle, Part F introduces internal and external fields and the concept of tuning using an external space field, Part G gives a brief introduction to conventional special relativity while Part H introduces special relativity in the presence of an APF, Part I describes the Lorentz contraction and the time dilation in the presence of an APF and Part J treats the special relativistic addition of velocities in the presence of an APF. After that, Section 2 develops the special relativistic transformations for the apparent space and time coordinate factors in the presence of an APF and the tuning of the external space fields that are required for superluminal flight, Section 3 develops the special relativistic apparent and intrinsic space and time coordinate factors and their transformations and the tuning of the APFs that are required for ultrafast subluminal flight, Section 4 considers the apparent and intrinsic special relativistic space and time coordinate factors and their transformations and the tuning of the APFs and the external space fields that are required for superluminal flight, and finally Section 5 describes some design concepts for ultrafast space vehicles. Appendix A introduces space and time coordinate factors into the description of the Lorentz contraction and the time dilation in the presence of an APF, Appendix B considers the introduction of space and time coordinate factors to the description of the addition of velocities in the presence of an APF and finally Appendix C gives a brief description of the liftoff conditions for a polyphase rocket.

A. Complex Number Space and Time Coordinates and the Space and Time Coordinate Factors of a Polyphase space vehicle at Rest in the Laboratory Frame of Reference.

The space and time coordinates in matter or vacuum that is located in an APF have been represented as complex numbers in an internal space.[5-12] A simple theory of the origin and nature of these complex number coordinates has already been developed.[9] Consider the constant real number spatial parameters L_α^c for the complex number space coordinates $\overline{\alpha}^{lab}$ in an APF in bulk matter, where L_α^c can be taken to be multiples of the Planck length. The real number constant spatial parameters L_α^c are divided into $\overline{\pi}_\alpha^{lab}$ parts of length $\Delta\overline{L}_\alpha^{lab}$ where the complex number spatial segmentation number $\overline{\pi}_\alpha^{lab}$ is represented as[9]

$$\overline{\pi}_\alpha^{lab} = \pi_\alpha^{lab}\exp(j\theta_{\pi\alpha}^{lab}) \tag{1}$$

so that

$$\Delta\overline{L}_\alpha^{lab} = \Delta L_\alpha^{lab}\exp(j\theta_{\Delta L\alpha}^{lab}) = L_\alpha^c/\overline{\pi}_\alpha^{lab} = L_\alpha^c/\pi_\alpha^{lab}\exp(-j\theta_{\pi\alpha}^{lab}) \tag{2}$$

and

$$\Delta L_\alpha^{lab} = L_\alpha^c/\pi_\alpha^{lab} \qquad \theta_{\Delta L\alpha}^{lab} = -\theta_{\pi\alpha}^{lab} \tag{3}$$

where $\alpha = x,y,z;\ r,z$ or ρ. Then the complex number space coordinate $\overline{\alpha}^{lab}$ can be written as

$$\overline{\alpha}^{lab} = \alpha^{lab}\exp(j\theta_\alpha^{lab}) = \overline{\eta}_\alpha^{lab}\Delta\overline{L}_\alpha^{lab} = \overline{\eta}_\alpha^{lab}L_\alpha^c/\overline{\pi}_\alpha^{lab} = \eta_\alpha^{lab}L_\alpha^c/\pi_\alpha^{lab}\exp[j(\theta_{\eta\alpha}^{lab}-\theta_{\pi\alpha}^{lab})] \tag{4}$$

where the complex number spatial counting number $\overline{\eta}_\alpha^{lab}$ is represented as

$$\overline{\eta}_\alpha^{lab} = \eta_\alpha^{lab}\exp(j\theta_{\eta\alpha}^{lab}) \tag{5}$$

so that

$$\alpha^{lab} = \eta_\alpha^{lab}L_\alpha^c/\pi_\alpha^{lab} = \eta_\alpha^{lab}\Delta L_\alpha^{lab} \qquad \theta_\alpha^{lab} = \theta_{\eta\alpha}^{lab} - \theta_{\pi\alpha}^{lab} \tag{6}$$

where η_α^{lab} and π_α^{lab} are not integers in the presence of an APF but are integers under zero APF conditions.

The time coordinate factor, which is analogous to that of the space coordinate factor, is developed as follows. Consider a real number constant L_t^c that is used to measure intervals of the complex number time coordinate $\overline{t}^{lab}$, where L_t^c can be taken to be multiples of the Planck time. In the presence of an APF in bulk matter the real number constant temporal parameter L_t^c is divided into $\overline{\pi}_t^{lab}$ parts of length $\Delta\overline{L}_t^{lab}$ where the complex number time segmentation number is represented as[9]

$$\overline{\pi}_t^{lab} = \pi_t^{lab} \exp(j\theta_{\pi t}^{lab}) \tag{7}$$

and where

$$\Delta\overline{L}_t^{lab} = \Delta L_t^{lab} \exp(j\theta_{\Delta Lt}^{lab}) = L_t^c / \overline{\pi}_t^{lab} = L_t^c / \pi_t^{lab} \exp(-j\theta_{\pi t}^{lab}) \tag{8}$$

and therefore

$$\Delta L_t^{lab} = L_t^c / \pi_t^{lab} \qquad \theta_{\Delta Lt}^{lab} = -\theta_{\pi t}^{lab} \tag{9}$$

The complex number time coordinate is then written as

$$\overline{t}^{lab} = t^{lab} \exp(j\theta_t^{lab}) = \overline{\eta}_t^{lab} \Delta\overline{L}_t^{lab} = \overline{\eta}_t^{lab} L_t^c / \overline{\pi}_t^{lab} = \eta_t^{lab} L_t^c / \pi_t^{lab} \exp[j(\theta_{\eta t}^{lab} - \theta_{\pi t}^{lab})] \tag{10}$$

where the complex number time counting number is written as

$$\overline{\eta}_t^{lab} = \eta_t^{lab} \exp(j\theta_{\eta t}^{lab}) \tag{11}$$

so that

$$t^{lab} = \eta_t^{lab} L_\alpha^c / \pi_t^{lab} = \eta_t^{lab} \Delta L_t^{lab} \qquad \theta_t^{lab} = \theta_{\eta t}^{lab} - \theta_{\pi t}^{lab} \tag{12}$$

where η_t^{lab} and π_t^{lab} are not integers when an APF is present in matter, and the complex number nature of the time coordinate is $\overline{t}^{lab}$ due only to the complex number time counting number $\overline{\eta}_t^{lab}$ and the complex number segmentation number $\overline{\pi}_t^{lab}$ because the constant L_t^c is a real number in equation (10). The real number constants L_α^c and L_t^c that appear in equations (4) and (10) can be taken to be multiples (generally very large multiples) of the Planck length and Planck time respectively because an APF is a field that induces internal phase angles in all physical quantities except the fundamental constants of nature (h, c and G). Equations (6) and (12) allow the possibility of $\theta_\alpha^{lab} \leq 0$ and $\theta_t^{lab} \leq 0$ which has been suggested is the case for the Earth's gravity field.[6-12]

The space and time coordinates in vacuum or matter that is located in an APF are also represented as complex numbers of the form[5-10]

$$\overline{\alpha}^{lab} = \alpha^{lab} \exp(j\theta_\alpha^{lab}) \qquad \overline{t}^{lab} = t^{lab} \exp(j\theta_t^{lab}) \tag{13}$$

where the linear coordinate magnitudes are α = x,y,z; r,z or ρ for Cartesian coordinates, cylindrical polar coordinates or spherical polar coordinates respectively. For instance equation (13) can describe the space and time coordinates within an atomic nucleus in an ambient phase γ ray field. The space and time coordinate magnitudes and internal phase angles for regions of space and time within matter and in an APF are given by

$$\alpha^{lab} = \alpha_o^{lab} C_{\sigma\alpha}^{lab} \qquad \theta_\alpha^{lab} = \theta_{\eta\alpha}^{lab} - \theta_{\pi\alpha}^{lab} \tag{14}$$

$$t^{lab} = t_o^{lab} C_{\sigma t}^{lab} \qquad \theta_t^{lab} = \theta_{\eta t}^{lab} - \theta_{\pi t}^{lab} \tag{15}$$

where α_o^{lab} and t_o^{lab} = values of the space and time coordinates for zero APF conditions within the polyphase space vehicle which is stationary in the fixed laboratory system, σ = 1,2,3 and where the space and time coordinate factors $C_{\sigma\alpha}^{lab}$ and $C_{\sigma t}^{lab}$ are given by[9]

$$C_{1\alpha}^{lab} = \cos\theta_{\eta\alpha}^{lab} \sec\theta_{\pi\alpha}^{lab} \qquad C_{1t}^{lab} = \cos\theta_{\eta t}^{lab} \sec\theta_{\pi t}^{lab} \qquad [hn_o^{lab}] \tag{16}$$

$$C_{2\alpha}^{lab} = \cos\theta_{\eta\alpha}^{lab} \cos\theta_{\pi\alpha}^{lab} \qquad C_{2t}^{lab} = \cos\theta_{\eta t}^{lab} \cos\theta_{\pi t}^{lab} \qquad [ln_o^{lab}] \tag{17}$$

$$C_{3\alpha}^{lab} = \sec\theta_{\eta\alpha}^{lab} \cos\theta_{\pi\alpha}^{lab} \qquad C_{3t}^{lab} = \sec\theta_{\eta t}^{lab} \cos\theta_{\pi t}^{lab} \qquad [ln_o^{lab}] \tag{18}$$

where $[hn_o^{lab}]$ = high-density polyphase space vehicle and $[ln_o^{lab}]$ = low-density polyphase space vehicle and where $\theta_{\eta\alpha}^{lab}, \theta_{\pi\alpha}^{lab}$ and $\theta_{\eta t}^{lab}, \theta_{\pi t}^{lab}$ are the internal phase angles of the complex number space and time counting and segmentation numbers. For bound matter in an APF

$$\theta_{\pi\alpha}^{lab} \geq \theta_{\eta\alpha}^{lab} \qquad \theta_{\pi t}^{lab} \geq \theta_{\eta t}^{lab} \tag{19}$$

and for the conditions in equation (19), equations (14) and (15) give[5-10]

$$\theta_{\alpha}^{lab} \leq 0 \qquad \theta_{t}^{lab} \leq 0 \tag{20}$$

which is true for bound systems. For spherical polar coordinates and for the radial coordinate $\alpha = \rho$ and equation (14) becomes

$$\alpha^{lab} = \alpha_o^{lab} C_{\sigma\alpha}^{lab} \qquad \theta_{\alpha}^{lab} = \theta_{\eta\alpha}^{lab} - \theta_{\pi\alpha}^{lab} \tag{21}$$

and the space coordinate factors $C_{\sigma\rho}^{lab}$ are[9]

$$C_{1\rho}^{lab} = \cos\theta_{\eta\rho}^{lab} \sec\theta_{\pi\rho}^{lab} \qquad [hn_o^{lab}] \tag{22}$$

$$C_{2\rho}^{lab} = \cos\theta_{\eta\rho}^{lab} \cos\theta_{\pi\rho}^{lab} \qquad [ln_o^{lab}] \tag{23}$$

$$C_{3\rho}^{lab} = \sec\theta_{\eta\rho}^{lab} \cos\theta_{\pi\rho}^{lab} \qquad [ln_o^{lab}] \tag{24}$$

The values of the space coordinate factors $C_{\sigma\rho}^{lab}$ for the radial coordinate of a spherical polar coordinate system can also be simply expressed in terms of the internal phase angles of the complex particle number and the complex particle number density.

It has been shown that for a finite sized element of solid, liquid or gaseous matter in an APF the complex number particle number density is written as[9]

$$\bar{n}^{lab} = n^{lab} \exp(j\theta_n^{lab}) = \bar{\aleph}_{lab}/\bar{V}_{lab} \qquad n^{lab} = \aleph^{lab} / V^{lab} \qquad \theta_n^{lab} = \theta_{\aleph}^{lab} - \theta_V^{lab} \tag{25}$$

where the complex number total particle number is written as

$$\bar{\aleph}_{lab} = \aleph^{lab} \exp(j\theta_{\aleph}^{lab}) \tag{26}$$

where

$$\mathfrak{N}^{lab} = N\cos\theta_{\mathfrak{N}}^{lab} \qquad [SN] \qquad (27)$$

$$\mathfrak{N}^{lab} = N\sec\theta_{\mathfrak{N}}^{lab} \qquad [LN] \qquad (28)$$

where [SN] = small N polyphase space vehicle, [LN] = large N polyphase space vehicle and where N = integer number of particles in the various systems of a polyphase space vehicle which interact with the APF. For a gravitational APF, N = total integer number of particles in a system, while for an electromagnetic APF, N = the net integer number of charged particles in a system.[6-10] The internal phase angle $\theta_{\mathfrak{N}}^{lab}$ can be calculated for an elastic body in a static electromagnetic field as[6-10]

$$\tan\theta_{\mathfrak{N}}^{lab} = (2\mathfrak{E}_{S}^{em} / G_o)^{1/2} \qquad (29)$$

where $\mathfrak{E}_{S}^{em}$ = energy density of a shearing electromagnetic APF and G_o = zero APF value of average shear modulus of polyphase space vehicle. If dynamical effects must be considered the replacement $G_o \rightarrow G_r$ can be made in equation (29) where G_r = effective dynamical shear modulus which has different values for an applied electromagnetic or gravitational APF and may correspond to a particular resonance frequency of a finite system.[6-10] The complex number volume is represented as

$$\overline{V}^{lab} = V^{lab}\exp(j\theta_V^{lab}) \qquad \theta_V^{lab} = \theta_{\mathfrak{N}}^{lab} - \theta_n^{lab} \qquad \theta_V^{lab} \leq 0 \qquad (30)$$

For homogeneous bound matter in an ambient 3-dimensional APF the internal phase angles of the space coordinates are then written as[6-10]

$$\theta_x^{lab} = \theta_y^{lab} = \theta_z^{lab} = \theta_r^{lab} = \theta_\rho^{lab} = 1/3\theta_V^{lab} = 1/3(\theta_{\mathfrak{N}}^{lab} - \theta_n^{lab}) \leq 0 \qquad (31)$$

while for a uniaxial APF in the x-direction

$$\theta_x^{lab} = \theta_V^{lab} = \theta_{\mathfrak{N}}^{lab} - \theta_n^{lab} \leq 0 \qquad (32)$$

From equations (25) through (30) it follows that[11]

$$n^{lab} = N / V^{lab}\cos\theta_{\mathfrak{N}}^{lab} \qquad [SN] \qquad (33)$$

$$n^{lab} = N / V^{lab}\sec\theta_{\mathfrak{N}}^{lab} \qquad [LN] \qquad (34)$$

$$\theta_n^{lab} = \theta_{\mathfrak{N}}^{lab} - \theta_V^{lab} = \theta_{\mathfrak{N}}^{lab} + \left|\theta_V^{lab}\right| \qquad \theta_n^{lab} \geq \theta_{\mathfrak{N}}^{lab} \qquad (35)$$

where for, for example, for spherical polar coordinates the volume element V^{lab} in the presence of a 3-dimensional homogeneous APF is given in terms of the volume element V_o^{lab} under zero APF conditions by[9]

$$V^{lab} / V_o^{lab} = (C_{1\rho}^{lab})^3 = \cos\theta_{\mathfrak{N}}^{lab}\sec\theta_n^{lab} \geq 1 \qquad [hn_o^{lab}, SN] \qquad (36)$$

$$V^{lab} / V_o^{lab} = (C_{1\rho}^{lab})^3 = \sec\theta_{\mathfrak{N}}^{lab}\sec\theta_n^{lab} \geq 1 \qquad [hn_o^{lab}, LN] \qquad (37)$$

$$V^{lab} / V_o^{lab} = (C_{2\rho}^{lab})^3 = \cos\theta_{\mathcal{N}}^{lab} \cos\theta_n^{lab} \leq 1 \qquad [ln_o^{lab}, SN] \tag{38}$$

$$V^{lab} / V_o^{lab} = (C_{2\rho}^{lab})^3 = \sec\theta_{\mathcal{N}}^{lab} \cos\theta_n^{lab} \leq 1 \qquad [ln_o^{lab}, LN] \tag{39}$$

$$V^{lab} / V_o^{lab} = (C_{3\rho}^{lab})^3 = \cos\theta_{\mathcal{N}}^{lab} \cos\theta_n^{lab} \leq 1 \qquad [ln_o^{lab}, SN] \tag{40}$$

$$V^{lab} / V_o^{lab} = (C_{3\rho}^{lab})^3 = \sec\theta_{\mathcal{N}}^{lab} \cos\theta_n^{lab} \leq 1 \qquad [ln_o^{lab}, LN] \tag{41}$$

where the space coordinate factors $C_{\sigma\rho}^{lab}$ for spherical polar coordinates are given by equations (22) through (24), and so because the right hand sides of equations (38) and (39) are the same as the right hand sides of equations (40) and (41) it follows that the space coordinate factors $C_{2\rho}^{lab}$ and $C_{3\rho}^{lab}$ of equations (23) and (24) give two possible solutions for the internal phase angles $\theta_{\eta\rho}^{lab}$ and $\theta_{\eta\rho}^{lab}$ of the counting and segmentation numbers for the [ld] case and therefore the [ld] case is degenerate.. Then equations (33), (34) and (36) through (41) give[11]

$$n^{lab} = n_o^{lab} \cos\theta_n^{lab} \leq n_o^{lab} \qquad [hn_o^{lab}] \tag{42}$$

$$n^{lab} = n_o^{lab} \sec\theta_n^{lab} \geq n_o^{lab} \qquad [ln_o^{lab}] \tag{43}$$

where n_o^{lab} = particle number density under zero APF conditions and which is given by

$$n_o^{lab} = N / V_o^{lab} \tag{44}$$

which is the conventional result.

Cartesian Coordinates.

For Cartesian coordinates the ratios of the volume elements with and without an APF present are given by

$$V^{lab} / V_o^{lab} = C_{1x}^{lab} C_{1y}^{lab} C_{1z}^{lab} = \cos\theta_{\mathcal{N}}^{lab} \sec\theta_n^{lab} \geq 1 \qquad [hn_o^{lab}, SN] \tag{45}$$

$$V^{lab} / V_o^{lab} = C_{1x}^{lab} C_{1y}^{lab} C_{1z}^{lab} = \sec\theta_{\mathcal{N}}^{lab} \sec\theta_n^{lab} \geq 1 \qquad [hn_o^{lab}, LN] \tag{46}$$

$$V^{lab} / V_o^{lab} = C_{\beta x}^{lab} C_{\beta y}^{lab} C_{\beta z}^{lab} = \cos\theta_{\mathcal{N}}^{lab} \cos\theta_n^{lab} \leq 1 \qquad [ln_o^{lab}, SN] \tag{47}$$

$$V^{lab} / V_o^{lab} = C_{\beta x}^{lab} C_{\beta y}^{lab} C_{\beta z}^{lab} = \sec\theta_{\mathcal{N}}^{lab} \cos\theta_n^{lab} \leq 1 \qquad [ln_o^{lab}, LN] \tag{48}$$

where $C_{1\alpha}^{lab}$ and $C_{\beta\alpha}^{lab}$ are given by equations (16) through (18) and where $\beta = 2,3$ describes the degeneracy of the $[ln_o^{lab}]$ state. For the case of a uniaxial APF in the x-direction, $C_{1y}^{lab} = 1$, $C_{1z}^{lab} = 1$, $C_{\beta y}^{lab} = 1$ and $C_{\beta z}^{lab} = 1$ with the result that equations (45) through (48) give

$$V^{lab} / V_o^{lab} = C_{1x}^{lab} = \cos\theta_{\mathcal{N}}^{lab} \sec\theta_n^{lab} \geq 1 \qquad [hn_o^{lab}, SN] \tag{49}$$

$$V^{lab} / V_o^{lab} = C_{1x}^{lab} = \sec\theta_{\mathcal{N}}^{lab} \sec\theta_n^{lab} \geq 1 \qquad [hn_o^{lab}, LN] \tag{50}$$

$$V^{lab} / V_o^{lab} = C_{\beta x}^{lab} = \cos\theta_{\mathcal{N}}^{lab} \cos\theta_n^{lab} \leq 1 \qquad [ln_o^{lab}, SN] \qquad (51)$$

$$V^{lab} / V_o^{lab} = C_{\beta x}^{lab} = \sec\theta_{\mathcal{N}}^{lab} \cos\theta_n^{lab} \leq 1 \qquad [ln_o^{lab}, LN] \qquad (52)$$

where $\beta = 2,3$.

Cylindrical Polar Coordinates.

For the case of cylindrical polar coordinates the ratios of the volume elements with and without an acting APF are given by

$$V^{lab} / V_o^{lab} = (C_{1r}^{lab})^2 C_{1z}^{lab} = \cos\theta_{\mathcal{N}}^{lab} \sec\theta_n^{lab} \geq 1 \qquad [hn_o^{lab}, SN] \qquad (53)$$

$$V^{lab} / V_o^{lab} = (C_{1r}^{lab})^2 C_{1z}^{lab} = \sec\theta_{\mathcal{N}}^{lab} \sec\theta_n^{lab} \geq 1 \qquad [hn_o^{lab}, LN] \qquad (54)$$

$$V^{lab} / V_o^{lab} = (C_{\beta r}^{lab})^2 C_{\beta z}^{lab} = \cos\theta_{\mathcal{N}}^{lab} \cos\theta_n^{lab} \leq 1 \qquad [ln_o^{lab}, SN] \qquad (55)$$

$$V^{lab} / V_o^{lab} = (C_{\beta r}^{lab})^2 C_{\beta z}^{lab} = \sec\theta_{\mathcal{N}}^{lab} \cos\theta_n^{lab} \leq 1 \qquad [ln_o^{lab}, LN] \qquad (56)$$

where $\beta = 2,3$. For the case of a uniaxial APF along the z-axis, $C_{1r}^{lab} = 1$ and $C_{\beta r}^{lab} = 1$ and equations (53) through (56) become equations (49) through (52) with x replaced by z. For the special case of a two-dimensional radial APF with $C_{1z}^{lab} = 1$ and $C_{\beta z}^{lab} = 1$ and equations (53) through (56) give

$$(V^{lab} / V_o^{lab})^{1/2} = C_{1r}^{lab} = \cos^{1/2}\theta_{\mathcal{N}}^{lab} \sec^{1/2}\theta_n^{lab} \geq 1 \qquad [hn_o^{lab}, SN] \qquad (57)$$

$$(V^{lab} / V_o^{lab})^{1/2} = C_{1r}^{lab} = \sec^{1/2}\theta_{\mathcal{N}}^{lab} \sec^{1/2}\theta_n^{lab} \geq 1 \qquad [hn_o^{lab}, LN] \qquad (58)$$

$$(V^{lab} / V_o^{lab})^{1/2} = C_{\beta r}^{lab} = \cos^{1/2}\theta_{\mathcal{N}}^{lab} \cos^{1/2}\theta_n^{lab} \leq 1 \qquad [ln_o^{lab}, SN] \qquad (59)$$

$$(V^{lab} / V_o^{lab})^{1/2} = C_{\beta r}^{lab} = \sec^{1/2}\theta_{\mathcal{N}}^{lab} \cos^{1/2}\theta_n^{lab} \leq 1 \qquad [ln_o^{lab}, LN] \qquad (60)$$

Spherical Polar Coordinates.

$$(V^{lab} / V_o^{lab})^{1/3} = C_{1\rho}^{lab} = \cos^{1/3}\theta_{\mathcal{N}}^{lab} \sec^{1/3}\theta_n^{lab} \geq 1 \qquad [hn_o^{lab}, SN] \qquad (61)$$

$$(V^{lab} / V_o^{lab})^{1/3} = C_{1\rho}^{lab} = \sec^{1/3}\theta_{\mathcal{N}}^{lab} \sec^{1/3}\theta_n^{lab} \geq 1 \qquad [hn_o^{lab}, LN] \qquad (62)$$

$$(V^{lab} / V_o^{lab})^{1/3} = C_{\beta\rho}^{lab} = \cos^{1/3}\theta_{\mathcal{N}}^{lab} \cos^{1/3}\theta_n^{lab} \leq 1 \qquad [ln_o^{lab}, SN] \qquad (63)$$

$$(V^{lab} / V_o^{lab})^{1/3} = C_{\beta\rho}^{lab} = \sec^{1/3}\theta_{\mathcal{N}}^{lab} \cos^{1/3}\theta_n^{lab} \leq 1 \qquad [ln_o^{lab}, LN] \qquad (64)$$

where $\beta = 1,2$. Note that equations (61) through (64) for spherical polar coordinates can also be obtained from equations (45) through (48) for a Cartesian coordinate system by taking $C_{1x}^{lab} = C_{1y}^{lab} = C_{1z}^{lab}$ and $C_{\beta x}^{lab} = C_{\beta y}^{lab} = C_{\beta z}^{lab}$ and from equations (53) through (56) for cylindrical polar coordinates by taking $C_{1r}^{lab} = C_{1z}^{lab}$ and $C_{\beta r}^{lab} = C_{\beta z}^{lab}$.

Consider now the case of a linear elastic system for which the internal phase angles θ_n^{lab} and $\theta_{\mathcal{N}}^{lab}$ are related by[9]

$$\cos\theta_n^{lab} = 1 - \tan\theta_{\mathcal{N}}^{lab} \qquad 0 \le \theta_n^{lab} \le \pi/2 \qquad 0 \le \theta_{\mathcal{N}}^{lab} \le \pi/4 \qquad [hn_o^{lab}] \tag{65}$$

$$\sec\theta_n^{lab} = 1 + \tan\theta_{\mathcal{N}}^{lab} \qquad 0 \le \theta_n^{lab} \le \pi/2 \qquad 0 \le \theta_{\mathcal{N}}^{lab} \le \pi/2 \qquad [ln_o^{lab}] \tag{66}$$

with $\theta_n^{lab} \ge \theta_{\mathcal{N}}^{lab}$, and which can now be used in equations (45) through (64) to describe the space coordinate factors for a linear elastic system in an APF.

<u>Cartesian Coordinates</u>

Combining equations (45) through (48) with equations (65) and (66) gives for the case of a 3-dimensional APF in Cartesian coordinates

$$C_{1x}^{lab}C_{1y}^{lab}C_{1z}^{lab} = \sec\theta_n^{lab}[1+(1-\cos\theta_n^{lab})^2]^{-1/2} = \cos\theta_{\mathcal{N}}^{lab}(1-\tan\theta_{\mathcal{N}}^{lab})^{-1} \qquad [hn_o^{lab}, SN] \tag{67}$$

$$C_{1x}^{lab}C_{1y}^{lab}C_{1z}^{lab} = \sec\theta_n^{lab}[1+(1-\cos\theta_n^{lab})^2]^{1/2} = \sec\theta_{\mathcal{N}}^{lab}(1-\tan\theta_{\mathcal{N}}^{lab})^{-1} \qquad [hn_o^{lab}, LN] \tag{68}$$

$$C_{\beta x}^{lab}C_{\beta y}^{lab}C_{\beta z}^{lab} = \cos\theta_n^{lab}[1+(\sec\theta_n^{lab}-1)^2]^{-1/2} = \cos\theta_{\mathcal{N}}^{lab}(1+\tan\theta_{\mathcal{N}}^{lab})^{-1} \qquad [ln_o^{lab}, SN] \tag{69}$$

$$C_{\beta x}^{lab}C_{\beta y}^{lab}C_{\beta z}^{lab} = \cos\theta_n^{lab}[1+(\sec\theta_n^{lab}-1)^2]^{1/2} = \sec\theta_{\mathcal{N}}^{lab}(1+\tan\theta_{\mathcal{N}}^{lab})^{-1} \qquad [ln_o^{lab}, LN] \tag{70}$$

where $\beta = 2,3$. Combining equations (49) through (52) with equations (65) and (66) gives for a uniaxial APF in a linear elastic body

$$C_{1x}^{lab} = \sec\theta_n^{lab}[1+(1-\cos\theta_n^{lab})^2]^{-1/2} = \cos\theta_{\mathcal{N}}^{lab}(1-\tan\theta_{\mathcal{N}}^{lab})^{-1} \qquad [hn_o^{lab}, SN] \tag{71}$$

$$C_{1x}^{lab} = \sec\theta_n^{lab}[1+(1-\cos\theta_n^{lab})^2]^{1/2} = \sec\theta_{\mathcal{N}}^{lab}(1-\tan\theta_{\mathcal{N}}^{lab})^{-1} \qquad [hn_o^{lab}, LN] \tag{72}$$

$$C_{\beta x}^{lab} = \cos\theta_n^{lab}[1+(\sec\theta_n^{lab}-1)^2]^{-1/2} = \cos\theta_{\mathcal{N}}^{lab}(1+\tan\theta_{\mathcal{N}}^{lab})^{-1} \qquad [ln_o^{lab}, SN] \tag{73}$$

$$C_{\beta x}^{lab} = \cos\theta_n^{lab}[1+(\sec\theta_n^{lab}-1)^2]^{1/2} = \sec\theta_{\mathcal{N}}^{lab}(1+\tan\theta_{\mathcal{N}}^{lab})^{-1} \qquad [ln_o^{lab}, LN] \tag{74}$$

for Cartesian coordinates.

<u>Cylindrical Polar Coordinates.</u>

Combining equations (53) through (56) with equations (65) and (66) gives for the case of a 3-dimensional APF

$$(C_{1r}^{lab})^2C_{1z}^{lab} = \sec\theta_n^{lab}[1+(1-\cos\theta_n^{lab})^2]^{-1/2} = \cos\theta_{\mathcal{N}}^{lab}(1-\tan\theta_{\mathcal{N}}^{lab})^{-1} \qquad [hn_o^{lab}, SN] \tag{75}$$

$$(C_{1r}^{lab})^2C_{1z}^{lab} = \sec\theta_n^{lab}[1+(1-\cos\theta_n^{lab})^2]^{1/2} = \sec\theta_{\mathcal{N}}^{lab}(1-\tan\theta_{\mathcal{N}}^{lab})^{-1} \qquad [hn_o^{lab}, LN] \tag{76}$$

$$(C^{lab}_{\beta r})^2 C^{lab}_{\beta z} = \cos\theta^{lab}_n[1+(\sec\theta^{lab}_n - 1)^2]^{-1/2} = \cos\theta^{lab}_{\mathcal{N}}(1+\tan\theta^{lab}_{\mathcal{N}})^{-1} \qquad [ln^{lab}_o, SN] \qquad (77)$$

$$(C^{lab}_{\beta r})^2 C^{lab}_{\beta z} = \cos\theta^{lab}_n[1+(\sec\theta^{lab}_n - 1)^2]^{1/2} = \sec\theta^{lab}_{\mathcal{N}}(1+\tan\theta^{lab}_{\mathcal{N}})^{-1} \qquad [ln^{lab}_o, LN] \qquad (78)$$

For a uniaxial APF, $C^{lab}_{1r} = 1$ and $C^{lab}_{\beta r} = 1$ and equations (75) through (78) become just the uniaxial APF equations (71) through (74) but now for the z-axis. For the special case of a 2-dimensional radial APF, $C^{lab}_{1z} = 1$ and $C^{lab}_{\beta z} = 1$ and equations (75) through (78) become

$$C^{lab}_{1r} = \sec^{1/2}\theta^{lab}_n[1+(1-\cos\theta^{lab}_n)^2]^{-1/4} = \cos^{1/2}\theta^{lab}_{\mathcal{N}}(1-\tan\theta^{lab}_{\mathcal{N}})^{-1/2} \qquad [hn^{lab}_o, SN] \qquad (79)$$

$$C^{lab}_{1r} = \sec^{1/2}\theta^{lab}_n[1+(1-\cos\theta^{lab}_n)^2]^{1/4} = \sec^{1/2}\theta^{lab}_{\mathcal{N}}(1-\tan\theta^{lab}_{\mathcal{N}})^{-1/2} \qquad [hn^{lab}_o, LN] \qquad (80)$$

$$C^{lab}_{\beta r} = \cos^{1/2}\theta^{lab}_n[1+(\sec\theta^{lab}_n - 1)^2]^{-1/4} = \cos^{1/2}\theta^{lab}_{\mathcal{N}}(1+\tan\theta^{lab}_{\mathcal{N}})^{-1/2} \qquad [ln^{lab}_o, SN] \qquad (81)$$

$$C^{lab}_{\beta r} = \cos^{1/2}\theta^{lab}_n[1+(\sec\theta^{lab}_n - 1)^2]^{1/4} = \sec^{1/2}\theta^{lab}_{\mathcal{N}}(1+\tan\theta^{lab}_{\mathcal{N}})^{-1/2} \qquad [ln^{lab}_o, LN] \qquad (82)$$

where $\beta = 2,3$.

Spherical Polar Coordinates.

Combining equations (61) through (64) with equations (65) and (66) gives for the case of a 3-dimensional APF in spherical polar coordinates

$$C^{lab}_{1\rho} = \sec^{1/3}\theta^{lab}_n[1+(1-\cos\theta^{lab}_n)^2]^{-1/6} = \cos^{1/3}\theta^{lab}_{\mathcal{N}}(1-\tan\theta^{lab}_{\mathcal{N}})^{-1/3} \qquad [hn^{lab}_o, SN] \qquad (83)$$

$$C^{lab}_{1\rho} = \sec^{1/3}\theta^{lab}_n[1+(1-\cos\theta^{lab}_n)^2]^{1/6} = \sec^{1/3}\theta^{lab}_{\mathcal{N}}(1-\tan\theta^{lab}_{\mathcal{N}})^{-1/3} \qquad [hn^{lab}_o, LN] \qquad (84)$$

$$C^{lab}_{\beta\rho} = \cos^{1/3}\theta^{lab}_n[1+(\sec\theta^{lab}_n - 1)^2]^{-1/6} = \cos^{1/3}\theta^{lab}_{\mathcal{N}}(1+\tan\theta^{lab}_{\mathcal{N}})^{-1/3} \qquad [ln^{lab}_o, SN] \qquad (85)$$

$$C^{lab}_{\beta\rho} = \cos^{1/3}\theta^{lab}_n[1+(\sec\theta^{lab}_n - 1)^2]^{1/6} = \sec^{1/3}\theta^{lab}_{\mathcal{N}}(1+\tan\theta^{lab}_{\mathcal{N}})^{-1/3} \qquad [ln^{lab}_o, LN] \qquad (86)$$

where $\beta = 2,3$. Equations (83) through (86) can also be obtained from equations (67) through (70) for Cartesian coordinates by taking $C^{lab}_{1x} = C^{lab}_{1y} = C^{lab}_{1z}$ and $C^{lab}_{\beta x} = C^{lab}_{\beta y} = C^{lab}_{\beta z}$.

The time coordinate factor for a polyphase space vehicle which is located in the stationary laboratory system and as seen by an observer in the stationary laboratory frame of reference can be obtained from the definitions of a characteristic average linear velocity s^{sv}_o of a mechanical, acoustic or electromagnetic wave in the materials of the polyphase space vehicle under zero APF conditions

$$s^{lab}_o = d\alpha^{lab}_o / dt^{lab}_o \qquad (87)$$

and for the case when a constant mechanical, acoustic or electromagnetic APF is present

$$\overline{\alpha}^{lab} = \alpha^{lab}\exp(j\theta^{lab}_\alpha) \qquad \overline{t}^{lab} = t^{lab}\exp(j\theta^{lab}_t) \qquad (88)$$

$$\bar{s}^{lab} = s^{lab} \exp(j\theta_s^{lab}) = d\bar{\alpha}^{lab} / d\bar{t}^{lab} \tag{89}$$

$$s^{lab} = d\alpha^{lab} / dt^{lab} = (d\alpha_o^{lab} / dt_o^{lab})(C_{\sigma\alpha}^{lab} / C_{\sigma t}^{lab}) = s_o^{lab}(C_{\sigma\alpha}^{lab} / C_{\sigma t}^{lab}) \tag{90}$$

$$\theta_s^{lab} = \theta_\alpha^{lab} - \theta_t^{lab} = \theta_{\eta\alpha}^{lab} - \theta_{\pi\alpha}^{lab} - (\theta_{\eta t}^{lab} - \theta_{\pi t}^{lab}) \tag{91}$$

so that the constant time coordinate factor $C_{\sigma t}^{lab}$ is related to the constant space coordinate factor $C_{\sigma\alpha}^{lab}$ by

$$C_{\sigma t}^{lab} = (s_o^{lab} / s^{lab}) C_{\sigma\alpha}^{lab} \tag{92}$$

where $\sigma = 1,2,3$; $\alpha = x,y,z$ or r,z or ρ and where s_o^{lab} = characteristic velocity of a mechanical, acoustic or electromagnetic wave propagating in the polyphase space vehicle which is resting in the ststionary laboratory frame of reference under zero APF conditions and s^{lab} = magnitude of the average velocity of a mechanical, acoustic or electromagnetic wave in the polyphase space vehicle which is fixed in the stationary laboratory frame of reference and in the presence of an APF. The constant space and time coordinate factors $C_{\sigma\alpha}^{lab}$ and $C_{\sigma t}^{lab}$ are expected to depend only on the properties of the polyphase space vehicle through the constant internal phase angles that appear in equation (91).

B. Forms of the Apparent Space and Time Coordinate Factors for a Moving Polyphase space vehicle as Seen From the Perspective of a Stationary System of Coordinates.

The space and time coordinates within matter or vacuum that is located in an APF have been represented as complex numbers in an internal space.[5-12] A simple theory of the origin and nature of these complex number coordinates has already been developed in a previously published paper.[9] Consider the constant real number spatial parameters L_α^c for the complex number space coordinates $\bar{\alpha}$ in an APF in vacuum, where L_α^c can be taken to be multiples of the Planck length. The real number constant spatial parameters L_α^c are divided into $\bar{\pi}_\alpha$ parts of length $\Delta\bar{L}_\alpha$ where the complex number spatial segmentation number $\bar{\pi}_\alpha$ is represented as[9]

$$\bar{\pi}_\alpha = \pi_\alpha \exp(j\theta_{\pi\alpha}) \tag{93}$$

so that

$$\Delta\bar{L}_\alpha = \Delta L_\alpha \exp(j\theta_{\Delta L\alpha}) = L_\alpha^c / \bar{\pi}_\alpha = L_\alpha^c / \pi_\alpha \exp(-j\theta_{\pi\alpha}) \tag{94}$$

and

$$\Delta L_\alpha = L_\alpha^c / \pi_\alpha \qquad \theta_{\Delta L\alpha} = -\theta_{\pi\alpha} \tag{95}$$

where for linear coordinates, $\alpha = x,y,z$; r,z or ρ. Then the complex number space coordinate $\bar{\alpha}$ can be written as

$$\bar{\alpha} = \alpha \exp(j\theta_\alpha) = \bar{\eta}_\alpha \Delta\bar{L}_\alpha = \bar{\eta}_\alpha L_\alpha^c / \bar{\pi}_\alpha = \eta_\alpha L_\alpha^c / \pi_\alpha \exp[j(\theta_{\eta\alpha} - \theta_{\pi\alpha})] \tag{96}$$

where the complex number spatial counting number is represented as

$$\overline{\eta}_\alpha = \eta_\alpha \exp(j\theta_{\eta\alpha}) \qquad (97)$$

so that

$$\alpha = \eta_\alpha L^c_\alpha / \imath_\alpha = \eta_\alpha \Delta L_\alpha \qquad \theta_\alpha = \theta_{\eta\alpha} - \theta_{\imath\alpha} \qquad (98)$$

where η_α and $\imath_\alpha$ are not integers in the presence of an APF but are integers under zero APF conditions.

The time coordinate factor, which is analogous to that of the space coordinate factor, is developed as follows. Consider a real number constant L^c_t that is used to measure intervals of the complex number time coordinate $\overline{t}$, where L^c_t can be taken to be multiples of the Planck time. In the presence of an APF in vacuum the real number constant temporal parameter L^c_t is divided into $\overline{\imath}_t$ parts of length $\Delta\overline{L}_t$ where the complex number time segmentation number is represented as[9]

$$\overline{\imath}_t = \imath_t \exp(j\theta_{\imath t}) \qquad (99)$$

and where

$$\Delta\overline{L}_t = \Delta L_t \exp(j\theta_{\Delta Lt}) = L^c_t / \overline{\imath}_t = L^c_t / \imath_t \exp(-j\theta_{\imath t}) \qquad (100)$$

and therefore

$$\Delta L_t = L^c_t / \imath_t \qquad \theta_{\Delta Lt} = -\,\theta_{\imath t} \qquad (101)$$

The complex number time coordinate is then written as

$$\overline{t} = t\exp(j\theta_t) = \overline{\eta}_t \Delta\overline{L}_t = \overline{\eta}_t L^c_t / \overline{\imath}_t = \eta_t L^c_t / \imath_t \exp[j(\theta_{\eta t} - \theta_{\imath t})] \qquad (102)$$

where the complex number time counting number is written as

$$\overline{\eta}_t = \eta_t \exp(j\theta_{\eta t}) \qquad (103)$$

so that

$$t = \eta_t L^c_t / \imath_t = \eta_t \Delta L_t \qquad \theta_t = \theta_{\eta t} - \theta_{\imath t} \qquad (104)$$

where η_t and $\imath_t$ are not integers when an APF is present in vacuum, and the complex number nature of the time coordinate is $\overline{t}$ due only to the complex number time counting number $\overline{\eta}_t$ and the complex number segmentation number $\overline{\imath}_t$ because the constant L^c_t is a real number in equation (102). The real number constants L^c_α and L^c_t that appear in equations (96) and (102) can be taken to be multiples (generally very large multiples) of the Planck length and Planck time respectively because an APF is a field that induces internal phase angles in all physical quantities except the fundamental constants of nature (h, c and G). Equations (98) and (104) allow the possibility of $\theta_\alpha \leq 0$ and $\theta_t \leq 0$ which has been suggested is the case for the Earth's gravity field.[6-12]

The space and time coordinates in vacuum or matter that is located in an APF are also represented as complex numbers of the form[5-12]

$$\bar{\alpha} = \alpha \exp(j\theta_\alpha) \qquad \bar{t} = t\exp(j\theta_t) \tag{105}$$

where the coordinate magnitudes are α = x,y,z; r,z or ρ for linear Cartesian coordinates, cylindrical polar coordinates or spherical polar coordinates respectively. For instance equation (105) can describe the space and time coordinates within an atomic nucleus in an ambient phase γ ray field. The space and time coordinate magnitudes and internal phase angles for regions of space and time within vacuum and in an APF are given by

$$\alpha = \alpha_o C_{\sigma\alpha} \qquad \theta_\alpha = \theta_{\eta\alpha} - \theta_{\pi\alpha} \tag{106}$$

$$t = t_o C_{\sigma t} \qquad \theta_t = \theta_{\eta t} - \theta_{\pi t} \tag{107}$$

where α_o and t_o = values of the space and time coordinates for zero APF conditions, σ = 1,2,3 and where the space and time coordinate factors $C_{\sigma\alpha}$ and $C_{\sigma t}$ are given by[9]

$$C_{1\alpha} = \cos\theta_{\eta\alpha} \sec\theta_{\pi\alpha} \qquad C_{1t} = \cos\theta_{\eta t} \sec\theta_{\pi t} \qquad [hn_{po}] \qquad [hn^{o}_{Dst}] \tag{108}$$

$$C_{2\alpha} = \cos\theta_{\eta\alpha} \cos\theta_{\pi\alpha} \qquad C_{2t} = \cos\theta_{\eta t} \cos\theta_{\pi t} \qquad [ln_{po}] \qquad [ln^{o}_{Dst}] \tag{109}$$

$$C_{3\alpha} = \sec\theta_{\eta\alpha} \cos\theta_{\pi\alpha} \qquad C_{3t} = \sec\theta_{\eta t} \cos\theta_{\pi t} \qquad [ln_{po}] \qquad [ln^{o}_{Dst}] \tag{110}$$

where $[hn_{po}]$ and $[hn^{o}_{Dst}]$ = zero APF values of the high particle-antiparticle pair number density and high-dimensional number density of the vacuum of the vacuum respectively, and $[ln_{po}]$ and $[ln^{o}_{Dst}]$ = zero APF values of the low particle-antiparticle pair number density and low-dimensional number density of the vacuum of the vacuum resectively and where $\theta_{\eta\alpha}$, $\theta_{\pi\alpha}$ and $\theta_{\eta t}$, $\theta_{\pi t}$ are the internal phase angles of the complex number space and time counting and segmentation numbers. For bound matter in an APF

$$\theta_{\pi\alpha} \geq \theta_{\eta\alpha} \qquad \theta_{\pi t} \geq \theta_{\eta t} \tag{111}$$

and for the conditions in equation (111), equations (106) and (107) give[5-12]

$$\theta_\alpha \leq 0 \qquad \theta_t \leq 0 \tag{112}$$

The values of the space coordinate factors $C_{\sigma\alpha}$ can also be simply expressed in terms of either the internal phase angles of the complex particle-antiparticle pair number and pair number density or by the phase angles of the complex dimension number and dimension number density.[9,10]

It has been shown that for a finite sized element of vacuum in an APF the complex number particle-antiparticle number density $\bar{n}_p$ is written as[9]

$$\bar{n}_p = n_p \exp(j\theta_{np}) = \bar{\aleph}_p/\bar{V}_p \qquad n_p = \aleph_p / V_p \qquad \theta_{np} = \theta_{\aleph p} - \theta_{Vp} \tag{113}$$

where the complex number total particle-antiparticle pair number is written as

$$\overline{\mathfrak{N}_p} = \mathfrak{N}_p \exp(j\theta_{\mathfrak{N}p}) \tag{114}$$

where

$$\mathfrak{N}_p = N_p \cos\theta_{\mathfrak{N}p} \qquad [SN_p] \tag{115}$$

$$\mathfrak{N}_p = N_p \sec\theta_{\mathfrak{N}p} \qquad [LN_p] \tag{116}$$

where $[SN_p]$ = small N_p, $[LN_p]$ = large N_p and where N_p = integer number of particle-antiparticle pairs of a vacuum which interact with the APF. The complex number volume is represented as

$$\overline{V}_p = V_p \exp(j\theta_{Vp}) \qquad \theta_{Vp} = \theta_{\mathfrak{N}p} - \theta_{np} \qquad \theta_{Vp} \le 0 \tag{117}$$

and where the subscript p refers to pairs of particles and antiparticles. For homogeneous bound vacuum in a 3-dimensional APF the internal phase angles of the space coordinates are then written as

$$\theta_x = \theta_y = \theta_z = \theta_r = \theta_\rho = 1/3\theta_{Vp} = 1/3(\theta_{\mathfrak{N}p} - \theta_{np}) \le 0 \tag{118}$$

while for a uniaxial APF in the x-direction

$$\theta_x = \theta_{Vp} = \theta_{\mathfrak{N}p} - \theta_{np} \le 0 \tag{119}$$

From equations (113) through (117) it follows that[11]

$$n_p = N_p / V_p \cos\theta_{\mathfrak{N}p} \quad [SN_p] \tag{120}$$

$$n_p = N_p / V_p \sec\theta_{\mathfrak{N}p} \quad [LN_p] \tag{121}$$

$$\theta_{np} = \theta_{\mathfrak{N}p} - \theta_{Vp} = \theta_{\mathfrak{N}p} + \left|\theta_{Vp}\right| \qquad \theta_{np} \ge \theta_{\mathfrak{N}p} \tag{122}$$

where the volume element V_p in the presence of a 3-dimensional APF is given in terms of the volume element V_p^o under zero APF conditions by[9,10]

$$V_p / V_{po} = C_{1x}C_{1y}C_{1z} = C_{1r}^2 C_{1z} = C_{1\rho}^3 = \cos\theta_{\mathfrak{N}p} \sec\theta_{np} \ge 1 \qquad [hn_{po}, SN_p] \tag{123}$$

$$V_p / V_{po} = C_{1x}C_{1y}C_{1z} = C_{1r}^2 C_{1z} = C_{1\rho}^3 = \sec\theta_{\mathfrak{N}p} \sec\theta_{np} \ge 1 \qquad [hn_{po}, LN_p] \tag{124}$$

$$V_p / V_{po} = C_{2x}C_{2y}C_{2z} = C_{2r}^2 C_{2z} = C_{2\rho}^3 = \cos\theta_{\mathfrak{N}p} \cos\theta_{np} \le 1 \qquad [ln_{po}, SN_p] \tag{125}$$

$$V_p / V_{po} = C_{2x}C_{2y}C_{2z} = C_{2r}^2 C_{2z} = C_{2\rho}^3 = \sec\theta_{\mathfrak{N}p} \cos\theta_{np} \le 1 \qquad [ln_{po}, LN_p] \tag{126}$$

$$V_p / V_{po} = C_{3x}C_{3y}C_{3z} = C_{3r}^2 C_{3z} = C_{3\rho}^3 = \cos\theta_{\mathfrak{N}p} \cos\theta_{np} \le 1 \qquad [ln_{po}, SN_p] \tag{127}$$

$$V_p / V_{po} = C_{3x}C_{3y}C_{3z} = C_{3r}^2 C_{3z} = C_{3\rho}^3 = \sec\theta_{Np} \cos\theta_{np} \le 1 \qquad [ln_{po}, LN_p] \qquad (128)$$

where in the case of the vacuum, $[hn_{po}]$ = high particle-antiparticle pair number density of vacuum, $[ln_{po}]$ = low particle-antiparticle pair number density of vacuum and where $[SN_p]$ and $[LN_p]$ = small number of particle-antiparticle pairs and large number of particle-antiparticle pairs of vacuum respectively all at zero APF conditions. The space coordinate factors are given by equations (108) through (110), and so because the right hand sides of equations (125) and (126) are the same as the right hand sides of equations (127) and (128) it follows that the space coordinate factors $C_{2\alpha}$ and $C_{3\alpha}$ of equations (109) and (110) give two possible solutions for the internal phase angles $\theta_{\eta\alpha}$ and $\theta_{n\alpha}$ of the counting and segmentation numbers for the $[ln_{po}]$ case and therefore the $[ln_{po}]$ case is degenerate. Then combining equations (120), (121) and (123) through (128) gives for the vacuum[11]

$$n_p = n_{po} \cos\theta_{np} \le n_{po} \qquad [hn_{po}] \qquad (129)$$

$$n_p = n_{po} \sec\theta_{np} \ge n_{po} \qquad [ln_{po}] \qquad (130)$$

where n_p = magnitude of the particle-antiparticle pair number density for the vacuum in an APF and where n_{po} = particle-antiparticle pair number density under zero APF conditions and which is given by

$$n_{po} = N_p / V_{po} \qquad (131)$$

which is the conventional result.

Cartesian Coordinates.

For Cartesian coordinates the ratios of the volume elements with and without an APF present are given by

$$V_p / V_{po} = C_{1x}C_{1y}C_{1z} = \cos\theta_{Np} \sec\theta_{np} \ge 1 \qquad [hn_{po}, SN_p] \qquad (132)$$

$$V_p / V_{po} = C_{1x}C_{1y}C_{1z} = \sec\theta_{Np} \sec\theta_{np} \ge 1 \qquad [hn_{po}, LN_p] \qquad (133)$$

$$V_p / V_{po} = C_{\beta x}C_{\beta y}C_{\beta z} = \cos\theta_{Np} \cos\theta_{np} \le 1 \qquad [ln_{po}, SN_p] \qquad (134)$$

$$V_p / V_{po} = C_{\beta x}C_{\beta y}C_{\beta z} = \sec\theta_{Np} \cos\theta_{np} \le 1 \qquad [ln_{po}, LN_p] \qquad (135)$$

where $C_{1\alpha}$ and $C_{\beta\alpha}$ are given by equations (108) through (110) and where $\beta = 2,3$ describes the degeneracy of the $[ln_{po}]$ state. For the case of a uniaxial APF in the x-direction, $C_{1y} = 1$, $C_{1z} = 1$, $C_{\beta y} = 1$ and $C_{\beta z} = 1$ with the result that equations (132) through (135) give

$$V_p / V_{po} = C_{1x} = \cos\theta_{Np} \sec\theta_{np} \ge 1 \qquad [hn_{po}, SN_p] \qquad (136)$$

$$V_p / V_{po} = C_{1x} = \sec\theta_{Np} \sec\theta_{np} \ge 1 \qquad [hn_{po}, LN_p] \qquad (137)$$

$$V_p / V_{po} = C_{\beta x} = \cos\theta_{Np} \cos\theta_{np} \leq 1 \qquad [ln_{po}, SN_p] \qquad (138)$$

$$V_p / V_{po} = C_{\beta x} = \sec\theta_{Np} \cos\theta_{np} \leq 1 \qquad [ln_{po}, LN_p] \qquad (139)$$

where $\beta = 2,3$.

Cylindrical Polar Coordinates.

For the case of cylindrical polar coordinates the ratios of the vacuum volume elements with and without an APF are given by

$$V_p / V_{po} = C_{1r}^2 C_{1z} = \cos\theta_{Np} \sec\theta_{np} \geq 1 \qquad [hn_{po}, SN_p] \qquad (140)$$

$$V_p / V_{po} = C_{1r}^2 C_{1z} = \sec\theta_{Np} \sec\theta_{np} \geq 1 \qquad [hn_{po}, LN_p] \qquad (141)$$

$$V_p / V_{po} = C_{\beta r}^2 C_{\beta z} = \cos\theta_{Np} \cos\theta_{np} \leq 1 \qquad [ln_{po}, SN_p] \qquad (142)$$

$$V_p / V_{po} = C_{\beta r}^2 C_{\beta z} = \sec\theta_{Np} \cos\theta_{np} \leq 1 \qquad [ln_{po}, LN_p] \qquad (143)$$

where $\beta = 2,3$. For the case of a uniaxial APF along the z-axis, $C_{1r} = 1$ and $C_{\beta r} = 1$ and equations (140) through (143) become

$$V_p / V_{po} = C_{1z} = \cos\theta_{Np} \sec\theta_{np} \geq 1 \qquad [hn_{po}, SN_p] \qquad (144)$$

$$V_p / V_{po} = C_{1z} = \sec\theta_{Np} \sec\theta_{np} \geq 1 \qquad [hn_{po}, LN_p] \qquad (145)$$

$$V_p / V_{po} = C_{\beta z} = \cos\theta_{Np} \cos\theta_{np} \leq 1 \qquad [ln_{po}, SN_p] \qquad (146)$$

$$V_p / V_{po} = C_{\beta z} = \sec\theta_{Np} \cos\theta_{np} \leq 1 \qquad [ln_{po}, LN_p] \qquad (147)$$

where $\beta = 2,3$. For the special case of a two-dimensional radial APF, $C_{1z} = 1$ and $C_{\beta z} = 1$ and equations (140) through (143) give

$$(V_p / V_{po})^{1/2} = C_{1r} = \cos^{1/2}\theta_{Np} \sec^{1/2}\theta_{np} \geq 1 \qquad [hn_{po}, SN_p] \qquad (148)$$

$$(V_p / V_{po})^{1/2} = C_{1r} = \sec^{1/2}\theta_{Np} \sec^{1/2}\theta_{np} \geq 1 \qquad [hn_{po}, LN_p] \qquad (149)$$

$$(V_p / V_{po})^{1/2} = C_{\beta r} = \cos^{1/2}\theta_{Np} \cos^{1/2}\theta_{np} \leq 1 \qquad [ln_{po}, SN_p] \qquad (150)$$

$$(V_p / V_{po})^{1/2} = C_{\beta r} = \sec^{1/2}\theta_{Np} \cos^{1/2}\theta_{np} \leq 1 \qquad [ln_{po}, LN_p] \qquad (151)$$

Spherical Polar Coordinates.

$$V_p / V_{po} = C_{1\rho}^3 = \cos\theta_{Np} \sec\theta_{np} \geq 1 \qquad [hn_{po}, SN_p] \qquad (152)$$

$$V_p / V_{po} = C_{1\rho}^3 = \sec\theta_{\aleph p} \sec\theta_{np} \geq 1 \qquad [hn_{po}, LN_p] \qquad (153)$$

$$V_p / V_{po} = C_{\beta\rho}^3 = \cos\theta_{\aleph p} \cos\theta_{np} \leq 1 \qquad [ln_{po}, SN_p] \qquad (154)$$

$$V_p / V_{po} = C_{\beta\rho}^3 = \sec\theta_{\aleph p} \cos\theta_{np} \leq 1 \qquad [ln_{po}, LN_p] \qquad (155)$$

$$(V_p / V_{po})^{1/3} = C_{1\rho} = \cos^{1/3}\theta_{\aleph p} \sec^{1/3}\theta_{np} \geq 1 \qquad [hn_{po}, SN_p] \qquad (156)$$

$$(V_p / V_{po})^{1/3} = C_{1\rho} = \sec^{1/3}\theta_{\aleph p} \sec^{1/3}\theta_{np} \geq 1 \qquad [hn_{po}, LN_p] \qquad (157)$$

$$(V_p / V_{po})^{1/3} = C_{\beta\rho} = \cos^{1/3}\theta_{\aleph p} \cos^{1/3}\theta_{np} \leq 1 \qquad [ln_{po}, SN_p] \qquad (158)$$

$$(V_p / V_{po})^{1/3} = C_{\beta\rho} = \sec^{1/3}\theta_{\aleph p} \cos^{1/3}\theta_{np} \leq 1 \qquad [ln_{po}, LN_p] \qquad (159)$$

where $\beta = 1,2$. Note that equations (152) through (159) for spherical polar coordinates can also be obtained from equations (132) through (135) for a Cartesian coordinate system by taking $C_{1x} = C_{1y} = C_{1z}$ and $C_{\beta x} = C_{\beta y} = C_{\beta z}$ and from equations (140) through (143) for cylindrical polar coordinates by taking $C_{1r} = C_{1z}$ and $C_{\beta r} = C_{\beta z}$.

Consider now the case of a linear elastic vacuum for which the internal phase angles θ_{np} and $\theta_{\aleph p}$ are related by[9]

$$\cos\theta_{np} = 1 - \tan\theta_{\aleph p} \qquad 0 \leq \theta_{np} \leq \pi/2 \qquad 0 \leq \theta_{\aleph p} \leq \pi/4 \qquad [hn_{po}] \qquad (160)$$

$$\sec\theta_{np} = 1 + \tan\theta_{\aleph p} \qquad 0 \leq \theta_{np} \leq \pi/2 \qquad 0 \leq \theta_{\aleph p} \leq \pi/2 \qquad [ln_{po}] \qquad (161)$$

with $\theta_{np} \geq \theta_{\aleph p}$, and which can now be used in equations (132) through (159) to describe the space coordinate factors for a linear elastic vacuum in an APF.

<u>Cartesian Coordinates</u>

Combining equations (132) through (135) with equations (160) and (161) gives for the case of a 3-dimensional APF in Cartesian coordinates

$$C_{1x}C_{1y}C_{1z} = \sec\theta_{np}[1 + (1 - \cos\theta_{np})^2]^{-1/2} = \cos\theta_{\aleph p}(1 - \tan\theta_{\aleph p})^{-1} \qquad [hn_{po}, SN_p] \qquad (162)$$

$$C_{1x}C_{1y}C_{1z} = \sec\theta_{np}[1 + (1 - \cos\theta_{np})^2]^{1/2} = \sec\theta_{\aleph p}(1 - \tan\theta_{\aleph p})^{-1} \qquad [hn_{po}, LN_p] \qquad (163)$$

$$C_{\beta x}C_{\beta y}C_{\beta z} = \cos\theta_{np}[1 + (\sec\theta_{np} - 1)^2]^{-1/2} = \cos\theta_{\aleph p}(1 + \tan\theta_{\aleph p})^{-1} \qquad [ln_{po}, SN_p] \qquad (164)$$

$$C_{\beta x}C_{\beta y}C_{\beta z} = \cos\theta_{np}[1 + (\sec\theta_{np} - 1)^2]^{1/2} = \sec\theta_{\aleph p}(1 + \tan\theta_{\aleph p})^{-1} \qquad [ln_{po}, LN_p] \qquad (165)$$

where $\beta = 2,3$. Combining equations (136) through (139) with equations (160) and (161) gives for a uniaxial APF along the x-coordinate in a linear elastic body

$$C_{1x} = \sec\theta_{np}[1+(1-\cos\theta_{np})^2]^{-1/2} = \cos\theta_{Np}(1-\tan\theta_{Np})^{-1} \quad [hn_{po}, SN_p] \quad (166)$$

$$C_{1x} = \sec\theta_{np}[1+(1-\cos\theta_{np})^2]^{1/2} = \sec\theta_{Np}(1-\tan\theta_{Np})^{-1} \quad [hn_{po}, LN_p] \quad (167)$$

$$C_{\beta x} = \cos\theta_{np}[1+(\sec\theta_{np}-1)^2]^{-1/2} = \cos\theta_{Np}(1+\tan\theta_{Np})^{-1} \quad [ln_{po}, SN_p] \quad (168)$$

$$C_{\beta x} = \cos\theta_{np}[1+(\sec\theta_{np}-1)^2]^{1/2} = \sec\theta_{Np}(1+\tan\theta_{Np})^{-1} \quad [ln_{po}, LN_p] \quad (169)$$

for Cartesian coordinates.

Cylindrical Polar Coordinates.

Combining equations (140) through (143) with equations (160) and (161) gives for the case of a 3-dimensional APF

$$C_{1r}^2 C_{1z} = \sec\theta_{np}[1+(1-\cos\theta_{np})^2]^{-1/2} = \cos\theta_{Np}(1-\tan\theta_{Np})^{-1} \quad [hn_{po}, SN_p] \quad (170)$$

$$C_{1r}^2 C_{1z} = \sec\theta_{np}[1+(1-\cos\theta_{np})^2]^{1/2} = \sec\theta_{Np}(1-\tan\theta_{Np})^{-1} \quad [hn_{po}, LN_p] \quad (171)$$

$$C_{\beta r}^2 C_{\beta z} = \cos\theta_{np}[1+(\sec\theta_{np}-1)^2]^{-1/2} = \cos\theta_{Np}(1+\tan\theta_{Np})^{-1} \quad [ln_{po}, SN_p] \quad (172)$$

$$C_{\beta r}^2 C_{\beta z} = \cos\theta_{np}[1+(\sec\theta_{np}-1)^2]^{1/2} = \sec\theta_{Np}(1+\tan\theta_{Np})^{-1} \quad [ln_{po}, LN_p] \quad (173)$$

For a uniaxial APF, $C_{1r} = 1$ and $C_{\beta r} = 1$ and equations (170) through (173) become the uniaxial APF equations (166) through (169) but now for the z-axis. For the case of a 2-dimensional radial APF, $C_{1z} = 1$ and $C_{\beta z} = 1$ and equations (170) through (173) become

$$C_{1r} = \sec^{1/2}\theta_{np}[1+(1-\cos\theta_{np})^2]^{-1/4} = \cos^{1/2}\theta_{Np}(1-\tan\theta_{Np})^{-1/2} \quad [hn_{po}, SN_p] \quad (174)$$

$$C_{1r} = \sec^{1/2}\theta_{np}[1+(1-\cos\theta_{np})^2]^{1/4} = \sec^{1/2}\theta_{Np}(1-\tan\theta_{Np})^{-1/2} \quad [hn_{po}, LN_p] \quad (175)$$

$$C_{\beta r} = \cos^{1/2}\theta_{np}[1+(\sec\theta_{np}-1)^2]^{-1/4} = \cos^{1/2}\theta_{Np}(1+\tan\theta_{Np})^{-1/2} \quad [ln_{po}, SN_p] \quad (176)$$

$$C_{\beta r} = \cos^{1/2}\theta_{np}[1+(\sec\theta_{np}-1)^2]^{1/4} = \sec^{1/2}\theta_{Np}(1+\tan\theta_{Np})^{-1/2} \quad [ln_{po}, LN_p] \quad (177)$$

where $\beta = 2,3$.

Spherical Polar Coordinates.

Combining equations (156) through (159) with equations (160) and (161) gives for the case of a 3-dimensional APF in spherical polar coordinates

$$C_{1\rho} = \sec^{1/3}\theta_{np}[1+(1-\cos\theta_{np})^2]^{-1/6} = \cos^{1/3}\theta_{Np}(1-\tan\theta_{Np})^{-1/3} \quad [hn_{po}, SN_p] \quad (178)$$

$$C_{1\rho} = \sec^{1/3}\theta_{np}[1+(1-\cos\theta_{np})^2]^{1/6} = \sec^{1/3}\theta_{Np}(1-\tan\theta_{Np})^{-1/3} \quad [hn_{po}, LN_p] \quad (179)$$

$$C_{\beta\rho} = \cos^{1/3}\theta_{np}[1+(\sec\theta_{np}-1)^2]^{-1/6} = \cos^{1/3}\theta_{\mathrm{N}p}(1+\tan\theta_{\mathrm{N}p})^{-1/3} \quad [\ln_{po}, SN_p] \qquad (180)$$

$$C_{\beta\rho} = \cos^{1/3}\theta_{np}[1+(\sec\theta_{np}-1)^2]^{1/6} = \sec^{1/3}\theta_{\mathrm{N}p}(1+\tan\theta_{\mathrm{N}p})^{-1/3} \quad [\ln_{po}, LN_p] \qquad (181)$$

where $\beta = 2,3$. Equations (178) through (181) can also be obtained from equations (162) through (165) for Cartesian coordinates by taking $C_{1x} = C_{1y} = C_{1z}$ and $C_{\beta x} = C_{\beta y} = C_{\beta z}$.

The space and time coordinate factors for the vacuum that contains an APF can also be related to the internal phase angles of the complex number dimensions of space and time.[9,10] It can be conjectured that the complex number of spacetime dimensions $\overline{D}_{st}$ is linearly related to the complex number of virtual particle-antiparticle pairs $\overline{\mathrm{N}}_p$ and the complex number spacetime dimension number density is linearly related to the complex number density of virtual particle-antiparticle pairs in the manner

$$\overline{D}_{st} = \mu\overline{\mathrm{N}}_p \qquad D_{st} = \mu\mathrm{N}_p \qquad \theta_{Dst} = \theta_{\mathrm{N}p} \qquad (182)$$

$$\overline{n}_{Dst} = \mu\overline{n}_{\mathrm{N}p} \qquad n_{Dst} = \mu n_{\mathrm{N}p} \qquad \theta_{nDst} = \theta_{np} \qquad (183)$$

where the particle-antiparticle pair formalism is given in equations (113) through (181). It has been suggested that the complex number dimensions of space and time can be written as[8-10]

$$\overline{D}_t = D_t\exp(j\theta_{Dt}) \qquad \overline{D}_s = D_s\exp(j\theta_{Ds}) \qquad \overline{D}_{st} = D_{st}\exp(j\theta_{Dst}) \qquad (184)$$

where for a low dimensional spacetime

$$D_t = N_{Dt}\cos\theta_{Dt} \qquad D_s = N_{Ds}\cos\theta_{Ds} \qquad D_{st} = N_{Dst}\cos\theta_{Dst} \qquad [SN_{Dst}] \qquad (185)$$

and for a high dimensional spacetime

$$D_t = N_{Dt}\sec\theta_{Dt} \qquad D_s = N_{Ds}\sec\theta_{Ds} \qquad D_{st} = N_{Dst}\sec\theta_{Dst} \qquad [LN_{Dst}] \qquad (186)$$

where $[SN_{Dst}]$ and $[LN_{Dst}]$ = small number of dimensions and large number of dimensions respectively and where N_{Dt}, N_{Ds} and $N_{Dst} = N_{Dt} + N_{Ds}$ = zero APF values of the dimensions of time, space and spacetime which in the vacuum of are generally integers but can be nonintegers in matter.[8-10]

It has been shown that for the vacuum in an APF the complex number dimension number density is written as[8-10]

$$\overline{n}_{Dst} = n_{Dst}\exp(j\theta_{nDst}) = \overline{D}_{st}/\overline{V}_{Dst} \qquad n_{Dst} = D_{st}/V_{Dst} \qquad \theta_{nDst} = \theta_{Dst} - \theta_{VDst} \qquad (187)$$

where the complex number spacetime dimension number is given in equations (184) through (186). The complex number volume of dimensions is represented as

$$\overline{V}_{Dst} = V_{Dst}\exp(j\theta_{VDst}) \qquad \theta_{VDst} = \theta_{Dst} - \theta_{nDst} \qquad \theta_{VDst} \le 0 \qquad (188)$$

For a homogeneous bound vacuum in a 3-dimensional APF, the internal phase angles of the space coordinates are written as

$$\theta_x = \theta_y = \theta_z = \theta_r = \theta_\rho = 1/3\theta_{VDst} = 1/3(\theta_{Dst} - \theta_{nDst}) \le 0 \qquad (189)$$

and for a uniaxial APF in the x-direction

$$\theta_x = \theta_{VDst} = \theta_{Dst} - \theta_{nDst} \le 0 \qquad (190)$$

For a 3-dimensional APF in the vacuum[9,10]

$$V_{Dst} / V^o_{Dst} = C_{1x}C_{1y}C_{1z} = C^2_{1r}C_{1z} = C^3_{1\rho} = \cos\theta_{Dst}\sec\theta_{nDst} \ge 1 \qquad [hn^o_{Dst}, SN_{Dst}] \qquad (191)$$

$$V_{Dst} / V^o_{Dst} = C_{1x}C_{1y}C_{1z} = C^2_{1r}C_{1z} = C^3_{1\rho} = \sec\theta_{Dst}\sec\theta_{nDst} \ge 1 \qquad [hn^o_{Dst}, LN_{Dst}] \qquad (192)$$

$$V_{Dst} / V^o_{Dst} = C_{2x}C_{2y}C_{2z} = C^2_{2r}C_{2z} = C^3_{2\rho} = \cos\theta_{Dst}\cos\theta_{nDst} \le 1 \qquad [ln^o_{Dst}, SN_{Dst}] \qquad (193)$$

$$V_{Dst} / V^o_{Dst} = C_{2x}C_{2y}C_{2z} = C^2_{2r}C_{2z} = C^3_{2\rho} = \sec\theta_{Dst}\cos\theta_{nDst} \le 1 \qquad [ln^o_{Dst}, LN_{Dst}] \qquad (194)$$

$$V_{Dst} / V^o_{Dst} = C_{3x}C_{3y}C_{3z} = C^2_{3r}C_{3z} = C^3_{3\rho} = \cos\theta_{Dst}\cos\theta_{nDst} \le 1 \qquad [ln^o_{Dst}, SN_{Dst}] \qquad (195)$$

$$V_{Dst} / V^o_{Dst} = C_{3x}C_{3y}C_{3z} = C^2_{3r}C_{3z} = C^3_{3\rho} = \sec\theta_{Dst}\cos\theta_{nDst} \le 1 \qquad [ln^o_{Dst}, LN_{Dst}] \qquad (196)$$

where in the case of the vacuum, $[hn^o_{Dst}]$ = high dimensional density of vacuum, $[ln^o_{Dst}]$ = low dimensional density of vacuum and where $[SN_{Dst}]$ and $[LN_{Dst}]$ = small number of dimensions and large number of dimensions respectively and where the space coordinate factors are given by equations (108) through (110), and so because the right hand sides of equations (193) and (194) are the same as the right hand sides of equations (195) and (196) it follows that the space coordinate factors $C_{2\alpha}$ and $C_{3\alpha}$ of equations (109) and (110) give two possible solutions for the internal phase angles $\theta_{\eta\alpha}$ and $\theta_{n\alpha}$ of the counting and segmentation numbers for the $[ln^o_{Dst}]$ case and therefore the $[ln^o_{Dst}]$ case is degenerate. Combining equations (187) through (196) gives[9,10]

$$n_{Dst} = N_{Dst} / V_{Dst}\cos\theta_{Dst} \qquad [SN_{Dst}] \qquad (197)$$

$$n_{Dst} = N_{Dst} / V_{Dst}\sec\theta_{Dst} \qquad [LN_{Dst}] \qquad (198)$$

Then it is easily shown that[9,10]

$$n_{Dst} = n^o_{Dst}\cos\theta_{nDst} \le n^o_{Dst} \qquad [hn^o_{Dst}] \qquad (199)$$

$$n_{Dst} = n^o_{Dst}\sec\theta_{nDst} \ge n^o_{Dst} \qquad [ln^o_{Dst}] \qquad (200)$$

where

$$n^o_{Dst} = N_{Dst} / V^o_{Dst} \qquad (201)$$

is the zero APF value of the dimensional number density. Equations (187) through (201) are the vacuum space and time dimension analogs of the vacuum particle-antiparticle equations (113) through (131).

Cartesian Coordinates.

For Cartesian coordinates the ratios of the volume elements with and without an APF present are given by

$$V_{Dst}/V_{Dst}^{o} = C_{1x}C_{1y}C_{1z} = \cos\theta_{Dst}\sec\theta_{nDst} \geq 1 \qquad [hn_{Dst}^{o}, SN_{Dst}] \qquad (202)$$

$$V_{Dst}/V_{Dst}^{o} = C_{1x}C_{1y}C_{1z} = \sec\theta_{Dst}\sec\theta_{nDst} \geq 1 \qquad [hn_{Dst}^{o}, LN_{Dst}] \qquad (203)$$

$$V_{Dst}/V_{Dst}^{o} = C_{2x}C_{2y}C_{2z} = \cos\theta_{Dst}\cos\theta_{nDst} \leq 1 \qquad [ln_{Dst}^{o}, SN_{Dst}] \qquad (204)$$

$$V_{Dst}/V_{Dst}^{o} = C_{2x}C_{2y}C_{2z} = \sec\theta_{Dst}\cos\theta_{nDst} \leq 1 \qquad [ln_{Dst}^{o}, LN_{Dst}] \qquad (205)$$

$$V_{Dst}/V_{Dst}^{o} = C_{3x}C_{3y}C_{3z} = \cos\theta_{Dst}\cos\theta_{nDst} \leq 1 \qquad [ln_{Dst}^{o}, SN_{Dst}] \qquad (206)$$

$$V_{Dst}/V_{Dst}^{o} = C_{3x}C_{3y}C_{3z} = \sec\theta_{Dst}\cos\theta_{nDst} \leq 1 \qquad [ln_{Dst}^{o}, LN_{Dst}] \qquad (207)$$

where $C_{1\alpha}$ and $C_{\beta\alpha}$ are given by equations (108) through (110) and where $\beta = 2,3$ refers to the degeneracy of the $[ln_{Dst}^{o}]$ state. For the case of a uniaxial APF in the x-direction, $C_{1y} = 1$, $C_{1z} = 1$, $C_{\beta y} = 1$ and $C_{\beta z} = 1$ so that equations (202) through (207) give

$$V_{Dst}/V_{Dst}^{o} = C_{1x} = \cos\theta_{Dst}\sec\theta_{nDst} \geq 1 \qquad [hn_{Dst}^{o}, SN_{Dst}] \qquad (208)$$

$$V_{Dst}/V_{Dst}^{o} = C_{1x} = \sec\theta_{Dst}\sec\theta_{nDst} \geq 1 \qquad [hn_{Dst}^{o}, LN_{Dst}] \qquad (209)$$

$$V_{Dst}/V_{Dst}^{o} = C_{\beta x} = \cos\theta_{Dst}\cos\theta_{nDst} \leq 1 \qquad [ln_{Dst}^{o}, SN_{Dst}] \qquad (210)$$

$$V_{Dst}/V_{Dst}^{o} = C_{\beta x} = \sec\theta_{Dst}\cos\theta_{nDst} \leq 1 \qquad [ln_{Dst}^{o}, LN_{Dst}] \qquad (211)$$

where $\beta = 2,3$.

Cylindrical Polar Coordinates.

For the case of cylindrical polar coordinates the ratios of the volume elements with and without an APF are given by

$$V_{Dst}/V_{Dst}^{o} = C_{1r}^{2}C_{1z} = \cos\theta_{Dst}\sec\theta_{nDst} \geq 1 \qquad [hn_{Dst}^{o}, SN_{Dst}] \qquad (212)$$

$$V_{Dst}/V_{Dst}^{o} = C_{1r}^{2}C_{1z} = \sec\theta_{Dst}\sec\theta_{nDst} \geq 1 \qquad [hn_{Dst}^{o}, LN_{Dst}] \qquad (213)$$

$$V_{Dst} / V^o_{Dst} = C^2_{\beta r} C_{\beta z} = \cos\theta_{Dst} \cos\theta_{nDst} \leq 1 \qquad [\ln^o_{Dst}, SN_{Dst}] \qquad (214)$$

$$V_{Dst} / V^o_{Dst} = C^2_{\beta r} C_{\beta z} = \sec\theta_{Dst} \cos\theta_{nDst} \leq 1 \qquad [\ln^o_{Dst}, LN_{Dst}] \qquad (215)$$

where $\beta = 2,3$. For the case of a uniaxial APF along the z-axis, $C_{1r} = 1$ and $C_{\beta r} = 1$ and equations (212) through (215) become

$$V_{Dst} / V^o_{Dst} = C_{1z} = \cos\theta_{Dst} \sec\theta_{nDst} \geq 1 \qquad [hn^o_{Dst}, SN_{Dst}] \qquad (216)$$

$$V_{Dst} / V^o_{Dst} = C_{1z} = \sec\theta_{Dst} \sec\theta_{nDst} \geq 1 \qquad [hn^o_{Dst}, LN_{Dst}] \qquad (217)$$

$$V_{Dst} / V^o_{Dst} = C_{\beta z} = \cos\theta_{Dst} \cos\theta_{nDst} \leq 1 \qquad [\ln^o_{Dst}, SN_{Dst}] \qquad (218)$$

$$V_{Dst} / V^o_{Dst} = C_{\beta z} = \sec\theta_{Dst} \cos\theta_{nDst} \leq 1 \qquad [\ln^o_{Dst}, LN_{Dst}] \qquad (219)$$

where $\beta = 2,3$. For the special case of a two-dimensional radial APF, $C_{1z} = 1$ and $C_{\beta z} = 1$ and equations (212) through (215) give

$$(V_{Dst} / V^o_{Dst})^{1/2} = C_{1r} = \cos^{1/2}\theta_{Dst} \sec^{1/2}\theta_{nDst} \geq 1 \qquad [hn^o_{Dst}, SN_{Dst}] \qquad (220)$$

$$(V_{Dst} / V^o_{Dst})^{1/2} = C_{1r} = \sec^{1/2}\theta_{Dst} \sec^{1/2}\theta_{nDst} \geq 1 \qquad [hn^o_{Dst}, LN_{Dst}] \qquad (221)$$

$$(V_{Dst} / V^o_{Dst})^{1/2} = C_{\beta r} = \cos^{1/2}\theta_{Dst} \cos^{1/2}\theta_{nDst} \leq 1 \qquad [\ln^o_{Dst}, SN_{Dst}] \qquad (222)$$

$$(V_{Dst} / V^o_{Dst})^{1/2} = C_{\beta r} = \sec^{1/2}\theta_{Dst} \cos^{1/2}\theta_{nDst} \leq 1 \qquad [\ln^o_{Dst}, LN_{Dst}] \qquad (223)$$

where $\beta = 2,3$.

Spherical Polar Coordinates.

For an isotropic APF in the vacuum,

$$V_{Dst} / V^o_{Dst} = C^3_{1\rho} = \cos\theta_{Dst} \sec\theta_{nDst} \geq 1 \qquad [hn^o_{Dst}, SN_{Dst}] \qquad (224)$$

$$V_{Dst} / V^o_{Dst} = C^3_{1\rho} = \sec\theta_{Dst} \sec\theta_{nDst} \geq 1 \qquad [hn^o_{Dst}, LN_{Dst}] \qquad (225)$$

$$V_{Dst} / V^o_{Dst} = C^3_{\beta\rho} = \cos\theta_{Dst} \cos\theta_{nDst} \leq 1 \qquad [\ln^o_{Dst}, SN_{Dst}] \qquad (226)$$

$$V_{Dst} / V^o_{Dst} = C^3_{\beta\rho} = \sec\theta_{Dst} \cos\theta_{nDst} \leq 1 \qquad [\ln^o_{Dst}, LN_{Dst}] \qquad (227)$$

or

$$(V_{Dst} / V^o_{Dst})^{1/3} = C_{1\rho} = \cos^{1/3}\theta_{Dst} \sec^{1/3}\theta_{nDst} \geq 1 \qquad [hn^o_{Dst}, SN_{Dst}] \qquad (228)$$

$$(V_{Dst} / V^o_{Dst})^{1/3} = C_{1\rho} = \sec^{1/3}\theta_{Dst} \sec^{1/3}\theta_{nDst} \geq 1 \qquad [hn^o_{Dst}, LN_{Dst}] \qquad (229)$$

$$(V_{Dst} / V^{o}_{Dst})^{1/3} = C_{\beta\rho} = \cos^{1/3}\theta_{Dst} \cos^{1/3}\theta_{nDst} \leq 1 \qquad [\ln^{o}_{Dst}, SN_{Dst}] \qquad (230)$$

$$(V_{Dst} / V^{o}_{Dst})^{1/3} = C_{\beta\rho} = \sec^{1/3}\theta_{Dst} \cos^{1/3}\theta_{nDst} \leq 1 \qquad [\ln^{o}_{Dst}, LN_{Dst}] \qquad (231)$$

where $\beta = 2,3$ describes the degeneracy of the $[\ln^{vo}_{Dst}]$ state.and where the space coordinate factors $C_{\sigma\rho}$ for spherical polar coordinates are given by equations (108) through (110) with $\alpha = \rho$. Note that equations (228) through (231) for spherical polar coordinates can also be obtained from equations (202) through (207) for a Cartesian coordinate system by taking $C_{1x} = C_{1y} = C_{1z}$ and $C_{\beta x} = C_{\beta y} = C_{\beta z}$ and from equations (212) through (215) for cylindrical polar coordinates by taking $C_{1r} = C_{1z}$ and $C_{\beta r} = C_{\beta z}$.

Consider now the case of a linear elastic vacuum for which the internal phase angles θ_{nDst} and θ_{Dst} are related by[9]

$$\cos\theta_{nDst} = 1 - \tan\theta_{Dst} \qquad 0 \leq \theta_{nDst} \leq \pi/2 \qquad 0 \leq \theta_{Dst} \leq \pi/4 \qquad [hn^{o}_{Dst}] \qquad (232)$$

$$\sec\theta_{nDst} = 1 + \tan\theta_{Dst} \qquad 0 \leq \theta_{nDst} \leq \pi/2 \qquad 0 \leq \theta_{Dst} \leq \pi/2 \qquad [\ln^{o}_{Dst}] \qquad (233)$$

with $\theta_{nDst} \geq \theta_{Dst}$, and which can now be used in equations (202) through (231) to describe the space coordinate factors for a linear elastic vacuum in an APF.

<u>Cartesian Coordinates</u>

Combining equations (202) through (207) with equations (232) and (233) gives for the case of a 3-dimensional APF in Cartesian coordinates

$$C_{1x}C_{1y}C_{1z} = \sec\theta_{nDst}[1 + (1 - \cos\theta_{nDst})^2]^{-1/2} = \cos\theta_{Dst}(1 - \tan\theta_{Dst})^{-1} \quad [hn^{o}_{Dst}, SN_{Dst}] \; (234)$$

$$C_{1x}C_{1y}C_{1z} = \sec\theta_{nDst}[1 + (1 - \cos\theta_{nDst})^2]^{1/2} = \sec\theta_{Dst}(1 - \tan\theta_{Dst})^{-1} \quad [hn^{o}_{Dst}, LN_{Dst}] \; (235)$$

$$C_{\beta x}C_{\beta y}C_{\beta z} = \cos\theta_{nDst}[1 + (\sec\theta_{nDst} - 1)^2]^{-1/2} = \cos\theta_{Dst}(1 + \tan\theta_{Dst})^{-1} \quad [\ln^{o}_{Dst}, SN_{Dst}] \; (236)$$

$$C_{\beta x}C_{\beta y}C_{\beta z} = \cos\theta_{nDst}[1 + (\sec\theta_{nDst} - 1)^2]^{1/2} = \sec\theta_{Dst}(1 + \tan\theta_{Dst})^{-1} \quad [\ln^{o}_{Dst}, LN_{Dst}] \; (237)$$

where $\beta = 2,3$. Combining equations (208) through (211) with equations (232) and (233) gives for a uniaxial APF in a linear elastic body

$$C_{1x} = \sec\theta_{nDst}[1 + (1 - \cos\theta_{nDst})^2]^{-1/2} = \cos\theta_{Dst}(1 - \tan\theta_{Dst})^{-1} \qquad [hn^{o}_{Dst}, SN_{Dst}] \qquad (238)$$

$$C_{1x} = \sec\theta_{nDst}[1 + (1 - \cos\theta_{nDst})^2]^{1/2} = \sec\theta_{Dst}(1 - \tan\theta_{Dst})^{-1} \qquad [hn^{o}_{Dst}, LN_{Dst}] \qquad (239)$$

$$C_{\beta x} = \cos\theta_{nDst}[1 + (\sec\theta_{nDst} - 1)^2]^{-1/2} = \cos\theta_{Dst}(1 + \tan\theta_{Dst})^{-1} \qquad [\ln^{o}_{Dst}, SN_{Dst}] \qquad (240)$$

$$C_{\beta x} = \cos\theta_{nDst}[1 + (\sec\theta_{nDst} - 1)^2]^{1/2} = \sec\theta_{Dst}(1 + \tan\theta_{Dst})^{-1} \qquad [\ln^{o}_{Dst}, LN_{Dst}] \qquad (241)$$

for Cartesian coordinates.

Cylindrical Polar Coordinates.

Combining equations (212) through (215) with equations (232) and (233) gives for the case of a 3-dimensional APF

$$C_{1r}^2 C_{1z} = \sec\theta_{nDst}[1+(1-\cos\theta_{nDst})^2]^{-1/2} = \cos\theta_{Dst}(1-\tan\theta_{Dst})^{-1} \quad [hn^o_{Dst}, SN_{Dst}] \quad (242)$$

$$C_{1r}^2 C_{1z} = \sec\theta_{nDst}[1+(1-\cos\theta_{nDst})^2]^{1/2} = \sec\theta_{Dst}(1-\tan\theta_{Dst})^{-1} \quad [hn^o_{Dst}, LN_{Dst}] \quad (243)$$

$$C_{\beta r}^2 C_{\beta z} = \cos\theta_{nDst}[1+(\sec\theta_{nDst}-1)^2]^{-1/2} = \cos\theta_{Dst}(1+\tan\theta_{Dst})^{-1} \quad [ln^o_{Dst}, SN_{Dst}] \quad (244)$$

$$C_{\beta r}^2 C_{\beta z} = \cos\theta_{nDst}[1+(\sec\theta_{nDst}-1)^2]^{1/2} = \sec\theta_{Dst}(1+\tan\theta_{Dst})^{-1} \quad [ln^o_{Dst}, LN_{Dst}] \quad (245)$$

For a uniaxial APF, $C_{1r} = 1$ and $C_{\beta r} = 1$ and equations (242) through (245) become just the uniaxial APF equations (238) through (241) but now for the z-axis. For the special case of a 2-dimensional radial APF, $C_{1z} = 1$ and $C_{\beta z} = 1$ and equations (242) through (245) become

$$C_{1r} = \sec^{1/2}\theta_{nDst}[1+(1-\cos\theta_{nDst})^2]^{-1/4} = \cos^{1/2}\theta_{Dst}(1-\tan\theta_{Dst})^{-1/2} \quad [hn^o_{Dst}, SN_{Dst}] \quad (246)$$

$$C_{1r} = \sec^{1/2}\theta_{nDst}[1+(1-\cos\theta_{nDst})^2]^{1/4} = \sec^{1/2}\theta_{Dst}(1-\tan\theta_{Dst})^{-1/2} \quad [hn^o_{Dst}, LN_{Dst}] \quad (247)$$

$$C_{\beta r} = \cos^{1/2}\theta_{nDst}[1+(\sec\theta_{nDst}-1)^2]^{-1/4} = \cos^{1/2}\theta_{Dst}(1+\tan\theta_{Dst})^{-1/2} \quad [ln^o_{Dst}, SN_{Dst}] \quad (248)$$

$$C_{\beta r} = \cos^{1/2}\theta_{nDst}[1+(\sec\theta_{nDst}-1)^2]^{1/4} = \sec^{1/2}\theta_{Dst}(1+\tan\theta_{Dst})^{-1/2} \quad [ln^o_{Dst}, LN_{Dst}] \quad (249)$$

where $\beta = 2,3$.

Spherical Polar Coordinates.

Combining equations (228) through (231) with equations (232) and (233) gives for the case of a 3-dimensional APF in spherical polar coordinates

$$C_{1\rho} = \sec^{1/3}\theta_{nDst}[1+(1-\cos\theta_{nDst})^2]^{-1/6} = \cos^{1/3}\theta_{Dst}(1-\tan\theta_{Dst})^{-1/3} \quad [hn^o_{Dst}, SN_{Dst}] \quad (250)$$

$$C_{1\rho} = \sec^{1/3}\theta_{nDst}[1+(1-\cos\theta_{nDst})^2]^{1/6} = \sec^{1/3}\theta_{Dst}(1-\tan\theta_{Dst})^{-1/3} \quad [hn^o_{Dst}, LN_{Dst}] \quad (251)$$

$$C_{\beta\rho} = \cos^{1/3}\theta_{nDst}[1+(\sec\theta_{nDst}-1)^2]^{-1/6} = \cos^{1/3}\theta_{Dst}(1+\tan\theta_{Dst})^{-1/3} \quad [ln^o_{Dst}, SN_{Dst}] \quad (252)$$

$$C_{\beta\rho} = \cos^{1/3}\theta_{nDst}[1+(\sec\theta_{nDst}-1)^2]^{1/6} = \sec^{1/3}\theta_{Dst}(1+\tan\theta_{Dst})^{-1/3} \quad [ln^o_{Dst}, LN_{Dst}] \quad (253)$$

where $\beta = 2,3$. Equations (250) through (253) can also be obtained from equations (234) through (237) for Cartesian coordinates by taking $C_{1x} = C_{1y} = C_{1z}$ and $C_{\beta x} = C_{\beta y} = C_{\beta z}$.

The time coordinate factor for a body (polyphase space vehicle) moving relative to a fixed frame of reference can be obtained from the definitions of linear velocity for uniform motion of a body under zero APF conditions

$$w_o = d\alpha_o / dt_o \tag{254}$$

and for the case when a constant APF is present

$$\overline{w} = w \exp(j\theta_w) = d\overline{\alpha} / d\overline{t} \tag{255}$$

$$w = d\alpha / dt = (d\alpha_o / dt_o)(C_{\sigma\alpha} / C_{\sigma t}) = w_o (C_{\sigma\alpha} / C_{\sigma t}) \tag{256}$$

$$\theta_w = \theta_\alpha - \theta_t = \theta_{\eta\alpha} - \theta_{\pi\alpha} - (\theta_{\eta t} - \theta_{\pi t}) \tag{257}$$

so that the constant time coordinate factor $C_{\sigma t}$ is related to the constant space coordinate factor $C_{\sigma\alpha}$ by

$$C_{\sigma t} = (w_o / w) C_{\sigma\alpha} \tag{258}$$

where $\sigma = 1,2,3$; $\alpha = x,y,z$ or r,z or ρ and where w_o = velocity of a body relative to a stationary frame of reference under zero APF conditions and w = magnitude of the velocity of a body relative to a stationary reference frame surrounded by an APF which is in the vacuum and in the body. Because of the large distance between the moving polyphase space vehicle and the observer in the stationary system, the constant space and time coordinate factors $C_{\sigma\alpha}$ and $C_{\sigma t}$ relative to a stationary frame of reference are expected to depend predominantly on the constant APF of the vacuum through the constant internal phase angles that appear in equation (257).

C. Forms of the Space and Time Coordinate Factors for Vacuum Distance $\overline{vt}$.

The space and time coordinates in matter or vacuum that is located in an APF have been represented as complex numbers in an internal space.[5-12] A simple theory of the origin and nature of these complex number coordinates has already been developed.[9] Consider the constant real number spatial parameters L^c_α for the complex number space coordinates $\overline{\alpha}^v$ in an APF in the vacuum, where L^c_α can be taken to be multiples of the Planck length. The real number constant spatial parameters L^c_α are divided into $\overline{n}^v_\alpha$ parts of length $\Delta\overline{L}^v_\alpha$ where the complex number spatial segmentation number $\overline{n}^v_\alpha$ is represented as[9]

$$\overline{n}^v_\alpha = n^v_\alpha \exp(j\theta^v_{n\alpha}) \tag{259}$$

so that

$$\Delta\overline{L}^v_\alpha = \Delta L^v_\alpha \exp(j\theta^v_{\Delta L\alpha}) = L^c_\alpha / \overline{n}^v_\alpha = L^c_\alpha / n^v_\alpha \exp(-j\theta^v_{n\alpha}) \tag{260}$$

and

$$\Delta L^v_\alpha = L^c_\alpha / n^v_\alpha \qquad \theta^v_{\Delta L\alpha} = -\theta^v_{n\alpha} \tag{261}$$

where $\alpha = x,y,z; r,z$ or ρ. Then the complex number space coordinate $\overline{\alpha}^{v}$ can be written as

$$\overline{\alpha}^{v} = \alpha^{v}\exp(j\theta_{\alpha}^{v}) = \overline{\eta}_{\alpha}^{v}\Delta\overline{L}_{\alpha}^{v} = \overline{\eta}_{\alpha}^{v}L_{\alpha}^{c}/\overline{\pi}_{\alpha}^{v} = \eta_{\alpha}^{v}L_{\alpha}^{c}/\pi_{\alpha}^{v}\exp[j(\theta_{\eta\alpha}^{v} - \theta_{\pi\alpha}^{v})] \quad (262)$$

where the complex number spatial counting number $\overline{\eta}_{\alpha}^{v}$ is represented as

$$\overline{\eta}_{\alpha}^{v} = \eta_{\alpha}^{v}\exp(j\theta_{\eta\alpha}^{v}) \quad (263)$$

so that

$$\alpha^{v} = \eta_{\alpha}^{v}L_{\alpha}^{c}/\pi_{\alpha}^{v} = \eta_{\alpha}^{v}\Delta L_{\alpha}^{v} \qquad \theta_{\alpha}^{v} = \theta_{\eta\alpha}^{v} - \theta_{\pi\alpha}^{v} \quad (264)$$

where η_{α}^{v} and π_{α}^{v} are not integers in the presence of an APF but are integers under zero APF conditions.

The time coordinate factor, which is analogous to that of the space coordinate factor, is developed as follows. Consider a real number constant L_{t}^{c} that is used to measure intervals of the complex number time coordinate $\overline{t}^{v}$, where L_{t}^{c} can be taken to be multiples of the Planck time. In the presence of an APF in bulk matter the real number constant temporal parameter L_{t}^{c} is divided into $\overline{\pi}_{t}^{v}$ parts of length $\Delta\overline{L}_{t}^{v}$ where the complex number time segmentation number is represented as[9]

$$\overline{\pi}_{t}^{v} = \pi_{t}^{v}\exp(j\theta_{\pi t}^{v}) \quad (265)$$

and where

$$\Delta\overline{L}_{t}^{v} = \Delta L_{t}^{v}\exp(j\theta_{\Delta Lt}^{v}) = L_{t}^{c}/\overline{\pi}_{t}^{v} = L_{t}^{c}/\pi_{t}^{v}\exp(-j\theta_{\pi t}^{v}) \quad (266)$$

and therefore

$$\Delta L_{t}^{v} = L_{t}^{c}/\pi_{t}^{v} \qquad \theta_{\Delta Lt}^{v} = -\theta_{\pi t}^{v} \quad (267)$$

The complex number time coordinate is then written as

$$\overline{t}^{v} = t^{v}\exp(j\theta_{t}^{v}) = \overline{\eta}_{t}^{v}\Delta\overline{L}_{t}^{v} = \overline{\eta}_{t}^{v}L_{t}^{c}/\overline{\pi}_{t}^{v} = \eta_{t}^{v}L_{t}^{c}/\pi_{t}^{v}\exp[j(\theta_{\eta t}^{v} - \theta_{\pi t}^{v})] \quad (268)$$

where the complex number time counting number is written as

$$\overline{\eta}_{t}^{v} = \eta_{t}^{v}\exp(j\theta_{\eta t}^{v}) \quad (269)$$

so that

$$t^{v} = \eta_{t}^{v}L_{\alpha}^{c}/\pi_{t}^{v} = \eta_{t}^{v}\Delta L_{t}^{v} \qquad \theta_{t}^{v} = \theta_{\eta t}^{v} - \theta_{\pi t}^{v} \quad (270)$$

where η_{t}^{v} and π_{t}^{v} are not integers when an APF is present in matter, and the complex number nature of the time coordinate is $\overline{t}^{v}$ due only to the complex number time counting number $\overline{\eta}_{t}^{v}$ and the complex number segmentation number $\overline{\pi}_{t}^{v}$ because the constant L_{t}^{c} is a real number in equation (268). The real number constants L_{α}^{c} and L_{t}^{c}

that appear in equations (262) and (268) can be taken to be multiples (generally very large multiples) of the Planck length and Planck time respectively because an APF is a field that induces internal phase angles in all physical quantities except the fundamental constants of nature (h, c and G). Equations (264) and (270) allow the possibility of $\theta^{v}_{\alpha} \leq 0$ and $\theta^{v}_{t} \leq 0$ which has been suggested is the case for the Earth's gravity field.

If an APF exists in the vacuum the complex number space and time coordinates are written as

$$\bar{\alpha}^{v} = \alpha^{v}\exp(j\theta^{v}_{\alpha}) \qquad \bar{t}^{v} = t^{v}\exp(j\theta^{v}_{t}) \tag{271}$$

where α = x,y,z or r,z or ρ and where α^{v} and t^{v} are related to their corresponding zero APF values by

$$\alpha^{v} = \alpha^{v}_{o}C^{v}_{\sigma\alpha} \qquad t^{v} = t^{v}_{o}C^{v}_{\sigma t} \tag{272}$$

where σ = 1,2,3 and where

$$C^{v}_{1\alpha} = \cos\theta^{v}_{\eta\alpha}\sec\theta^{v}_{\pi\alpha} \qquad C^{v}_{1t} = \cos\theta^{v}_{\eta t}\sec\theta^{v}_{\pi t} \qquad [hn^{vo}_{p}] \qquad [hn^{vo}_{Dst}] \tag{273}$$

$$C^{v}_{2\alpha} = \cos\theta^{v}_{\eta\alpha}\cos\theta^{v}_{\pi\alpha} \qquad C^{v}_{2t} = \cos\theta^{v}_{\eta t}\cos\theta^{v}_{\pi t} \qquad [ln^{vo}_{p}] \qquad [ln^{vo}_{Dst}] \tag{274}$$

$$C^{v}_{3\alpha} = \sec\theta^{v}_{\eta\alpha}\cos\theta^{v}_{\pi\alpha} \qquad C^{v}_{3t} = \sec\theta^{v}_{\eta t}\cos\theta^{v}_{\pi t} \qquad [ln^{vo}_{p}] \qquad [ln^{vo}_{Dst}] \tag{275}$$

where $[hn^{vo}_{p}]$ and $[hn^{vo}_{Dst}]$ = zero APF values of the high particle-antiparticle pair number density and high-dimensional number density of the vacuum of the vacuum respectively, and $[ln^{vo}_{p}]$ and $[ln^{vo}_{Dst}]$ = zero APF values of the low particle-antiparticle pair number density and low-dimensional number density of the vacuum of the vacuum and where $\theta^{v}_{\eta\alpha}, \theta^{v}_{\pi\alpha}$ and $\theta^{v}_{\eta t}, \theta^{v}_{\pi t}$ are the internal phase angles of the complex number counting and segmentation numbers of the space and time and time coordinates respectively.[9] Then the internal phase angles θ^{v}_{α} and θ^{v}_{t} of the space and time coordinates that span a vacuum in which an APF is present, and which appear in equation (271), are given by[9]

$$\theta^{v}_{\alpha} = \theta^{v}_{\eta\alpha} - \theta^{v}_{\pi\alpha} \qquad \theta^{v}_{\alpha} = \theta^{v}_{\eta\alpha} - \theta^{v}_{\pi\alpha} \tag{276}$$

which allows these angles to have negative values.

It has been shown that for a finite sized element of vacuum in an APF the complex number particle-antiparticle number density $\bar{n}^{v}_{p}$ is written as[9]

$$\bar{n}^{v}_{p} = n^{v}_{p}\exp(j\theta_{np}) = \bar{\mathfrak{N}}^{v}_{p}/\bar{V}^{v}_{p} \qquad n^{v}_{p} = \mathfrak{N}^{v}_{p}/V^{v}_{p} \qquad \theta^{v}_{np} = \theta^{v}_{\mathfrak{N}p} - \theta^{v}_{Vp} \tag{277}$$

where the complex number total particle-antiparticle pair number is written as

$$\bar{\mathfrak{N}}^{v}_{p} = \mathfrak{N}^{v}_{p}\exp(j\theta^{v}_{\mathfrak{N}p}) \tag{278}$$

where

$$\mathfrak{N}^{v}_{p} = N_{p}\cos\theta^{v}_{\mathfrak{N}p} \qquad [SN_{p}] \tag{279}$$

$$\mathbb{N}_p^v = N_p \sec\theta_{\mathbb{N}p}^v \qquad [LN_p] \tag{280}$$

where $[SN_p]$ = small N_p, $[LN_p]$ = large N_p and where N_p = integer number of particle-antiparticle pairs of a vacuum which interact with the APF. The complex number volume is represented as

$$\overline{V}_p^v = V_p^v \exp(j\theta_{Vp}^v) \qquad \theta_{Vp}^v = \theta_{\mathbb{N}p}^v - \theta_{np}^v \qquad \theta_{Vp}^v \le 0 \tag{281}$$

and where the subscript p refers to pairs of particles and antiparticles. For a bound homogeneous vacuum in a 3-dimensional APF the internal phase angles of the space coordinates are then written as

$$\theta_x^v = \theta_y^v = \theta_z^v = \theta_r^v = \theta_\rho^v = 1/3\theta_{Vp}^v = 1/3(\theta_{\mathbb{N}p}^v - \theta_{np}^v) \le 0 \tag{282}$$

while for a uniaxial APF in the x-direction

$$\theta_x^v = \theta_{Vp}^v = \theta_{\mathbb{N}p}^v - \theta_{np}^v \le 0 \tag{283}$$

From equations (277) through (281) it follows that[11]

$$n_p^v = N_p / V_p^v \cos\theta_{\mathbb{N}p}^v \qquad [SN_p] \tag{284}$$

$$n_p^v = N_p / V_p^v \sec\theta_{\mathbb{N}p}^v \qquad [LN_p] \tag{285}$$

$$\theta_{np}^v = \theta_{\mathbb{N}p}^v - \theta_{Vp}^v = \theta_{\mathbb{N}p}^v + \left|\theta_{Vp}^v\right| \qquad \theta_{np}^v \ge \theta_{\mathbb{N}p}^v \tag{286}$$

where the volume element V_p^v in the presence of a 3-dimensional APF is given in terms of the volume element V_p^{vo} under zero APF conditions by[9,10]

$$V_p^v / V_p^{vo} = C_{1x}^v C_{1y}^v C_{1z}^v = (C_{1r}^v)^2 C_{1z}^v = (C_{1\rho}^v)^3 = \cos\theta_{\mathbb{N}p}^v \sec\theta_{np}^v \ge 1 \qquad [hn_p^{vo}, SN_p] \tag{287}$$

$$V_p^v / V_p^{vo} = C_{1x}^v C_{1y}^v C_{1z}^v = (C_{1r}^v)^2 C_{1z}^v = (C_{1\rho}^v)^3 = \sec\theta_{\mathbb{N}p}^v \sec\theta_{np}^v \ge 1 \qquad [hn_p^{vo}, LN_p] \tag{288}$$

$$V_p^v / V_p^{vo} = C_{2x}^v C_{2y}^v C_{2z}^v = (C_{2r}^v)^2 C_{2z}^v = (C_{2\rho}^v)^3 = \cos\theta_{\mathbb{N}p}^v \cos\theta_{np}^v \le 1 \qquad [ln_p^{vo}, SN_p] \tag{289}$$

$$V_p^v / V_p^{vo} = C_{2x}^v C_{2y}^v C_{2z}^v = (C_{2r}^v)^2 C_{2z}^v = (C_{2\rho}^v)^3 = \sec\theta_{\mathbb{N}p}^v \cos\theta_{np}^v \le 1 \qquad [ln_p^{vo}, LN_p] \tag{290}$$

$$V_p^v / V_p^{vo} = C_{3x}^v C_{3y}^v C_{3z}^v = (C_{3r}^v)^2 C_{3z}^v = (C_{3\rho}^v)^3 = \cos\theta_{\mathbb{N}p}^v \cos\theta_{np}^v \le 1 \qquad [ln_p^{vo}, SN_p] \tag{291}$$

$$V_p^v / V_p^{vo} = C_{3x}^v C_{3y}^v C_{3z}^v = (C_{3r}^v)^2 C_{3z}^v = (C_{3\rho}^v)^3 = \sec\theta_{\mathbb{N}p}^v \cos\theta_{np}^v \le 1 \qquad [ln_p^{vo}, LN_p] \tag{292}$$

where in the case of the vacuum, $[hn_p^{vo}]$ = high particle-antiparticle pair number density of vacuum, $[ln_p^{vo}]$ = low particle-antiparticle pair number density of vacuum and where $[SN_p]$ and $[LN_p]$ = small number of particle-antiparticle pairs and large number of particle-antiparticle pairs of vacuum respectively all at zero APF conditions. The space coordinate factors are given by equations (273) through (275), and so because the right hand sides of equations (289) and (290) are the same as the right hand sides of equations (291) and (292) it follows that the space coordinate factors $C^{v}_{2\alpha}$ and $C^{v}_{3\alpha}$ of equations (274) and (275) give two possible solutions for the internal phase angles $\theta^{v}_{\eta\alpha}$ and $\theta^{v}_{n\alpha}$ of the counting and segmentation numbers for the $[ln^{vo}_{p}]$ case and therefore the $[ln^{vo}_{p}]$ case is degenerate. Then equations (284), (285) and (287) through (292) give[11]

$$n^{v}_{p} = n^{vo}_{p} \cos\theta^{v}_{np} \le n^{vo}_{p} \qquad [hn^{vo}_{p}] \tag{293}$$

$$n^{v}_{p} = n^{vo}_{p} \sec\theta^{v}_{np} \ge n^{vo}_{p} \qquad [ln^{vo}_{p}] \tag{294}$$

where n^{v}_{p} = particle-antiparticle pair number density for the vacuum in an APF and where n^{vo}_{p} = particle-antiparticle pair number density under zero APF conditions and which is given by

$$n^{vo}_{p} = N_{p} / V^{vo}_{p} \tag{295}$$

which is the conventional result.

Cartesian Coordinates.

For Cartesian coordinates the ratios of the volume elements with and without an APF present are given by

$$V^{v}_{p} / V^{vo}_{p} = C^{v}_{1x} C^{v}_{1y} C^{v}_{1z} = \cos\theta^{v}_{\mathrm{N}p} \sec\theta^{v}_{np} \ge 1 \qquad [hn^{vo}_{p}, SN_{p}] \tag{296}$$

$$V^{v}_{p} / V^{vo}_{p} = C^{v}_{1x} C^{v}_{1y} C^{v}_{1z} = \sec\theta^{v}_{\mathrm{N}p} \sec\theta^{v}_{np} \ge 1 \qquad [hn^{vo}_{p}, LN_{p}] \tag{297}$$

$$V^{v}_{p} / V^{vo}_{p} = C^{v}_{\beta x} C^{v}_{\beta y} C^{v}_{\beta z} = \cos\theta^{v}_{\mathrm{N}p} \cos\theta^{v}_{np} \le 1 \qquad [ln^{vo}_{p}, SN_{p}] \tag{298}$$

$$V^{v}_{p} / V^{vo}_{p} = C^{v}_{\beta x} C^{v}_{\beta y} C^{v}_{\beta z} = \sec\theta^{v}_{\mathrm{N}p} \cos\theta^{v}_{np} \le 1 \qquad [ln^{vo}_{p}, LN_{p}] \tag{299}$$

where $C^{v}_{1\alpha}$ and $C^{v}_{\beta\alpha}$ are given by equations (273) through (275) and where $\beta = 2,3$ describes the degeneracy of the $[ln^{vo}_{p}]$ state. For the case of a uniaxial APF in the x-direction, $C^{v}_{1y} = 1, C^{v}_{1z} = 1, C^{v}_{\beta y} = 1$ and $C^{v}_{\beta z} = 1$ with the result that equations (296) through (299) give

$$V^{v}_{p} / V^{vo}_{p} = C^{v}_{1x} = \cos\theta^{v}_{\mathrm{N}p} \sec\theta^{v}_{np} \ge 1 \qquad [hn^{vo}_{p}, SN_{p}] \tag{300}$$

$$V^{v}_{p} / V^{vo}_{p} = C^{v}_{1x} = \sec\theta^{v}_{\mathrm{N}p} \sec\theta^{v}_{np} \ge 1 \qquad [hn^{vo}_{p}, LN_{p}] \tag{301}$$

$$V^{v}_{p} / V^{vo}_{p} = C^{v}_{\beta x} = \cos\theta^{v}_{\mathrm{N}p} \cos\theta^{v}_{np} \le 1 \qquad [ln^{vo}_{p}, SN_{p}] \tag{302}$$

$$V^{v}_{p} / V^{vo}_{p} = C^{v}_{\beta x} = \sec\theta^{v}_{\mathrm{N}p} \cos\theta^{v}_{np} \le 1 \qquad [ln^{vo}_{p}, LN_{p}] \tag{303}$$

where $\beta = 2,3$.

Cylindrical Polar Coordinates.

For the case of cylindrical polar coordinates the ratios of the vacuum volume elements with and without an APF are given by

$$V_p^v / V_p^{vo} = (C_{1r}^v)^2 C_{1z}^v = \cos\theta_{\tilde{N}p}^v \sec\theta_{np}^v \geq 1 \qquad [hn_p^{vo}, SN_p] \tag{304}$$

$$V_p^v / V_p^{vo} = (C_{1r}^v)^2 C_{1z}^v = \sec\theta_{\tilde{N}p}^v \sec\theta_{np}^v \geq 1 \qquad [hn_p^{vo}, LN_p] \tag{305}$$

$$V_p^v / V_p^{vo} = (C_{\beta r}^v)^2 C_{\beta z}^v = \cos\theta_{\tilde{N}p}^v \cos\theta_{np}^v \leq 1 \qquad [ln_p^{vo}, SN_p] \tag{306}$$

$$V_p^v / V_p^{vo} = (C_{\beta r}^v)^2 C_{\beta z}^v = \sec\theta_{\tilde{N}p}^v \cos\theta_{np}^v \leq 1 \qquad [ln_p^{vo}, LN_p] \tag{307}$$

where $\beta = 2,3$. For the case of a uniaxial APF along the z-axis, $C_{1r}^v = 1$ and $C_{\beta r}^v = 1$ and equations (304) through (307) become

$$V_p^v / V_p^{vo} = C_{1z}^v = \cos\theta_{\tilde{N}p}^v \sec\theta_{np}^v \geq 1 \qquad [hn_p^{vo}, SN_p] \tag{308}$$

$$V_p^v / V_p^{vo} = C_{1z}^v = \sec\theta_{\tilde{N}p}^v \sec\theta_{np}^v \geq 1 \qquad [hn_p^{vo}, LN_p] \tag{309}$$

$$V_p^v / V_p^{vo} = C_{\beta z}^v = \cos\theta_{\tilde{N}p}^v \cos\theta_{np}^v \leq 1 \qquad [ln_p^{vo}, SN_p] \tag{310}$$

$$V_p^v / V_p^{vo} = C_{\beta z}^v = \sec\theta_{\tilde{N}p}^v \cos\theta_{np}^v \leq 1 \qquad [ln_p^{vo}, LN_p] \tag{311}$$

where $\beta = 2,3$. For the special case of a two-dimensional radial APF, $C_{1z}^v = 1$ and $C_{\beta z}^v = 1$ and equations (304) through (307) give

$$(V_p^v / V_p^{vo})^{1/2} = C_{1r}^v = \cos^{1/2}\theta_{\tilde{N}p}^v \sec^{1/2}\theta_{np}^v \geq 1 \qquad [hn_p^{vo}, SN_p] \tag{312}$$

$$(V_p^v / V_p^{vo})^{1/2} = C_{1r}^v = \sec^{1/2}\theta_{\tilde{N}p}^v \sec^{1/2}\theta_{np}^v \geq 1 \qquad [hn_p^{vo}, LN_p] \tag{313}$$

$$(V_p^v / V_p^{vo})^{1/2} = C_{\beta r}^v = \cos^{1/2}\theta_{\tilde{N}p}^v \cos^{1/2}\theta_{np}^v \leq 1 \qquad [ln_p^{vo}, SN_p] \tag{314}$$

$$(V_p^v / V_p^{vo})^{1/2} = C_{\beta r}^v = \sec^{1/2}\theta_{\tilde{N}p}^v \cos^{1/2}\theta_{np}^v \leq 1 \qquad [ln_p^{vo}, LN_p] \tag{315}$$

Spherical Polar Coordinates.

$$V_p^v / V_p^{vo} = (C_{1\rho}^v)^3 = \cos\theta_{\tilde{N}p}^v \sec\theta_{np}^v \geq 1 \qquad [hn_p^{vo}, SN_p] \tag{316}$$

$$V_p^v / V_p^{vo} = (C_{1\rho}^v)^3 = \sec\theta_{\tilde{N}p}^v \sec\theta_{np}^v \geq 1 \qquad [hn_p^{vo}, LN_p] \tag{317}$$

$$V_p^v / V_p^{vo} = (C_{\beta\rho}^v)^3 = \cos\theta_{\tilde{N}p}^v \cos\theta_{np}^v \leq 1 \qquad [ln_p^{vo}, SN_p] \tag{318}$$

$$V_p^v / V_p^{vo} = (C_{\beta\rho}^v)^3 = \sec\theta_{\tilde{N}p}^v \cos\theta_{np}^v \leq 1 \qquad [ln_p^{vo}, LN_p] \tag{319}$$

$$(V_p^v / V_p^{vo})^{1/3} = C_{1\rho}^v = \cos^{1/3}\theta_{Np}^v \sec^{1/3}\theta_{np}^v \geq 1 \qquad [hn_p^{vo}, SN_p] \qquad (320)$$

$$(V_p^v / V_p^{vo})^{1/3} = C_{1\rho}^v = \sec^{1/3}\theta_{Np}^v \sec^{1/3}\theta_{np}^v \geq 1 \qquad [hn_p^{vo}, LN_p] \qquad (321)$$

$$(V_p^v / V_p^{vo})^{1/3} = C_{\beta\rho}^v = \cos^{1/3}\theta_{Np}^v \cos^{1/3}\theta_{np}^v \leq 1 \qquad [ln_p^{vo}, SN_p] \qquad (322)$$

$$(V_p^v / V_p^{vo})^{1/3} = C_{\beta\rho}^v = \sec^{1/3}\theta_{Np}^v \cos^{1/3}\theta_{np}^v \leq 1 \qquad [ln_p^{vo}, LN_p] \qquad (323)$$

where $\beta = 1,2$. Note that equations (316) through (323) for spherical polar coordinates can also be obtained from equations (296) through (299) for a Cartesian coordinate system by taking $C_{1x}^v = C_{1y}^v = C_{1z}^v$ and $C_{\beta x}^v = C_{\beta y}^v = C_{\beta z}^v$ and from equations (304) through (307) for cylindrical polar coordinates by taking $C_{1r}^v = C_{1z}^v$ and $C_{\beta r}^v = C_{\beta z}^v$.

Consider now the case of a linear elastic vacuum for which the internal phase angles θ_{np}^v and θ_{Np}^v are related by[9]

$$\cos\theta_{np}^v = 1 - \tan\theta_{Np}^v \qquad 0 \leq \theta_{np}^v \leq \pi/2 \qquad 0 \leq \theta_{Np}^v \leq \pi/4 \qquad [hn_p^{vo}] \qquad (324)$$

$$\sec\theta_{np}^v = 1 + \tan\theta_{Np}^v \qquad 0 \leq \theta_{np}^v \leq \pi/2 \qquad 0 \leq \theta_{Np}^v \leq \pi/2 \qquad [ln_p^{vo}] \qquad (325)$$

with $\theta_{np}^v \geq \theta_{Np}^v$, and which can now be used in equations (296) through (323) to describe the space coordinate factors for a linear elastic vacuum in an APF.

Cartesian Coordinates

Combining equations (296) through (299) with equations (324) and (325) gives for the case of a 3-dimensional APF in Cartesian coordinates

$$C_{1x}^v C_{1y}^v C_{1z}^v = \sec\theta_{np}^v [1 + (1 - \cos\theta_{np}^v)^2]^{-1/2} = \cos\theta_{Np}^v (1 - \tan\theta_{Np}^v)^{-1} \qquad [hn_p^{vo}, SN_p] \qquad (326)$$

$$C_{1x}^v C_{1y}^v C_{1z}^v = \sec\theta_{np}^v [1 + (1 - \cos\theta_{np}^v)^2]^{1/2} = \sec\theta_{Np}^v (1 - \tan\theta_{Np}^v)^{-1} \qquad [hn_p^{vo}, LN_p] \qquad (327)$$

$$C_{\beta x}^v C_{\beta y}^v C_{\beta z}^v = \cos\theta_{np}^v [1 + (\sec\theta_{np}^v - 1)^2]^{-1/2} = \cos\theta_{Np}^v (1 + \tan\theta_{Np}^v)^{-1} \qquad [ln_p^{vo}, SN_p] \qquad (328)$$

$$C_{\beta x}^v C_{\beta y}^v C_{\beta z}^v = \cos\theta_{np}^v [1 + (\sec\theta_{np}^v - 1)^2]^{1/2} = \sec\theta_{Np}^v (1 + \tan\theta_{Np}^v)^{-1} \qquad [ln_p^{vo}, LN_p] \qquad (329)$$

where $\beta = 2,3$. Combining equations (300) through (303) with equations (324) and (325) gives for a uniaxial APF along the x-coordinate in a linear elastic body

$$C_{1x}^v = \sec\theta_{np}^v [1 + (1 - \cos\theta_{np}^v)^2]^{-1/2} = \cos\theta_{Np}^v (1 - \tan\theta_{Np}^v)^{-1} \qquad [hn_p^{vo}, SN_p] \qquad (330)$$

$$C_{1x}^v = \sec\theta_{np}^v [1 + (1 - \cos\theta_{np}^v)^2]^{1/2} = \sec\theta_{Np}^v (1 - \tan\theta_{Np}^v)^{-1} \qquad [hn_p^{vo}, LN_p] \qquad (331)$$

$$C_{\beta x}^v = \cos\theta_{np}^v [1 + (\sec\theta_{np}^v - 1)^2]^{-1/2} = \cos\theta_{Np}^v (1 + \tan\theta_{Np}^v)^{-1} \qquad [ln_p^{vo}, SN_p] \qquad (332)$$

$$C^{v}_{\beta x} = \cos\theta^{v}_{np}[1+(\sec\theta^{v}_{np}-1)^2]^{1/2} = \sec\theta^{v}_{Np}(1+\tan\theta^{v}_{Np})^{-1} \qquad [\ln^{vo}_{p}, LN_{p}] \qquad (333)$$

for Cartesian coordinates.

Cylindrical Polar Coordinates.

Combining equations (304) through (307) with equations (324) and (325) gives for the case of a 3-dimensional APF

$$(C^{v}_{1r})^2 C^{v}_{1z} = \sec\theta^{v}_{np}[1+(1-\cos\theta^{v}_{np})^2]^{-1/2} = \cos\theta^{v}_{Np}(1-\tan\theta^{v}_{Np})^{-1} \qquad [hn^{vo}_{p}, SN_{p}] \qquad (334)$$

$$(C^{v}_{1r})^2 C^{v}_{1z} = \sec\theta^{v}_{np}[1+(1-\cos\theta^{v}_{np})^2]^{1/2} = \sec\theta^{v}_{Np}(1-\tan\theta^{v}_{Np})^{-1} \qquad [hn^{vo}_{p}, LN_{p}] \qquad (335)$$

$$(C^{v}_{\beta r})^2 C^{v}_{\beta z} = \cos\theta^{v}_{np}[1+(\sec\theta^{v}_{np}-1)^2]^{-1/2} = \cos\theta^{v}_{Np}(1+\tan\theta^{v}_{Np})^{-1} \qquad [\ln^{vo}_{p}, SN_{p}] \qquad (336)$$

$$(C^{v}_{\beta r})^2 C^{v}_{\beta z} = \cos\theta^{v}_{np}[1+(\sec\theta^{v}_{np}-1)^2]^{1/2} = \sec\theta^{v}_{Np}(1+\tan\theta^{v}_{Np})^{-1} \qquad [\ln^{vo}_{p}, LN_{p}] \qquad (337)$$

For a uniaxial APF, $C^{v}_{1r} = 1$ and $C^{v}_{\beta r} = 1$ and equations (334) through (337) become just the uniaxial APF equations (330) through (333) but now for the z-axis. For the special case of a 2-dimensional radial APF, $C^{v}_{1z} = 1$ and $C^{v}_{\beta z} = 1$ and equations (334) through (337) become

$$C^{v}_{1r} = \sec^{1/2}\theta^{v}_{np}[1+(1-\cos\theta^{v}_{np})^2]^{-1/4} = \cos^{1/2}\theta^{v}_{Np}(1-\tan\theta^{v}_{Np})^{-1/2} \qquad [hn^{vo}_{p}, SN_{p}] \qquad (338)$$

$$C^{v}_{1r} = \sec^{1/2}\theta^{v}_{np}[1+(1-\cos\theta^{v}_{np})^2]^{1/4} = \sec^{1/2}\theta^{v}_{Np}(1-\tan\theta^{v}_{Np})^{-1/2} \qquad [hn^{vo}_{p}, LN_{p}] \qquad (339)$$

$$C^{v}_{\beta r} = \cos^{1/2}\theta^{v}_{np}[1+(\sec\theta^{v}_{np}-1)^2]^{-1/4} = \cos^{1/2}\theta^{v}_{Np}(1+\tan\theta^{v}_{Np})^{-1/2} \qquad [\ln^{vo}_{p}, SN_{p}] \qquad (340)$$

$$C^{v}_{\beta r} = \cos^{1/2}\theta^{v}_{np}[1+(\sec\theta^{v}_{np}-1)^2]^{1/4} = \sec^{1/2}\theta^{v}_{Np}(1+\tan\theta^{v}_{Np})^{-1/2} \qquad [\ln^{vo}_{p}, LN_{p}] \qquad (341)$$

where $\beta = 2,3$.

Spherical Polar Coordinates.

Combining equations (320) through (323) with equations (324) and (325) gives for the case of a 3-dimensional APF in spherical polar coordinates

$$C^{v}_{1\rho} = \sec^{1/3}\theta^{v}_{np}[1+(1-\cos\theta^{v}_{np})^2]^{-1/6} = \cos^{1/3}\theta^{v}_{Np}(1-\tan\theta^{v}_{Np})^{-1/3} \qquad [hn^{vo}_{p}, SN_{p}] \qquad (342)$$

$$C^{v}_{1\rho} = \sec^{1/3}\theta^{v}_{np}[1+(1-\cos\theta^{v}_{np})^2]^{1/6} = \sec^{1/3}\theta^{v}_{Np}(1-\tan\theta^{v}_{Np})^{-1/3} \qquad [hn^{vo}_{p}, LN_{p}] \qquad (343)$$

$$C^{v}_{\beta\rho} = \cos^{1/3}\theta^{v}_{np}[1+(\sec\theta^{v}_{np}-1)^2]^{-1/6} = \cos^{1/3}\theta^{v}_{Np}(1+\tan\theta^{v}_{Np})^{-1/3} \qquad [\ln^{vo}_{p}, SN_{p}] \qquad (344)$$

$$C^{v}_{\beta\rho} = \cos^{1/3}\theta^{v}_{np}[1+(\sec\theta^{v}_{np}-1)^2]^{1/6} = \sec^{1/3}\theta^{v}_{Np}(1+\tan\theta^{v}_{Np})^{-1/3} \qquad [\ln^{vo}_{p}, LN_{p}] \qquad (345)$$

where $\beta = 2,3$. Equations (342) through (345) can also be obtained from equations (326) through (329) for Cartesian coordinates by taking $C^{v}_{1x} = C^{v}_{1y} = C^{v}_{1z}$ and $C^{v}_{\beta x} = C^{v}_{\beta y} = C^{v}_{\beta z}$.

The space and time coordinate factors for the vacuum that contains an APF can also be related to the internal phase angles of the complex number dimensions of space and time.[9,10] It can be conjectured that the complex number of spacetime dimensions $\overline{D}^{v}_{Dst}$ is linearly related to the complex number of virtual particle-antiparticle pairs $\overline{\mathfrak{N}}^{v}_{p}$ and the complex number spacetime dimension number density is linearly related to the complex number density of virtual particle-antiparticle pairs in the manner

$$\overline{D}^{v}_{st} = \mu\overline{\mathfrak{N}}^{v}_{p} \qquad D^{v}_{st} = \mu\mathfrak{N}^{v}_{p} \qquad \theta^{v}_{Dst} = \theta^{v}_{\mathfrak{N}p} \tag{346}$$

$$\overline{n}^{v}_{Dst} = \mu\overline{n}^{v}_{p} \qquad n^{v}_{Dst} = \mu n^{v}_{p} \qquad \theta^{v}_{nDst} = \theta^{v}_{np} \tag{347}$$

where the particle-antiparticle pair formalism is given in equations (277) through (295).

The space and time coordinate factors for the vacuum that contains an APF can also be related to the internal phase angles of the complex number dimensions of space and time.[9,10] It has been suggested that the complex number dimensions of space and time can be written as[8-10]

$$\overline{D}^{v}_{t} = D^{v}_{t}\exp(j\theta^{v}_{Dt}) \qquad \overline{D}^{v}_{s} = D^{v}_{s}\exp(j\theta^{v}_{Ds}) \qquad \overline{D}^{v}_{st} = D^{v}_{st}\exp(j\theta^{v}_{Dst}) \tag{348}$$

where for a low dimensional spacetime

$$D^{v}_{t} = N_{Dt}\cos\theta^{v}_{Dt} \qquad D^{v}_{s} = N_{Ds}\cos\theta^{v}_{Ds} \qquad D^{v}_{st} = N_{Dst}\cos\theta^{v}_{Dst} \qquad [SN_{Dst}] \tag{349}$$

and for a high dimensional spacetime

$$D^{v}_{t} = N_{Dt}\sec\theta^{v}_{Dt} \qquad D^{v}_{s} = N_{Ds}\sec\theta^{v}_{Ds} \qquad D^{v}_{st} = N_{Dst}\sec\theta^{v}_{Dst} \qquad [LN_{Dst}] \tag{350}$$

where $[SN_{Dst}]$ and $[LN_{Dst}]$ = small number of dimensions and large number of dimensions respectively and where N_{Dt}, N_{Ds} and $N_{Dst} = N_{Dt} + N_{Ds}$ = zero APF values of the dimensions of time, space and spacetime which in the vacuum of are generally integers but can be nonintegers in matter.[8-10] The value of the internal phase angle θ^{v}_{Dt} is given is given for a constant amplitude electromagnetic wave APF by

$$\tan\theta^{v}_{Dt} = (2\mathfrak{E}^{em}/G^{em}_{v})^{1/2} \qquad \mathfrak{E}^{em} = 1/2G^{em}_{v}\tan^{2}\theta^{v}_{Dt} \tag{351}$$

where $\mathfrak{E}^{em}$ = energy density of an electromagnetic APF and G^{em}_{v} = dynamic shear modulus of the vacuum in the presence of an electromagnetic APF.[8-10] If the APF is a constant amplitude gravitational wave in nature the governing relation for the internal phase angles is

$$\tan\theta^{v}_{Dst} = (2\mathfrak{E}^{g}/G^{g}_{v})^{1/2} \qquad \mathfrak{E}^{g} = 1/2G^{g}_{v}\tan^{2}\theta^{v}_{Dst} \tag{352}$$

where $\mathfrak{E}^{g}$ = energy density of an gravitational wave APF and G^{g}_{v} = dynamic shear modulus of the vacuum in the presence of an gravitational APF. Equations (351) and (352) are supplemented by an addition law for the complex number time, space and spacetime dimensions which can be used to determine all of the internal phase angles $\theta^v_{Dt}, \theta^v_{Ds}$ and θ^v_{Dst}.[8-10]

It has been shown that for the vacuum in an APF the complex number dimension number density is written as[8-10]

$$\bar{n}^v_{Dst} = n^v_{Dst}\exp(j\theta^v_{nDst}) = \bar{D}^v_{st}/\bar{V}^v_{Dst} \qquad n^v_{Dst} = D^v_{st}/V^v_{Dst} \qquad \theta_{nDst} = \theta_{Dst} - \theta_{VDst} \quad (353)$$

where the complex number spacetime dimension number is given in equations (348) through (350). The complex number volume of the spacetime dimensions is represented as

$$\bar{V}^v_{Dst} = V^v_{Dst}\exp(j\theta^v_{VDst}) \qquad \theta^v_{VDst} = \theta^v_{Dst} - \theta^v_{nDst} \qquad \theta^v_{VDst} \le 0 \quad (354)$$

For a bound homogeneous vacuum in a 3-dimensional APF, the internal phase angles of the space coordinates are written as

$$\theta^v_x = \theta^v_y = \theta^v_z = \theta^v_r = \theta^v_\rho = 1/3\theta^v_{VDst} = 1/3(\theta^v_{Dst} - \theta^v_{nDst}) \le 0 \quad (355)$$

and for a uniaxial APF in the x-direction

$$\theta^v_x = \theta^v_{VDst} = \theta^v_{Dst} - \theta^v_{nDst} \le 0 \quad (356)$$

For a 3-dimensional APF in the vacuum[9,10]

$$V^v_{Dst}/V^{vo}_{Dst} = C^v_{1x}C^v_{1y}C^v_{1z} = (C^v_{1r})^2C^v_{1z} = (C^v_{1\rho})^3 = \cos\theta^v_{Dst}\sec\theta^v_{nDst} \ge 1 \quad [hn^{vo}_{Dst}, SN_{Dst}] \quad (357)$$

$$V^v_{Dst}/V^{vo}_{Dst} = C^v_{1x}C^v_{1y}C^v_{1z} = (C^v_{1r})^2C^v_{1z} = (C^v_{1\rho})^3 = \sec\theta^v_{Dst}\sec\theta^v_{nDst} \ge 1 \quad [hn^{vo}_{Dst}, LN_{Dst}] \quad (358)$$

$$V^v_{Dst}/V^{vo}_{Dst} = C^v_{2x}C^v_{2y}C^v_{2z} = (C^v_{2r})^2C^v_{2z} = (C^v_{2\rho})^3 = \cos\theta^v_{Dst}\cos\theta^v_{nDst} \le 1 \quad [ln^{vo}_{Dst}, SN_{Dst}] \quad (359)$$

$$V^v_{Dst}/V^{vo}_{Dst} = C^v_{2x}C^v_{2y}C^v_{2z} = (C^v_{2r})^2C^v_{2z} = (C^v_{2\rho})^3 = \sec\theta^v_{Dst}\cos\theta^v_{nDst} \le 1 \quad [ln^{vo}_{Dst}, LN_{Dst}] \quad (360)$$

$$V^v_{Dst}/V^{vo}_{Dst} = C^v_{3x}C^v_{3y}C^v_{3z} = (C^v_{3r})^2C^v_{3z} = (C^v_{3\rho})^3 = \cos\theta^v_{Dst}\cos\theta^v_{nDst} \le 1 \quad [ln^{vo}_{Dst}, SN_{Dst}] \quad (361)$$

$$V^v_{Dst}/V^{vo}_{Dst} = C^v_{3x}C^v_{3y}C^v_{3z} = (C^v_{3r})^2C^v_{3z} = (C^v_{3\rho})^3 = \sec\theta^v_{Dst}\cos\theta^v_{nDst} \le 1 \quad [ln^{vo}_{Dst}, LN_{Dst}] \quad (362)$$

where in the case of the vacuum, $[hn^{vo}_{Dst}]$ = high dimensional density of vacuum, $[ln^{vo}_{Dst}]$ = low dimensional density of vacuum and where $[SN_{Dst}]$ and $[LN_{Dst}]$ = small number of dimensions and large number of dimensions respectively and where the space coordinate factors are given by equations (273) through (275), and so because the right hand sides of equations (359) and (360) are the same as the right hand sides of equations (361) and (362) it follows that the space coordinate factors $C^v_{2\alpha}$ and $C^v_{3\alpha}$ of equations (274) and (275) give two possible solutions for the internal phase angles $\theta^v_{\eta\alpha}$ and $\theta^v_{\pi\alpha}$ of the counting and segmentation numbers for the $[ln^{vo}_{Dst}]$ case and therefore the $[ln^{vo}_{Dst}]$ case is degenerate. Combining equations (348) through (350) with (353) and (354) gives the following results[9,10]

$$n^{v}_{Dst} = N_{Dst} / V^{v}_{Dst} \cos\theta^{v}_{Dst} \qquad [SN_{Dst}] \qquad (363)$$

$$n^{v}_{Dst} = N_{Dst} / V^{v}_{Dst} \sec\theta^{v}_{Dst} \qquad [LN_{Dst}] \qquad (364)$$

Then using equations (357) through (362) it is easily shown that[9,10]

$$n^{v}_{Dst} = n^{vo}_{Dst} \cos\theta^{v}_{nDst} \le n^{vo}_{Dst} \qquad [hn^{vo}_{Dst}] \qquad (365)$$

$$n^{v}_{Dst} = n^{vo}_{Dst} \sec\theta^{v}_{nDst} \ge n^{vo}_{Dst} \qquad [ln^{vo}_{Dst}] \qquad (366)$$

where

$$n^{vo}_{Dst} = N_{Dst} / V^{vo}_{Dst} \qquad (367)$$

is the zero APF value of the dimensional number density. Equations (348) through (367) are the vacuum space and time dimension analogs of the vacuum particle-antiparticle equations (277) through (295).

Cartesian Coordinates.

For Cartesian coordinates the ratios of the volume elements with and without an APF present are given by

$$V^{v}_{Dst} / V^{vo}_{Dst} = C^{v}_{1x} C^{v}_{1y} C^{v}_{1z} = \cos\theta^{v}_{Dst} \sec\theta^{v}_{nDst} \ge 1 \qquad [hn^{vo}_{Dst}, SN_{Dst}] \qquad (368)$$

$$V^{v}_{Dst} / V^{vo}_{Dst} = C^{v}_{1x} C^{v}_{1y} C^{v}_{1z} = \sec\theta^{v}_{Dst} \sec\theta^{v}_{nDst} \ge 1 \qquad [hn^{vo}_{Dst}, LN_{Dst}] \qquad (369)$$

$$V^{v}_{Dst} / V^{vo}_{Dst} = C^{v}_{\beta x} C^{v}_{\beta y} C^{v}_{\beta z} = \cos\theta^{v}_{Dst} \cos\theta^{v}_{nDst} \le 1 \qquad [ln^{vo}_{Dst}, SN_{Dst}] \qquad (370)$$

$$V^{v}_{Dst} / V^{vo}_{Dst} = C^{v}_{\beta x} C^{v}_{\beta y} C^{v}_{\beta z} = \sec\theta^{v}_{Dst} \cos\theta^{v}_{nDst} \le 1 \qquad [ln^{vo}_{Dst}, LN_{Dst}] \qquad (371)$$

where $C^{v}_{1\alpha}$ and $C^{v}_{\beta\alpha}$ are given by equations (273) through (275) and where $\beta = 2,3$ describes the degeneracy of the $[ln^{vo}_{Dst}]$ state. For the case of a uniaxial APF in the x-direction, $C^{v}_{1y} = 1, C^{v}_{1z} = 1, C^{v}_{\beta y} = 1$ and $C^{v}_{\beta z} = 1$ with the result that equations (368) through (371) give

$$V^{v}_{Dst} / V^{vo}_{Dst} = C^{v}_{1x} = \cos\theta^{v}_{Dst} \sec\theta^{v}_{nDst} \ge 1 \qquad [hn^{vo}_{Dst}, SN_{Dst}] \qquad (372)$$

$$V^{v}_{Dst} / V^{vo}_{Dst} = C^{v}_{1x} = \sec\theta^{v}_{Dst} \sec\theta^{v}_{nDst} \ge 1 \qquad [hn^{vo}_{Dst}, LN_{Dst}] \qquad (373)$$

$$V^{v}_{Dst} / V^{vo}_{Dst} = C^{v}_{\beta x} = \cos\theta^{v}_{Dst} \cos\theta^{v}_{nDst} \le 1 \qquad [ln^{vo}_{Dst}, SN_{Dst}] \qquad (374)$$

$$V^{v}_{Dst} / V^{vo}_{Dst} = C^{v}_{\beta x} = \sec\theta^{v}_{Dst} \cos\theta^{v}_{nDst} \le 1 \qquad [ln^{vo}_{Dst}, LN_{Dst}] \qquad (375)$$

where $\beta = 2,3$.

Cylindrical Polar Coordinates.

For the case of cylindrical polar coordinates the ratios of the volume elements with and without an APF are given by

$$V^{v}_{Dst} / V^{vo}_{Dst} = (C^{v}_{1r})^2 C^{v}_{1z} = \cos\theta^{v}_{Dst} \sec\theta^{v}_{nDst} \geq 1 \qquad [hn^{vo}_{Dst}, SN_{Dst}] \qquad (376)$$

$$V^{v}_{Dst} / V^{vo}_{Dst} = (C^{v}_{1r})^2 C^{v}_{1z} = \sec\theta^{v}_{Dst} \sec\theta^{v}_{nDst} \geq 1 \qquad [hn^{vo}_{Dst}, LN_{Dst}] \qquad (377)$$

$$V^{v}_{Dst} / V^{vo}_{Dst} = (C^{v}_{\beta r})^2 C^{v}_{\beta z} = \cos\theta^{v}_{Dst} \cos\theta^{v}_{nDst} \leq 1 \qquad [ln^{vo}_{Dst}, SN_{Dst}] \qquad (378)$$

$$V^{v}_{Dst} / V^{vo}_{Dst} = (C^{v}_{\beta r})^2 C^{v}_{\beta z} = \sec\theta^{v}_{Dst} \cos\theta^{v}_{nDst} \leq 1 \qquad [ln^{vo}_{Dst}, LN_{Dst}] \qquad (379)$$

where $\beta = 2,3$. For the case of a uniaxial APF along the z-axis, $C^{v}_{1r} = 1$ and $C^{v}_{\beta r} = 1$ and equations (376) through (379) become

$$V^{v}_{Dst} / V^{vo}_{Dst} = C^{v}_{1z} = \cos\theta^{v}_{Dst} \sec\theta^{v}_{nDst} \geq 1 \qquad [hn^{vo}_{Dst}, SN_{Dst}] \qquad (380)$$

$$V^{v}_{Dst} / V^{vo}_{Dst} = C^{v}_{1z} = \sec\theta^{v}_{Dst} \sec\theta^{v}_{nDst} \geq 1 \qquad [hn^{vo}_{Dst}, LN_{Dst}] \qquad (381)$$

$$V^{v}_{Dst} / V^{vo}_{Dst} = C^{v}_{\beta z} = \cos\theta^{v}_{Dst} \cos\theta^{v}_{nDst} \leq 1 \qquad [ln^{vo}_{Dst}, SN_{Dst}] \qquad (382)$$

$$V^{v}_{Dst} / V^{vo}_{Dst} = C^{v}_{\beta z} = \sec\theta^{v}_{Dst} \cos\theta^{v}_{nDst} \leq 1 \qquad [ln^{vo}_{Dst}, LN_{Dst}] \qquad (383)$$

where $\beta = 2,3$. For the special case of a two-dimensional radial APF, $C^{v}_{1z} = 1$ and $C^{v}_{\beta z} = 1$ and equations (376) through (379) give

$$(V^{v}_{Dst} / V^{vo}_{Dst})^{1/2} = C^{v}_{1r} = \cos^{1/2}\theta^{v}_{Dst} \sec^{1/2}\theta^{v}_{nDst} \geq 1 \qquad [hn^{vo}_{Dst}, SN_{Dst}] \qquad (384)$$

$$(V^{v}_{Dst} / V^{vo}_{Dst})^{1/2} = C^{v}_{1r} = \sec^{1/2}\theta^{v}_{Dst} \sec^{1/2}\theta^{v}_{nDst} \geq 1 \qquad [hn^{vo}_{Dst}, LN_{Dst}] \qquad (385)$$

$$(V^{v}_{Dst} / V^{vo}_{Dst})^{1/2} = C^{v}_{\beta r} = \cos^{1/2}\theta^{v}_{Dst} \cos^{1/2}\theta^{v}_{nDst} \leq 1 \qquad [ln^{vo}_{Dst}, SN_{Dst}] \qquad (386)$$

$$(V^{v}_{Dst} / V^{vo}_{Dst})^{1/2} = C^{v}_{\beta r} = \sec^{1/2}\theta^{v}_{Dst} \cos^{1/2}\theta^{v}_{nDst} \leq 1 \qquad [ln^{vo}_{Dst}, LN_{Dst}] \qquad (387)$$

where $\beta = 2,3$.

Spherical Polar Coordinates.

For an isotropic APF in the vacuum,

$$V^{v}_{Dst} / V^{vo}_{Dst} = (C^{v}_{1\rho})^3 = \cos\theta^{v}_{Dst} \sec\theta^{v}_{nDst} \geq 1 \qquad [hn^{vo}_{Dst}, SN_{Dst}] \qquad (388)$$

$$V^{v}_{Dst} / V^{vo}_{Dst} = (C^{v}_{1\rho})^3 = \sec\theta^{v}_{Dst} \sec\theta^{v}_{nDst} \geq 1 \qquad [hn^{vo}_{Dst}, LN_{Dst}] \qquad (389)$$

$$V^{v}_{Dst} / V^{vo}_{Dst} = (C^{v}_{\beta\rho})^3 = \cos\theta^{v}_{Dst} \cos\theta^{v}_{nDst} \leq 1 \qquad [\ln^{vo}_{Dst}, SN_{Dst}] \qquad (390)$$

$$V^{v}_{Dst} / V^{vo}_{Dst} = (C^{v}_{\beta\rho})^3 = \sec\theta^{v}_{Dst} \cos\theta^{v}_{nDst} \leq 1 \qquad [\ln^{vo}_{Dst}, LN_{Dst}] \qquad (391)$$

or

$$(V^{v}_{Dst} / V^{vo}_{Dst})^{1/3} = C^{v}_{1\rho} = \cos^{1/3}\theta^{v}_{Dst} \sec^{1/3}\theta^{v}_{nDst} \geq 1 \qquad [hn^{vo}_{Dst}, SN_{Dst}] \qquad (392)$$

$$(V^{v}_{Dst} / V^{vo}_{Dst})^{1/3} = C^{v}_{1\rho} = \sec^{1/3}\theta^{v}_{Dst} \sec^{1/3}\theta^{v}_{nDst} \geq 1 \qquad [hn^{vo}_{Dst}, LN_{Dst}] \qquad (393)$$

$$(V^{v}_{Dst} / V^{vo}_{Dst})^{1/3} = C^{v}_{\beta\rho} = \cos^{1/3}\theta^{v}_{Dst} \cos^{1/3}\theta^{v}_{nDst} \leq 1 \qquad [\ln^{vo}_{Dst}, SN_{Dst}] \qquad (394)$$

$$(V^{v}_{Dst} / V^{vo}_{Dst})^{1/3} = C^{v}_{\beta\rho} = \sec^{1/3}\theta^{v}_{Dst} \cos^{1/3}\theta^{v}_{nDst} \leq 1 \qquad [\ln^{vo}_{Dst}, LN_{Dst}] \qquad (395)$$

where the index $\beta = 2,3$ describes the degeneracy of the $[\ln^{vo}_{Dst}]$ state and where the space coordinate factors $C^{v}_{\sigma\rho}$ for spherical polar coordinates in the vacuum are given by equations (273) through (275) with $\alpha = \rho$. Note that equations (392) through (395) for spherical polar coordinates can also be obtained from equations (368) through (377) for a Cartesian coordinate system by choosing $C^{v}_{1x} = C^{v}_{1y} = C^{v}_{1z}$ and $C^{v}_{\beta x} = C^{v}_{\beta y} = C^{v}_{\beta z}$ and from equations (376) through (379) for cylindrical polar coordinates by requiring the condition $C^{v}_{1r} = C^{v}_{1z}$ and $C^{v}_{\beta r} = C^{v}_{\beta z}$.

Consider now the case of a linear elastic vacuum for which the internal phase angles θ^{v}_{nDst} and θ^{v}_{Dst} are related by[9]

$$\cos\theta^{v}_{nDst} = 1 - \tan\theta^{v}_{Dst} \qquad 0 \leq \theta^{v}_{nDst} \leq \pi/2 \qquad 0 \leq \theta^{v}_{Dst} \leq \pi/4 \qquad [hn^{vo}_{Dst}] \qquad (396)$$

$$\sec\theta^{v}_{nDst} = 1 + \tan\theta^{v}_{Dst} \qquad 0 \leq \theta^{v}_{nDst} \leq \pi/2 \qquad 0 \leq \theta^{v}_{Dst} \leq \pi/2 \qquad [\ln^{vo}_{Dst}] \qquad (397)$$

with $\theta^{v}_{nDst} \geq \theta^{v}_{Dst}$, and which can now be used in equations (368) through (395) to describe the space coordinate factors for a linear elastic vacuum in an APF.

Cartesian Coordinates

Combining equations (368) through (371) with equations (396) and (397) gives for the case of a 3-dimensional APF in Cartesian coordinates

$$C^{v}_{1x}C^{v}_{1y}C^{v}_{1z} = \sec\theta^{v}_{nDst}[1 + (1 - \cos\theta^{v}_{nDst})^2]^{-1/2} = \cos\theta^{v}_{Dst}(1 - \tan\theta^{v}_{Dst})^{-1} \quad [hn^{vo}_{Dst}, SN_{Dst}] \ (398)$$

$$C^{v}_{1x}C^{v}_{1y}C^{v}_{1z} = \sec\theta^{v}_{nDst}[1 + (1 - \cos\theta^{v}_{nDst})^2]^{1/2} = \sec\theta^{v}_{Dst}(1 - \tan\theta^{v}_{Dst})^{-1} \quad [hn^{vo}_{Dst}, LN_{Dst}] \ (399)$$

$$C^{v}_{\beta x}C^{v}_{\beta y}C^{v}_{\beta z} = \cos\theta^{v}_{nDst}[1 + (\sec\theta^{v}_{nDst} - 1)^2]^{-1/2} = \cos\theta^{v}_{Dst}(1 + \tan\theta^{v}_{Dst})^{-1} \quad [\ln^{vo}_{Dst}, SN_{Dst}] \ (400)$$

$$C^{v}_{\beta x}C^{v}_{\beta y}C^{v}_{\beta z} = \cos\theta^{v}_{nDst}[1 + (\sec\theta^{v}_{nDst} - 1)^2]^{1/2} = \sec\theta^{v}_{Dst}(1 + \tan\theta^{v}_{Dst})^{-1} \quad [\ln^{vo}_{Dst}, LN_{Dst}] \ (401)$$

where $\beta = 2,3$. Combining equations (372) through (375) with equations (396) and (397) gives for a uniaxial APF in the x-direction for a linear elastic body

$$C^{v}_{1x} = \sec\theta^{v}_{nDst}[1+(1-\cos\theta^{v}_{nDst})^2]^{-1/2} = \cos\theta^{v}_{Dst}(1-\tan\theta^{v}_{Dst})^{-1} \qquad [hn^{vo}_{Dst}, SN_{Dst}] \quad (402)$$

$$C^{v}_{1x} = \sec\theta^{v}_{nDst}[1+(1-\cos\theta^{v}_{nDst})^2]^{1/2} = \sec\theta^{v}_{Dst}(1-\tan\theta^{v}_{Dst})^{-1} \qquad [hn^{vo}_{Dst}, LN_{Dst}] \quad (403)$$

$$C^{v}_{\beta x} = \cos\theta^{v}_{nDst}[1+(\sec\theta^{v}_{nDst}-1)^2]^{-1/2} = \cos\theta^{v}_{Dst}(1+\tan\theta^{v}_{Dst})^{-1} \qquad [ln^{vo}_{Dst}, SN_{Dst}] \quad (404)$$

$$C^{v}_{\beta x} = \cos\theta^{v}_{nDst}[1+(\sec\theta^{v}_{nDst}-1)^2]^{1/2} = \sec\theta^{v}_{Dst}(1+\tan\theta^{v}_{Dst})^{-1} \qquad [ln^{vo}_{Dst}, LN_{Dst}] \quad (405)$$

for Cartesian coordinates.

Cylindrical Polar Coordinates.

Combining equations (376) through (379) with equations (396) and (397) gives for the case of a 3-dimensional APF

$$(C^{v}_{1r})^2 C^{v}_{1z} = \sec\theta^{v}_{nDst}[1+(1-\cos\theta^{v}_{nDst})^2]^{-1/2} = \cos\theta^{v}_{Dst}(1-\tan\theta^{v}_{Dst})^{-1} \quad [hn^{vo}_{Dst}, SN_{Dst}] \quad (406)$$

$$(C^{v}_{1r})^2 C^{v}_{1z} = \sec\theta^{v}_{nDst}[1+(1-\cos\theta^{v}_{nDst})^2]^{1/2} = \sec\theta^{v}_{Dst}(1-\tan\theta^{v}_{Dst})^{-1} \quad [hn^{vo}_{Dst}, LN_{Dst}] \quad (407)$$

$$(C^{v}_{1r})^2 C^{v}_{\beta z} = \cos\theta^{v}_{nDst}[1+(\sec\theta^{v}_{nDst}-1)^2]^{-1/2} = \cos\theta^{v}_{Dst}(1+\tan\theta^{v}_{Dst})^{-1} \quad [ln^{vo}_{Dst}, SN_{Dst}] \quad (408)$$

$$(C^{v}_{1r})^2 C^{v}_{\beta z} = \cos\theta^{v}_{nDst}[1+(\sec\theta^{v}_{nDst}-1)^2]^{1/2} = \sec\theta^{v}_{Dst}(1+\tan\theta^{v}_{Dst})^{-1} \quad [ln^{vo}_{Dst}, LN_{Dst}] \quad (409)$$

For a uniaxial APF, $C^{v}_{1r} = 1$ and $C^{v}_{\beta r} = 1$ and equations (406) through (409) become just the uniaxial APF equations (402) through (405) but now for the z-axis. For the special case of a 2-dimensional radial APF, $C^{v}_{1z} = 1$ and $C^{v}_{\beta z} = 1$ and equations (406) through (409) become

$$C^{v}_{1r} = \sec^{1/2}\theta^{v}_{nDst}[1+(1-\cos\theta^{v}_{nDst})^2]^{-1/4} = \cos^{1/2}\theta^{v}_{Dst}(1-\tan\theta^{v}_{Dst})^{-1/2} \quad [hn^{vo}_{Dst}, SN_{Dst}] \quad (410)$$

$$C^{v}_{1r} = \sec^{1/2}\theta^{v}_{nDst}[1+(1-\cos\theta^{v}_{nDst})^2]^{1/4} = \sec^{1/2}\theta^{v}_{Dst}(1-\tan\theta^{v}_{Dst})^{-1/2} \quad [hn^{vo}_{Dst}, LN_{Dst}] \quad (411)$$

$$C^{v}_{\beta r} = \cos^{1/2}\theta^{v}_{nDst}[1+(\sec\theta^{v}_{nDst}-1)^2]^{-1/4} = \cos^{1/2}\theta^{v}_{Dst}(1+\tan\theta^{v}_{Dst})^{-1/2} \quad [ln^{vo}_{Dst}, SN_{Dst}] \quad (412)$$

$$C^{v}_{\beta r} = \cos^{1/2}\theta^{v}_{nDst}[1+(\sec\theta^{v}_{nDst}-1)^2]^{1/4} = \sec^{1/2}\theta^{v}_{Dst}(1+\tan\theta^{v}_{Dst})^{-1/2} \quad [ln^{vo}_{Dst}, LN_{Dst}] \quad (413)$$

where $\beta = 2,3$ describes the degeneracy of the $[ln^{vo}_{Dst}]$ state.

Spherical Polar Coordinates.

Combining equations (392) through (395) with equations (396) and (397) gives for the case of a 3-dimensional APF in spherical polar coordinates

$$C^{v}_{1\rho} = \sec^{1/3}\theta^{v}_{nDst}[1+(1-\cos\theta^{v}_{nDst})^2]^{-1/6} = \cos^{1/3}\theta^{v}_{Dst}(1-\tan\theta^{v}_{Dst})^{-1/3} \quad [hn^{vo}_{Dst}, SN_{Dst}] \quad (414)$$

$$C^{v}_{1\rho} = \sec^{1/3}\theta^{v}_{nDst}[1+(1-\cos\theta^{v}_{nDst})^2]^{1/6} = \sec^{1/3}\theta^{v}_{Dst}(1-\tan\theta^{v}_{Dst})^{-1/3} \quad [hn^{vo}_{Dst}, LN_{Dst}] \quad (415)$$

$$C^{v}_{\beta\rho} = \cos^{1/3}\theta^{v}_{nDst}[1+(\sec\theta^{v}_{nDst}-1)^2]^{-1/6} = \cos^{1/3}\theta^{v}_{Dst}(1+\tan\theta^{v}_{Dst})^{-1/3} \quad [ln^{vo}_{Dst}, SN_{Dst}] \quad (416)$$

$$C^{v}_{\beta\rho} = \cos^{1/3}\theta^{v}_{nDst}[1+(\sec\theta^{v}_{nDst}-1)^2]^{1/6} = \sec^{1/3}\theta^{v}_{Dst}(1+\tan\theta^{v}_{Dst})^{-1/3} \quad [ln^{vo}_{Dst}, LN_{Dst}] \quad (417)$$

where $\beta = 2,3$. Equations (414) through (417) can also be obtained from equations (398) through (401) for Cartesian coordinates by taking $C^{v}_{1x} = C^{v}_{1y} = C^{v}_{1z}$ and $C^{v}_{\beta x} = C^{v}_{\beta y} = C^{v}_{\beta z}$.

The time coordinate factor for a frame of reference that is moving through the vacuum relative to a fixed frame of reference can be obtained from the definitions of linear velocity for uniform motion of a body under zero APF conditions in the vacuum

$$v_o = d\alpha^{v}_{o} / dt^{v}_{o} \quad (418)$$

and for the case when a constant APF is present in the vacuum

$$\bar{v} = v\exp(j\theta_v) = d\bar{\alpha}^{v} / d\bar{t}^{v} \quad (419)$$

$$v = d\alpha^{v} / dt^{v} = (d\alpha^{v}_{o} / dt^{v}_{o})(C^{v}_{\sigma\alpha} / C^{v}_{\sigma t}) = v_o (C^{v}_{\sigma\alpha} / C^{v}_{\sigma t}) \quad (420)$$

$$\theta_v = \theta^{v}_{\alpha} - \theta^{v}_{t} = \theta^{v}_{\eta\alpha} - \theta^{v}_{\pi\alpha} - (\theta^{v}_{\eta t} - \theta^{v}_{\pi t}) \quad (421)$$

so that the time coordinate factor $C^{v}_{\sigma t}$ is related to the space coordinate factor $C^{v}_{\sigma\alpha}$ by

$$C^{v}_{\sigma t} = (v_o / v) C^{v}_{\sigma\alpha} \quad (422)$$

where $\sigma = 1,2,3$; $\alpha = x,y,z$ or r,z or ρ and where v_o = velocity of a reference frame moving in the vacuum relative to a stationary frame of reference under zero APF conditions, and v = magnitude of the velocity of a reference frame that is moving through the vacuum relative to a stationary reference frame when an APF exists in the vacuum. Note that no body is present, stationary or moving, and v and v_o refer to the relative velocity of two frames of reference through the vacuum with and without an APF. The constant space and time coordinate factors $C^{v}_{\sigma\alpha}$ and $C^{v}_{\sigma t}$ that describe the space and time coordinates of the vacuum in the presence of an APF are expected to depend only on the properties of the vacuum as given by the constant phase angles $\theta^{v}_{\eta\alpha}, \theta^{v}_{\pi\alpha}$ and $\theta^{v}_{\eta t}, \theta^{v}_{\pi t}$ or equivalently θ^{v}_{Dst} and θ^{v}_{nDst} or θ^{v}_{Np} and θ^{v}_{np} and the ratio v/v_o.

D. Forms of the Space and Time Coordinate Factors for a Moving Polyphase space vehicle From the Perspective of a Moving Coordinate System.

The space and time coordinates in matter or vacuum that is located in an APF have been represented as complex numbers in an internal space.[5-12] A simple theory of the origin and nature of these complex number coordinates has already been developed.[9] Consider the constant real number spatial parameters L^{c}_{α} for the complex number space coordinates $\bar{\alpha}'$ in an APF in bulk matter, where L^{c}_{α} can be taken to be multiples of the Planck length. The real number constant spatial parameters L^{c}_{α} are divided into $\bar{n}'_{\alpha}$ parts of length $\Delta\overline{L}'_\alpha$ where the complex number spatial segmentation number $\overline{n}'_\alpha$ is represented as[9]

$$\overline{n}'_\alpha = n'_\alpha \exp(j\theta'_{n\alpha}) \tag{423}$$

so that

$$\Delta\overline{L}'_\alpha = \Delta L'_\alpha \exp(j\theta'_{\Delta L\alpha}) = L^c_\alpha / \overline{n}'_\alpha = L^c_\alpha / n'_\alpha \exp(-j\theta'_{n\alpha}) \tag{424}$$

and

$$\Delta L'_\alpha = L^c_\alpha / n'_\alpha \qquad \theta'_{\Delta L\alpha} = -\theta'_{n\alpha} \tag{425}$$

where α = x,y,z; r,z or ρ. Then the complex number space coordinate $\overline{\alpha}'$ can be written as

$$\overline{\alpha}' = \alpha' \exp(j\theta'_\alpha) = \overline{\eta}'_\alpha \Delta\overline{L}'_\alpha = \overline{\eta}'_\alpha L^c_\alpha / \overline{n}'_\alpha = \eta'_\alpha L^c_\alpha / n'_\alpha \exp[j(\theta'_{\eta\alpha} - \theta'_{n\alpha})] \tag{426}$$

where the complex number spatial counting number is represented as

$$\overline{\eta}'_\alpha = \eta'_\alpha \exp(j\theta'_{\eta\alpha}) \tag{427}$$

so that

$$\alpha' = \eta'_\alpha L^c_\alpha / n'_\alpha = \eta'_\alpha \Delta L'_\alpha \qquad \theta'_\alpha = \theta'_{\eta\alpha} - \theta'_{n\alpha} \tag{428}$$

where η'_α and n'_α are not integers in the presence of an APF but are integers under zero APF conditions.

The time coordinate factor, is developed as follows. Consider a real number constant L^c_t that is used to measure intervals of the complex number time coordinate $\overline{t}'$, where L^c_t can be taken to be multiples of the Planck time. In the presence of an APF in bulk matter the real number constant temporal parameter L^c_t is divided into $\overline{n}'_t$ parts of length $\Delta\overline{L}'_t$ where the complex number time segmentation number is represented as[9]

$$\overline{n}'_t = n'_t \exp(j\theta'_{nt}) \tag{429}$$

and where

$$\Delta\overline{L}'_t = \Delta L'_t \exp(j\theta'_{\Delta Lt}) = L^c_t / \overline{n}'_t = L^c_t / n'_t \exp(-j\theta'_{nt}) \tag{430}$$

and therefore

$$\Delta L'_t = L^c_t / n'_t \qquad \theta'_{\Delta Lt} = -\theta'_{nt} \tag{431}$$

The complex number time coordinate is then written as

$$\overline{t}' = t' \exp(j\theta'_t) = \overline{\eta}'_t \Delta\overline{L}'_t = \overline{\eta}'_t L^c_t / \overline{n}'_t = \eta'_t L^c_t / n'_t \exp[j(\theta'_{\eta t} - \theta'_{nt})] \tag{432}$$

where the complex number time counting number is written as

$$\overline{\eta}'_t = \eta'_t \exp(j\theta'_{\eta t}) \tag{433}$$

so that

$$t' = \eta'_t L^c_t / \imath'_t = \eta'_t \Delta L'_t \qquad \theta'_t = \theta'_{\eta t} - \theta'_{\imath t} \tag{434}$$

where η'_t and $\imath'_t$ are not integers when an APF is present in matter, and the complex number nature of the time coordinate is $\bar{t}'$ due only to the complex number time counting number $\overline{\eta}_t$ and the complex number segmentation number $\overline{\imath}'_t$ because the constant L^c_t is a real number in equation (432). The real number constants L^c_α and L^c_t that appear in equations (426) and (432) can be taken to be multiples (generally very large multiples) of the Planck length and Planck time respectively because an APF is a field that induces internal phase angles in all physical quantities except the fundamental constants of nature (h, c and G). Equations (428) and (434) allow the possibility of $\theta'_\alpha \leq 0$ and $\theta'_t \leq 0$.

Part D delineates the space and time coordinate factors for several types of moving coordinate systems. The space and time coordinates in vacuum or matter that is located in an APF are represented as complex numbers of the form[5-10]

$$\overline{\alpha}' = \alpha' \exp(j\theta'_\alpha) \qquad \bar{t}' = t' \exp(j\theta'_t) \tag{435}$$

where the coordinate magnitudes are $\alpha' = x', y', z'$ or r', z' or ρ' for Cartesian coordinates, cylindrical polar coordinates or spherical polar coordinates respectively. The space and time coordinate magnitudes and internal phase angles for regions of space and time within matter and in an APF are given by

$$\alpha' = \alpha'_o C'_{\sigma\alpha} \qquad \theta'_\alpha = \theta'_{\eta\alpha} - \theta'_{\imath\alpha} \tag{436}$$

$$t' = t'_o C'_{\sigma t} \qquad \theta'_t = \theta'_{\eta t} - \theta'_{\imath t} \tag{437}$$

where α'_o and t'_o = values of the space and time coordinates for zero APF conditions, σ = 1,2,3 and where the space and time coordinate factors $C'_{\sigma\alpha}$ and $C'_{\sigma t}$ for the space and time coordinates of the moving system are given by[9]

$$C'_{1\alpha} = \cos\theta'_{\eta\alpha} \sec\theta'_{\imath\alpha} \qquad C'_{1t} = \cos\theta'_{\eta t} \sec\theta'_{\imath t} \qquad [hn'_{po}] \tag{438}$$

$$C'_{2\alpha} = \cos\theta'_{\eta\alpha} \cos\theta'_{\imath\alpha} \qquad C'_{2t} = \cos\theta'_{\eta t} \cos\theta'_{\imath t} \qquad [ln'_o] \tag{439}$$

$$C'_{3\alpha} = \sec\theta'_{\eta\alpha} \cos\theta'_{\imath\alpha} \qquad C'_{3t} = \sec\theta'_{\eta t} \cos\theta'_{\imath t} \qquad [ln'_o] \tag{440}$$

where $[hn'_o]$ = high-density matter of the polyphase space vehicle and $[ln'_o]$ = low-density matter of the polyphase space vehicle and where $\theta'_{\eta\alpha}, \theta'_{\imath\alpha}$ and $\theta'_{\eta t}, \theta'_{\imath t}$ are the apparent internal phase angles of the complex number space and time counting and segmentation numbers as seen by an observer in the moving coordinate system. For bound matter in an APF

$$\theta'_{\imath\alpha} \geq \theta'_{\eta\alpha} \qquad \theta'_{\imath t} \geq \theta'_{\eta t} \tag{441}$$

and for the conditions in equation (441), equations (436) and (437) give

$$\theta'_{\alpha} \le 0 \qquad \theta'_{t} \le 0 \tag{442}$$

The values of the space coordinate factors $C'_{\sigma\alpha}$ for the radial coordinate of a spherical polar coordinate system can also be simply expressed in terms of the internal phase angles of the complex particle number density.[9]

For a finite sized element of vacuum in an APF the apparent complex number particle-antiparticle number density $\bar{n}_p$ as seen from a moving coordiate system is written as[9]

$$\bar{n}'_p = n'_p \exp(j\theta_{np}) = \overline{\mathbb{N}'_p}/\overline{V'_p} \qquad n'_p = \mathbb{N}'_p / V'_p \qquad \theta'_{np} = \theta'_{\mathbb{N}p} - \theta'_{Vp} \tag{443}$$

where the complex number total particle-antiparticle pair number is written as

$$\overline{\mathbb{N}'_p} = \mathbb{N}'_p \exp(j\theta'_{\mathbb{N}p}) \tag{444}$$

where

$$\mathbb{N}'_p = N_p \cos\theta'_{\mathbb{N}p} \qquad [SN_p] \tag{445}$$

$$\mathbb{N}'_p = N_p \sec\theta'_{\mathbb{N}p} \qquad [LN_p] \tag{446}$$

where $[SN_p]$ = small N_p, $[LN_p]$ = large N_p and where N_p = integer number of particle-antiparticle pairs of a vacuum which interact with the APF. The complex number volume is represented as

$$\overline{V'_p} = V'_p \exp(j\theta'_{Vp}) \qquad \theta_{Vp} = \theta_{\mathbb{N}p} - \theta_{np} \qquad \theta_{Vp} \le 0 \tag{447}$$

and where the subscript p refers to pairs of particles and antiparticles. For a bound homogeneous vacuum in a 3-dimensional APF the internal phase angles of the space coordinates are then written as

$$\theta'_x = \theta'_y = \theta'_z = \theta'_r = \theta'_\rho = 1/3\theta'_{Vp} = 1/3(\theta'_{\mathbb{N}p} - \theta'_{np}) \le 0 \tag{448}$$

while for a uniaxial APF in the x-direction

$$\theta_x = \theta_{Vp} = \theta_{\mathbb{N}p} - \theta_{np} \le 0 \tag{449}$$

From equations (443) through (447) it follows that[11]

$$n'_p = N_p / V'_p \cos\theta'_{\mathbb{N}p} \quad [SN_p] \tag{450}$$

$$n'_p = N_p / V'_p \sec\theta'_{\mathbb{N}p} \quad [LN_p] \tag{451}$$

$$\theta'_{np} = \theta'_{\mathbb{N}p} - \theta'_{Vp} = \theta'_{\mathbb{N}p} + |\theta'_{Vp}| \qquad \theta'_{np} \ge \theta'_{\mathbb{N}p} \tag{452}$$

where the apparent volume element V'_p of vacuum in the presence of a 3-dimensional APF is given in terms of the apparent volume element of vacuum V'_{po} under zero APF conditions by[9,10]

$$V'_p / V'_{po} = C'_{1x} C'_{1y} C'_{1z} = C'^2_{1r} C'_{1z} = C'^3_{1\rho} = \cos\theta'_{\text{Ꞥ}p} \sec\theta'_{np} \geq 1 \qquad [hn'_{po}, SN_p] \qquad (453)$$

$$V'_p / V'_{po} = C'_{1x} C'_{1y} C'_{1z} = C'^2_{1r} C'_{1z} = C'^3_{1\rho} = \sec\theta'_{\text{Ꞥ}p} \sec\theta'_{np} \geq 1 \qquad [hn'_{po}, LN_p] \qquad (454)$$

$$V'_p / V'_{po} = C'_{2x} C'_{2y} C'_{2z} = C'^2_{2r} C'_{2z} = C'^3_{2\rho} = \cos\theta'_{\text{Ꞥ}p} \cos\theta'_{np} \leq 1 \qquad [ln'_{po}, SN_p] \qquad (455)$$

$$V'_p / V'_{po} = C'_{2x} C'_{2y} C'_{2z} = C'^2_{2r} C'_{2z} = C'^3_{2\rho} = \sec\theta'_{\text{Ꞥ}p} \cos\theta'_{np} \leq 1 \qquad [ln'_{po}, LN_p] \qquad (456)$$

$$V'_p / V'_{po} = C'_{3x} C'_{3y} C'_{3z} = C'^2_{3r} C_{3z} = C'^3_{3\rho} = \cos\theta'_{\text{Ꞥ}p} \cos\theta'_{np} \leq 1 \qquad [ln'_{po}, SN_p] \qquad (457)$$

$$V'_p / V'_{po} = C'_{3x} C'_{3y} C'_{3z} = C'^2_{3r} C_{3z} = C'^3_{3\rho} = \sec\theta'_{\text{Ꞥ}p} \cos\theta'_{np} \leq 1 \qquad [ln'_{po}, LN_p] \qquad (458)$$

where in the case of the vacuum, $[hn'_{po}]$ indicates apparent high particle-antiparticle pair number density of vacuum, $[ln'_{po}]$ indicates apparent low particle-antiparticle pair number density of vacuum and where $[SN_p]$ and $[LN_p]$ = small integer number of particle-antiparticle pairs and large integer number of particle-antiparticle pairs of vacuum respectively all at zero APF conditions. The space coordinate factors are given by equations (438) through (440), and so because the right hand sides of equations (455) and (456) are the same as the right hand sides of equations (457) and (458) it follows that the apparent space coordinate factors $C'_{2\alpha}$ and $C'_{3\alpha}$ of equations (439) and (440) give two possible solutions for the internal phase angles θ'_{np} and $\theta'_{\eta\rho}$ of the apparent counting and segmentation numbers for the $[ln'_{po}]$ case and therefore the $[ln'_{po}]$ case is degenerate. Then combining equations (450), (451) and (453) through (458) gives[11]

$$n'_p = n'_{po} \cos\theta'_{np} \leq n'_{po} \qquad [hn'_{po}] \qquad (459)$$

$$n'_p = n'_{po} \sec\theta'_{np} \geq n'_{po} \qquad [ln'_{po}] \qquad (460)$$

where n'_p = apparent particle-antiparticle pair number density for the vacuum in an APF and where n'_{po} = apparent particle-antiparticle pair number density under zero APF conditions and which is given by

$$n'_{po} = N_p / V'_{po} \qquad (461)$$

which is the conventional result.

<u>Cartesian Coordinates.</u>

As seen from a moving Cartesian coordinate system, the ratios of the volume elements of the vacuum with and without an APF present are given by

$$V'_p / V'_{po} = C'_{1x} C'_{1y} C'_{1z} = \cos\theta'_{\text{Ꞥ}p} \sec\theta'_{np} \geq 1 \qquad [hn'_{po}, SN_p] \qquad (462)$$

$$V'_p / V'_{po} = C'_{1x} C'_{1y} C'_{1z} = \sec\theta'_{\text{N}p} \sec\theta'_{np} \geq 1 \qquad [hn'_{po}, LN_p] \tag{463}$$

$$V'_p / V'_{po} = C'_{\beta x} C'_{\beta y} C'_{\beta z} = \cos\theta'_{\text{N}p} \cos\theta'_{np} \leq 1 \qquad [ln'_{po}, SN_p] \tag{464}$$

$$V'_p / V'_{po} = C'_{\beta x} C'_{\beta y} C'_{\beta z} = \sec\theta'_{\text{N}p} \cos\theta'_{np} \leq 1 \qquad [ln'_{po}, LN_p] \tag{465}$$

where $C'_{1\alpha}$ and $C'_{\beta\alpha}$ are given by equations (438) through (440) and where $\beta = 2,3$ describes the degeneracy of the $[ln_{po}]$ state of the vacuum. For the case of a uniaxial APF in the x-direction, $C'_{1y} = 1$, $C'_{1z} = 1$, $C'_{\beta y} = 1$ and $C'_{\beta z} = 1$ with the result that, equations (462) through (465) give

$$V'_p / V'_{po} = C'_{1x} = \cos\theta'_{\text{N}p} \sec\theta'_{np} \geq 1 \qquad [hn'_{po}, SN_p] \tag{466}$$

$$V'_p / V'_{po} = C'_{1x} = \sec\theta'_{\text{N}p} \sec\theta'_{np} \geq 1 \qquad [hn'_{po}, LN_p] \tag{467}$$

$$V'_p / V'_{po} = C'_{\beta x} = \cos\theta'_{\text{N}p} \cos\theta'_{np} \leq 1 \qquad [ln'_{po}, SN_p] \tag{468}$$

$$V'_p / V'_{po} = C'_{\beta x} = \sec\theta'_{\text{N}p} \cos\theta'_{np} \leq 1 \qquad [ln'_{po}, LN_p] \tag{469}$$

where $\beta = 2,3$.

Cylindrical Polar Coordinates.

For the case of cylindrical polar coordinates the ratios of the vacuum volume elements with and without an APF are given by

$$V'_p / V'_{po} = C'^2_{1r} C'_{1z} = \cos\theta'_{\text{N}p} \sec\theta'_{np} \geq 1 \qquad [hn'_{po}, SN_p] \tag{470}$$

$$V'_p / V'_{po} = C'^2_{1r} C'_{1z} = \sec\theta'_{\text{N}p} \sec\theta'_{np} \geq 1 \qquad [hn'_{po}, LN_p] \tag{471}$$

$$V'_p / V'_{po} = C'^2_{\beta r} C'_{\beta z} = \cos\theta'_{\text{N}p} \cos\theta'_{np} \leq 1 \qquad [ln'_{po}, SN_p] \tag{472}$$

$$V'_p / V'_{po} = C'^2_{\beta r} C'_{\beta z} = \sec\theta'_{\text{N}p} \cos\theta'_{np} \leq 1 \qquad [ln'_{po}, LN_p] \tag{473}$$

where $\beta = 2,3$. For the case of a uniaxial APF along the z-axis, $C'_{1r} = 1$ and $C'_{\beta r} = 1$ and equations (470) through (473) become

$$V_p / V_{po} = C_{1z} = \cos\theta_{\text{N}p} \sec\theta_{np} \geq 1 \qquad [hn_{po}, SN_p] \tag{474}$$

$$V_p / V_{po} = C_{1z} = \sec\theta_{\text{N}p} \sec\theta_{np} \geq 1 \qquad [hn_{po}, LN_p] \tag{475}$$

$$V_p / V_{po} = C_{\beta z} = \cos\theta_{\text{N}p} \cos\theta_{np} \leq 1 \qquad [ln_{po}, SN_p] \tag{476}$$

$$V_p / V_{po} = C_{\beta z} = \sec\theta_{\text{N}p} \cos\theta_{np} \leq 1 \qquad [ln_{po}, LN_p] \tag{477}$$

where $\beta = 2,3$. For the special case of a two-dimensional radial APF, $C'_{1z} = 1$ and equations (470) through (473) give

$$(V'_p / V'_{po})^{1/2} = C'_{1r} = \cos^{1/2}\theta'_{\aleph p} \sec^{1/2}\theta'_{np} \geq 1 \qquad [hn'_{po}, SN_p] \qquad (478)$$

$$(V'_p / V'_{po})^{1/2} = C'_{1r} = \sec^{1/2}\theta'_{\aleph p} \sec^{1/2}\theta'_{np} \geq 1 \qquad [hn'_{po}, LN_p] \qquad (479)$$

$$(V'_p / V'_{po})^{1/2} = C'_{\beta r} = \cos^{1/2}\theta'_{\aleph p} \cos^{1/2}\theta'_{np} \leq 1 \qquad [ln'_{po}, SN_p] \qquad (480)$$

$$(V'_p / V'_{po})^{1/2} = C'_{\beta r} = \sec^{1/2}\theta'_{\aleph p} \cos^{1/2}\theta'_{np} \leq 1 \qquad [ln'_{po}, LN_p] \qquad (481)$$

Spherical Polar Coordinates.

$$V'_p / V'_{po} = C'^3_{1\rho} = \cos\theta'_{\aleph p} \sec\theta'_{np} \geq 1 \qquad [hn'_{po}, SN_p] \qquad (482)$$

$$V'_p / V'_{po} = C'^3_{1\rho} = \sec\theta'_{\aleph p} \sec\theta'_{np} \geq 1 \qquad [hn'_{po}, LN_p] \qquad (483)$$

$$V'_p / V'_{po} = C'^3_{\beta\rho} = \cos\theta'_{\aleph p} \cos\theta'_{np} \leq 1 \qquad [ln'_{po}, SN_p] \qquad (484)$$

$$V'_p / V'_{po} = C'^3_{\beta\rho} = \sec\theta'_{\aleph p} \cos\theta'_{np} \leq 1 \qquad [ln'_{po}, LN_p] \qquad (485)$$

or

$$(V'_p / V'_{po})^{1/3} = C'_{1\rho} = \cos^{1/3}\theta'_{\aleph p} \sec^{1/3}\theta'_{np} \geq 1 \qquad [hn'_{po}, SN_p] \qquad (486)$$

$$(V'_p / V'_{po})^{1/3} = C'_{1\rho} = \sec^{1/3}\theta'_{\aleph p} \sec^{1/3}\theta'_{np} \geq 1 \qquad [hn'_{po}, LN_p] \qquad (487)$$

$$(V'_p / V'_{po})^{1/3} = C'_{\beta\rho} = \cos^{1/3}\theta'_{\aleph p} \cos^{1/3}\theta'_{np} \leq 1 \qquad [ln'_{po}, SN_p] \qquad (488)$$

$$(V'_p / V'_{po})^{1/3} = C'_{\beta\rho} = \sec^{1/3}\theta'_{\aleph p} \cos^{1/3}\theta'_{np} \leq 1 \qquad [ln'_{po}, LN_p] \qquad (489)$$

where $\beta = 1,2$. Note that equations (482) through (489) for spherical polar coordinates can also be obtained from equations (462) through (465) for a Cartesian coordinate system by taking $C'_{1x} = C'_{1y} = C'_{1z}$ and $C'_{\beta x} = C'_{\beta y} = C'_{\beta z}$ and from equations (470) through (473) for cylindrical polar coordinates by taking $C'_{1r} = C'_{1z}$ and $C'_{\beta r} = C'_{\beta z}$.

Consider now the case of a linear elastic vacuum for which the internal phase angles θ_{np} and $\theta_{\aleph p}$ are related by[9]

$$\cos\theta'_{np} = 1 - \tan\theta'_{\aleph p} \qquad 0 \leq \theta'_{np} \leq \pi/2 \qquad 0 \leq \theta'_{\aleph p} \leq \pi/4 \qquad [hn'_{po}] \qquad (490)$$

$$\sec\theta'_{np} = 1 + \tan\theta'_{\aleph p} \qquad 0 \leq \theta'_{np} \leq \pi/2 \qquad 0 \leq \theta'_{\aleph p} \leq \pi/2 \qquad [ln'_{po}] \qquad (491)$$

with $\theta'_{np} \geq \theta'_{\aleph p}$, and which can now be used in equations (462) through (489) to describe the space coordinate factors for a linear elastic vacuum in an APF.

Cartesian Coordinates

Combining equations (462) through (465) with equations (490) and (491) gives for the case of a 3-dimensional APF in the vacuum and for Cartesian coordinates

$$C'_{1x}C'_{1y}C'_{1z} = \sec\theta'_{np}[1+(1-\cos\theta'_{np})^2]^{-1/2} = \cos\theta'_{Np}(1-\tan\theta'_{Np})^{-1} \quad [hn'_{po}, SN_p] \qquad (492)$$

$$C'_{1x}C'_{1y}C'_{1z} = \sec\theta'_{np}[1+(1-\cos\theta'_{np})^2]^{1/2} = \sec\theta'_{Np}(1-\tan\theta'_{Np})^{-1} \quad [hn'_{po}, LN_p] \qquad (493)$$

$$C'_{\beta x}C'_{\beta y}C'_{\beta z} = \cos\theta'_{np}[1+(\sec\theta'_{np}-1)^2]^{-1/2} = \cos\theta'_{Np}(1+\tan\theta'_{Np})^{-1} \quad [ln'_{po}, SN_p] \qquad (494)$$

$$C'_{\beta x}C'_{\beta y}C'_{\beta z} = \cos\theta'_{np}[1+(\sec\theta'_{np}-1)^2]^{1/2} = \sec\theta'_{Np}(1+\tan\theta'_{Np})^{-1} \quad [ln'_{po}, LN_p] \qquad (495)$$

where $\beta = 2,3$. Combining equations (466) through (469) with equations (490) and (491) gives for a uniaxial APF along the x-coordinate in a linear elastic vacuum

$$C'_{1x} = \sec\theta'_{np}[1+(1-\cos\theta'_{np})^2]^{-1/2} = \cos\theta'_{Np}(1-\tan\theta'_{Np})^{-1} \quad [hn'_{po}, SN_p] \qquad (496)$$

$$C'_{1x} = \sec\theta'_{np}[1+(1-\cos\theta'_{np})^2]^{1/2} = \sec\theta'_{Np}(1-\tan\theta'_{Np})^{-1} \quad [hn'_{po}, LN_p] \qquad (497)$$

$$C'_{\beta x} = \cos\theta'_{np}[1+(\sec\theta'_{np}-1)^2]^{-1/2} = \cos\theta'_{Np}(1+\tan\theta'_{Np})^{-1} \quad [ln'_{po}, SN_p] \qquad (498)$$

$$C'_{\beta x} = \cos\theta'_{np}[1+(\sec\theta'_{np}-1)^2]^{1/2} = \sec\theta'_{Np}(1+\tan\theta'_{Np})^{-1} \quad [ln'_{po}, LN_p] \qquad (499)$$

for Cartesian coordinates.

Cylindrical Polar Coordinates.

Combining equations (470) through (473) with equations (490) and (491) gives for the case of a 3-dimensional APF in the vacuum

$$C'^2_{1r}C'_{1z} = \sec\theta'_{np}[1+(1-\cos\theta'_{np})^2]^{-1/2} = \cos\theta'_{Np}(1-\tan\theta'_{Np})^{-1} \quad [hn'_{po}, SN_p] \qquad (500)$$

$$C'^2_{1r}C'_{1z} = \sec\theta'_{np}[1+(1-\cos\theta'_{np})^2]^{1/2} = \sec\theta'_{Np}(1-\tan\theta'_{Np})^{-1} \quad [hn'_{po}, LN_p] \qquad (501)$$

$$C'^2_{\beta r}C'_{\beta z} = \cos\theta'_{np}[1+(\sec\theta'_{np}-1)^2]^{-1/2} = \cos\theta'_{Np}(1+\tan\theta'_{Np})^{-1} \quad [ln'_{po}, SN_p] \qquad (502)$$

$$C'^2_{\beta r}C'_{\beta z} = \cos\theta'_{np}[1+(\sec\theta'_{np}-1)^2]^{1/2} = \sec\theta'_{Np}(1+\tan\theta'_{Np})^{-1} \quad [ln'_{po}, LN_p] \qquad (503)$$

For a uniaxial APF, $C'_{1r} = 1$ and $C'_{\beta r} = 1$ and equations (500) through (503) become just the uniaxial APF equations (496) through (499) but now for the z-axis. For the special case of a 2-dimensional radial APF in the vacuum, $C'_{1z} = 1$ and $C'_{\beta z} = 1$ and equations (500) through (503) become

$$C'_{1r} = \sec^{1/2}\theta'_{np}[1+(1-\cos\theta'_{np})^2]^{-1/4} = \cos^{1/2}\theta'_{Np}(1-\tan\theta'_{Np})^{-1/2} \quad [hn'_{po}, SN_p] \qquad (504)$$

$$C'_{1r} = \sec^{1/2}\theta'_{np}[1+(1-\cos\theta'_{np})^2]^{1/4} = \sec^{1/2}\theta'_{\mathbb{N}p}(1-\tan\theta'_{\mathbb{N}p})^{-1/2} \qquad [hn'_{po}, LN_p] \qquad (505)$$

$$C'_{\beta r} = \cos^{1/2}\theta'_{np}[1+(\sec\theta'_{np}-1)^2]^{-1/4} = \cos^{1/2}\theta'_{\mathbb{N}p}(1+\tan\theta'_{\mathbb{N}p})^{-1/2} \qquad [ln'_{po}, SN_p] \qquad (506)$$

$$C'_{\beta r} = \cos^{1/2}\theta'_{np}[1+(\sec\theta'_{np}-1)^2]^{1/4} = \sec^{1/2}\theta'_{\mathbb{N}p}(1+\tan\theta'_{\mathbb{N}p})^{-1/2} \qquad [ln'_{po}, LN_p] \qquad (507)$$

where $\beta = 2,3$.

<u>Spherical Polar Coordinates.</u>

Combining equations (486) through (489) with equations (490) and (491) gives for the case of a 3-dimensional APF in spherical polar coordinates

$$C'_{1\rho} = \sec^{1/3}\theta'_{np}[1+(1-\cos\theta'_{np})^2]^{-1/6} = \cos^{1/3}\theta'_{\mathbb{N}p}(1-\tan\theta'_{\mathbb{N}p})^{-1/3} \qquad [hn'_{po}, SN_p] \qquad (508)$$

$$C'_{1\rho} = \sec^{1/3}\theta'_{np}[1+(1-\cos\theta'_{np})^2]^{1/6} = \sec^{1/3}\theta'_{\mathbb{N}p}(1-\tan\theta'_{\mathbb{N}p})^{-1/3} \qquad [hn'_{po}, LN_p] \qquad (509)$$

$$C'_{\beta\rho} = \cos^{1/3}\theta'_{np}[1+(\sec\theta'_{np}-1)^2]^{-1/6} = \cos^{1/3}\theta'_{\mathbb{N}p}(1+\tan\theta'_{\mathbb{N}p})^{-1/3} \qquad [ln'_{po}, SN_p] \qquad (510)$$

$$C'_{\beta\rho} = \cos^{1/3}\theta'_{np}[1+(\sec\theta'_{np}-1)^2]^{1/6} = \sec^{1/3}\theta'_{\mathbb{N}p}(1+\tan\theta'_{\mathbb{N}p})^{-1/3} \qquad [ln'_{po}, LN_p] \qquad (511)$$

where $\beta = 2,3$. Equations (508) through (511) can also be obtained from equations (492) through (495) for Cartesian coordinates by taking $C'_{1x} = C'_{1y} = C'_{1z}$ and $C'_{\beta x} = C'_{\beta y} = C'_{\beta z}$.

The space and time coordinate factors for the vacuum that contains an APF can also be related to the internal phase angles of the complex number dimensions of space and time.[9,10] It can be conjectured that the complex number of spacetime dimensions $\overline{D}_{st}$ is linearly related to the complex number of virtual particle-antiparticle pairs $\overline{\mathbb{N}}_p$ and the complex number spacetime dimension number density is linearly related to the complex number density of virtual particle-antiparticle pairs as seen from the moving (primed) system in the manner

$$\overline{D}'_{st} = \mu\overline{\mathbb{N}}'_p \qquad D'_{st} = \mu\mathbb{N}'_p \qquad \theta'_{Dst} = \theta'_{\mathbb{N}p} \qquad (512)$$

$$\overline{n}'_{Dst} = \mu\overline{n}'_{\mathbb{N}p} \qquad n'_{Dst} = \mu n'_{\mathbb{N}p} \qquad \theta'_{nDst} = \theta'_{np} \qquad (513)$$

where the particle-antiparticle pair formalism is given in equations (443) through (461). It has been suggested that the complex number dimensions of space and time can be written as[8-10]

$$\overline{D}'_t = D'_t\exp(j\theta'_{Dt}) \qquad \overline{D}'_s = D'_s\exp(j\theta'_{Ds}) \qquad \overline{D}'_{st} = D'_{st}\exp(j\theta'_{Dst}) \qquad (514)$$

where for a low dimensional spacetime

$$D'_t = N_{Dt}\cos\theta'_{Dt} \qquad D'_s = N_{Ds}\cos\theta'_{Ds} \qquad D'_{st} = N_{Dst}\cos\theta'_{Dst} \qquad [SN_{Dst}] \qquad (515)$$

and for a high dimensional spacetime

$$D'_t = N_{Dt} \sec\theta'_{Dt} \qquad D'_s = N_{Ds} \sec\theta'_{Ds} \qquad D'_{st} = N_{Dst} \sec\theta'_{Dst} \qquad [LN_{Dst}] \qquad (516)$$

where [SN_{Dst}] and [LN_{Dst}] = small number of dimensions and large number of dimensions respectively and where N_{Dt}, N_{Ds} and $N_{Dst} = N_{Dt} + N_{Ds}$ = zero APF values of the dimensions of time, space and spacetime which in the vacuum of are generally integers but can be nonintegers in matter.[8-10]

From the perspective of the moving coordinate system for the vacuum in an APF, the complex number dimension number density of the vacuum is written as[8-10]

$$\bar{n}'_{Dst} = n'_{Dst} \exp(j\theta'_{nDst}) = \bar{D}'_{st}/\bar{V}'_{Dst} \qquad n' = D'_{st} / V'_{Dst} \qquad \theta'_{nDst} = \theta'_{Dst} - \theta'_{VDst} \qquad (517)$$

where the complex number spacetime dimension number is given in equations (444) through (446). The complex number volume of dimensions is represented as

$$\bar{V}'_{Dst} = V'_{Dst} \exp(j\theta'_{VDst}) \qquad \theta'_{VDst} = \theta'_{Dst} - \theta'_{nDst} \qquad \theta'_{VDst} \le 0 \qquad (518)$$

For a bound homogeneous vacuum in a 3-dimensional APF, the internal phase angles of the space coordinates are written as

$$\theta'_x = \theta'_y = \theta'_z = \theta'_r = \theta'_\rho = 1/3\theta'_{VDst} = 1/3(\theta'_{Dst} - \theta'_{nDst}) \le 0 \qquad (519)$$

and for a uniaxial APF in the x-direction

$$\theta_x = \theta_{VDst} = \theta_{Dst} - \theta_{nDst} \le 0 \qquad (520)$$

For a 3-dimensional APF in the vacuum[9,10]

$$V'_{Dst} / V'^{o}_{Dst} = C'_{1x}C'_{1y}C'_{1z} = C'^2_{1r}C'_{1z} = C'^3_{1\rho} = \cos\theta'_{Dst} \sec\theta'_{nDst} \ge 1 \qquad [hn'^{o}_{Dst}, SN_{Dst}] \qquad (521)$$

$$V'_{Dst} / V'^{o}_{Dst} = C'_{1x}C'_{1y}C'_{1z} = C'^2_{1r}C'_{1z} = C'^3_{1\rho} = \sec\theta'_{Dst} \sec\theta'_{nDst} \ge 1 \qquad [hn'^{o}_{Dst}, LN_{Dst}] \qquad (522)$$

$$V'_{Dst} / V'^{o}_{Dst} = C'_{2x}C'_{2y}C'_{2z} = C'^2_{2r}C'_{2z} = C'^3_{2\rho} = \cos\theta'_{Dst} \cos\theta'_{nDst} \le 1 \qquad [ln'^{o}_{Dst}, SN_{Dst}] \qquad (523)$$

$$V'_{Dst} / V'^{o}_{Dst} = C'_{2x}C'_{2y}C'_{2z} = C'^2_{2r}C'_{2z} = C'^3_{2\rho} = \sec\theta'_{Dst} \cos\theta'_{nDst} \le 1 \qquad [ln'^{o}_{Dst}, LN_{Dst}] \qquad (524)$$

$$V'_{Dst} / V'^{o}_{Dst} = C'_{3x}C'_{3y}C'_{3z} = C'^2_{3r}C'_{3z} = C'^3_{3\rho} = \cos\theta'_{Dst} \cos\theta'_{nDst} \le 1 \qquad [ln'^{o}_{Dst}, SN_{Dst}] \qquad (525)$$

$$V'_{Dst} / V'^{o}_{Dst} = C'_{3x}C'_{3y}C'_{3z} = C'^2_{3r}C'_{3z} = C'^3_{3\rho} = \sec\theta'_{Dst} \cos\theta'_{nDst} \le 1 \qquad [ln'^{o}_{Dst}, LN_{Dst}] \qquad (526)$$

where in the case of the vacuum, $[hn'^{o}_{Dst}]$ = high dimensional density of vacuum, $[ln'^{o}_{Dst}]$ = low dimensional density of vacuum and where [SN_{Dst}] and [LN_{Dst}] = small number of dimensions and large number of dimensions respectively and where the space coordinate factors are given by equations (438) through (440), and so because the right hand sides of equations (523) and (524) are the same as the right hand sides of equations (525) and

(526) it follows that the space coordinate factors $C'_{2\alpha}$ and $C'_{3\alpha}$ of equations (439) and (440) give two possible solutions for the internal phase angles $\theta'_{\eta\rho}$ and $\theta'_{\eta\rho}$ of the counting and segmentation numbers for the $[\ln'^{o}_{Dst}]$ case and therefore the $[\ln'^{o}_{Dst}]$ case is degenerate. Combining equations (514) through (518) gives[9,10]

$$n'_{Dst} = N_{Dst} / V'_{Dst} \cos\theta'_{Dst} \qquad [SN_{Dst}] \tag{527}$$

$$n'_{Dst} = N_{Dst} / V'_{Dst} \sec\theta'_{Dst} \qquad [LN_{Dst}] \tag{528}$$

Then it is easily shown by equations (521) through (528) that[9,10]

$$n'_{Dst} = n'^{o}_{Dst} \cos\theta'_{nDst} \le n'^{o}_{Dst} \qquad [hn'^{o}_{Dst}] \tag{529}$$

$$n'_{Dst} = n'^{o}_{Dst} \sec\theta'_{nDst} \ge n'^{o}_{Dst} \qquad [\ln'^{o}_{Dst}] \tag{530}$$

where

$$n'^{o}_{Dst} = N_{Dst} / V'^{o}_{Dst} \tag{531}$$

is the zero APF value of the dimensional number density. Equations (514) through (531) are the vacuum space and time dimension analogs of the vacuum particle-antiparticle equations (443) through (461).

Cartesian Coordinates.

For Cartesian coordinates the ratios of the vacuum volume elements with and without an APF present are given by

$$V'_{Dst} / V'^{o}_{Dst} = C'_{1x} C'_{1y} C'_{1z} = \cos\theta'_{Dst} \sec\theta'_{nDst} \ge 1 \qquad [hn'^{o}_{Dst}, SN_{Dst}] \tag{532}$$

$$V'_{Dst} / V'^{o}_{Dst} = C'_{1x} C'_{1y} C'_{1z} = \sec\theta'_{Dst} \sec\theta'_{nDst} \ge 1 \qquad [hn'^{o}_{Dst}, LN_{Dst}] \tag{533}$$

$$V'_{Dst} / V'^{o}_{Dst} = C'_{2x} C'_{2y} C'_{2z} = \cos\theta'_{Dst} \cos\theta'_{nDst} \le 1 \qquad [\ln'^{o}_{Dst}, SN_{Dst}] \tag{534}$$

$$V'_{Dst} / V'^{o}_{Dst} = C'_{2x} C'_{2y} C'_{2z} = \sec\theta'_{Dst} \cos\theta'_{nDst} \le 1 \qquad [\ln'^{o}_{Dst}, LN_{Dst}] \tag{535}$$

$$V'_{Dst} / V'^{o}_{Dst} = C'_{3x} C'_{3y} C'_{3z} = \cos\theta'_{Dst} \cos\theta'_{nDst} \le 1 \qquad [\ln'^{o}_{Dst}, SN_{Dst}] \tag{536}$$

$$V'_{Dst} / V'^{o}_{Dst} = C'_{3x} C'_{3y} C'_{3z} = \sec\theta'_{Dst} \cos\theta'_{nDst} \le 1 \qquad [\ln'^{o}_{Dst}, LN_{Dst}] \tag{537}$$

where $C'_{1\alpha}$ and $C'_{\beta\alpha}$ are given by equations (438) through (440) and where $\beta = 2,3$ describes the degeneracy of the [ld] state. For the case of a uniaxial APF in the x-direction, $C'_{1y} = 1, C'_{1z} = 1, C'_{\beta y} = 1$ and $C'_{\beta z} = 1$ with the result that, equations (532) through (537) give

$$V'_{Dst} / V'^{o}_{Dst} = C'_{1x} = \cos\theta'_{Dst} \sec\theta'_{nDst} \ge 1 \qquad [hn'^{o}_{Dst}, SN_{Dst}] \tag{538}$$

$$V'_{Dst} / V'^{o}_{Dst} = C'_{1x} = \sec\theta'_{Dst} \sec\theta'_{nDst} \geq 1 \qquad [hn'^{o}_{Dst}, LN_{Dst}] \qquad (539)$$

$$V'_{Dst} / V'^{o}_{Dst} = C'_{\beta x} = \cos\theta'_{Dst} \cos\theta'_{nDst} \leq 1 \qquad [ln'^{o}_{Dst}, SN_{Dst}] \qquad (122)$$

$$V'_{Dst} / V'^{o}_{Dst} = C'_{\beta x} = \sec\theta'_{Dst} \cos\theta'_{nDst} \leq 1 \qquad [ln'^{o}_{Dst}, LN_{Dst}] \qquad (123)$$

where $\beta = 2,3$.

Cylindrical Polar Coordinates.

For the case of cylindrical polar coordinates the ratios of the vacuum volume elements with and without an APF are given by

$$V'_{Dst} / V'^{o}_{Dst} = C'^{2}_{1r} C'_{1z} = \cos\theta'_{Dst} \sec\theta'_{nDst} \geq 1 \qquad [hn'^{o}_{Dst}, SN_{Dst}] \qquad (542)$$

$$V'_{Dst} / V'^{o}_{Dst} = C'^{2}_{1r} C'_{1z} = \sec\theta'_{Dst} \sec\theta'_{nDst} \geq 1 \qquad [hn'^{o}_{Dst}, LN_{Dst}] \qquad (543)$$

$$V'_{Dst} / V'^{o}_{Dst} = C'^{2}_{\beta r} C'_{\beta z} = \cos\theta'_{Dst} \cos\theta'_{nDst} \leq 1 \qquad [ln'^{o}_{Dst}, SN_{Dst}] \qquad (544)$$

$$V'_{Dst} / V'^{o}_{Dst} = C'^{2}_{\beta r} C'_{\beta z} = \sec\theta'_{Dst} \cos\theta'_{nDst} \leq 1 \qquad [ln'^{o}_{Dst}, LN_{Dst}] \qquad (545)$$

where $\beta = 2,3$. For the case of a uniaxial APF along the z-axis, $C'_{1r} = 1$ and $C'_{\beta r} = 1$ and equations (542) through (545) become

$$V'_{Dst} / V'^{o}_{Dst} = C'_{1z} = \cos\theta'_{Dst} \sec\theta'_{nDst} \geq 1 \qquad [hn'^{o}_{Dst}, SN_{Dst}] \qquad (546)$$

$$V'_{Dst} / V'^{o}_{Dst} = C'_{1z} = \sec\theta'_{Dst} \sec\theta'_{nDst} \geq 1 \qquad [hn'^{o}_{Dst}, LN_{Dst}] \qquad (547)$$

$$V'_{Dst} / V'^{o}_{Dst} = C'_{\beta z} = \cos\theta'_{Dst} \cos\theta'_{nDst} \leq 1 \qquad [ln'^{o}_{Dst}, SN_{Dst}] \qquad (548)$$

$$V'_{Dst} / V'^{o}_{Dst} = C'_{\beta z} = \sec\theta'_{Dst} \cos\theta'_{nDst} \leq 1 \qquad [ln'^{o}_{Dst}, LN_{Dst}] \qquad (549)$$

where $\beta = 2,3$. For the special case of a two-dimensional radial APF, $C'_{1z} = 1$ and equations (542) through (545) give

$$(V'_{Dst} / V'^{o}_{Dst})^{1/2} = C'_{1r} = \cos^{1/2}\theta'_{Dst} \sec^{1/2}\theta'_{nDst} \geq 1 \qquad [hn'^{o}_{Dst}, SN_{Dst}] \qquad (550)$$

$$(V'_{Dst} / V'^{o}_{Dst})^{1/2} = C'_{1r} = \sec^{1/2}\theta'_{Dst} \sec^{1/2}\theta'_{nDst} \geq 1 \qquad [hn'^{o}_{Dst}, LN_{Dst}] \qquad (551)$$

$$(V'_{Dst} / V'^{o}_{Dst})^{1/2} = C'_{\beta r} = \cos^{1/2}\theta'_{Dst} \cos^{1/2}\theta'_{nDst} \leq 1 \qquad [ln'^{o}_{Dst}, SN_{Dst}] \qquad (552)$$

$$(V'_{Dst} / V'^{o}_{Dst})^{1/2} = C'_{\beta r} = \sec^{1/2}\theta'_{Dst} \cos^{1/2}\theta'_{nDst} \leq 1 \qquad [ln'^{o}_{Dst}, LN_{Dst}] \qquad (553)$$

where $\beta = 2,3$ indicates the degeneracy associated with the possible low-dimensional states of the vacuum.

Spherical Polar Coordinates.

For an isotropic APF in the vacuum, and for spherical coordinates, the ratios of the volumes for the case when the APF is present to the case when the APF is absent are given by

$$V'_{Dst} / V'^{o}_{Dst} = C'^{3}_{1\rho} = \cos\theta'_{Dst} \sec\theta'_{nDst} \geq 1 \qquad [hn'^{o}_{Dst}, SN_{Dst}] \qquad (554)$$

$$V'_{Dst} / V'^{o}_{Dst} = C'^{3}_{1\rho} = \sec\theta'_{Dst} \sec\theta'_{nDst} \geq 1 \qquad [hn'^{o}_{Dst}, LN_{Dst}] \qquad (555)$$

$$V'_{Dst} / V'^{o}_{Dst} = C'^{3}_{\beta\rho} = \cos\theta'_{Dst} \cos\theta'_{nDst} \leq 1 \qquad [ln'^{o}_{Dst}, SN_{Dst}] \qquad (556)$$

$$V'_{Dst} / V'^{o}_{Dst} = C'^{3}_{\beta\rho} = \sec\theta'_{Dst} \cos\theta'_{nDst} \leq 1 \qquad [ln'^{o}_{Dst}, LN_{Dst}] \qquad (557)$$

or

$$(V'_{Dst} / V'^{o}_{Dst})^{1/3} = C'_{1\rho} = \cos^{1/3}\theta'_{Dst} \sec^{1/3}\theta'_{nDst} \geq 1 \qquad [hn'^{o}_{Dst}, SN_{Dst}] \qquad (558)$$

$$(V'_{Dst} / V'^{o}_{Dst})^{1/3} = C'_{1\rho} = \sec^{1/3}\theta'_{Dst} \sec^{1/3}\theta'_{nDst} \geq 1 \qquad [hn'^{o}_{Dst}, LN_{Dst}] \qquad (559)$$

$$(V'_{Dst} /' V^{o}_{Dst})^{1/3} = C'_{\beta\rho} = \cos^{1/3}\theta'_{Dst} \cos^{1/3}\theta'_{nDst} \leq 1 \qquad [ln'^{o}_{Dst}, SN_{Dst}] \qquad (560)$$

$$(V'_{Dst} /' V^{o}_{Dst})^{1/3} = C'_{\beta\rho} = \sec^{1/3}\theta'_{Dst} \cos^{1/3}\theta'_{nDst} \leq 1 \qquad [ln'^{o}_{Dst}, LN_{Dst}] \qquad (561)$$

where $\beta = 2,3$ indicates the degeneracy of the $[ln'^{o}_{Dst}]$ case and where the space coordinate factors $C'_{\sigma\rho}$ for spherical polar coordinates are given by equations (438) through (440) with $\alpha = \rho$. Note that equations (558) through (561) for spherical polar coordinates can also be obtained from equations (532) through (537) for a Cartesian coordinate system by taking $C'_{1x} = C'_{1y} = C'_{1z}$ and $C'_{\beta x} = C'_{\beta y} = C'_{\beta z}$ and from equations (542) through (545) for cylindrical polar coordinates by taking $C'_{1r} = C'_{1z}$ and $C'_{\beta r} = C'_{\beta z}$.

Consider now the case of a linear elastic vacuum for which the internal phase angles θ'_{nDst} and θ'_{Dst} are related by[9]

$$\cos\theta'_{nDst} = 1 - \tan\theta'_{Dst} \qquad 0 \leq \theta'_{nDst} \leq \pi/2 \qquad 0 \leq \theta'_{Dst} \leq \pi/4 \qquad [hn'^{o}_{Dst}] \qquad (562)$$

$$\sec\theta'_{nDst} = 1 + \tan\theta'_{Dst} \qquad 0 \leq \theta'_{nDst} \leq \pi/2 \qquad 0 \leq \theta'_{Dst} \leq \pi/2 \qquad [ln'^{o}_{Dst}] \qquad (563)$$

where the trigonometrical functions in these two equations show that for high or low dimensional vacuums it is always true that $\theta'_{nDst} \geq \theta'_{Dst}$, and which can now be used in equations (532) through (561) to describe the space coordinate factors for the possibility of a linear elastic vacuum of high or low dimensions in the presence of an APF however small.

Cartesian Coordinates

Combining equations (532) through (537) with equations (562) and (563) gives for the case of a 3-dimensional APF in the vacuumand described by Cartesian coordinates

$$C'_{1x}C'_{1y}C'_{1z} = \sec\theta'_{nDst}[1+(1-\cos\theta'_{nDst})^2]^{-1/2} = \cos\theta'_{Dst}(1-\tan\theta'_{Dst})^{-1} \quad [hn'^{o}_{Dst}, SN_{Dst}] \quad (564)$$

$$C'_{1x}C'_{1y}C'_{1z} = \sec\theta'_{nDst}[1+(1-\cos\theta'_{nDst})^2]^{1/2} = \sec\theta'_{Dst}(1-\tan\theta'_{Dst})^{-1} \quad [hn'^{o}_{Dst}, LN_{Dst}] \quad (565)$$

$$C'_{\beta x}C'_{\beta y}C'_{\beta z} = \cos\theta'_{nDst}[1+(\sec\theta'_{nDst}-1)^2]^{-1/2} = \cos\theta'_{Dst}(1+\tan\theta'_{Dst})^{-1} \quad [ln'^{o}_{Dst}, SN_{Dst}] \quad (566)$$

$$C'_{\beta x}C'_{\beta y}C'_{\beta z} = \cos\theta'_{nDst}[1+(\sec\theta'_{nDst}-1)^2]^{1/2} = \sec\theta'_{Dst}(1+\tan\theta'_{Dst})^{-1} \quad [ln'^{o}_{Dst}, LN_{Dst}] \quad (567)$$

where $\beta = 2,3$. Combining $C'_{1y} = C'_{1z} = 1$ and $C'_{\beta y} = C'_{\beta z} = 1$ with equations (564) through (567) gives for a uniaxial APF in the x-direction for a linear elastic vacuum

$$C'_{1x} = \sec\theta'_{nDst}[1+(1-\cos\theta'_{nDst})^2]^{-1/2} = \cos\theta'_{Dst}(1-\tan\theta'_{Dst})^{-1} \quad [hn'^{o}_{Dst}, SN_{Dst}] \quad (568)$$

$$C'_{1x} = \sec\theta'_{nDst}[1+(1-\cos\theta'_{nDst})^2]^{1/2} = \sec\theta'_{Dst}(1-\tan\theta'_{Dst})^{-1} \quad [hn'^{o}_{Dst}, LN_{Dst}] \quad (569)$$

$$C'_{\beta x} = \cos\theta'_{nDst}[1+(\sec\theta'_{nDst}-1)^2]^{-1/2} = \cos\theta'_{Dst}(1+\tan\theta'_{Dst})^{-1} \quad [ln'^{o}_{Dst}, SN_{Dst}] \quad (570)$$

$$C'_{\beta x} = \cos\theta'_{nDst}[1+(\sec\theta'_{nDst}-1)^2]^{1/2} = \sec\theta'_{Dst}(1+\tan\theta'_{Dst})^{-1} \quad [ln'^{o}_{Dst}, LN_{Dst}] \quad (571)$$

for Cartesian coordinates.

Cylindrical Polar Coordinates.

Combining equations (542) through (545) with equations (562) and (563) gives for the case of a 3-dimensional APF in a linear elastic vacuum

$$C'^2_{1r}C'_{1z} = \sec\theta'_{nDst}[1+(1-\cos\theta'_{nDst})^2]^{-1/2} = \cos\theta'_{Dst}(1-\tan\theta'_{Dst})^{-1} \quad [hn'^{o}_{Dst}, SN_{Dst}] \quad (572)$$

$$C'^2_{1r}C'_{1z} = \sec\theta'_{nDst}[1+(1-\cos\theta'_{nDst})^2]^{1/2} = \sec\theta'_{Dst}(1-\tan\theta'_{Dst})^{-1} \quad [hn'^{o}_{Dst}, LN_{Dst}] \quad (573)$$

$$C'^2_{\beta r}C'_{\beta z} = \cos\theta'_{nDst}[1+(\sec\theta'_{nDst}-1)^2]^{-1/2} = \cos\theta'_{Dst}(1+\tan\theta'_{Dst})^{-1} \quad [ln'^{o}_{Dst}, SN_{Dst}] \quad (574)$$

$$C'^2_{\beta r}C'_{\beta z} = \cos\theta'_{nDst}[1+(\sec\theta'_{nDst}-1)^2]^{1/2} = \sec\theta'_{Dst}(1+\tan\theta'_{Dst})^{-1} \quad [ln'^{o}_{Dst}, LN_{Dst}] \quad (575)$$

For a uniaxial APF, $C'_{1r} = 1$ and $C'_{\beta r} = 1$ and equations (572) through (575) become just the uniaxial APF equations (568) through (571) but now for the z-axis. For the special case of a 2-dimensional radial APF, $C'_{1z} = 1$ and $C'_{\beta z} = 1$ and equations (572) through (575) become

$$C'_{1r} = \sec^{1/2}\theta'_{nDst}[1+(1-\cos\theta'_{nDst})^2]^{-1/4} = \cos^{1/2}\theta'_{Dst}(1-\tan\theta'_{Dst})^{-1/2} \quad [hn'^{o}_{Dst}, SN_{Dst}] \quad (576)$$

$$C'_{1r} = \sec^{1/2}\theta'_{nDst}[1+(1-\cos\theta'_{nDst})^2]^{1/4} = \sec^{1/2}\theta'_{Dst}(1-\tan\theta'_{Dst})^{-1/2} \quad [hn'^{o}_{Dst}, LN_{Dst}] \quad (577)$$

$$C'_{\beta r} = \cos^{1/2}\theta'_{nDst}[1+(\sec\theta'_{nDst}-1)^2]^{-1/4} = \cos^{1/2}\theta'_{Dst}(1+\tan\theta'_{Dst})^{-1/2} \quad [ln'^{o}_{Dst}, SN_{Dst}] \quad (578)$$

$$C'_{\beta r} = \cos^{1/2}\theta'_{nDst}[1+(\sec\theta'_{nDst}-1)^2]^{1/4} = \sec^{1/2}\theta'_{Dst}(1+\tan\theta'_{Dst})^{-1/2} \quad [ln'^{o}_{Dst}, LN_{Dst}] \quad (579)$$

where $\beta = 2,3$.

Spherical Polar Coordinates.

Combining equations (558) through (561) with equations (562) and (563) gives for the case of a 3-dimensional APF in spherical polar coordinates

$$C'_{1\rho} = \sec^{1/3}\theta'_{nDst}[1+(1-\cos\theta'_{nDst})^2]^{-1/6} = \cos^{1/3}\theta'_{Dst}(1-\tan\theta'_{Dst})^{-1/3} \quad [hn'^{o}_{Dst}, SN_{Dst}] \quad (580)$$

$$C'_{1\rho} = \sec^{1/3}\theta'_{nDst}[1+(1-\cos\theta'_{nDst})^2]^{1/6} = \sec^{1/3}\theta'_{Dst}(1-\tan\theta'_{Dst})^{-1/3} \quad [hn'^{o}_{Dst}, LN_{Dst}] \quad (581)$$

$$C'_{\beta\rho} = \cos^{1/3}\theta'_{nDst}[1+(\sec\theta'_{nDst}-1)^2]^{-1/6} = \cos^{1/3}\theta'_{Dst}(1+\tan\theta'_{Dst})^{-1/3} \quad [ln'^{o}_{Dst}, SN_{Dst}] \quad (582)$$

$$C'_{\beta\rho} = \cos^{1/3}\theta'_{nDst}[1+(\sec\theta'_{nDst}-1)^2]^{1/6} = \sec^{1/3}\theta'_{Dst}(1+\tan\theta'_{Dst})^{-1/3} \quad [ln'^{o}_{Dst}, LN_{Dst}] \quad (583)$$

where $\beta = 2,3$. Equations (580) through (583) can also be obtained from equations (568) through (571) for Cartesian coordinates by taking $C'_{1x} = C'_{1y} = C'_{1z}$ and $C'_{\beta x} = C'_{\beta y} = C'_{\beta z}$.

The time coordinate factor for a body moving relative to a moving frame of reference can be obtained from the definitions of linear velocity for uniform motion of a body under zero APF conditions as

$$u_o = d\alpha'_o / dt'_o \quad (584)$$

and for the case when a constant APF is present

$$\bar{u} = u\exp(j\theta_u) = d\bar{\alpha}'/d\bar{t}' \quad (585)$$

$$u = d\alpha'/dt' = (d\alpha'_o/dt'_o)(C'_{\sigma\alpha}/C'_{\sigma t}) = u_o(C'_{\sigma\alpha}/C'_{\sigma t}) \quad (586)$$

$$\theta_u = \theta'_\alpha - \theta'_t = \theta'_{\eta\alpha} - \theta'_{n\alpha} - (\theta'_{\eta t} - \theta'_{nt}) \quad (587)$$

so that the time coordinate factor $C'_{\sigma t}$ is related to the space coordinate factor $C'_{\sigma\alpha}$ by

$$C'_{\sigma t} = (u_o/u)C'_{\sigma\alpha} \quad (588)$$

where $\sigma = 1,2,3$; $\alpha' = x', y', z'$ or r', z' or ρ' and where u_o = velocity of a body relative to a stationary frame of reference under zero APF conditions and u = magnitude of the velocity of a body relative to a moving reference frame surrounded by a constant APF which is in the vacuum and in the body. For astronomical distances, the constant space and time coordinate factors $C'_{\sigma\alpha}$ and $C'_{\sigma t}$ for a moving frame of reference are expected to depend essentially on the properties of the APF in the vacuum through the constant internal phase angles that appear in equation (587) because the dominant part of the distance x′ for a polyphase space vehicle is in the vacuum.

E. Forms of the Space Coordinate Factors for a Coordinate System that is Attached to the Rear of a Polyphase Space Vehicle.

The space and time coordinates in matter or vacuum that is situated in an APF have been represented as complex numbers in an internal space.[5-12] A simple theory of the origin and nature of these complex number coordinates has already been developed.[9] Consider the constant real number spatial parameters L^c_α for the complex number space coordinates $\overline{\alpha}^{sv}$ in an APF in bulk matter, where L^c_α can be taken to be multiples of the Planck length. The real number constant spatial parameters L^c_α are divided into $\overline{\text{n}}^{sv}_\alpha$ parts of length $\Delta\overline{L}^{sv}_\alpha$ where the complex number spatial segmentation number $\overline{\text{n}}^{sv}_\alpha$ is represented as[9]

$$\overline{\text{n}}^{sv}_\alpha = \text{n}^{sv}_\alpha \exp(j\theta^{sv}_{\text{n}\alpha}) \tag{589}$$

so that

$$\Delta\overline{L}^{sv}_\alpha = \Delta L^{sv}_\alpha \exp(j\theta^{sv}_{\Delta L\alpha}) = L^c_\alpha / \overline{\text{n}}^{sv}_\alpha = L^c_\alpha / \text{n}^{sv}_\alpha \exp(-j\theta^{sv}_{\text{n}\alpha}) \tag{590}$$

and

$$\Delta L^{sv}_\alpha = L^c_\alpha / \text{n}^{sv}_\alpha \qquad \theta^{sv}_{\Delta L\alpha} = -\theta^{sv}_{\text{n}\alpha} \tag{591}$$

where $\alpha = x,y,z;\ r,z$ or ρ. Then the complex number space coordinate $\overline{\alpha}^{sv}$ can be written as

$$\overline{\alpha}^{sv} = \alpha^{sv} \exp(j\theta^{sv}_\alpha) = \overline{\eta}^{sv}_\alpha \Delta\overline{L}^{sv}_\alpha = \overline{\eta}^{sv}_\alpha L^c_\alpha / \overline{\text{n}}^{sv}_\alpha = \eta^{sv}_\alpha L^c_\alpha / \text{n}^{sv}_\alpha \exp[j(\theta^{sv}_{\eta\alpha} - \theta^{sv}_{\text{n}\alpha})] \tag{592}$$

where the complex number spatial counting number $\overline{\eta}^{sv}_\alpha$ is represented as

$$\overline{\eta}^{sv}_\alpha = \eta^{sv}_\alpha \exp(j\theta^{sv}_{\eta\alpha}) \tag{593}$$

so that

$$\alpha^{sv} = \eta^{sv}_\alpha L^c_\alpha / \text{n}^{sv}_\alpha = \eta^{sv}_\alpha \Delta L^{sv}_\alpha \qquad \theta^{sv}_\alpha = \theta^{sv}_{\eta\alpha} - \theta^{sv}_{\text{n}\alpha} \tag{594}$$

where η^{sv}_α and n^{sv}_α are not integers in the presence of an APF but are integers under zero APF conditions.

The time coordinate factor, which is analogous to that of the space coordinate factor, is developed as follows. Consider a real number constant L^c_t that is used to measure intervals of the complex number time coordinate $\overline{t}^{sv}$, where L^c_t can be taken to be multiples of the Planck time. In the presence of an APF in bulk matter the real number constant temporal parameter L^c_t is divided into $\overline{\text{n}}^{sv}_t$ parts of length $\Delta\overline{L}^{sv}_t$ where the complex number time segmentation number is represented as[9]

$$\overline{n}_t^{sv} = n_t^{sv}\exp(j\theta_{nt}^{sv}) \qquad (595)$$

and where

$$\Delta\overline{L}_t^{sv} = \Delta L_t^{sv}\exp(j\theta_{\Delta Lt}^{sv}) = L_t^c/\overline{n}_t^{sv} = L_t^c/n_t^{sv}\exp(-j\theta_{nt}^{sv}) \qquad (596)$$

and therefore

$$\Delta L_t^{sv} = L_t^c/n_t^{sv} \qquad \theta_{\Delta Lt}^{sv} = -\theta_{nt}^{sv} \qquad (597)$$

The complex number time coordinate is then written as

$$\overline{t}^{sv} = t^{sv}\exp(j\theta_t^{sv}) = \overline{\eta}_t^{sv}\Delta\overline{L}_t^{sv} = \overline{\eta}_t^{sv}L_t^c/\overline{n}_t^{sv} = \eta_t^{sv}L_t^c/n_t^{sv}\exp[j(\theta_{\eta t}^{sv} - \theta_{nt}^{sv})] \qquad (598)$$

where the complex number time counting number is written as

$$\overline{\eta}_t^{sv} = \eta_t^{sv}\exp(j\theta_{\eta t}^{sv}) \qquad (599)$$

so that

$$t^{sv} = \eta_t^{sv}L_\alpha^c/n_t^{sv} = \eta_t^{sv}\Delta L_t^{sv} \qquad \theta_t^{sv} = \theta_{\eta t}^{sv} - \theta_{nt}^{sv} \qquad (600)$$

where η_t^{sv} and n_t^{sv} are not integers when an APF is present in matter, and the complex number nature of the time coordinate is $\overline{t}^{sv}$ due only to the complex number time counting number $\overline{\eta}_t^{sv}$ and the complex number segmentation number $\overline{n}_t^{sv}$ because the constant L_t^c is a real number in equation (598). The real number constants L_α^c and L_t^c that appear in equations (592) and (598) can be taken to be multiples (generally very large multiples) of the Planck length and Planck time respectively because an APF is a field that induces internal phase angles in all physical quantities except the fundamental constants of nature (h, c and G). Equations (594) and (600) allow the possibility of $\theta_\alpha^{sv} \le 0$ and $\theta_t^{sv} \le 0$ which has been suggested is the case for the Earth's gravity field.[5-10]

Part E investigates the space and time coordinate factors for several types of coordinate systems that are attached to the polyphase space vehicle itself, that is they are co-moving with the polyphase space vehicle. The space and time coordinate systems whose origin is situated in a polyphase space vehicle that is located in an APF are represented as complex numbers of the form[5-12]

$$\overline{\alpha}^{sv} = \alpha^{sv}\exp(j\theta_\alpha^{sv}) \qquad \overline{t}^{sv} = t^{sv}\exp(j\theta_t^{sv}) \qquad (601)$$

where the coordinate magnitudes are $\alpha^{sv} = x^{sv}, y^{sv}, z^{sv}$ or r^{sv}, z^{sv} or ρ^{sv} for Cartesian coordinates, cylindrical polar coordinates or spherical polar coordinates respectively. The space and time coordinate magnitudes and internal phase angles for regions of a polyphase space vehicle in an APF are given by

$$\alpha^{sv} = \alpha_o^{sv}C_{\sigma\alpha}^{sv} \qquad \theta_\alpha^{sv} = \theta_{\eta\alpha}^{sv} - \theta_{n\alpha}^{sv} \qquad (602)$$

$$t^{sv} = t_o^{sv}C_{\sigma t}^{sv} \qquad \theta_t^{sv} = \theta_{\eta t}^{sv} - \theta_{nt}^{sv} \qquad (603)$$

where α_o^{sv} and t_o^{sv} = values of the space and time coordinates within the polyphase space vehicle for zero APF conditions, $\sigma = 1,2,3$ and where the space and time coordinate factors $C_{\sigma\alpha}^{sv}$ and $C_{\sigma t}^{sv}$ are given by[9]

$$C_{1\alpha}^{sv} = \cos\theta_{\eta\alpha}^{sv} \sec\theta_{n\alpha}^{sv} \qquad C_{1t}^{sv} = \cos\theta_{\eta t}^{sv} \sec\theta_{nt}^{sv} \qquad [hn_o^{sv}] \tag{604}$$

$$C_{2\alpha}^{sv} = \cos\theta_{\eta\alpha}^{sv} \cos\theta_{n\alpha}^{sv} \qquad C_{2t}^{sv} = \cos\theta_{\eta t}^{sv} \cos\theta_{nt}^{sv} \qquad [ln_o^{sv}] \tag{605}$$

$$C_{3\alpha}^{sv} = \sec\theta_{\eta\alpha}^{sv} \cos\theta_{n\alpha}^{sv} \qquad C_{3t}^{sv} = \sec\theta_{\eta t}^{sv} \cos\theta_{nt}^{sv} \qquad [ln_o^{sv}] \tag{606}$$

where $[hn_o^{sv}]$ = high-density polyphase space vehicle and $[ln_o^{sv}]$ = low-density polyphase space vehicle and where $\theta_{\eta\alpha}^{sv}, \theta_{n\alpha}^{sv}$ and $\theta_{\eta t}^{sv}, \theta_{nt}^{sv}$ are the internal phase angles of the complex number space and time counting and segmentation numbers. For bound matter in an APF

$$\theta_{n\alpha}^{sv} \geq \theta_{\eta\alpha}^{sv} \qquad \theta_{nt}^{sv} \geq \theta_{\eta t}^{sv} \tag{607}$$

and for the conditions in equation (607), equations (602) and (603) give

$$\theta_{\alpha}^{sv} \leq 0 \qquad \theta_{t}^{sv} \leq 0 \tag{608}$$

which is true for bound systems. For spherical polar coordinates and for the radial coordinate $\alpha = \rho$ and equation (602) becomes

$$\alpha^{sv} = \alpha_o^{sv} C_{\sigma\alpha}^{sv} \qquad \theta_{\alpha}^{sv} = \theta_{\eta\alpha}^{sv} - \theta_{n\alpha}^{sv} \tag{609}$$

and the space coordinate factors $C_{\sigma\rho}^{sv}$ are[9]

$$C_{1\rho}^{sv} = \cos\theta_{\eta\rho}^{sv} \sec\theta_{n\rho}^{sv} \qquad [hn_o^{sv}] \tag{610}$$

$$C_{2\rho}^{sv} = \cos\theta_{\eta\rho}^{sv} \cos\theta_{n\rho}^{sv} \qquad [ln_o^{sv}] \tag{611}$$

$$C_{3\rho}^{sv} = \sec\theta_{\eta\rho}^{sv} \cos\theta_{n\rho}^{sv} \qquad [ln_o^{sv}] \tag{612}$$

The values of the space coordinate factors $C_{\sigma\rho}^{sv}$ for the radial coordinate of a spherical polar coordinate system can also be simply expressed in terms of the internal phase angles of the complex particle number and the complex particle number density.[9]

For a polyphase space vehicle in an APF, the average complex number particle number density is written as[9]

$$\overline{n}^{sv} = n^{sv} \exp(j\theta_n^{sv}) = \overline{\mathcal{N}}_{sv}/\overline{V}_{sv} \qquad n^{sv} = \mathcal{N}^{sv}/V^{sv} \qquad \theta_n^{sv} = \theta_{\mathcal{N}}^{sv} - \theta_V^{sv} \tag{613}$$

where the complex number total particle number of the polyphase space vehicle is written as

$$\overline{\mathcal{N}}_{sv} = \mathcal{N}^{sv} \exp(j\theta_{\mathcal{N}}^{sv}) \tag{614}$$

where

$$\mathfrak{N}^{sv} = N\cos\theta^{sv}_{\mathfrak{N}} \qquad [SN] \qquad (615)$$

$$\mathfrak{N}^{sv} = N\sec\theta^{sv}_{\mathfrak{N}} \qquad [LN] \qquad (616)$$

where [SN] = small N, [LN] = large N and where N = integer number of particles which interact with the APF. For a gravitational APF, $N = N^{tot}$ = total integer number of molecules in the polyphase space vehicle, while for an electromagnetic APF, $N = ZN^{tot}$ where Z = average integer atomic number of the material in the polyphase space vehicle. For practical reasons, the dominant APF in a polyphase space vehicle may be electromagnetic in nature. The angle $\theta^{sv}_{\mathfrak{N}}$ can be calculated for a linear static elastic model an as

$$\tan\theta^{sv}_{\mathfrak{N}} = (2\mathfrak{E}^{em}/G^{sv}_{o})^{1/2} \qquad \mathfrak{E}^{em} = 1/2G^{sv}_{o}\tan^{2}\theta^{sv}_{\mathfrak{N}} \qquad (617)$$

where $\mathfrak{E}^{em}$ = energy density of an electromagnetic APF and G^{sv}_{o} = zero APF value of the average shear modulus of the polyphase space vehicle. If dynamical effects must be considered the replacement $G^{sv}_{o} \to G^{sv}_{r}$ can be made in equation (617) where G^{sv}_{r} = an effective dynamical shear modulus which has different values for an applied electromagnetic or gravitational APF and may correspond to a particular resonance frequency of a finite system.[6-10] The complex number volume is represented as

$$\overline{V}^{sv} = V^{sv}\exp(j\theta^{sv}_{V}) \qquad \theta^{sv}_{V} = \theta^{sv}_{\mathfrak{N}} - \theta^{sv}_{n} \qquad \theta^{sv}_{V} \le 0 \qquad (618)$$

For homogeneous bound matter in a 3-dimensional APF the internal phase angles of the space coordinates are then written as

$$\theta^{sv}_{x} = \theta^{sv}_{y} = \theta^{sv}_{z} = \theta^{sv}_{r} = \theta^{sv}_{\rho} = 1/3\theta^{sv}_{V} = 1/3(\theta^{sv}_{\mathfrak{N}} - \theta^{sv}_{n}) \le 0 \qquad (619)$$

while for a uniaxial field along the x diection

$$\theta^{sv}_{x} = \theta^{sv}_{V} = \theta^{sv}_{\mathfrak{N}} - \theta^{sv}_{n} \le 0 \qquad (620)$$

From equations (613) through (618) it follows that[11]

$$n^{sv} = N/V^{sv}\cos\theta^{sv}_{\mathfrak{N}} \qquad [SN] \qquad (621)$$

$$n^{sv} = N/V^{sv}\sec\theta^{sv}_{\mathfrak{N}} \qquad [LN] \qquad (622)$$

$$\theta^{sv}_{n} = \theta^{sv}_{\mathfrak{N}} - \theta^{sv}_{V} = \theta^{sv}_{\mathfrak{N}} + \left|\theta^{sv}_{V}\right| \qquad \theta^{sv}_{n} \ge \theta^{sv}_{\mathfrak{N}} \qquad (623)$$

where for, for example, for spherical polar coordinates the volume element V^{sv} in the presence of a 3-dimensional homogeneous APF is given in terms of the volume element V^{sv}_{o} under zero APF conditions by[9]

$$V^{sv}/V^{sv}_{o} = (C^{sv}_{1\rho})^{3} = \cos\theta^{sv}_{\mathfrak{N}}\sec\theta^{sv}_{n} \ge 1 \qquad [hn^{sv}_{o}, SN] \qquad (624)$$

$$V^{sv}/V_o^{sv} = (C_{1\rho}^{sv})^3 = \sec\theta_{\mathcal{N}}^{sv}\sec\theta_n^{sv} \geq 1 \qquad [hn_o^{sv}, LN] \qquad (625)$$

$$V^{sv}/V_o^{sv} = (C_{2\rho}^{sv})^3 = \cos\theta_{\mathcal{N}}^{sv}\cos\theta_n^{sv} \leq 1 \qquad [ln_o^{sv}, SN] \qquad (626)$$

$$V^{sv}/V_o^{sv} = (C_{2\rho}^{sv})^3 = \sec\theta_{\mathcal{N}}^{sv}\cos\theta_n^{sv} \leq 1 \qquad [ln_o^{sv}, LN] \qquad (627)$$

$$V^{sv}/V_o^{sv} = (C_{3\rho}^{sv})^3 = \cos\theta_{\mathcal{N}}^{sv}\cos\theta_n^{sv} \leq 1 \qquad [ln_o^{sv}, SN] \qquad (628)$$

$$V^{sv}/V_o^{sv} = (C_{3\rho}^{sv})^3 = \sec\theta_{\mathcal{N}}^{sv}\cos\theta_n^{sv} \leq 1 \qquad [ln_o^{sv}, LN] \qquad (629)$$

where the space coordinate factors $C_{\sigma\rho}^{sv}$ for spherical polar coordinates are given by equations (610) through (612), and so because the right hand sides of equations (626) and (627) are the same as the right hand sides of equations (628) and (629) it follows that the space coordinate factors $C_{2\rho}^{sv}$ and $C_{3\rho}^{sv}$ of equations (611) and (612) give two possible solutions for the internal phase angles $\theta_{\eta\rho}^{sv}$ and $\theta_{\eta\rho}^{sv}$ of the counting and segmentation numbers for the $[ln_o^{sv}]$ case and therefore the $[ln_o^{sv}]$ case is degenerate. Then equations (621), (622) and (624) through (629) give[11]

$$n^{sv} = n_o^{sv}\cos\theta_n^{sv} \leq n_o^{sv} \qquad [hn_o^{sv}] \qquad (630)$$

$$n^{sv} = n_o^{sv}\sec\theta_n^{sv} \geq n_o^{sv} \qquad [ln_o^{sv}] \qquad (632)$$

where n_o^{sv} = particle number density under zero APF conditions and which is given by

$$n_o^{sv} = N/V_o^{sv} \qquad (632)$$

which is the conventional result.

Cartesian Coordinates.

For Cartesian coordinates the ratios of the volume elements with and without an APF present are given by

$$V^{sv}/V_o^{sv} = C_{1x}^{sv}C_{1y}^{sv}C_{1z}^{sv} = \cos\theta_{\mathcal{N}}^{sv}\sec\theta_n^{sv} \geq 1 \qquad [hn_o^{sv}, SN] \qquad (633)$$

$$V^{sv}/V_o^{sv} = C_{1x}^{sv}C_{1y}^{sv}C_{1z}^{sv} = \sec\theta_{\mathcal{N}}^{sv}\sec\theta_n^{sv} \geq 1 \qquad [hn_o^{sv}, LN] \qquad (634)$$

$$V^{sv}/V_o^{sv} = C_{\beta x}^{sv}C_{\beta y}^{sv}C_{\beta z}^{sv} = \cos\theta_{\mathcal{N}}^{sv}\cos\theta_n^{sv} \leq 1 \qquad [ln_o^{sv}, SN] \qquad (635)$$

$$V^{sv}/V_o^{sv} = C_{\beta x}^{sv}C_{\beta y}^{sv}C_{\beta z}^{sv} = \sec\theta_{\mathcal{N}}^{sv}\cos\theta_n^{sv} \leq 1 \qquad [ln_o^{sv}, LN] \qquad (636)$$

where $C_{1\alpha}^{sv}$ and $C_{\beta\alpha}^{sv}$ are given by equations (604) through (606) and where $\beta = 2,3$ describes the degeneracy of the $[ln_o^{sv}]$ state. For the case of a uniaxial APF in the x-direction, $C_{1y}^{sv} = 1, C_{1z}^{sv} = 1, C_{\beta y}^{sv} = 1$ and $C_{\beta z}^{sv} = 1$ with the result that, equations (633) through (636) give

$$V^{sv} / V_o^{sv} = C_{1x}^{sv} = \cos\theta_{\mathcal{N}}^{sv} \sec\theta_n^{sv} \geq 1 \qquad [hn_o^{sv}, SN] \qquad (637)$$

$$V^{sv} / V_o^{sv} = C_{1x}^{sv} = \sec\theta_{\mathcal{N}}^{sv} \sec\theta_n^{sv} \geq 1 \qquad [hn_o^{sv}, LN] \qquad (638)$$

$$V^{sv} / V_o^{sv} = C_{\beta x}^{sv} = \cos\theta_{\mathcal{N}}^{sv} \cos\theta_n^{sv} \leq 1 \qquad [ln_o^{sv}, SN] \qquad (639)$$

$$V^{sv} / V_o^{sv} = C_{\beta x}^{sv} = \sec\theta_{\mathcal{N}}^{sv} \cos\theta_n^{sv} \leq 1 \qquad [ln_o^{sv}, LN] \qquad (640)$$

where $\beta = 2,3$.

Cylindrical Polar Coordinates.

For the case of cylindrical polar coordinates the ratios of the volume elements with and without an acting APF are given by

$$V^{sv} / V_o^{sv} = (C_{1r}^{sv})^2 C_{1z}^{sv} = \cos\theta_{\mathcal{N}}^{sv} \sec\theta_n^{sv} \geq 1 \qquad [hn_o^{sv}, SN] \qquad (641)$$

$$V^{sv} / V_o^{sv} = (C_{1r}^{sv})^2 C_{1z}^{sv} = \sec\theta_{\mathcal{N}}^{sv} \sec\theta_n^{sv} \geq 1 \qquad [hn_o^{sv}, LN] \qquad (642)$$

$$V^{sv} / V_o^{sv} = (C_{\beta r}^{sv})^2 C_{\beta z}^{sv} = \cos\theta_{\mathcal{N}}^{sv} \cos\theta_n^{sv} \leq 1 \qquad [ln_o^{sv}, SN] \qquad (643)$$

$$V^{sv} / V_o^{sv} = (C_{\beta r}^{sv})^2 C_{\beta z}^{sv} = \sec\theta_{\mathcal{N}}^{sv} \cos\theta_n^{sv} \leq 1 \qquad [ln_o^{sv}, LN] \qquad (644)$$

where $\beta = 2,3$. For the case of a uniaxial APF along the z-axis, $C_{1r}^{sv} = 1$ and $C_{\beta r}^{sv} = 1$ and equations (641) through (644) become

$$V^{sv} / V_o^{sv} = C_{1z}^{sv} = \cos\theta_{\mathcal{N}}^{sv} \sec\theta_n^{sv} \geq 1 \qquad [hn_o^{sv}, SN] \qquad (645)$$

$$V^{sv} / V_o^{sv} = C_{1z}^{sv} = \sec\theta_{\mathcal{N}}^{sv} \sec\theta_n^{sv} \geq 1 \qquad [hn_o^{sv}, LN] \qquad (646)$$

$$V^{sv} / V_o^{sv} = C_{\beta z}^{sv} = \cos\theta_{\mathcal{N}}^{sv} \cos\theta_n^{sv} \leq 1 \qquad [ln_o^{sv}, SN] \qquad (647)$$

$$V^{sv} / V_o^{sv} = C_{\beta z}^{sv} = \sec\theta_{\mathcal{N}}^{sv} \cos\theta_n^{sv} \leq 1 \qquad [ln_o^{sv}, LN] \qquad (648)$$

where $\beta = 2,3$. For the special case of a two-dimensional radial APF, $C_{1z}^{sv} = 1$ and equations (641) through (644) give

$$(V^{sv} / V_o^{sv})^{1/2} = C_{1r}^{sv} = \cos^{1/2}\theta_{\mathcal{N}}^{sv} \sec^{1/2}\theta_n^{sv} \geq 1 \qquad [hn_o^{sv}, SN] \qquad (649)$$

$$(V^{sv} / V_o^{sv})^{1/2} = C_{1r}^{sv} = \sec^{1/2}\theta_{\mathcal{N}}^{sv} \sec^{1/2}\theta_n^{sv} \geq 1 \qquad [hn_o^{sv}, LN] \qquad (650)$$

$$(V^{sv} / V_o^{sv})^{1/2} = C_{\beta r}^{sv} = \cos^{1/2}\theta_{\mathcal{N}}^{sv} \cos^{1/2}\theta_n^{sv} \leq 1 \qquad [ln_o^{sv}, SN] \qquad (651)$$

$$(V^{sv} / V_o^{sv})^{1/2} = C_{\beta r}^{sv} = \sec^{1/2}\theta_{\mathcal{N}}^{sv} \cos^{1/2}\theta_n^{sv} \leq 1 \qquad [ln_o^{sv}, LN] \qquad (652)$$

where $\beta = 2,3$.

Spherical Polar Coordinates.

$$(V^{sv}/V_o^{sv})^{1/3} = C_{1\rho}^{sv} = \cos^{1/3}\theta_{\aleph}^{sv}\sec^{1/3}\theta_n^{sv} \geq 1 \qquad [hn_o^{sv}, SN] \qquad (653)$$

$$(V^{sv}/V_o^{sv})^{1/3} = C_{1\rho}^{sv} = \sec^{1/3}\theta_{\aleph}^{sv}\sec^{1/3}\theta_n^{sv} \geq 1 \qquad [hn_o^{sv}, LN] \qquad (654)$$

$$(V^{sv}/V_o^{sv})^{1/3} = C_{\beta\rho}^{sv} = \cos^{1/3}\theta_{\aleph}^{sv}\cos^{1/3}\theta_n^{sv} \leq 1 \qquad [ln_o^{sv}, SN] \qquad (655)$$

$$(V^{sv}/V_o^{sv})^{1/3} = C_{\beta\rho}^{sv} = \sec^{1/3}\theta_{\aleph}^{sv}\cos^{1/3}\theta_n^{sv} \leq 1 \qquad [ln_o^{sv}, LN] \qquad (656)$$

where $\beta = 1,2$. Note that equations (653) through (656) for spherical polar coordinates can also be obtained from equations (633) through (636) for a Cartesian coordinate system by taking $C_{1x}^{sv} = C_{1y}^{sv} = C_{1z}^{sv}$ and $C_{\beta x}^{sv} = C_{\beta y}^{sv} = C_{\beta z}^{sv}$ and from equations (641) through (644) for cylindrical polar coordinates by taking $C_{1r}^{sv} = C_{1z}^{sv}$ and $C_{\beta r}^{sv} = C_{\beta z}^{sv}$.

Consider now the case of a polyphase space vehicle treated as a linear elastic system for which the internal phase angles θ_n^{sv} and $\theta_{\aleph}^{sv}$ are related by[9]

$$\cos\theta_n^{sv} = 1 - \tan\theta_{\aleph}^{sv} \qquad 0 \leq \theta_n^{sv} \leq \pi/2 \qquad 0 \leq \theta_{\aleph}^{sv} \leq \pi/4 \qquad [hn_o^{sv}] \qquad (657)$$

$$\sec\theta_n^{sv} = 1 + \tan\theta_{\aleph}^{sv} \qquad 0 \leq \theta_n^{sv} \leq \pi/2 \qquad 0 \leq \theta_{\aleph}^{sv} \leq \pi/2 \qquad [ln_o^{sv}] \qquad (658)$$

with $\theta_n^{sv} \geq \theta_{\aleph}^{sv}$, and which can now be used in equations (633) through (656) to describe the space coordinate factors for a linear elastic system in an APF.

Cartesian Coordinates

Combining equations (633) through (636) with equations (657) and (658) gives for the case of a 3-dimensional APF in Cartesian coordinates

$$C_{1x}^{sv}C_{1y}^{sv}C_{1z}^{sv} = \sec\theta_n^{sv}[1 + (1 - \cos\theta_n^{sv})^2]^{-1/2} = \cos\theta_{\aleph}^{sv}(1 - \tan\theta_{\aleph}^{sv})^{-1} \qquad [hn_o^{sv}, SN] \qquad (659)$$

$$C_{1x}^{sv}C_{1y}^{sv}C_{1z}^{sv} = \sec\theta_n^{sv}[1 + (1 - \cos\theta_n^{sv})^2]^{1/2} = \sec\theta_{\aleph}^{sv}(1 - \tan\theta_{\aleph}^{sv})^{-1} \qquad [hn_o^{sv}, LN] \qquad (660)$$

$$C_{\beta x}^{sv}C_{\beta y}^{sv}C_{\beta z}^{sv} = \cos\theta_n^{sv}[1 + (\sec\theta_n^{sv} - 1)^2]^{-1/2} = \cos\theta_{\aleph}^{sv}(1 + \tan\theta_{\aleph}^{sv})^{-1} \qquad [ln_o^{sv}, SN] \qquad (661)$$

$$C_{\beta x}^{sv}C_{\beta y}^{sv}C_{\beta z}^{sv} = \cos\theta_n^{sv}[1 + (\sec\theta_n^{sv} - 1)^2]^{1/2} = \sec\theta_{\aleph}^{sv}(1 + \tan\theta_{\aleph}^{sv})^{-1} \qquad [ln_o^{sv}, LN] \qquad (662)$$

where $\beta = 2,3$. Combining equations (637) through (640) with equations (657) and (658) gives for a uniaxial APF in a linear elastic polyphase space vehicle

$$C_{1x}^{sv} = \sec\theta_n^{sv}[1 + (1 - \cos\theta_n^{sv})^2]^{-1/2} = \cos\theta_{\aleph}^{sv}(1 - \tan\theta_{\aleph}^{sv})^{-1} \qquad [hn_o^{sv}, SN] \qquad (663)$$

$$C_{1x}^{sv} = \sec\theta_n^{sv}[1 + (1 - \cos\theta_n^{sv})^2]^{1/2} = \sec\theta_{\aleph}^{sv}(1 - \tan\theta_{\aleph}^{sv})^{-1} \qquad [hn_o^{sv}, LN] \qquad (664)$$

$$C^{sv}_{\beta x} = \cos\theta^{sv}_{n}[1+(\sec\theta^{sv}_{n}-1)^2]^{-1/2} = \cos\theta^{sv}_{\mathcal{N}}(1+\tan\theta^{sv}_{\mathcal{N}})^{-1} \quad [\ln^{sv}_{o}, SN] \quad (665)$$

$$C^{sv}_{\beta x} = \cos\theta^{sv}_{n}[1+(\sec\theta^{sv}_{n}-1)^2]^{1/2} = \sec\theta^{sv}_{\mathcal{N}}(1+\tan\theta^{sv}_{\mathcal{N}})^{-1} \quad [\ln^{sv}_{o}, LN] \quad (666)$$

for Cartesian coordinates.

Cylindrical Polar Coordinates.

Combining equations (641) through (644) with equations (657) and (658) gives for the case of a 3-dimensional APF

$$(C^{sv}_{1r})^2 C^{sv}_{1z} = \sec\theta^{sv}_{n}[1+(1-\cos\theta^{sv}_{n})^2]^{-1/2} = \cos\theta^{sv}_{\mathcal{N}}(1-\tan\theta^{sv}_{\mathcal{N}})^{-1} \quad [hn^{sv}_{o}, SN] \quad (667)$$

$$(C^{sv}_{1r})^2 C^{sv}_{1z} = \sec\theta^{sv}_{n}[1+(1-\cos\theta^{sv}_{n})^2]^{1/2} = \sec\theta^{sv}_{\mathcal{N}}(1-\tan\theta^{sv}_{\mathcal{N}})^{-1} \quad [hn^{sv}_{o}, LN] \quad (668)$$

$$(C^{sv}_{\beta r})^2 C^{sv}_{\beta z} = \cos\theta^{sv}_{n}[1+(\sec\theta^{sv}_{n}-1)^2]^{-1/2} = \cos\theta^{sv}_{\mathcal{N}}(1+\tan\theta^{sv}_{\mathcal{N}})^{-1} \quad [\ln^{sv}_{o}, SN] \quad (669)$$

$$(C^{sv}_{\beta r})^2 C^{sv}_{\beta z} = \cos\theta^{sv}_{n}[1+(\sec\theta^{sv}_{n}-1)^2]^{1/2} = \sec\theta^{sv}_{\mathcal{N}}(1+\tan\theta^{sv}_{\mathcal{N}})^{-1} \quad [\ln^{sv}_{o}, LN] \quad (670)$$

For a uniaxial APF, $C^{sv}_{1r} = 1$ and $C^{sv}_{\beta r} = 1$ and equations (667) through (670) become the uniaxial APF equations (663) through (666) but now for the z-axis. For the special case of a 2-dimensional radial APF, $C^{sv}_{1z} = 1$ and $C^{sv}_{\beta z} = 1$ and equations (667) through (670) become

$$C^{sv}_{1r} = \sec^{1/2}\theta^{sv}_{n}[1+(1-\cos\theta^{sv}_{n})^2]^{-1/4} = \cos^{1/2}\theta^{sv}_{\mathcal{N}}(1-\tan\theta^{sv}_{\mathcal{N}})^{-1/2} \quad [hn^{sv}_{o}, SN] \quad (671)$$

$$C^{sv}_{1r} = \sec^{1/2}\theta^{sv}_{n}[1+(1-\cos\theta^{sv}_{n})^2]^{1/4} = \sec^{1/2}\theta^{sv}_{\mathcal{N}}(1-\tan\theta^{sv}_{\mathcal{N}})^{-1/2} \quad [hn^{sv}_{o}, LN] \quad (672)$$

$$C^{sv}_{\beta r} = \cos^{1/2}\theta^{sv}_{n}[1+(\sec\theta^{sv}_{n}-1)^2]^{-1/4} = \cos^{1/2}\theta^{sv}_{\mathcal{N}}(1+\tan\theta^{sv}_{\mathcal{N}})^{-1/2} \quad [\ln^{sv}_{o}, SN] \quad (673)$$

$$C^{sv}_{\beta r} = \cos^{1/2}\theta^{sv}_{n}[1+(\sec\theta^{sv}_{n}-1)^2]^{1/4} = \sec^{1/2}\theta^{sv}_{\mathcal{N}}(1+\tan\theta^{sv}_{\mathcal{N}})^{-1/2} \quad [\ln^{sv}_{o}, LN] \quad (674)$$

where $\beta = 2,3$.

Spherical Polar Coordinates.

Combining equations (653) through (656) with equations (657) and (658) gives for the case of a 3-dimensional APF in spherical polar coordinates

$$C^{sv}_{1\rho} = \sec^{1/3}\theta^{sv}_{n}[1+(1-\cos\theta^{sv}_{n})^2]^{-1/6} = \cos^{1/3}\theta^{sv}_{\mathcal{N}}(1-\tan\theta^{sv}_{\mathcal{N}})^{-1/3} \quad [hn^{sv}_{o}, SN] \quad (675)$$

$$C^{sv}_{1\rho} = \sec^{1/3}\theta^{sv}_{n}[1+(1-\cos\theta^{sv}_{n})^2]^{1/6} = \sec^{1/3}\theta^{sv}_{\mathcal{N}}(1-\tan\theta^{sv}_{\mathcal{N}})^{-1/3} \quad [hn^{sv}_{o}, LN] \quad (676)$$

$$C^{sv}_{\beta\rho} = \cos^{1/3}\theta^{sv}_{n}[1+(\sec\theta^{sv}_{n}-1)^2]^{-1/6} = \cos^{1/3}\theta^{sv}_{\mathcal{N}}(1+\tan\theta^{sv}_{\mathcal{N}})^{-1/3} \quad [\ln^{sv}_{o}, SN] \quad (677)$$

$$C^{sv}_{\beta\rho} = \cos^{1/3}\theta^{sv}_{n}[1+(\sec\theta^{sv}_{n}-1)^2]^{1/6} = \sec^{1/3}\theta^{sv}_{\mathbb{N}}(1+\tan\theta^{sv}_{\mathbb{N}})^{-1/3} \qquad [\ln^{sv}_{o}, LN] \qquad (678)$$

where $\beta = 2,3$. Equations (675) through (678) can also be obtained from equations (659) through (662) for Cartesian coordinates by taking $C^{sv}_{1x} = C^{sv}_{1y} = C^{sv}_{1z}$ and $C^{sv}_{\beta x} = C^{sv}_{\beta y} = C^{sv}_{\beta z}$.

According to Einstein covariance of the laws of physics for two inertial systems, the space and time coordinate factors for the polyphase space vehicle at rest in the laboratory frame of reference on Earth as described in Part A must be equal to the corresponding quantities measured in the frame of reference attached to the space vehicle moving at constant relative speed as described in Part E, and therefore

$$C^{sv}_{\sigma\alpha} = C^{lab}_{\sigma\alpha} \qquad (679)$$

$$C^{sv}_{\sigma t} = C^{lab}_{\sigma t} \qquad (680)$$

where $\sigma = 1, 2, 3$ and where $\alpha = x, y, z$ for Cartesian coordinates, $\alpha = r, z$ for cylindrical coordinates and $\alpha = \rho$ for spherical polar coordinates. It is shown in Part F that for superluminal flight a tuned manmade external space field must be generated in the polyphase space vehicle so that $C^{sv}_{\sigma t}$ is an imaginary number in external space and is given by equation (719) as

$$C^{sv}_{\sigma t} = (s^{sv}_{\alpha o}/s^{sv}_{\alpha G})C^{sv}_{\sigma\alpha} = [s^{sv}_{\alpha o}/(is^{sv}_{m\alpha G})]C^{sv}_{\sigma\alpha} \qquad (681)$$

where $s^{sv}_{\alpha o}$ = constant phase velocity of a mechanical, acoustic or electromagnetic wave in the various components of the space vehicle and $s^{sv}_{m\alpha G}$ = magnitude of the of the tuned phase velocity $s^{sv}_{\alpha G}$ of the waves whose phase angles have been tuned so that $s^{sv}_{\alpha G}$ is an imaginary number. Then by equation (680) it follows that

$$C^{lab}_{\sigma t} = [s^{lab}_{\alpha o}/(is^{lab}_{m\alpha G})]C^{lab}_{\sigma\alpha} \qquad (682)$$

where

$$s^{lab}_{\alpha o} = s^{sv}_{\alpha o} \qquad s^{lab}_{m\alpha G} = s^{sv}_{m\alpha G} \qquad (683)$$

so that tuning methods developed in laboratories on Earth are expected to be operational in a polyphase space vehicle traveling at $v_o \geq c$ a long distance from Earth.

F. The Time Coordinate Factor for the Case of an Observer Attached to a Polyphase Space Vehicle Traveling at Subluminal and Superluminal Speeds and the Tuning Condition on the Ordinary External Space Field that is Required for Superluminal Flight of a Polyphase SpaceVehicle.

The time coordinate factor for a polyphase space vehicle as seen by an observer in a frame of reference attached to the rear of the polyphase space vehicle can be obtained from the definitions of a characteristic average constant linear velocity $s^{sv}_{\alpha o}$ of a mechanical, acoustic or electromagnetic wave in the materials of the polyphase space vehicle under zero APF conditions

$$s^{sv}_{\alpha o} = d\alpha^{sv}_{o}/dt^{sv}_{o} \qquad (684)$$

For the case of a constant APF which induces constant internal phase angles in all physical quantities except the fundamental constants of nature (and as used for the case of ultrafast subluminal flight in Section 3) the complex number speed of a mechanical, acoustic or electromagnetic wave is calculated from a choice of complex number space and time coordinates as follows

$$\bar{\alpha}_{L}^{sv} = \alpha_{L}^{sv} \exp(j\theta_{\alpha L}^{sv}) \qquad \bar{t}_{L}^{sv} = t_{L}^{sv} \exp(j\theta_{tL}^{sv}) \tag{685}$$

$$\alpha_{L}^{sv} = \alpha_{o}^{sv} C_{\sigma\alpha}^{sv} \qquad t_{L}^{sv} = t_{o}^{sv} C_{\sigma t}^{sv} \tag{686}$$

$$\bar{s}_{\alpha L}^{sv} = s_{\alpha L}^{sv} \exp(j\theta_{s\alpha L}^{sv}) = d\bar{\alpha}_{L}^{sv} / d\bar{t}_{L}^{sv} \tag{687}$$

$$s_{\alpha L}^{sv} = d\alpha_{L}^{sv} / dt_{L}^{sv} = (d\alpha_{o}^{sv} / dt_{o}^{sv})(C_{\sigma\alpha}^{sv} / C_{\sigma t}^{sv}) = s_{\alpha o}^{sv}(C_{\sigma\alpha}^{sv} / C_{\sigma t}^{sv}) \tag{688}$$

$$\theta_{s\alpha L}^{sv} = \theta_{\alpha L}^{sv} - \theta_{tL}^{sv} = \theta_{\eta\alpha L}^{sv} - \theta_{\pi\alpha L}^{sv} - (\theta_{\eta tL}^{sv} - \theta_{\pi tL}^{sv}) \tag{689}$$

where the j refers to a complex number in an internal space and where the index L refers to a polyphase space vehicle with a speed less than or equal to the light speed in vacuo. The time coordinate factor $C_{\sigma t}^{sv}$ is related to the space coordinate factor $C_{\sigma\alpha}^{sv}$ by

$$C_{\sigma t}^{sv} = (s_{\alpha o}^{sv} / s_{\alpha L}^{sv}) C_{\sigma\alpha}^{sv} \tag{690}$$

where $\sigma = 1,2,3$; $\alpha = x,y,z$ or r,z or ρ and where $s_{\alpha o}^{sv}$ = velocity of a mechanical, acoustic or electromagnetic wave propagating in the polyphase space vehicle relative to a frame of reference attached to the polyphase space vehicle under zero APF conditions, $s_{\alpha L}^{sv}$ = magnitude of the average velocity of a mechanical, acoustic or electromagnetic wave in the materials of the polyphase space vehicle relative to a frame of reference that is attached to the polyphase space vehicle in the presence of an APF. The time coordinate factor $C_{\sigma t}^{sv}$ is a real number for subluminal flight. Equation (690) is used in equations (854) and (855) and in equations (858) through (879) of Section 3 to show that ultrafast flight up to and including the light speed in vacuo is possible for a polyphase space vehicle with a properly tuned APF that cancels γ_{oL} of equation (725) of Part G.

For superluminal flight (see Section 4) an ordinary wave component in external space must be tuned so as to induce in the materials of the polyphase space vehicle a set of complex number space and time coordinates of a general form that can be engineered for the case $v_o \geq c$ so as to cancel the $-i$ of the term $-i\gamma_{oG}$ term that appears in the special relativity equation (725) of Part G. Then the manmade tuned APF component fields in internal space of the wave field are used to can cancel out the effects of γ_{oG} as explained in Section 4. The general forms of the coordinates required for this are

$$\bar{\alpha}_{G}^{sv} = \alpha_{G}^{sv} \exp(j\theta_{\alpha G}^{sv}) \exp(i\bar{\varphi}_{\alpha}^{sv}) \qquad \bar{t}_{G}^{sv} = t_{G}^{sv} \exp(j\theta_{tG}^{sv}) \exp(i\bar{\varphi}_{t}^{sv}) \tag{691}$$

where j refers to an internal space and i refers to the ordinary external space, where the index G refers to a space vehicle traveling at a speed greater than the light speed in vacuo and finally where, for constant phase angles, the velocity of the wave is

$$\bar{s}_{\alpha G}^{sv} = s_{\alpha G}^{sv} \exp(j\theta_{s\alpha G}^{sv}) = d\bar{\alpha}_{G}^{sv} / d\bar{t}_{G}^{sv} \tag{692}$$

$$\bar{s}_{\alpha G}^{sv} = s_{m\alpha G}^{sv} \exp(j\theta_{s\alpha G}^{sv}) \exp(i\bar{\varphi}_{s\alpha G}^{sv}) = d\bar{\alpha}_{G}^{sv} / d\bar{t}_{G}^{sv} \tag{693}$$

$$s_{\alpha G}^{sv} = d\alpha_{G}^{sv} / dt_{G}^{sv} = s_{\alpha o}^{sv} (C_{\sigma\alpha}^{sv} / C_{\sigma t}^{sv}) \tag{694}$$

where $s_{m\alpha G}^{sv}$ is the magnitude of the complex number wave velocity $s_{\alpha G}^{sv}$ which is a complex number in external space for superluminal flight, so that

$$s_{m\alpha G}^{sv} = \left| s_{\alpha G}^{sv} \right| = s_{\alpha o}^{sv} (C_{\sigma\alpha}^{sv} / \left| C_{\sigma t}^{sv} \right|) \tag{695}$$

where the time coordinate factor $C_{\sigma t}^{sv}$ is a complex number in external space for superluminal flight and which is written as

$$C_{\sigma t}^{sv} = \left| C_{\sigma t}^{sv} \right| \exp(i\varphi_{C\sigma t}^{sv}) \tag{696}$$

and as described in Section 4, the tuned value for superluminal flight is given by

$$C_{\sigma t}^{sv} = -i\left| C_{\sigma t}^{sv} \right| = \left| C_{\sigma t}^{sv} \right| \exp(-i\pi/2) \qquad \varphi_{C\sigma t}^{sv} = -\pi/2 \tag{697}$$

and where

$$\theta_{s\alpha G}^{sv} = \theta_{\alpha G}^{sv} - \theta_{tG}^{sv} \qquad \bar{\varphi}_{s\alpha G}^{sv} = \bar{\varphi}_{\alpha}^{sv} - \bar{\varphi}_{t}^{sv} - \varphi_{C\sigma t}^{sv} \tag{698}$$

and where the external phase angles for the wave are complex numbers in an internal space because of the presence of an APF and are written as

$$\bar{\varphi}_{\alpha}^{sv} = \bar{k}^{sv}\bar{\delta}_{\alpha}^{sv} \qquad \bar{\varphi}_{t}^{sv} = \bar{\omega}^{sv}\bar{\delta}_{t}^{sv} \qquad \bar{\varphi}_{s\alpha}^{sv} = \bar{k}^{sv}\bar{\delta}_{\alpha}^{sv} - \bar{\omega}^{sv}\bar{\delta}_{t}^{sv} \tag{699}$$

where $\bar{\delta}_{\alpha}^{sv}$ and $\bar{\delta}_{t}^{sv}$ are the distance and the time coordinates of the external wave that propagates in the space vehicle in the presence of an APF and $\bar{k}^{sv}$ and $\bar{\omega}^{sv}$ are the wave number and the frequency of the external wave which is propagating in the presence of an APF, and these quantities are represented as

$$\bar{\delta}_{\alpha}^{sv} = \delta_{\alpha o}^{sv} \exp(j\theta_{\delta\alpha}^{sv}) \qquad \bar{\delta}_{t}^{sv} = \delta_{to}^{sv} \exp(j\theta_{\delta t}^{sv}) \tag{700}$$

$$\bar{k}^{sv} = k_{o}^{sv} \exp(j\theta_{k}^{sv}) \qquad \bar{\omega}^{sv} = \omega_{o}^{sv} \exp(j\theta_{\omega}^{sv}) \tag{701}$$

The tuned complex number external phase angle of the velocity $\bar{\varphi}_{s\alpha}^{sv}$ is written for superluminal flight as

$$\bar{\varphi}_{s\alpha G}^{sv} = k_{o}^{sv}\delta_{\alpha o}^{sv} \exp[j(\theta_{k}^{sv} + \theta_{\delta\alpha}^{sv})] - \omega_{o}^{sv}\delta_{to}^{sv} \exp[j(\theta_{\omega}^{sv} + \theta_{\delta t}^{sv})] - \varphi_{C\sigma t}^{sv} = \pi/2 \tag{702}$$

which can also be written as

$$\bar{\varphi}_{s\alpha G}^{sv} = \varphi_{\delta\alpha R}^{sv} - \varphi_{\delta t R}^{sv} - \varphi_{C\sigma t}^{sv} + j(\varphi_{\delta\alpha I}^{sv} - \varphi_{\delta t I}^{sv}) = \pi/2 \tag{703}$$

where

$$\varphi_{\delta\alpha R}^{sv} = k_{o}^{sv}\delta_{\alpha o}^{sv} \cos(\theta_{k}^{sv} + \theta_{\delta\alpha}^{sv}) \qquad \varphi_{\delta t R}^{sv} = \omega_{o}^{sv}\delta_{to}^{sv} \cos(\theta_{\omega}^{sv} + \theta_{\delta t}^{sv}) \tag{704}$$

$$\varphi^{sv}_{\delta\alpha I} = k^{sv}_{o}\delta^{sv}_{\alpha o}\sin(\theta^{sv}_{k} + \theta^{sv}_{\delta\alpha}) \qquad \varphi^{sv}_{\delta t I} = \omega^{sv}_{o}\delta^{sv}_{to}\sin(\theta^{sv}_{\omega} + \theta^{sv}_{\delta t}) \tag{705}$$

The first condition for superluminal flight is that $\overline{\varphi}^{sv}_{s\alpha G}$ given by equations (702) and (703) must be a real number in internal space and given by

$$\overline{\varphi}^{sv}_{s\alpha G} = \varphi^{sv}_{s\alpha G} = \pi/2 \tag{706}$$

so that the conventional part of the wave field induces only an external space phase angle and does not change the internal space phase angles created by the APF portion of the wave field which will be subsequently tuned in Section 4, so that the conventional external part is tuned to the condition

$$\varphi^{sv}_{\delta\alpha I} - \varphi^{sv}_{\delta t I} = 0 \tag{707}$$

which, after using equation (705), is equivalent to

$$k^{sv}_{o}\delta^{sv}_{\alpha o}\sin(\theta^{sv}_{k} + \theta^{sv}_{\delta\alpha}) = \omega^{sv}_{o}\delta^{sv}_{to}\sin(\theta^{sv}_{\omega} + \theta^{sv}_{\delta t}) \tag{708}$$

and equation (703) becomes

$$\overline{\varphi}^{sv}_{s\alpha G} = \varphi^{sv}_{\delta\alpha R} - \varphi^{sv}_{\delta t R} - \varphi^{sv}_{C\sigma t} = \pi/2 \tag{709}$$

where this angle of value $\pi/2$ is an engineered or tuned value and equation (709) is only valid for this tuned condition. The form of the complex number number space and time coordinates within the polyphase space vehicle that are tuned for superluminal flight are

$$\overline{\alpha}^{sv}_{G} = \alpha^{sv}_{G}\exp(j\theta^{sv}_{\alpha G}) \qquad \overline{t}^{sv}_{G} = t^{sv}_{G}\exp(j\theta^{sv}_{tG}) \tag{710}$$

$$\alpha^{sv}_{G} = \alpha^{sv}_{o}C^{sv}_{\sigma\alpha}\exp(i\varphi^{sv}_{\delta\alpha R}) \qquad t^{sv}_{G} = t^{sv}_{o}C^{sv}_{\sigma t}\exp(i\varphi^{sv}_{\delta t R}) = t^{sv}_{o}\left|C^{sv}_{\sigma t}\right|\exp[i(\varphi^{sv}_{\delta t R} + \varphi^{sv}_{C\sigma t})] \tag{711}$$

where α^{sv}_{G} and t^{sv}_{G} are real numbers in an internal space and complex numbers in external space. Therefore the space and time coordinates within a polyphase space vehicle must be complex numbers in both internal and external space. Within the condition that the APF in each section of the space vehicle is independent of the space and time coordinates in each section so that all phase angles and all space and time coordinate factors are constants within each section of the space vehicle, the complex number velocity of the wave in each section of the space vehicle is written as

$$\overline{s}^{sv}_{\alpha G} = s^{sv}_{\alpha G}\exp(j\theta^{sv}_{s\alpha G}) = d\overline{\alpha}^{sv}_{G}/d\overline{t}^{sv}_{G} \tag{712}$$

$$\begin{aligned} s^{sv}_{\alpha G} &= d\alpha^{sv}_{G}/dt^{sv}_{G} = s^{sv}_{\alpha o}(C^{sv}_{\sigma\alpha}/\left|C^{sv}_{\sigma t}\right|)\exp(i\varphi^{sv}_{s\alpha G}) \\ &= s^{sv}_{\alpha o}(C^{sv}_{\sigma\alpha}/\left|C^{sv}_{\sigma t}\right|)\exp[i(\varphi^{sv}_{\delta\alpha R} - \varphi^{sv}_{\delta t R} - \varphi^{sv}_{C\sigma t})] \end{aligned} \tag{713}$$

The magnitude $s^{sv}_{m\alpha G}$ of the complex number $s^{sv}_{\alpha G}$ is

$$s^{sv}_{m\alpha G} = s^{sv}_{\alpha o}(C^{sv}_{\sigma\alpha}/\left|C^{sv}_{\sigma t}\right|) \tag{714}$$

where $s^{sv}_{\alpha o}$ is given by equation (684), so that equation (713) can be written as

$$s^{sv}_{\alpha G} = s^{sv}_{m\alpha G} \exp(i\varphi^{sv}_{s\alpha G}) = s^{sv}_{m\alpha G} \exp[i(\varphi^{sv}_{\delta\alpha R} - \varphi^{sv}_{\delta t R} - \varphi^{sv}_{C\sigma t})] \tag{715}$$

where $s^{sv}_{m\alpha G}$ = magnitude of the velocity of a mechanical, acoustic or electromagnetic wave in the α direction of a polyphase space vehicle. For the special case when the conventional part of the wave field in each section of the polyphase space vehicle is tuned to the condition in equation (706), then the following condition is valid

$$\varphi^{sv}_{s\alpha G} = \varphi^{sv}_{\delta\alpha R} - \varphi^{sv}_{\delta t R} - \varphi^{sv}_{C\sigma t} = \pi/2 \qquad \varphi^{sv}_{\delta\alpha R} - \varphi^{sv}_{\delta t R} = 0 \qquad \varphi^{sv}_{C\sigma t} = -\pi/2 \tag{716}$$

which according to equations (703) and (704) is equivalent to

$$k^{sv}_{o}\delta^{sv}_{\alpha o} \cos(\theta^{sv}_{k} + \theta^{sv}_{\delta\alpha}) - \omega^{sv}_{o}\delta^{sv}_{to} \cos(\theta^{sv}_{\omega} + \theta^{sv}_{\delta t}) = 0 \tag{717}$$

(see Section 4 for a complete treatment), which is always possible for the conventional portion of the wave field, then equation (715) for the wave velocity becomes

$$s^{sv}_{\alpha G} = is^{sv}_{m\alpha G} \qquad \left|s^{sv}_{\alpha G}\right| = s^{sv}_{m\alpha G} \qquad \varphi^{sv}_{s\alpha G} = \pi/2 \tag{718}$$

and for this constant intensity APF and for the tuned condition in equation (716), the time coordinate factor is given by

$$C^{sv}_{\sigma t} = (s^{sv}_{\alpha o} / s^{sv}_{\alpha G})C^{sv}_{\sigma\alpha} = [s^{sv}_{\alpha o} /(is^{sv}_{m\alpha G})]C^{sv}_{\sigma\alpha} \qquad \left|C^{sv}_{\sigma t}\right| = (s^{sv}_{\alpha o} / s^{sv}_{m\alpha G})C^{sv}_{\sigma\alpha} \qquad \varphi^{sv}_{C\sigma t} = -\pi/2 \tag{719}$$

and it is these expressions that are used in equations (886) and (887) and (890) through (911) of Section 4 to remove the imaginary index i from the equations, and that at least mathematically allows the superluminal flight of a polyphase space vehicle. Combining equations (708) and (717) gives

$$\theta^{sv}_{k} + \theta^{sv}_{\delta\alpha} = \theta^{sv}_{\omega} + \theta^{sv}_{\delta t} \tag{720}$$

which is the requirement for the tuning condition in equation (706).

The space and time coordinate factors $C^{sv}_{\sigma\alpha}$ and $C^{sv}_{\sigma t}$ relative to a frame of reference that is attached to the polyphase space vehicle are expected to depend only on the properties of the polyphase space vehicle. It should be pointed out that according to the Einstein principle of covariance of the laws of physics, all quantities Q^{lab} ascribed to the polyphase space vehicle at rest in the laboratory system on Earth as described in Part A are equal to the corresponding quantities Q^{sv} of the polyphase space vehicle as seen by an observer attached to the polyphase space vehicle that is moving uniformly relative to the laboratory on Earth as in Part E, so that $Q^{lab} = Q^{sv}$. It is clear from a comparison of equations (685), (691), (710) and (711) that different forms of the complex number space and time coordinates are required for subluminal and superluminal inertial systems, but this does not violate any law of nature. Rather, the difference in form of the space and time coordinates is due to a pilot throwing a switch that changes an APF into an APF/conventional field combination and vice versa at the point $v_o = c$. But the switch can also be thrown in the space vehicle when it is stationary in the laboratory system so that the rule $Q^{lab} = Q^{sv}$ is also valid for the complex number coordinates.

G. Special Relativity and Einstein Covariance for Real Number Space and Time Coordinates that Correspond to Zero APF.

Special relativity has a long history.[13] The primary postulates of the special theory of relativity is that of Einstein invariance of the laws of physics and the constancy of the speed of light, which, for the propagation of light states that the quantity $c^2t_o^2 - x_o^2$ is invariant of the coordinate frame of reference chosen, and therefore[3,4,13]

$$c^2t_o^2 - x_o^2 = c^2(t'_o)^2 - (x'_o)^2 = 0 = \text{invariant} \tag{721}$$

which is an invariant that is independent of the motion of the observer, where one observer is located at the origin of the stationary (unprimed) coordinate system and the second observer is located at the origin of the moving (primed) coordinate system which is moving relative to the stationary-coordinate system with velocity v_o. The subscript "o" refers to zero APF condition under which special relativity and Einstein invariance were historically developed. Equation (721) gives rise to the Lorentz transformation of coordinates that relates the space and time coordinates of a moving body as observed from the moving and stationary-coordinate systems and which is written as[3,4,13]

$$x'_o = \gamma_o(x_o - v_o t_o) \qquad x_o = \gamma_o(x'_o + v_o t'_o) \tag{722}$$

$$t'_o = \gamma_o(t_o - v_o c^{-2} x_o) \qquad t_o = \gamma_o(t'_o + v_o c^{-2} x'_o) \tag{723}$$

which leads to the following expression for the mass of a moving body as see from the rest frame of reference

$$m = \gamma_o m_r \tag{724}$$

where m_r = rest mass (proper mass) and where the Lorentz-Einstein factor γ_o is given by

$$\gamma_o = (1-\beta_o^2)^{-1/2} = \begin{cases} \gamma_{oL} & 0 \le \beta_o < 1 \\ \infty & \beta_o = 1 \\ -i\gamma_{oG} & 1 < \beta_o \le \infty \end{cases} \tag{725}$$

where $\beta_o = v_o / c$ and where

$$\gamma_{oL} = (1-\beta_o^2)^{-1/2} \qquad 0 \le \beta_o < 1 \qquad 1 \le \gamma_{oL} < \infty \tag{726}$$

$$\gamma_{oG} = (\beta_o^2 - 1)^{-1/2} \qquad 1 < \beta_o \le \infty \qquad 0 \le \gamma_{oG} < \infty \tag{727}$$

Equation (725) shows that special relativity for the case of zero APF requires that the speed of a body be less than the light speed, that is $\beta_o < 1$ because, as shown in equation (725), the Lorentz-Einstein factor γ_o has a singularity for $\beta_o = 1$ and the case $\beta_o > 1$ leads to an imaginary value of the Lorentz-Einstein factor γ_o and this in turn requires that the space and time coordinates in equations (722) and (723) and the mass in equation (724) are imaginary numbers which does not make sense. Thus for matter, only $\beta_o < 1$ and $\gamma_o = \gamma_{oL}$ are possible for special relativity under zero APF conditions.

The standard expressions for the Lorentz contraction and the time dilation correspond to zero APF and are obtained from equations (722) and (723) as[3,4,13]

$$\Delta x_o = \gamma_{oL}^{-1} \Delta x'_o \tag{728}$$

$$\Delta t_o = \gamma_{oL} \Delta t'_o \tag{729}$$

where $\Delta x'_o$ and $\Delta t'_o$ are the values of a distance interval and a time interval as measured by an observer attached to the moving coordinate system. Finally, the conventional special relativistic velocity addition formulas correspond to zero APF and are easily obtained from the Lorentz transformations in equations (722) and (723) as[3,4,13]

$$w_o = (u_o + v_o)(1 + u_o v_o c^{-2})^{-1} \tag{730}$$

$$u_o = (w_o - v_o)(1 - w_o v_o c^{-2})^{-1} \tag{731}$$

where v_o = relative velocity of the moving frame of reference and the stationary frame of reference, u_o = velocity of the moving body as measured from the moving frame of reference and w_o = velocity of the moving body as measured from stationary frame of reference. This very brief summary of conventional special relativity is sufficient for the reader to proceed. The standard special relativistic equations (721) through (731) are valid for the case of zero APF and will now be generalized to the case of nonzero cosmic and manmade APFs.

H. Special Relativity and Einstein Covariance for Complex Number Space and Time Coordinates in the Presence of an APF.

About fifteen years ago it was suggested that space and time coordinates of spacetime may have, under some conditions, a broken symmetry in the sense that these coordinates are complex numbers in an internal space.[12] Einstein covariance and the constancy of the speed of light are assumed to be valid in this complex number spacetime, and in particular was applied to light propagation in the vacuum with an APF as follows[5]

$$c^2 \bar{t}^2 - \bar{x}^2 = c^2 (\bar{t}')^2 - (\bar{x}')^2 = 0 = \text{invariant} \tag{732}$$

Equation (732) gives rise to the complex number Lorentz transformations

$$\bar{x}' = \bar{\gamma}(\bar{x} - \bar{v}\bar{t}) \qquad \bar{x} = \bar{\gamma}(\bar{x}' + \bar{v}\bar{t}') \tag{733}$$

$$\bar{t}' = \bar{\gamma}(\bar{t} - \bar{v}\bar{x}c^{-2}) \qquad \bar{t} = \bar{\gamma}(\bar{t}' + \bar{v}\bar{x}'c^{-2}) \tag{734}$$

where

$$\bar{\gamma} = (1 - \bar{\beta}^2)^{-1/2} \qquad \bar{\beta} = \bar{v}/c \tag{735}$$

where in c = real number light speed in the vacuum which is a constant of nature and is assumed to have the same real number value when observed from the stationary or moving inertial systems even in the presence of a cosmic APF which induces internal phase angles in the space and time coordinates and in the kinematical and dynamical quantities such as the relative velocity $\overline{v}$.[5] In the presence of an APF the space and time coordinates that appear in the Lorentz transformation equations (733) and (734) are represented as[5]

$$\overline{x} = x\exp(j\theta_x) \qquad \overline{t} = t\exp(j\theta_t) \qquad \overline{x}' = x'\exp(j\theta'_x) \qquad \overline{t}' = t'\exp(\theta'_t) \qquad (736)$$

while the kinematic quantities $\overline{v}$, $\overline{\beta}$ and $\overline{\gamma}$ are written as

$$\overline{v} = v\exp(j\theta_v) \qquad \overline{\beta} = \beta\exp(j\theta_\beta) \qquad \overline{\gamma} = \gamma\exp(j\theta_\gamma) \qquad (737)$$

Then equations (735) and (737) give the following magnitudes and internal phase angles[5]

$$\beta = v/c \qquad \theta_\beta = \theta_v \qquad (738)$$

$$\gamma = [1 - 2\beta^2\cos(2\theta_\beta) + \beta^4]^{-1/4} \qquad \tan(2\theta_\gamma) = \beta^2\sin(2\theta_\beta)[1 - \beta^2\cos(2\theta_\beta)]^{-1} \qquad (739)$$

An examination of equation (739) shows that the magnitude γ has no singular points and therefore special relativity in a spacetime with broken internal symmetries that are induced by the presence of an APF places no restrictions on the velocity of a body, and a body can go from subluminal flight to superluminal flight without any mathematical disasters.[5] Instead of a singularity at $\beta = 1$ the function γ has the value[5]

$$\gamma_1 = [2\sin\theta_\beta]^{-1/2} \qquad \beta = 1 \qquad (740)$$

The maximum value of γ is given by[5]

$$\gamma_{max} = [\sin(2\theta_\beta)]^{-1/2} \qquad \beta_{max} = [\cos(2\theta_\beta)]^{1/2} \qquad (741)$$

and occurs at the point $\beta = \beta_{max}$. The complex number relativistic mass corresponding to the the complex number boost $\overline{\gamma}$ is given by

$$\overline{m} = m\exp(j\theta_m) = \overline{\gamma}m_r \qquad m = \gamma m_r \qquad \theta_m = \theta_\gamma \qquad (742)$$

However in the present theory γ_o, corresponding to a zero value of an APF, is not discarded as a useless artifact and still plays an important role in the theory of ultrafast flight. In fact, it will be shown in Section 3 and Section 4 of this paper that the Lorentz-Einstein factor γ_o, with its singularity at $v_o = c$, as described by equation (725), requires that natural objects cannot obtain subluminal flight up to and including $v = c$ and superluminal flight with $v > c$. In particular it will be shown in Sections 2 through 4 that the Lorentz-Einstein factors $\gamma_o = \gamma_{oL}$ and $\gamma_o = -i\gamma_{oG}$ given in equation (725) forbid ultrafast flight in natural objects and that only polyphase space vehicles which contain an artificial adjustable APF of a specific form to cancel the singularity effects of γ_{oL} can undergo ultrafast subluminal flight with $v \le c$, and only the joint action of an artificial and adjustable external space field and an artificial and adjustable APF of a specific form can cancel the joint effects of $-i\gamma_{oG}$ (which includes the double effects of the $-$ i term and the singularity of the γ_{oG} term) and can allow a polyphase space vehicle to undergo superluminal flight with $v \geq c$.

The real and imaginary parts of the complex number Lorentz transformations in equations (733) and (734) can be written as[5]

$$x'\cos\theta'_x = \gamma[x\cos(\theta_x + \theta_\gamma) - vt\cos(\theta_v + \theta_t + \theta_\gamma)] \equiv \gamma E \tag{743}$$

$$x'\sin\theta'_x = \gamma[x\sin(\theta_x + \theta_\gamma) - vt\sin(\theta_v + \theta_t + \theta_\gamma)] \equiv \gamma F \tag{744}$$

$$t'\cos\theta'_t = \gamma[t\cos(\theta_t + \theta_\gamma) - vxc^{-2}\cos(\theta_v + \theta_x + \theta_\gamma)] \equiv \gamma G \tag{745}$$

$$t'\sin\theta'_t = \gamma[t\sin(\theta_t + \theta_\gamma) - vxc^{-2}\sin(\theta_v + \theta_x + \theta_\gamma)] \equiv \gamma H \tag{746}$$

$$x\cos\theta_x = \gamma[x'\cos(\theta'_x + \theta_\gamma) + vt'\cos(\theta_v + \theta'_t + \theta_\gamma)] \equiv \gamma K \tag{747}$$

$$x\sin\theta_x = \gamma[x'\sin(\theta'_x + \theta_\gamma) + vt'\sin(\theta_v + \theta'_t + \theta_\gamma)] \equiv \gamma L \tag{748}$$

$$t\cos\theta_t = \gamma[t'\cos(\theta'_t + \theta_\gamma) + vx'c^{-2}\cos(\theta_v + \theta'_x + \theta_\gamma)] \equiv \gamma M \tag{749}$$

$$t\sin\theta_t = \gamma[t'\sin(\theta'_t + \theta_\gamma) + vx'c^{-2}\sin(\theta_v + \theta'_x + \theta_\gamma)] \equiv \gamma N \tag{750}$$

where

$$E = x\cos(\theta_x + \theta_\gamma) - vt\cos(\theta_v + \theta_t + \theta_\gamma) \tag{751}$$

$$F = x\sin(\theta_x + \theta_\gamma) - vt\sin(\theta_v + \theta_t + \theta_\gamma) \tag{752}$$

$$G = t\cos(\theta_t + \theta_\gamma) - vxc^{-2}\cos(\theta_v + \theta_x + \theta_\gamma) \tag{753}$$

$$H = t\sin(\theta_t + \theta_\gamma) - vxc^{-2}\sin(\theta_v + \theta_x + \theta_\gamma) \tag{754}$$

$$K = x'\cos(\theta'_x + \theta_\gamma) + vt'\cos(\theta_v + \theta'_t + \theta_\gamma) \tag{755}$$

$$L = x'\sin(\theta'_x + \theta_\gamma) + vt'\sin(\theta_v + \theta'_t + \theta_\gamma) \tag{756}$$

$$M = t'\cos(\theta'_t + \theta_\gamma) + vx'c^{-2}\cos(\theta_v + \theta'_x + \theta_\gamma) \tag{757}$$

$$N = t'\sin(\theta'_t + \theta_\gamma) + vx'c^{-2}\sin(\theta_v + \theta'_x + \theta_\gamma) \tag{758}$$

and where γ and θ_γ are given by equation (739). Then it is easy to calculate $x', \theta'_x, t', \theta'_t$ and their inverses x, θ_x, t, θ_t as follows[5]

$$x' = \gamma I_x \qquad t' = \gamma I_t \qquad x = \gamma I'_x \qquad t = \gamma I'_t \tag{759}$$

where

$$I_x = (E^2 + F^2)^{1/2} \tag{760}$$

$$= [x^2 + v^2t^2 - 2vtx\cos(\theta_t + \theta_v - \theta_x)]^{1/2}$$

$$= [(x - vt)^2 + 2vtx\Psi_x]^{1/2}$$

$$I_t = (G^2 + H^2)^{1/2} \tag{761}$$

$$= [t^2 + v^2x^2c^{-4} - 2vxtc^{-2}\cos(\theta_x + \theta_v - \theta_t)]^{1/2}$$

$$= [(t - vxc^{-2})^2 + 2vxtc^{-2}\Psi_t]^{1/2}$$

$$I'_x = (K^2 + L^2)^{1/2} \tag{762}$$

$$= [(x')^2 + v^2(t')^2 + 2vt'x'\cos(\theta'_t + \theta_v - \theta'_x)]^{1/2}$$

$$= [(x' + vt')^2 - 2vt'x'\Psi'_x]^{1/2}$$

$$I'_t = (M^2 + N^2)^{1/2} \tag{763}$$

$$= [(t')^2 + v^2(x')^2c^{-4} + 2vx't'c^{-2}\cos(\theta'_x + \theta_v - \theta'_t)]^{1/2}$$

$$= [(t' + vx'c^{-2})^2 - 2vx't'c^{-2}\Psi'_t]^{1/2}$$

and

$$\tan\theta'_x = F/E \qquad \tan\theta'_t = H/G \qquad \tan\theta_x = L/K \qquad \tan\theta_t = N/M \tag{764}$$

where E,F,G,H,K,L,M and N are defined by equations (751) through (758), and where

$$\Psi_x = 1 - \cos(\theta_t + \theta_v - \theta_x) \qquad \Psi_t = 1 - \cos(\theta_x + \theta_v - \theta_t) \tag{765}$$

$$\Psi'_x = 1 - \cos(\theta'_t + \theta_v - \theta'_x) \qquad \Psi'_t = 1 - \cos(\theta'_x + \theta_v - \theta'_t) \tag{766}$$

For the special case where $x' = 0$ when the moving body is attached to the moving axis, equations (743), (744), (751), (752), (759), (760) and (765) give

$$I_x = 0 \qquad x = vt \qquad \theta_x = \theta^v_x \qquad \theta_t = \theta^v_t \qquad \theta^v_x = \theta^v_t + \theta_v \tag{767}$$

$$\Psi_x = \Psi^v_x = 0 \qquad \Psi_t = \Psi^v_t = 1 - \cos(2\theta_v) \qquad E = 0 \qquad F = 0 \tag{768}$$

where the superscript "v" refers to the vacuum state, and where θ_v = internal phase angle of the relative velocity of the stationary and moving axis and which also refers to the vacuum. As will be described in more detail in Section 2, the quantities Ψ'_x and Ψ'_t depend on the properties of the moving body for this special case, and in particular on the cosmic APF which induces an internal phase angle θ_v to the relative velocity of the polyphase space vehicle and a stationary observer on Earth.

I. Lorentz Contraction and the Time Dilation for Spacetime with an APF.

The forms of the Lorentz transformations of spacetime coordinates that are located in an APF and that are best to investigate the Lorentz contraction and time dilation respectively are taken from equations (733) and (734) as

$$\bar{x}' = \bar{\gamma}(\bar{x} - \bar{v}\bar{t}) \tag{769}$$

$$\bar{t}' = \bar{\gamma}(\bar{t} - \bar{v}\bar{x}c^{-2}) \tag{770}$$

For the study of the Lorentz contraction, consider a set of complex number space and time coordinates that satisfy equation (769), then

$$\bar{x}_1' = \bar{\gamma}(\bar{x}_1 - \bar{v}\bar{t}_1) \qquad \bar{x}_2' = \bar{\gamma}(\bar{x}_2 - \bar{v}\bar{t}_2) \tag{771}$$

Arrange that the measurements occur at $\bar{t}_1 = \bar{t}_2$ in the stationary frame of reference and then obtain the following from equation (771)

$$\Delta\bar{x}' = \bar{x}_2' - \bar{x}_1' = \bar{\gamma}(\bar{x}_2 - \bar{x}_1) = \bar{\gamma}\Delta\bar{x} \tag{772}$$

But a length as seen by an observer attached to the moving coordinate system is given by

$$\Delta\bar{x}' = \Delta x' \exp(j\theta_{\Delta x}') = \bar{x}_2' - \bar{x}_1' \tag{773}$$

and a length as seen from the stationary system of coordinates is given by

$$\Delta\bar{x} = \Delta x \exp(j\theta_{\Delta x}) = \bar{x}_2 - \bar{x}_1 \tag{774}$$

Then equation (772) can be written as

$$\Delta\bar{x} = (\bar{\gamma})^{-1}\Delta\bar{x}' \qquad \Delta x = \gamma^{-1}\Delta x' \qquad \theta_{\Delta x} = \theta_{\Delta x}' - \theta_\gamma \tag{775}$$

The relationship between the measured lengths is then given by the real part of the complex number relation in equation (775)

$$\Delta x_m = \gamma^{-1}\Delta x_m' \sec\theta_{\Delta x}' \cos\theta_{\Delta x} = \gamma^{-1}\Delta x_m' \sec\theta_{\Delta x}' \cos(\theta_{\Delta x}' - \theta_\gamma) \tag{776}$$

where the measured lengths in the stationary and moving coordinate systems are respectively

$$\Delta x_m = \Delta x \cos\theta_{\Delta x} \qquad \Delta x_m' = \Delta x' \cos\theta_{\Delta x}' \tag{777}$$

and therefore the Lorentz contraction $\Delta x_m < \Delta x_m'$ as seen by an observer in the stationary-coordinate system is valid only if

$$\gamma > \sec\theta_{\Delta x}' \cos(\theta_{\Delta x}' - \theta_\gamma) \tag{778}$$

which is generally true except for very large values of v > c where γ drops below $\gamma = 1$ and eventually $\gamma \rightarrow 0$ as $\beta \rightarrow \infty$ as shown in Reference 5. Therefore at very large values of v > c, equation (776) shows that $\Delta x_m > \Delta x'_m$ and there is a spatial dilation. Equations (775) and (776) describe the Lorentz contraction in the presence of an APF and should be compared with the result in equation (728) which is valid for the case of zero APF. Note that the results in equations (775) and (776) assume that $t_1 = t_2$ and $\theta_{t1} = \theta_{t2}$. The space coordinate factors are introduced into the theory of the Lorentz-Fitzgerald contraction in Appendix A.

The description of the time dilation in the presence of an APF is obtained from equation (770) by assuming that the moving coordinate system, with its attached observer, is itself attached to the moving body so that $\bar{x}' = 0$ and $\bar{x} = \bar{v}\bar{t}$, so that

$$\bar{t}' = \bar{\gamma}\bar{t}(1 - \bar{\beta}^2) = (\bar{\gamma})^{-1}\bar{t} \quad \text{or} \quad \bar{t} = \bar{\gamma}\bar{t}' \tag{779}$$

The same result is true for some chosen interval of time, thus

$$\Delta\bar{t} = \bar{\gamma}\Delta\bar{t}' \tag{780}$$

where the interval of time as seen by an observer attached to the moving body is written as

$$\Delta\bar{t}' = \Delta t' \exp(j\theta'_{\Delta t}) \tag{781}$$

and the interval of time as seen by an observer in the stationary-coordinate system is given by

$$\Delta\bar{t} = \Delta t \exp(j\theta_{\Delta t}) \tag{782}$$

Then equation (780) can be written as

$$\Delta t = \gamma\Delta t' \qquad \theta_{\Delta t} = \theta'_{\Delta t} + \theta_\gamma \tag{783}$$

The measured values of the two time intervals are given by

$$\Delta t_m = \Delta t \cos\theta_{\Delta t} \qquad \Delta t'_m = \Delta t' \cos\theta'_{\Delta t} \tag{784}$$

Then combining equations (783) and (784) gives the relationship between the measured time intervals as seen by an observer in the stationary system of coordinates and by an observer in the moving system of coordinates which are attached to the moving body, as

$$\Delta t_m = \gamma\Delta t'_m \sec\theta'_{\Delta t} \cos\theta_{\Delta t} = \gamma\Delta t'_m \sec\theta'_{\Delta t} \cos(\theta'_{\Delta t} + \theta_\gamma) \tag{785}$$

where γ and θ_γ are given by equation (739) and γ contains no singularities but has a maximum value given by equation (741). Equation (785) shows that the time dilation $\Delta t_m > \Delta t'_m$ as seen by an observer in the stationary-coordinate system is valid only if

$$\gamma > \cos\theta'_{\Delta t} \sec(\theta'_{\Delta t} + \theta_\gamma) \tag{786}$$

which is generally true except for large values of $\beta = v/c$ with $\beta > 1$ where the factor γ decreases to values below $\gamma = 1$ and eventually $\gamma \rightarrow 0$ as $\beta \rightarrow \infty$ as shown in Reference 5. Therefore at very large values of $v > c$, equation (785) shows that $\Delta t_m < \Delta t'_m$ and there is a time contraction so that the rates of processes would appear to be larger. A full description of the time dilation requires the introduction of the time coordinate factors and this is done in Appendix A.

J. Special Relativistic Addition of Velocities in the Presence of an APF.

In the presence of an APF, the special relativistic velocity addition equations (730) and (731) have been generalized to the following expressions[5]

$$\overline{w} = (\overline{u} + \overline{v})(1 + \overline{uv}c^{-2})^{-1} \tag{787}$$

$$\overline{u} = (\overline{w} - \overline{v})(1 - \overline{wv}c^{-2})^{-1} \tag{788}$$

where $\overline{u}$ = complex number speed of a body as seen by an observer on the moving frame of reference, $\overline{v}$ = complex number relative velocity of the moving and stationary frames of reference, $\overline{w}$ = complex number velocity as seen by an observer in the stationary frame of reference and where c = light speed in the vacuum which is a real number because it is a constant of nature.[5] A full description of the velocity addition formulas in equations (787) and (788) requires the use of the space and time coordinate factors and this will be accomplished in Appendix B.

This brief Introduction is sufficient to allow the use of the original Einstein's theory special relativity, and its modification to the case of the presence of an APF, for the calculation of the relative space and time coordinate factors of a moving polyphase space vehicle from the perspective of stationary and moving observers and to show that ultrafast flight, including superluminal flight, may be possible for a polyphase space vehicle when the following conditions prevail

a. Einstein's theory of special relativity is valid for the case of a zero cosmic APF and a zero manmade APF.
b. There exists a cosmic APF, however small, in all the vacuum of spacetime, which requires that all physical quantities, including space and time coordinates but with the exception of the fundamental constants of nature, are complex numbers in an internal space.
c. Einsteins idea of the form covariance of the laws of physics is valid for matter and the vacuum in the presence of an APF.
d. A series of manmade mechanical, acoustic or electromagnetic APFs can be generated within the sections of a polyphase space vehicle, which require that all physical quantities including space and time coordinates, but with the exception of the fundamental constants of nature, are complex numbers in an internal space.
e. For subluminal and superluminal flight the manmade APFs inside the polyphase space vehicle can be tuned to cancel the infinities associated with γ_{oL} and γ_{oG} respectively at the point $v_o = c$.

f. For superluminal flight, a manmade ordinary external space mechanical, acoustic or electromagnetic field is generated within the polyphase space vehicle that can eliminate the imaginary index i that appears in $\gamma_o = -i\gamma_{oG}$ for $v_o \geq c$.

This will now be examined in Sections 2 through 4.

2. SPECIAL RELATIVISTIC TRANSFORMATIONS FOR THE SPACE AND TIME COORDINATE FACTORS. This Section derives the transformation formulas for the space and time coordinate factors that are associated with the complex number Lorentz transformations of the space and time coordinates that are located in an APF. Special relativistic expressions for the relationship between the space and time coordinate factors of a moving polyphase space vehicle as seen by an observer in the stationary frame of reference and a moving frame of reference are derived.

A. General Transformations of the Space and Time Coordinate Factors as Obtained From the Lorentz Transformations.

In the presence of an APF, the Lorentz transformations of the space and time coordinate magnitudes shown in equation (759) gives the connection between the magnitudes x′, t′ of the complex number space and time coordinates of an event as seen by an observer in the moving system and the magnitudes x, t of the complex number space and time coordinates of the event as seen by an observer in the stationary-coordinate system as

$$x' = \gamma I_{\sigma x}(x, t, v, \theta_x, \theta_t, \theta_v) \qquad t' = \gamma I_{\sigma t}(x, t, v, \theta_x, \theta_t, \theta_v) \tag{789}$$

$$x = \gamma I'_{\sigma x}(x', t', v, \theta'_x, \theta'_t, \theta_v) \qquad t = \gamma I'_{\sigma t}(x', t', v, \theta'_x, \theta'_t, \theta_v) \tag{790}$$

where the functions $I_{\sigma x}, I_{\sigma t}, I'_{\sigma x}$ and $I'_{\sigma t}$ are obtained from equations (760) through (763). The corresponding phase angles $\theta'_x, \theta'_t, \theta_x$ and θ_t are given by equation (764). The magnitudes x′, t′, x and t of the complex number space and time coordinates that appear in the (789) and (790) can be related to their zero APF values after using the coordinate factor equations (106), (107), (436) and (437) and the zero APF Lorentz transformations (722) and (723) with the result that

$$x' = C'_{\sigma x} x'_o = C'_{\sigma x} \gamma_o (x_o - v_o t_o) \tag{791}$$

$$t' = C'_{\sigma t} t'_o = C'_{\sigma t} \gamma_o (t_o - v_o x_o c^{-2}) \tag{792}$$

$$x = C_{\sigma x} x_o = C_{\sigma x} \gamma_o (x'_o + v_o t'_o) \tag{793}$$

$$t = C_{\sigma t} t_o = C_{\sigma t} \gamma_o (t'_o + v_o x'_o c^{-2}) \tag{794}$$

where $x_o, t_o, x'_o,$ and t'_o for zero APF conditions are related by the standard Lorentz transformations given in equations (722) through (727), and where γ_o is defined by equations (725) through (727). Then the functions $I_{\sigma x}, I_{\sigma t}, I'_{\sigma x}$ and $I'_{\sigma t}$ can be written in terms of x_o, t_o, x'_o and t'_o and $C_{\sigma x}, C_{\sigma t}, C'_{\sigma x}$ and $C'_{\sigma t}$ as

$$I_{\sigma x} = [(C_{\sigma x}x_o - vC_{\sigma t}t_o)^2 + 2vt_o x_o C_{\sigma t}C_{\sigma x}\Psi_x]^{1/2} \tag{795}$$

$$= \{[C_{\sigma x}x_o - v_o t_o (C^v_{\sigma x}/C^v_{\sigma t})C_{\sigma t}]^2 + 2v_o t_o x_o (C^v_{\sigma x}/C^v_{\sigma t})C_{\sigma t}C_{\sigma x}\Psi_x\}^{1/2}$$

$$= C_{\sigma x}\{[x_o - v_o t_o (C^v_{\sigma x}/C^v_{\sigma t})(C_{\sigma t}/C_{\sigma x})]^2 + 2v_o t_o x_o (C^v_{\sigma x}/C^v_{\sigma t})(C_{\sigma t}/C_{\sigma x})\Psi_x\}^{1/2}$$

$$I_{\sigma t} = [(C_{\sigma t}t_o - vC_{\sigma x}x_o c^{-2})^2 + 2vt_o x_o C_{\sigma t}C_{\sigma x}c^{-2}\Psi_t]^{1/2} \tag{796}$$

$$= \{[C_{\sigma t}t_o - v_o x_o c^{-2}(C^v_{\sigma x}/C^v_{\sigma t})C_{\sigma x}]^2 + 2v_o t_o x_o c^{-2}(C^v_{\sigma x}/C^v_{\sigma t})C_{\sigma t}C_{\sigma x}\Psi_t\}^{1/2}$$

$$= C_{\sigma t}\{[t_o - v_o x_o c^{-2}(C^v_{\sigma x}/C^v_{\sigma t})(C_{\sigma x}/C_{\sigma t})]^2 + 2v_o t_o x_o c^{-2}(C^v_{\sigma x}/C^v_{\sigma t})(C_{\sigma x}/C_{\sigma t})\Psi_t\}^{1/2}$$

$$I'_{\sigma x} = [(C'_{\sigma x}x'_o + vC'_{\sigma t}t'_o)^2 - 2vt'_o x'_o C'_{\sigma t}C'_{\sigma x}\Psi'_x]^{1/2} \tag{797}$$

$$= \{[C'_{\sigma x}x'_o + v_o t'_o (C^v_{\sigma x}/C^v_{\sigma t})C'_{\sigma t}]^2 - 2v_o t'_o x'_o (C^v_{\sigma x}/C^v_{\sigma t})C'_{\sigma t}C'_{\sigma x}\Psi'_x\}^{1/2}$$

$$= C'_{\sigma x}\{[x'_o + v_o t'_o (C^v_{\sigma x}/C^v_{\sigma t})(C'_{\sigma t}/C'_{\sigma x})]^2 - 2v_o t'_o x'_o (C^v_{\sigma x}/C^v_{\sigma t})(C'_{\sigma t}/C'_{\sigma x})\Psi'_x\}^{1/2}$$

$$I'_{\sigma t} = [(C'_{\sigma t}t'_o + vC'_{\sigma x}x'_o c^{-2})^2 - 2vt'_o x'_o C'_{\sigma t}C'_{\sigma x}c^{-2}\Psi'_t]^{1/2} \tag{798}$$

$$= \{[C'_{\sigma t}t'_o + v_o x'_o c^{-2}(C^v_{\sigma x}/C^v_{\sigma t})C'_{\sigma x}]^2 - 2v_o t'_o x'_o c^{-2}(C^v_{\sigma x}/C^v_{\sigma t})C'_{\sigma t}C'_{\sigma x}\Psi'_t\}^{1/2}$$

$$= C'_{\sigma t}\{[t'_o + v_o x'_o c^{-2}(C^v_{\sigma x}/C^v_{\sigma t})(C'_{\sigma x}/C'_{\sigma t})]^2 - 2v_o t'_o x'_o c^{-2}(C^v_{\sigma x}/C^v_{\sigma t})(C'_{\sigma x}/C'_{\sigma t})\Psi'_t\}^{1/2}$$

where Ψ_x, Ψ_t, Ψ'_x and Ψ'_t are given by equations (765) and (766), and where C^v_x and C^v_t refer to the relative velocity parameters v and v_o and do not refer to either the stationary or moving coordinate systems and where $\sigma = 1$ in the case of a $[hn^o_{Dst}]$ vacuum and $\sigma = 2, 3$ in the case of a $[ln^o_{Dst}]$ vacuum (see Parts B, C and D of Section 1).

From the previous results it is easy to calculate the space and time coordinate factors $C_{\sigma x}, C_{\sigma t}, C'_{\sigma x}$ and $C'_{\sigma t}$. Combining equations (789), (790) and (791) through (794) gives

$$C'_{\sigma x} = \gamma\gamma_o^{-1}(x_o - v_o t_o)^{-1}I_{\sigma x} = \gamma(x'_o)^{-1}I_{\sigma x} = \gamma\gamma_o^{-1}W_{\sigma x} \tag{799}$$

$$C'_{\sigma t} = \gamma\gamma_o^{-1}(t_o - v_o x_o c^{-2})^{-1}I_{\sigma t} = \gamma(t'_o)^{-1}I_{\sigma t} = \gamma\gamma_o^{-1}W_{\sigma t} \tag{800}$$

$$C_{\sigma x} = \gamma\gamma_o^{-1}(x'_o + v_o t'_o)^{-1}I'_{\sigma x} = \gamma\gamma_o^{-1}W'_{\sigma x} \tag{801}$$

$$C_{\sigma t} = \gamma\gamma_o^{-1}(t'_o + v_o x'_o c^{-2})^{-1}I'_{\sigma t} = \gamma\gamma_o^{-1}W'_{\sigma t} \tag{802}$$

where $W_{\sigma x}, W_{\sigma t}, W'_{\sigma x}$ and $W'_{\sigma t}$ are given by equations (799) through (802) to be

$$W_{\sigma x} = (x_o - v_o t_o)^{-1}I_{\sigma x} \qquad W_{\sigma t} = (t_o - v_o x_o c^{-2})^{-1}I_{\sigma t} \tag{803}$$

$$W'_{\sigma x} = (x'_o + v_o t'_o)^{-1} I'_{\sigma x} \qquad W'_{\sigma t} = (t'_o + v_o x'_o c^{-2})^{-1} I'_{\sigma t} \tag{804}$$

and where the Einstein-Lorentz factor γ_o is given by equations (725) through (727), the factor γ is defined by equation (739) and $I_{\sigma x}, I_{\sigma t}, I'_{\sigma x}$ and $I'_{\sigma t}$ are given by equations (795) through (798).

B. Case Where the Moving Axis is Attached to the Rear of the Polyphase Space Vehicle.

For the case when the rear of the polyphase space vehicle is attached to the moving coordinate system, the following conditions are valid

$$x'_o = x_o^{sv} = 0 \qquad t'_o = t_o^{sv} \qquad x_o = v_o t_o \tag{805}$$

$$t_o = \gamma_o t'_o \qquad u_o = 0 \qquad w_o = v_o \tag{806}$$

$$\bar{x}' = \bar{x}^{sv} = 0 \qquad x' = x^{sv} = 0 \qquad \theta'_x = \theta_x^{sv} \tag{807}$$

$$\bar{t} = \bar{\gamma}\bar{t}' \qquad t = \gamma t' \qquad \theta_t = \theta_\gamma + \theta'_t \tag{808}$$

$$\bar{x} = \bar{v}\bar{t} \qquad x = vt \qquad \theta_x = \theta_v + \theta_t \tag{809}$$

$$\bar{x}_v = \bar{x} \qquad x_v = x \qquad \theta_x^v = \theta_x \tag{810}$$

$$\bar{t}_v = \bar{t} \qquad t_v = t \qquad \theta_t^v = \theta_t \tag{811}$$

$$\bar{x}_v = \bar{v}\bar{t}_v \qquad x_v = vt_v \qquad \theta_x^v = \theta_v + \theta_t^v \tag{812}$$

$$\bar{u} = 0 \qquad u = 0 \qquad \theta_u = 0 \tag{813}$$

$$\bar{w} = \bar{v} \qquad w = v \qquad \theta_w = \theta_v \tag{814}$$

$$C'_{\sigma x} = C_{\sigma x}^{sv} \qquad C'_{\sigma t} = C_{\sigma t}^{sv} \qquad \Psi_x = \Psi_x^v = 0 \tag{815}$$

$$E = 0 \qquad F = 0 \tag{816}$$

Then combining equations (795) through equations (798) with equations (805) through (816) gives

$$I_{\sigma x} = \gamma^{-1} C_{\sigma x}^{sv} x_o^{sv} = C_{\sigma x}^{v} x_o^{v} (2\Psi_x^v)^{1/2} = 0 \tag{817}$$

$$I_{\sigma t} = \gamma^{-1} C_{\sigma t}^{sv} t_o^{sv} = C_{\sigma t}^{v} t_o^{v} H_{\sigma t}^{v} \tag{819}$$

$$I'_{\sigma x} = I_{\sigma x}^{sv} = v_o t_o^{sv} C_{\sigma t}^{sv} (C_{\sigma x}^{v} / C_{\sigma t}^{v}) = v t_o^{sv} C_{\sigma t}^{sv} \tag{820}$$

$$I'_{\sigma t} = I_{\sigma t}^{sv} = C_{\sigma t}^{sv} t_o^{sv} \tag{820}$$

where

$$H^{v}_{\sigma t} = [(1 - v^2 c^{-2})^2 + 2v^2 c^{-2} \Psi^{v}_{t}]^{1/2} \tag{821}$$

$$= \{[1 - v_o^2 c^{-2} (C^{v}_{\sigma x} / C^{v}_{\sigma t})^2]^2 + 2v_o^2 c^{-2} (C^{v}_{\sigma x} / C^{v}_{\sigma t})^2 \Psi^{v}_{t}\}^{1/2}$$

where Ψ^{v}_{x} and Ψ^{v}_{t} are given by equation (768), where $C^{v}_{\sigma x}$ and $C^{v}_{\sigma t}$ are the space and time coordinate factors of the vacuum directly behind the moving coordinate system (and now also directly behind the polyphase space vehicle for this special case) and where $C^{sv}_{\sigma x}$ and $C^{sv}_{\sigma t}$ are the space and time coordinate factors of the polyphase space vehicle itself and $\sigma = 1$ in the case of a $[hn^{o}_{Dst}]$ vacuum and $\sigma = 2, 3$ for the degenerate case of a $[ln^{o}_{Dst}]$ vacuum (see Parts B, C, D and E of Section 1). For the case at hand with $x'_o = 0$ and $\bar{x}' = 0$, equations (803) and (804) combined with equations (817) through (821) give the following general results

$$W_{\sigma x} = \gamma_o \gamma^{-1} C^{sv}_{\sigma x} \tag{822}$$

$$W_{\sigma t} = \gamma_o^2 C^{v}_{\sigma t} H^{v}_{\sigma t} = \gamma_o^2 \gamma^{-2} C_{\sigma t} = \gamma_o \gamma^{-1} C^{sv}_{\sigma t} \tag{823}$$

$$W'_{\sigma x} = W^{sv}_{\sigma x} = (C^{v}_{\sigma x} / C^{v}_{\sigma t}) C^{sv}_{\sigma t} = (v / v_o) C^{sv}_{\sigma t} = \xi^{sv}_{x} C^{sv}_{\sigma x} \tag{824}$$

$$W'_{\sigma t} = W^{sv}_{\sigma t} = C^{sv}_{\sigma t} = \gamma \gamma_o^{-1} W_{\sigma t} \tag{825}$$

where

$$\xi^{sv}_{x} = (v / v_o)(s^{sv}_{xo} / s^{sv}_{x}) \tag{826}$$

where the zero APF value of the characteristic velocity of a mechanical, acoustic or electromagnetic waves in the polyphase space vehicle is

$$s^{sv}_{xo} = dx^{sv}_{o} / dt^{sv}_{o} \tag{827}$$

and where s^{sv}_{x} is the characteristic velocity of a mechanical, acoustic or electromagnetic wave in the polyphase space vehicle with an internal APF. Consider now the special cases of subluminal and superluminal flight respectively

$$s^{sv}_{x} = s^{sv}_{xL} \qquad \xi^{sv}_{x} = \xi^{sv}_{xL} \qquad v \le c \tag{828}$$

$$s^{sv}_{x} = s^{sv}_{xG} \qquad \xi^{sv}_{x} = \xi^{sv}_{xG} \qquad v \ge c \tag{829}$$

where equation (688) gives for subluminal flight

$$s^{sv}_{xL} = dx^{sv}_{L} / dt^{sv}_{L} = (dx^{sv}_{o} / dt^{sv}_{o})(C^{sv}_{\sigma x} / C^{sv}_{\sigma t}) = s^{sv}_{xo} (C^{sv}_{\sigma x} / C^{sv}_{\sigma t}) \tag{830}$$

where s^{sv}_{xL} = real number characteristic velocity of a mechanical, acoustic or electromagnetic wave in the polyphase space vehicle which is moving with $v_o \le c$, and where also for subluminal flight the velocity ratio function ξ^{sv}_{xL} is given by

$$\xi^{sv}_{x} = \xi^{sv}_{xL} = (v/v_o)(s^{sv}_{xo}/s^{sv}_{xL}) = \text{real number} \tag{831}$$

For superluminal flight, equations (712) through (718) give the tuned relation

$$s^{sv}_{xG} = dx^{sv}_{G} / dt^{sv}_{G} = is^{sv}_{xo}(C^{sv}_{\sigma x} / |C^{sv}_{\sigma t}|) = is^{sv}_{mxG} = \text{ imaginary number} \qquad (832)$$

which is obtained from the first tuning condition on the external phase angle of a wave propagating in the material of the space vehicle which is obtained from equation (716) as

$$\varphi^{sv}_{sxG} = \varphi^{sv}_{\delta xR} - \varphi^{sv}_{\delta tR} - \varphi^{sv}_{C\sigma t} = \pi / 2 \qquad \varphi^{sv}_{\delta xR} - \varphi^{sv}_{\delta tR} = 0 \qquad \varphi^{sv}_{C\sigma t} = -\pi / 2 \qquad (833)$$

where s^{sv}_{xG} = imaginary number characteristic velocity of a mechanical, acoustic or electromagnetic wave in the space vehicle for a polyphase space vehicle moving with velocity $v_o \geq c$ and where the magnitude of the imaginary velocity is

$$s^{sv}_{mxG} = s^{sv}_{xo}(C^{sv}_{\sigma x} / |C^{sv}_{\sigma t}|) \qquad (834)$$

and where for superluminal flight with the requirement of the tuning condition in equation (833) the velocity ratio function is

$$\xi^{sv}_{xG} = (v/v_o)(s^{sv}_{xo}/s^{sv}_{xG}) = (v / v_o)[s^{sv}_{xo} /(is^{sv}_{mxG})] = -i\xi^{sv}_{mxG} = \text{imaginary number} \qquad (835)$$

where

$$\xi^{sv}_{mxG} = (v / v_o)[s^{sv}_{xo} /(s^{sv}_{mxG})] \qquad (836)$$

where v and v_o are the relative velocities of the space vehicle with and without a cosmic APF present, and where, as in equations (684) and (690), s^{sv}_{xo} = characteristic average velocity of a mechanical, acoustic or electromagnetic wave propagating in the x-direction in the polyphase space vehicle relative to a frame of reference co-moving with and attached to the rear of the polyphase space vehicle under zero APF conditions and s^{sv}_{xL} = real number average velocity of a mechanical, acoustic or electromagnetic wave in the polyphase space vehicle relative to a frame of reference attached to the rear of the space vehicle traveling at $v_o \leq c$ and in which there is a local artificial APF, where s^{sv}_{xG} = imaginary number average velocity of a mechanical, acoustic or electromagnetic wave in the polyphase space vehicle which is traveling at a relative speed $v_o \geq c$, where s^{sv}_{mxG} = magnitude of the average velocity of a mechanical, acoustic or electromagnetic wave in a polyphase space vehicle which is traveling at a relative speed $v_o \geq c$, x^{sv}_{L} = coordinate variable in the x-direction within the space vehicle traveling at $v_o \leq c$, t^{sv}_{L} = time coordinate variable within the polyphase space vehicle traveling at $v_o \leq c$, x^{sv}_{G} = characteristic length of a space vehicle traveling at a speed $v_o \geq c$ in the x-direction and t^{sv}_{G} = characteristic time within the polyphase space vehicle traveling at a speed $v_o \geq c$.

Combining equations (799) through (802) with equations (805) through (816) and (822) through (825) gives the following general results for the special case when the moving frame of reference is attached to the rear of the polyphase space vehicle

$$C'_{\sigma x} = C^{sv}_{\sigma x} \qquad (837)$$

$$C'_{\sigma t} = C^{sv}_{\sigma t} = \gamma(t^{sv}_{o})^{-1} I_{\sigma t} = \gamma\gamma_o(t^{v}_{o})^{-1} I_{\sigma t} = \gamma\gamma_o C^{v}_{\sigma t} H^{v}_{\sigma t} = \gamma_o \gamma^{-1} C_{\sigma t} \qquad (838)$$

$$C_{\sigma x} = \gamma\gamma_o^{-1}(v_o t_o^{sv})^{-1} I_{\sigma x}^{sv} = \gamma\gamma_o^{-1} W_{\sigma x}^{sv} = \gamma\gamma_o^{-1}(v/v_o) C_{\sigma t}^{sv} = \gamma\gamma_o^{-1}(C_{\sigma x}^{v}/C_{\sigma t}^{v}) C_{\sigma t}^{sv} = \gamma\gamma_o^{-1}\xi_x^{sv} C_{\sigma x}^{sv} \quad (839)$$

$$C_{\sigma t} = \gamma\gamma_o^{-1}(t_o^{sv})^{-1} I_{\sigma t}^{sv} = \gamma\gamma_o^{-1} W_{\sigma t}^{sv} = \gamma^2 C_{\sigma t}^{v} H_{\sigma t}^{v} = \gamma\gamma_o^{-1}(s_{xo}^{sv}/s_x^{sv}) C_{\sigma x}^{sv} = \gamma\gamma_o^{-1} C_{\sigma t}^{sv} \quad (840)$$

where the velocity factor ξ_x^{sv} is given by equations (826) through (836). Thus the case $x_o' = 0$ corresponds to a moving frame of reference that is attached to the rear of the polyphase space vehicle and gives $C_{\sigma x}' = C_{\sigma x}^{sv}$, where $C_{\sigma x}^{sv}$ is a known quantity given by equations (589) through (640). Note that $C_{\sigma t}^{v}$ and $C_{\sigma t}^{sv}$ are obtained from equations (422) and (690) respectively. It is also clear from equations (839) and (840) that

$$C_{\sigma x}/C_{\sigma t} = C_{\sigma x}^{v}/C_{\sigma t}^{v} = v/v_o \quad (841)$$

so that for the case $x_o' = 0$ and $\bar{x}' = 0$, it follows that $C_{\sigma x} \neq C_{\sigma x}^{v}$ and $C_{\sigma t} \neq C_{\sigma t}^{v}$ but the proportionality in equation (841) is valid. From equations (838) through (840) it follows that in general

$$I_{\sigma t} = \gamma^{-1} t_o^{sv}(s_{xo}^{sv}/s_x^{sv}) C_{\sigma x}^{sv} = \gamma^{-1}\gamma_o^{-1} t_o^{v}(s_{xo}^{sv}/s_x^{sv}) C_{\sigma x}^{sv} = \gamma^{-1} I_{\sigma t}^{sv} \quad (842)$$

$$I_{\sigma x}^{sv} = v_o t_o^{sv}\xi_x^{sv} C_{\sigma x}^{sv} = v t_o^{sv}(s_{xo}^{sv}/s_x^{sv}) C_{\sigma x}^{sv} = v\gamma_o^{-1} t_o^{v}(s_{xo}^{sv}/s_x^{sv}) C_{\sigma x}^{sv} \quad (843)$$

$$W_{\sigma x}^{sv} = \xi_x^{sv} C_{\sigma x}^{sv} = (v/v_o)(s_{xo}^{sv}/s_x^{sv}) C_{\sigma x}^{sv} = (v/v_o) C_{\sigma t}^{sv} = (v/v_o) W_{\sigma t}^{sv} \quad (844)$$

$$C_{\sigma t}^{sv} = (s_{xo}^{sv}/s_x^{sv}) C_{\sigma x}^{sv} \quad (845)$$

$$I_{\sigma t}^{sv} = t_o^{sv}(s_{xo}^{sv}/s_x^{sv}) C_{\sigma x}^{sv} = \gamma_o^{-1} t_o^{v}(s_{xo}^{sv}/s_x^{sv}) C_{\sigma x}^{sv} = v I_{\sigma x}^{sv} \quad (846)$$

$$W_{\sigma t}^{sv} = (s_{xo}^{sv}/s_x^{sv}) C_{\sigma x}^{sv} = C_{\sigma t}^{sv} \quad (847)$$

$$C_{\sigma t}^{v} H_{\sigma t}^{v} = \gamma^{-1}\gamma_o^{-1}(s_{xo}^{sv}/s_x^{sv}) C_{\sigma x}^{sv} = \gamma^{-1}\gamma_o^{-1} C_{\sigma t}^{sv} \quad (848)$$

For the special cases of superluminal motion with $v_o \geq c, \gamma_o = -i\gamma_{oG}$ and $s_{xG}^{sv} = i s_{mxG}^{sv}$, and in agreement with equation (832), the following quantities are imaginary numbers, $C_{\sigma t}^{sv}$, $W_{\sigma x}^{sv}$, $W_{\sigma t}^{sv}$ and t_o^{sv}.

The general case of a moving axis which is attached to the rear of a polyphase space vehicle is described for a nonzero APF by $\bar{x}' = \bar{x}^{sv} = 0$ and equations (805) through (831) and (837) through (848). For the case of a zero APF, $x_o' = x_o^{sv} = 0$ is valid and these equations become the following limiting equations for subluminal flight

$$I_{\sigma x} = x_o^{sv}\gamma_o^{-1} W_{\sigma x} = 0 \qquad I_{\sigma t} = \gamma_o^{-2} t_o^{v} \qquad H_{\sigma t}^{v} = \gamma_o^{-2} = 1 - v_o^2 c^{-2} \quad (849)$$

$$I_{\sigma x}^{sv} = v_o t_o^{sv} \qquad I_{\sigma t}^{sv} = t_o^{sv} \qquad I_{\sigma x}^{sv} = v_o I_{\sigma t}^{sv} \quad (850)$$

$$W_{\sigma x} = 1 \qquad W_{\sigma t} = 1 \qquad W_{\sigma x}^{sv} = 1 \qquad W_{\sigma t}^{sv} = 1 \quad (851)$$

$$C_{\sigma x}^{sv} = 1 \qquad C_{\sigma t}^{sv} = \gamma_o^{-1}(t_o^{sv})^{-1} t_o^{v} = 1 \quad (852)$$

$$C_{\sigma x} = 1 \qquad C_{\sigma t} = 1 \qquad C_{\sigma x}^{v} = 1 \qquad C_{\sigma t}^{v} = 1 \quad (853)$$

as expected. Note that in equation (852) the result $C^{sv}_{\sigma t} = 1$ requires the use of the time dilation equation (729) for zero APF.

The apparent space and time coordinate factors $C_{\sigma x}$ and $C_{\sigma t}$ of the coordinates of a space vehicle as seen from the stationary-coordinate system and as represented by equations (93) through (258) and are represented in terms of the the space and time coordinate factors of the space vehicle $C^{sv}_{\sigma x}$ and $C^{sv}_{\sigma t}$ as given by equations (589) through (678) and (684) through (720) and the space and time coordinate factors of the vacuum $C^{v}_{\sigma x}$ and $C^{v}_{\sigma t}$ as given by equations (259) through (422). It will be shown in Section 3 that only equations (828), (830) and (831) are applicable to ultrafast subluminal flight and it will be demonstrated in Section 4 that only equations (829) and (832) through (836) are applicable to superluminal flight.

3. <u>THE SPECIAL RELATIVISTIC APPARENT SPACE AND TIME COORDINATE FACTOR EQUATIONS AND THE TUNING CONDITIONS FOR ULTRAFAST SUBLUMINAL FLIGHT.</u> This Section gives the conditions on the engineered artificial APF that is generated in a polyphase space vehicle and that is required for ultrafast subluminal flight with $v \le c$, and also shows that natural objects cannot undergo ultrafast subluminal flight even if they contain a natural APF.

A. The Apparent Space and Time Coordinate Factor Equations and the Tuning Conditions for the Ultrafast Subluminal Flight of Polyphase Space Vehicles with $v \le c$.

For subluminal flight the zero APF Einstein-Lorentz factor γ_o is given by $\gamma_o = \gamma_{oL}$ where γ_{oL} is given by equation (726). For the case when a cosmic APF is assumed to exist throughout spacetime, the Einstein-Lorentz factor γ_{oL} transforms into γ for $v \le c$ which is given by equation (739) and which does not have a singularity at $v = c$. This section shows that the fact that γ is a well-behaved function of relative velocity is not sufficient to show that ultrafast subluminal motion $v \le c$ is possible, and in fact it will be shown that even in the presence of a cosmic field only manmade polyphase space vehicles can travel at speeds up to and including the light speed. In the presence of a cosmic APF, only a polyphase space vehicle with a properly engineered internal APF in a polyphase space vehicle can travel up to and including the light speed. These conclusions derive from equations (839) and (840) giving the corresponding apparent space and time coordinate factors of the space vehicle as

$$C_{\sigma x} = \gamma\gamma^{-1}_{oL}(v/v_o)C^{sv}_{\sigma t} = \gamma\gamma^{-1}_{oL}\xi^{sv}_{xL}C^{sv}_{\sigma x} \tag{854}$$

$$C_{\sigma t} = \gamma\gamma^{-1}_{oL}(s^{sv}_{xo}/s^{sv}_{xL})C^{sv}_{\sigma x} = \gamma\gamma^{-1}_{oL}C^{sv}_{\sigma t} \tag{855}$$

where for subluminal flight the characteristic wave velocity s^{sv}_{xL} of an electromagnetic, acoustic or mechanical APF wave in a polyphase space vehicle is given by equations (830) and where also for subluminal flight the velocity ratio function ξ^{sv}_{xL} is given by equation (831). Equations (854) and (855) contain both γ and γ_{oL}, and therefore the Einstein-Lorentz factor γ_{oL} with its singularity is still there to haunt the space vehicle design engineer.

Assume first that the long tube of vacuum in which the space vehicle is moving in a weak uniaxial cosmic APF in the x-direction is a small volume of low-density vacuum $[\ln_{po}, SN_p]$ when compared to the surrounding universe which is a large and relatively empty object $[\ln_o^{un}, LN^{un}]$, so that the apparent space and time coordinate factors $C_{\sigma x}$ and $C_{\sigma t}$ of the space vehicle are given by equations (138) and (258) that represent the vacuum in the space between the distant fixed observer and the space vehicle (see Part B of Section 1). The space vehicle itself is assumed to be a small, high-density object $[hn_o^{sv}, SN]$ containing a manmade uniaxial APF in the x-direction so that the space and time coordinate factors of the space vehicle $C_{\sigma x}^{sv}$ and $C_{\sigma t}^{sv}$ are given by equations (637) and (690) (see Part E of Section 1). The polyphase space vehicle is a small, high-density $[hn_o^{sv}, SN]$ object in the sense that the surrounding universe is a very large and nearly empty and is an $[\ln_o^{un}, LN^{un}]$, object. For ultrafast subluminal flight, the polyphase space vehicle is composed of sections made of solids, liquids and gases all of which can be described as $[hn_o^{sv}, SN]$ structures, so that the angles $\theta_{\mathcal{N}}^{sv}$ and θ_n^{sv} are taken to be averages over the entire volume of the space vehicle and over each solid, liquid and gas structure as follows

$$\theta_{\mathcal{N}}^{sv} = (V^{sv})^{-1} \sum_k \theta_{\mathcal{N}k}^{sv} V_k^{sv} \qquad \theta_n^{sv} = (V^{sv})^{-1} \sum_k \theta_{nk}^{sv} V_k^{sv} \tag{856}$$

where the total volume V^{sv} of the polyphase space vehicle is given by

$$V^{sv} = \sum_k V_k^{sv} \tag{857}$$

where the volume of the k'th component of the polyphase space vehicle is V_k^{sv}. Within these assumptions equation (854) can be written as

$$\cos\theta_{\mathcal{N}p} \cos\theta_{np} = \gamma\gamma_{oL}^{-1} \xi_{xL}^{sv} \cos\theta_{\mathcal{N}}^{sv} \sec\theta_n^{sv} \qquad [\ln_{po}, SN_p],\ [hn_o^{sv}, SN] \tag{858}$$

where ξ_{xL}^{sv} is given by equations (830) and (831). If the cosmic APF is 3-dimensional and isotropic and the artificial APF in the space vehicle is uniaxial, then after using equation (158), equation (858) is replaced by

$$\cos^{1/3}\theta_{\mathcal{N}p} \cos^{1/3}\theta_{np} = \gamma\gamma_{oL}^{-1} \xi_{xL}^{sv} \cos\theta_{\mathcal{N}}^{sv} \sec\theta_n^{sv} \qquad [\ln_{po}, SN_p],\ [hn_o^{sv}, SN] \tag{859}$$

If the cosmic APF is 3-dimensional and isotropic and the polyphase space vehicle is spherical in shape and contains an internally generated 3-dimensional isotropic APF, equations (854) can be written as

$$\cos^{1/3}\theta_{\mathcal{N}p} \cos^{1/3}\theta_{np} = \gamma\gamma_{oL}^{-1} \xi_{xL}^{sv} \cos^{1/3}\theta_{\mathcal{N}}^{sv} \sec^{1/3}\theta_n^{sv} \qquad [\ln_{po}, SN_p],\ [hn_o^{sv}, SN] \tag{860}$$

Equivalently, these equations can be written in terms of the internal phase angles of the dimensions of spacetime by using equations (210) and (230) as

$$\cos\theta_{Dst} \cos\theta_{nDst} = \gamma\gamma_{oL}^{-1} \xi_{xL}^{sv} \cos\theta_{\mathcal{N}}^{sv} \sec\theta_n^{sv} \qquad [\ln_{Dst}^{o}, SN_{Dst}],\ [hn_o^{sv}, SN] \tag{861}$$

$$\cos^{1/3}\theta_{Dst} \cos^{1/3}\theta_{nDst} = \gamma\gamma_{oL}^{-1} \xi_{xL}^{sv} \cos\theta_{\mathcal{N}}^{sv} \sec\theta_n^{sv} \qquad [\ln_{Dst}^{o}, SN_{Dst}],\ [hn_o^{sv}, SN] \tag{862}$$

$$\cos^{1/3}\theta_{Dst} \cos^{1/3}\theta_{nDst} = \gamma\gamma_{oL}^{-1} \xi_{xL}^{sv} \cos^{1/3}\theta_{\mathcal{N}}^{sv} \sec^{1/3}\theta_n^{sv} \qquad [\ln_{Dst}^{o}, SN_{Dst}],\ [hn_o^{sv}, SN] \tag{863}$$

Solutions to each of these equations are not in general possible if a space vehicle acted like a natural body with fixed values of the angles $\theta_{\mathfrak{N}}^{sv}$ and θ_{n}^{sv}, however for a polyphase space vehicle with an internal APF generator the magnitude of the uniaxial APF can adjusted to the situation that the angles $\theta_{\mathfrak{N}}^{sv}$ and θ_{n}^{sv} have the values for a uniaxial artificial APF determined by

$$\sec\theta_{\mathfrak{N}}^{sv}\cos\theta_{n}^{sv} = \gamma\gamma_{oL}^{-1}(s_{xo}^{sv}/s_{xL}^{sv}) \qquad s_{xL}^{sv} \geq s_{xo}^{sv} \qquad (864)$$

while for an isotropic spherical artificial APF

$$\sec^{1/3}\theta_{\mathfrak{N}}^{sv}\cos^{1/3}\theta_{n}^{sv} = \gamma\gamma_{oL}^{-1}(s_{xo}^{sv}/s_{xL}^{sv}) \qquad s_{xL}^{sv} \geq s_{xo}^{sv} \qquad (865)$$

where s_{xL}^{sv} is the characteristic velocity of mechanical, acoustic or electromagnetic waves in the polyphase space vehicle and is given by equation (830) and where $s_{xL}^{sv} \geq s_{xo}^{sv}$ for the $[hn_{o}^{sv}, SN]$ polyphase space vehicle, and in which case the remaining vacuum equations obtained from equations (861) through (863) are

$$\cos\theta_{\mathfrak{N}p}\cos\theta_{np} = v/v_{o} \qquad \cos^{1/3}\theta_{\mathfrak{N}p}\cos^{1/3}\theta_{np} = v/v_{o} \qquad (866)$$

or

$$\cos\theta_{Dst}\cos\theta_{nDst} = v/v_{o} \qquad \cos^{1/3}\theta_{Dst}\cos^{1/3}\theta_{nDst} = v/v_{o} \qquad (867)$$

which are the equations for the phase angles of the cosmic APF in the $[ln_{po}, SN_{p}]$ or $[ln_{Dst}^{o}, SN_{Dst}]$ vacuum with $v \leq v_{o}$. Equations (864) through (867) give the conditions for ultrafast subluminal flight. The angles $\theta_{\mathfrak{N}}^{sv}$ and θ_{n}^{sv} are related by equation (619) and also by equations (657) for a $[hn_{o}^{sv}, SN]$ linear elastic space vehicle so that these angles are not independent. Nevertheless, as a crude approximation it can be assumed that $\theta_{\mathfrak{N}}^{sv}$ and θ_{n}^{sv} are independent in equations (864) and (865), and this gives the following approximate conditions for ultrafast subluminal flight

$$\sec\theta_{n}^{sv} = \gamma_{oL} \qquad \sec\theta_{\mathfrak{N}}^{sv} = \gamma(s_{xo}^{sv}/s_{xL}^{sv}) \qquad s_{xL}^{sv} \geq s_{xo}^{sv} \qquad (868)$$

$$\sec^{1/3}\theta_{n}^{sv} = \gamma_{oL} \qquad \sec^{1/3}\theta_{\mathfrak{N}}^{sv} = \gamma(s_{xo}^{sv}/s_{xL}^{sv}) \qquad s_{xL}^{sv} \geq s_{xo}^{sv} \qquad (869)$$

These equations are only valid if the strength of the APF in the space vehicle, and therefore also the angles $\theta_{\mathfrak{N}}^{sv}$ and θ_{n}^{sv}, can be engineered so as to satisfy equations (864) or (865). Equations (864) through (869) show that the engineered APF within a high-density polyphase space vehicle cancels the effects of the singularity of γ_{oL} at $v_{o} = c$ by requiring

$$\theta_{n}^{sv} = \theta_{\mathfrak{N}}^{sv} + \left|\theta_{V}^{sv}\right| = \pi/2 \qquad \theta_{\mathfrak{N}}^{sv} = \pi/4 \qquad \theta_{V}^{sv} = -\pi/4 \qquad (870)$$

at this point, and therefore polyphase space vehicles can undergo ultrafast subluminal flight. The condition $\theta_{n}^{sv} = \pi/2$ and $\theta_{\mathfrak{N}}^{sv} = \pi/4$ for a solid-body, liquid or gas structure corresponds to an APF whose energy density for a stationary APF is given by $\mathfrak{E} = 1/2(K_{o}^{sv} + G_{o}^{sv})$ while for a time dependent APF (see Section 5) the energy density is

$\mathfrak{E} = K_o^{sv} + G_o^{sv}$, where K_o^{sv} and G_o^{sv} are the zero APF values of the bulk modulus and the shear modulus of a solid, liquid or gaseous portion of an elastic polyphase space vehicle.[9]

B. Conditions for the Ultrafast Subluminal Flight of a Linear Elastic Space Vehicle.

In this section it is assumed that the solids, liquids and gases that compose the polyphase space vehicle are linear elastic structures. Equations (858) through (860) can be rewritten for linear elastic space vehicles after using equations (663) and (675) as follows, for a uniaxial cosmic APF throughout all space and a uniaxial manmade APF in a polyphase space vehicle

$$\cos\theta_{\mathfrak{N}p}\cos\theta_{np} = \gamma\gamma_{oL}^{-1}\xi_{xL}^{sv}\sec\theta_n^{sv}[1+(1-\cos\theta_n^{sv})^2]^{-1/2} \qquad [ln_{po}, SN_p],\ [hn_o^{sv}, SN] \quad (871)$$
$$= \gamma\gamma_{oL}^{-1}\xi_{xL}^{sv}\cos\theta_{\mathfrak{N}}^{sv}(1-\tan\theta_{\mathfrak{N}}^{sv})^{-1}$$

and for a 3-dimensional isotropic cosmic APF and an artificial uniaxial APF in a cylindrical polyphase space vehicle

$$\cos^{1/3}\theta_{\mathfrak{N}p}\cos^{1/3}\theta_{np} = \gamma\gamma_{oL}^{-1}\xi_{xL}^{sv}\sec\theta_n^{sv}[1+(1-\cos\theta_n^{sv})^2]^{-1/2} \quad [ln_{po}, SN_p],[hn_o^{sv}, SN] \quad (872)$$
$$= \gamma\gamma_{oL}^{-1}\xi_{xL}^{sv}\cos\theta_{\mathfrak{N}}^{sv}(1-\tan\theta_{\mathfrak{N}}^{sv})^{-1}$$

and for a 3-dimensional isotropic cosmic APF and a 3-dimensional manmade APF within a spherical polyphase space vehicle

$$\cos^{1/3}\theta_{\mathfrak{N}p}\cos^{1/3}\theta_{np} = \gamma\gamma_{oL}^{-1}\xi_{xL}^{sv}\sec^{1/3}\theta_n^{sv}[1+(1-\cos\theta_n^{sv})^2]^{-1/6} \quad [ln_{po}, SN_p],[hn_o^{sv}, SN] (873)$$
$$= \gamma\gamma_{oL}^{-1}\xi_{xL}^{sv}\cos^{1/3}\theta_{\mathfrak{N}}^{sv}(1-\tan\theta_{\mathfrak{N}}^{sv})^{-1/3}$$

where ξ_{xL}^{sv} is given by equations (830) and (831) and corresponds to the artificial APF generated in the polyphase space vehicle. Equations (871) through (873) are the linear elastic equivalents of equations (858) through (863) and do not have solutions for arbitrary values of the angles $\theta_{\mathfrak{N}}^{sv}$ or θ_n^{sv} which are now related by equation (657) for a high-density $[hn_o^{sv}, SN]$ polyphase space vehicle. These equations can only be satisfied if the angles $\theta_{\mathfrak{N}}^{sv}$ or θ_n^{sv} are adjusted by the manmade APF in the polyphase space vehicle so as to satisfy the following conditions, first for a uniaxial artificial APF

$$\cos\theta_n^{sv}[1+(1-\cos\theta_n^{sv})^2]^{1/2} = \gamma\gamma_{oL}^{-1}(s_{xo}^{sv}/s_{xL}^{sv}) \qquad (874)$$

$$\sec\theta_{\mathfrak{N}}^{sv}(1-\tan\theta_{\mathfrak{N}}^{sv}) = \gamma\gamma_{oL}^{-1}(s_{xo}^{sv}/s_{xL}^{sv}) \qquad (875)$$

and for a spherical artificial APF

$$\cos^{1/3}\theta_n^{sv}[1+(1-\cos\theta_n^{sv})^2]^{1/6} = \gamma\gamma_{oL}^{-1}(s_{xo}^{sv}/s_{xL}^{sv}) \qquad (876)$$

$$\sec^{1/3}\theta_{\mathfrak{N}}^{sv}(1-\tan\theta_{\mathfrak{N}}^{sv})^{1/3} = \gamma\gamma_{oL}^{-1}(s_{xo}^{sv}/s_{xL}^{sv}) \qquad (877)$$

and as before the auxiliary vacuum equations corresponding to equations (874) through (877) are

$$\cos\theta_{\mathbb{N}p}\cos\theta_{np} = v/v_o \qquad \cos^{1/3}\theta_{\mathbb{N}p}\cos^{1/3}\theta_{np} = v/v_o \tag{878}$$

or equivalently as

$$\cos\theta_{Dst}\cos\theta_{nDst} = v/v_o \qquad \cos^{1/3}\theta_{Dst}\cos^{1/3}\theta_{nDst} = v/v_o \tag{879}$$

where s_{xL}^{sv} is the characteristic velocity of mechanical, acoustic or electromagnetic waves in a polyphase space vehicle and is given by equations (830) and (833). Equations (871) through (877) are valid for high-density matter provided that the following conditions are valid[9]

$$\mathfrak{E} \le 1/2(K_o^{sv} + G_o^{sv}) \qquad \text{static APF} \tag{880}$$

$$\mathfrak{E} \le K_o^{sv} + G_o^{sv} \qquad \text{time dependent APF} \tag{881}$$

Therefore ultrafast flight of an engineered linear elastic polyphase space vehicle is possible because the angles θ_n^{sv} and $\theta_{\mathbb{N}}^{sv}$ have the values $\theta_n^{sv} = \pi/2$ and $\theta_{\mathbb{N}}^{sv} = \pi/4$ at the point $v_o = c$ where $\gamma_{oL}^{-1} = 0$ and where $\mathfrak{E} = 1/2(K_o^{sv} + G_o^{sv})$ for static APF and $\mathfrak{E} = K_o^{sv} + G_o^{sv}$ for a time dependent APF. It will now be shown that ultrafast flight of a natural object with arbitrary values of the angles θ_n^{sv} and $\theta_{\mathbb{N}}^{sv}$ is not possible, even though γ has no singularity for v = c, because the Einstein-Lorentz factor γ_{oL} still enters the transformation equations for the apparent values of the space and time coordinate factors, and only a manmade space vehicle with an internally engineered artificial APF can cancel the effects of γ_{oL} and undergo ultrafast subluminal flight.

C. The Impossibility of of Ultrafast Subluminal Flight of Natural Objects.

Small high-density natural objects (NO) such as planets, stars and galaxies (which are small objects relative to the universe) that contain a natural APF satisfy the same form of equations (860) or (863) which are valid for a space vehicle and can be written as

$$\cos^{1/3}\theta_{\mathbb{N}p}\cos^{1/3}\theta_{np} = \gamma\gamma_{oL}^{-1}\xi_{xL}^{NO}\cos^{1/3}\theta_{\mathbb{N}}^{NO}\sec^{1/3}\theta_n^{NO} \qquad [ln_{po}, SN_p],\ [hn_o^{NO}, SN] \tag{882}$$

where ξ_{xL}^{NO} is given by

$$\xi_{xL}^{NO} = (v/v_o)(s_{xo}^{NO}/s_{xL}^{NO}) = \text{real number} \tag{883}$$

with

$$s_{xL}^{NO} = dx_L^{NO}/dt_L^{NO} = (dx_o^{NO}/dt_o^{NO})(C_{\sigma x}^{NO}/C_{\sigma t}^{NO}) = s_{xo}^{NO}(C_{\sigma x}^{NO}/C_{\sigma t}^{NO}) \tag{884}$$

where s_{xL}^{NO} = real number characteristic velocity of a mechanical, acoustic or electromagnetic wave in the natural object and where v and v_o are the relative velocities of the natural object with and without a cosmic APF present, and where, as in equation (684), s_{xo}^{NO} = average velocity of a mechanical, acoustic or electromagnetic wave propagating in the x-direction in the natural object relative to a frame of reference attached to the natural object under zero APF conditions and s_{xL}^{NO} = real number average velocity of a mechanical, acoustic or electromagnetic wave in the natural object relative to a frame of reference attached to the natural object in which there is a natural APF, and where x_L^{NO} = coordinate variable in the x-direction within the natural object, t_L^{NO} = time coordinate variable within the natural object. The quantity ξ_{xL}^{NO} corresponds to the propagation of a wave in the natural object, and it is clear that equation (882) has no real number solution for the case of constant or arbitrarily varying angles $\theta_{\mathbb{N}}^{NO}$ and θ_n^{NO} because the angles $\theta_{\mathbb{N}p}$ and θ_{np} of the vacuum are small and cannot agree with the singularity of the Einstein-Lorentz factor γ_{oL} at the point $v_o = c$. Therefore small high-density natural objects should not be expected to undergo ultrafast subluminal flight. Consider now a large high-density natural object such as the expanding universe, which is a high-density object compared to the vacuum and which is cutting a large swath through the vacuum. For this case equations (159) and (654) give

$$\sec^{1/3}\theta_{\mathbb{N}p}\cos^{1/3}\theta_{np} = \gamma\gamma_{oL}^{-1}\xi_{xL}\sec^{1/3}\theta_{\mathbb{N}}^{NO}\sec^{1/3}\theta_n^{NO} \qquad [\ln_{po}, LN_p],\ [hn_o^{NO}, LN] \qquad (885)$$

Because the angles $\theta_{\mathbb{N}p}$ and θ_{np} of the vacuum are small, equation (885) has no possible solutions that compensate for the singularity of γ_{oL} when $v_o = c$ for constant values or for any arbitrary or specified variation of the angles $\theta_{\mathbb{N}}^{NO}$ and θ_n^{NO}, and therefore large high-density natural objects also cannot travel at ultrafast subluminal speeds. Therefore even though γ has no singularity at $v = c$ for natural objects due to the presence of a cosmic APF, ultrafast motion is prohibited for natural objects by the singularity of the Einstein-Lorentz factor γ_{oL} at the point $v_o = c$ in equations (882) and (885).

4. THE RELATIVISTIC APPARENT SPACE AND TIME COORDINATE FACTOR EQUATIONS AND THE TUNING CONDITIONS FOR SUPERLUMINAL FLIGHT. This section gives the mathematical conditions on the engineered artificial APF that are required for superluminal flight of a polyphase space vehicle with $v \geq c$ and shows that natural objects cannot undergo superluminal flight even if they contain a natural APF.

A. Equations for the Apparent Space and Time Coordinate Factors and the Tuning Conditions Required by Polyphase Space Vehicles for Superluminal Flight with $v \geq c$.

Superluminal flight under zero cosmic APF conditions is impossible and the Einstein-Lorentz factor γ_o for this case is given by $\gamma_o = -i\gamma_{oG}$ where γ_{oG} is given by equation (727). The Einstein-Lorentz factor γ_{oG} has a singularity at $v_o = c$. The imaginary value of γ_o indicates that superluminal flight for a zero cosmic APF is impossible. For the case when a cosmic APF is assumed to exist throughout spacetime, the factor γ, which is given by equation (739), does not have a singularity at $v = c$ and is analytic in the range $v \geq c$. But the fact that γ is a well-behaved function of relative velocity is not sufficient to show that superluminal motion is possible because the factor γ_{oG} also appears in equations (839) and (840) as well as γ, and in fact it will be demonstrated that even in the presence of a cosmic field natural bodies cannot move faster than light in vacuo. However, it will be shown that in the presence of a cosmic APF a polyphase space vehicle, with a properly engineered mechanical, acoustic or electromagnetic APF within the vehicle, can travel faster than the light speed in vacuo. In order to see this one must look at equations (697), (835), (839) and (840) which, for the case at hand where the moving frame of reference is attached to the rear of the space vehicle, give the apparent space and time coordinate factors of the space vehicle as

$$C_{\sigma x} = i\gamma\gamma_{oG}^{-1}(v/v_o)C_{\sigma t}^{sv} = i\gamma\gamma_{oG}^{-1}\xi_{xG}^{sv}C_{\sigma x}^{sv} = \gamma\gamma_{oG}^{-1}\xi_{mxG}^{sv}C_{\sigma x}^{sv} \qquad (886)$$

$$C_{\sigma t} = i\gamma\gamma_{oG}^{-1}(s_{xo}^{sv}/s_{xG}^{sv})C_{\sigma x}^{sv} = i\gamma\gamma_{oG}^{-1}C_{\sigma t}^{sv} = \gamma\gamma_{oG}^{-1}\left|C_{\sigma t}^{sv}\right| \qquad (887)$$

where the tuned imaginary number characteristic velocity s_{xG}^{sv} of a mechanical, acoustic or electromagnetic wave in a polyphase space vehicle and the imaginary number velocity ratio function ξ_{xG}^{sv} are given by equations (832) and (835) respectively and cancel the i terms that appear in equations (886) and (887). Equations (886) and (887) contain both γ and γ_{oG}, and therefore the Einstein-Lorentz factor γ_{oG} with its singularity at $v_o = c$ is still there to haunt the space vehicle design engineer, but there is now no longer any imaginary numbers in external space to consider for all $v_o \geq c$. The imaginary value of the Einstein-Lorentz factor γ_o in external space occurs only because the zero APF state is unphysical for $v_o \geq c$, and it does not imply that the game is over for superluminal flight of a polyphase space vehicle even though the apparent space and time coordinate factors in equations (886) and (887) are real numbers.

Assume first that the vacuum in which the space vehicle is moving contains a weak uniaxial cosmic APF in the x-direction and that the volume of low-density vacuum traversed is small (on a cosmic scale) so that the apparent space and time coordinate factors $C_{\sigma x}$ and $C_{\sigma t}$ of the space vehicle are given by equations (138) and (258) that represent the vacuum in the space between the fixed observer and the space vehicle, and the polyphase space vehicle itself is assumed to be a small (on a cosmic scale) high-density object so that the space and time coordinate factors of the space vehicle $C_{\sigma x}^{sv}$ and $C_{\sigma t}^{sv}$ are given by equations (637) and (690). The polyphase space vehicle contains a manmade uniaxial APF in the x-direction within the space vehicle. For superluminal flight, the polyphase space vehicle is composed of sections made of solids, liquids and gases all of which can be described as $[hn_o^{sv}, SN]$ structures, so that the angles θ_N^{sv} and θ_n^{sv} and φ_{sxG}^{sv} are taken to be averages over the entire volume of the space vehicle and over each solid, liquid and gas structure as follows

$$\theta_N^{sv} = (V^{sv})^{-1}\sum_k \theta_{Nk}^{sv}V_k^{sv} \qquad \theta_n^{sv} = (V^{sv})^{-1}\sum_k \theta_{nk}^{sv}V_k^{sv} \qquad \varphi_{sxG}^{sv} = (V^{sv})^{-1}\sum_k \varphi_{sxGk}^{sv}V_k^{sv} \qquad (888)$$

where the total volume V^{sv} of the polyphase space vehicle is given by

$$V^{sv} = \sum_k V_k^{sv} \qquad (889)$$

where the volume of the k'th component of the polyphase space vehicle is V_k^{sv}. The angle φ_{sxG}^{sv} that is defined by equation (833) occurs only in superluminal flight. Within these assumptions and after using equation (835), equation (886) can be written as

$$\cos\theta_{Np}\cos\theta_{np} = \gamma\gamma_{oG}^{-1}\xi_{mxG}^{sv}\cos\theta_N^{sv}\sec\theta_n^{sv} \qquad [ln_{po}, SN_p],\ [hn_o^{sv}, SN] \qquad (890)$$

where ξ_{mxG}^{sv} is given by equation (836). If the cosmic APF is 3-dimensional and isotropic and the artificial APF in the space vehicle is uniaxial then, after using equation (158), equation (890) is replaced by

$$\cos^{1/3}\theta_{\text{Ɲp}}\cos^{1/3}\theta_{np} = \gamma\gamma_{oG}^{-1}\xi_{mxG}^{sv}\cos\theta_{\text{Ɲ}}^{sv}\sec\theta_{n}^{sv} \qquad [\ln_{po},SN_p],\ [hn_o^{sv},SN] \qquad (891)$$

If the cosmic APF is 3-dimensional and isotropic and the polyphase space vehicle is spherical in shape and contains an internally generated 3-dimensional isotropic APF, equations (886) can be written as

$$\cos^{1/3}\theta_{\text{Ɲp}}\cos^{1/3}\theta_{np} = \gamma\gamma_{oG}^{-1}\xi_{mxG}\cos^{1/3}\theta_{\text{Ɲ}}^{sv}\sec^{1/3}\theta_{n}^{sv} \qquad [\ln_{po},SN_p],\ [hn_o^{sv},SN] \qquad (892)$$

Equivalently, these equations can be written in terms of the internal phase angles of the dimensions of spacetime by using equations (204) and (230) as

$$\cos\theta_{Dst}\cos\theta_{nDst} = \gamma\gamma_{oG}^{-1}\xi_{mxG}^{sv}\cos\theta_{\text{Ɲ}}^{sv}\sec\theta_{n}^{sv} \qquad [\ln_{Dst}^{o},SN_{Dst}],\ [hn_o^{sv},SN] \qquad (893)$$

$$\cos^{1/3}\theta_{Dst}\cos^{1/3}\theta_{nDst} = \gamma\gamma_{oG}^{-1}\xi_{mxG}^{sv}\cos\theta_{\text{Ɲ}}^{sv}\sec\theta_{n}^{sv} \qquad [\ln_{Dst}^{o},SN_{Dst}],\ [hn_o^{sv},SN] \qquad (894)$$

$$\cos^{1/3}\theta_{Dst}\cos^{1/3}\theta_{nDst} = \gamma\gamma_{oG}^{-1}\xi_{mxG}^{sv}\cos^{1/3}\theta_{\text{Ɲ}}^{sv}\sec^{1/3}\theta_{n}^{sv} \qquad [\ln_{Dst}^{o},SN_{Dst}],\ [hn_o^{sv},SN] \qquad (895)$$

where ξ_{mxG}^{sv} is given by equation (836). Solutions to each of these equations are not in general possible if a space vehicle acted like a natural body with fixed or arbitrarily varying values of the angles $\theta_{\text{Ɲ}}^{sv}$ and θ_{n}^{sv} and φ_{sxG}^{sv}, however for a polyphase space vehicle with an internal APF generator and external space field generator, the external space phase angle φ_{sxG}^{sv}, defined in equation (833), can be adjusted to $\varphi_{sxG}^{sv} = \pi/2$ for all $v_o \geq c$, and the magnitude of the manmade uniaxial APF can be adjusted to the condition that the internal phase angles $\theta_{\text{Ɲ}}^{sv}$ and θ_{n}^{sv} have the values determined by

$$\sec\theta_{\text{Ɲ}}^{sv}\cos\theta_{n}^{sv} = i\gamma\gamma_{oG}^{-1}(s_{xo}^{sv}/s_{xG}^{sv}) = \gamma\gamma_{oG}^{-1}(s_{xo}^{sv}/s_{mxG}^{sv}) \qquad s_{mxG}^{sv} \geq s_{xo}^{sv} \qquad (896)$$

while for an isotropic spherical artificial APF in spherical polyphase space vehicle

$$\sec^{1/3}\theta_{\text{Ɲ}}^{sv}\cos^{1/3}\theta_{n}^{sv} = i\gamma\gamma_{oG}^{-1}(s_{xo}^{sv}/s_{xG}^{sv}) = \gamma\gamma_{oG}^{-1}(s_{xo}^{sv}/s_{mxG}^{sv}) \qquad s_{mxG}^{sv} \geq s_{xo}^{sv} \qquad (897)$$

where the imaginary number velocity s_{xG}^{sv} and the real number measured velocity s_{mxG}^{sv} are given by equations (832) through (834) and where $s_{mxG}^{sv} \geq s_{xo}^{sv}$ for the $[hn_o^{sv},SN]$ polyphase space vehicle, and in which case the auxiliary vacuum equations obtained from equations (893) through (895) are

$$\cos\theta_{\text{Ɲp}}\cos\theta_{np} = v/v_o \qquad \cos^{1/3}\theta_{\text{Ɲp}}\cos^{1/3}\theta_{np} = v/v_o \qquad (898)$$

or

$$\cos\theta_{Dst}\cos\theta_{nDst} = v/v_o \qquad \cos^{1/3}\theta_{Dst}\cos^{1/3}\theta_{nDst} = v/v_o \qquad (899)$$

which are the equations for the phase angles of the cosmic APF in the $[\ln_{po},SN_p]$ or $[\ln_{Dst}^{o},SN_{Dst}]$ vacuum with $v \leq v_o$. Equations (896) and (897) give the conditions for superluminal flight. The angles $\theta_{\text{Ɲ}}^{sv}$ and θ_{n}^{sv} are always related by equation (619) and also by equation (657) for a $[hn_o^{sv},SN]$ linear elastic space vehicle so that these angles are not independent. Nevertheless, as a crude approximation it can be assumed that $\theta_{\text{Ɲ}}^{sv}$ and θ_{n}^{sv} are independent in equations (896) and (897), and this gives the following approximate conditions for superluminal flight with $v_o \geq c$

$$\sec\theta_{n}^{sv} = \gamma_{oG} \qquad \sec\theta_{\mathbb{N}}^{sv} = i\gamma(s_{xo}^{sv}/s_{xG}^{sv}) = \gamma(s_{xo}^{sv}/s_{mxG}^{sv}) \qquad s_{mxG}^{sv} \geq s_{xo}^{sv} \qquad (900)$$

$$\sec^{1/3}\theta_{n}^{sv} = \gamma_{oG} \qquad \sec^{1/3}\theta_{\mathbb{N}}^{sv} = i\gamma(s_{xo}^{sv}/s_{xG}^{sv}) = \gamma(s_{xo}^{sv}/s_{mxG}^{sv}) \qquad s_{mxG}^{sv} \geq s_{xo}^{sv} \qquad (901)$$

These equations are only valid if the amplitudes and frequencies of the APFs and the external space conventional fields within the space vehicle, and therefore also the angles $\theta_{\mathbb{N}}^{sv}$ and θ_{n}^{sv} and φ_{sxG}^{sv}, can be engineered so as to satisfy equations (896) or (897), (900) and (901) and for which the imaginary index i in these equations is also canceled by the imaginary number s_{xG}^{sv} given by equations (832) through (834). Equations (896) through (901) show that a number of engineered APFs and external field generators within the high-density solid-body, liquid or gaseous structures of a polyphase space vehicle can cancel the effects of the singularity of γ_{oG} at $v_o = c$ and can cancel the effects of the imaginary index i by requiring the average phase angles

$$\theta_{n}^{sv} = \theta_{\mathbb{N}}^{sv} + \left|\theta_{V}^{sv}\right| = \pi/2 \qquad \theta_{\mathbb{N}}^{sv} = \pi/4 \qquad \theta_{V}^{sv} = -\pi/4 \qquad \varphi_{sxG}^{sv} = \pi/2 \qquad (902)$$

at this point, and therefore a properly tuned polyphase space vehicles can undergo superluminal flight. The condition $\theta_{n}^{sv} = \pi/2$ and $\theta_{\mathbb{N}}^{sv} = \pi/4$ for a solid-body, liquid or gaseous structure corresponds to an APF whose energy density for a stationary APF is given by $\mathfrak{E} = 1/2(K_{o}^{sv} + G_{o}^{sv})$ while for a time dependent APF the energy density is $\mathfrak{E} = K_{o}^{sv} + G_{o}^{sv}$, where K_{o}^{sv} and G_{o}^{sv} are the zero APF values of the bulk modulus and the shear modulus of a solid, liquid or gaseous portion of a polyphase space vehicle (see Section 5).[9]

B. Conditions for the Superluminal Flight of a Linear Elastic Space Vehicle.

Equations (890) through (892) can be rewritten for linear elastic space vehicles and for a uniaxial cosmic APF throughout all space and a uniaxial manmade mechanical, acoustic or electromagnetic APF in the polyphase space vehicle and after using equations (663) and (675) as follows for $v_o \geq c$

$$\cos\theta_{\mathbb{N}p}\cos\theta_{np} = \gamma\gamma_{oG}^{-1}\xi_{mxG}^{sv}\sec\theta_{n}^{sv}[1+(1-\cos\theta_{n}^{sv})^{2}]^{-1/2} \qquad [\ln_{po}, SN_{p}],\ [hn_{o}^{sv}, SN] \qquad (903)$$
$$= \gamma\gamma_{oG}^{-1}\xi_{mxG}^{sv}\cos\theta_{\mathbb{N}}^{sv}(1-\tan\theta_{\mathbb{N}}^{sv})^{-1}$$

and for a 3-dimensional isotropic cosmic APF throughout all space and an artificial uniaxial APF in a cylindrical polyphase space vehicle

$$\cos^{1/3}\theta_{\mathbb{N}p}\cos^{1/3}\theta_{np} = \gamma\gamma_{oG}^{-1}\xi_{mxG}^{sv}\sec\theta_{n}^{sv}[1+(1-\cos\theta_{n}^{sv})^{2}]^{-1/2} \quad [\ln_{po}, SN_{p}],[hn_{o}^{sv}, SN] \quad (904)$$
$$= \gamma\gamma_{oG}^{-1}\xi_{mxG}^{sv}\cos\theta_{\mathbb{N}}^{sv}(1-\tan\theta_{\mathbb{N}}^{sv})^{-1}$$

and for a 3-dimensional isotropic cosmic APF throughout all space and a 3-dimensional manmade APF within a spherical polyphase space vehicle

$$\cos^{1/3}\theta_{\mathbb{N}p}\cos^{1/3}\theta_{np} = \gamma\gamma_{oG}^{-1}\xi_{mxG}^{sv}\sec^{1/3}\theta_{n}^{sv}[1+(1-\cos\theta_{n}^{sv})^{2}]^{-1/6} \ [\ln_{po}, SN_{p}],[hn_{o}^{sv}, SN] \ (905)$$
$$= \gamma\gamma_{oG}^{-1}\xi_{mxG}^{sv}\cos^{1/3}\theta_{\mathbb{N}}^{sv}(1-\tan\theta_{\mathbb{N}}^{sv})^{-1/3}$$

where where ξ^{sv}_{mxG} is given by equation (836). Equations (903) through (905) are the linear elastic equivalents of equations (890) through (892) and do not have analytic solutions for arbitrary values of the angles $\theta^{sv}_{\mathfrak{N}}$ or θ^{sv}_{n} which are now related by equation (657) for a high-density polyphase space vehicle. These equations can only be satisfied if the angles $\theta^{sv}_{\mathfrak{N}}$ or θ^{sv}_{n} are adjusted by the manmade APF in the polyphase space vehicle so as to satisfy the following conditions for $v_o \geq c$, first for a uniaxial artificial APF

$$\cos\theta^{sv}_{n}[1+(1-\cos\theta^{sv}_{n})^2]^{1/2} = i\gamma\gamma^{-1}_{oG}(s^{sv}_{xo}/s^{sv}_{xG}) = \gamma\gamma^{-1}_{oG}(s^{sv}_{xo}/s^{sv}_{mxG}) \tag{906}$$

$$\sec\theta^{sv}_{\mathfrak{N}}(1-\tan\theta^{sv}_{\mathfrak{N}}) = i\gamma\gamma^{-1}_{oG}(s^{sv}_{xo}/s^{sv}_{xG}) = \gamma\gamma^{-1}_{oG}(s^{sv}_{xo}/s^{sv}_{mxG}) \tag{907}$$

and for a spherical artificial APF

$$\cos^{1/3}\theta^{sv}_{n}[1+(1-\cos\theta^{sv}_{n})^2]^{1/6} = i\gamma\gamma^{-1}_{oG}(s^{sv}_{xo}/s^{sv}_{xG}) = \gamma\gamma^{-1}_{oG}(s^{sv}_{xo}/s^{sv}_{mxG}) \tag{908}$$

$$\sec^{1/3}\theta^{sv}_{\mathfrak{N}}(1-\tan\theta^{sv}_{\mathfrak{N}})^{1/3} = i\gamma\gamma^{-1}_{oG}(s^{sv}_{xo}/s^{sv}_{xG}) = \gamma\gamma^{-1}_{oG}(s^{sv}_{xo}/s^{sv}_{mxG}) \tag{909}$$

and as before the auxiliary vacuum equations corresponding to equations (906) through (909) are

$$\cos\theta_{\mathfrak{N}p}\cos\theta_{np} = v/v_o \qquad \cos^{1/3}\theta_{\mathfrak{N}p}\cos^{1/3}\theta_{np} = v/v_o \tag{910}$$

or equivalently as

$$\cos\theta_{Dst}\cos\theta_{nDst} = v/v_o \qquad \cos^{1/3}\theta_{Dst}\cos^{1/3}\theta_{nDst} = v/v_o \tag{911}$$

Equations (903) through (909) are valid for high-density matter provided that the following conditions are valid[9]

$$\mathfrak{E} \leq 1/2(K^{sv}_{o} + G^{sv}_{o}) \qquad \text{static APF} \tag{912}$$

$$\mathfrak{E} \leq K^{sv}_{o} + G^{sv}_{o} \qquad \text{time dependent APF} \tag{913}$$

In addition the imaginary number i is also canceled out of equations (903) through (909) by the quantities s^{sv}_{xG} and ξ^{sv}_{xG} that have been engineered so as to have imaginary number values with $\varphi^{sv}_{sxG} = \pi/2$ for all $v_o \geq c$. Therefore superluminal flight of an engineered and properly tuned linear elastic polyphase space vehicle is possible not only because the angles θ^{sv}_{n} and $\theta^{sv}_{\mathfrak{N}}$ have the engineered values $\theta^{sv}_{n} = \pi/2$ and $\theta^{sv}_{\mathfrak{N}} = \pi/4$ at the point $v_o = c$ where $\gamma^{-1}_{oG} = 0$ and where $\mathfrak{E} = 1/2(K^{sv}_{o} + G^{sv}_{o})$ for static APF and $\mathfrak{E} = K^{sv}_{o} + G^{sv}_{o}$ for a time dependent APF, but also because of the required values of s^{sv}_{xG} and ξ^{sv}_{xG} compensate for the imaginary index i that arises from the Lorentz factor. Superluminal flight of a natural object with arbitrary values of the angles θ^{sv}_{n} and $\theta^{sv}_{\mathfrak{N}}$ is not possible, even though γ has no singularity for v = c, because the Einstein-Lorentz factor γ_{oG} still enters the transformation equations for the apparent values of the space and time coordinate factors, and only a manmade space vehicle with an engineered internal APF to cancel the effects of γ_{oG} and with an engineered ordinary external space field to cancel the i in the term $\gamma_o = -i\gamma_{oG}$ can undergo superluminal flight.

C. The Impossibility of the Superluminal Flight of Natural Objects.

Small high-density natural objects (NO) such as planets, stars and galaxies (which are small objects relative to the universe), and which may have a natural internal APF that satisfies the same form of equations (892) or (895) which are valid for a space vehicle and can be written for natural objects as

$$\cos^{1/3}\theta_{\mathbb{N}p}\cos^{1/3}\theta_{np} = i\gamma\gamma_{oG}^{-1}\xi_{xG}^{NO}\cos^{1/3}\theta_{\mathbb{N}}^{NO}\sec^{1/3}\theta_{n}^{NO} \quad [\ln_{po}, SN_p], [hn_o^{NO}, SN] \tag{914}$$

where

$$\xi_{xG}^{NO} = (v/v_o)(s_{xo}^{NO}/s_{xG}^{NO}) \tag{915}$$

with

$$s_{xG}^{NO} = s_{mxG}^{NO}\exp(i\varphi_{sx}^{NO}) = s_{mxG}^{NO}\exp[i(\varphi_{\delta x}^{NO} - \varphi_{\delta t}^{NO} - \varphi_{C\sigma t}^{NO})] \tag{916}$$

where s_{xG}^{NO} = complex number characteristic velocity of a mechanical, acoustic or electromagnetic wave in the natural object for an APF within the natural object, and where the magnitude s_{mxG}^{NO} of the velocity s_{xG}^{NO} is

$$s_{mxG}^{NO} = s_{xo}^{NO}(C_{\sigma x}^{NO}/|C_{\sigma t}^{NO}|) \tag{917}$$

where v and v_o are the relative velocities of the natural object with and without a cosmic APF present, and where, as in equation (684), s_{xo}^{NO} = characteristic average velocity of a mechanical, acoustic or electromagnetic wave propagating in the x-direction in the natural object relative to a frame of reference attached to the natural object under zero APF conditions and where s_{xG}^{NO} = complex number average velocity of a mechanical, acoustic or electromagnetic wave in the natural object, where s_{mxG}^{NO} = magnitude of the average velocity of a mechanical, acoustic or electromagnetic wave in a natural object. For these conditions, equation (914) would be impossible because it sets a real number equal to a complex number. Even if by some curious and rare chance the condition $\varphi_{sx}^{sv} = \varphi_{\delta x}^{NO} - \varphi_{\delta t}^{NO} - \varphi_{C\sigma t}^{NO} = \pi/2$ were valid in a natural object and equation (915) became

$$\xi_{xG}^{NO} = (v/v_o)(s_{xo}^{NO}/s_{xG}^{NO}) = (v/v_o)[s_{xo}^{NO}/(is_{mxG}^{NO})] \tag{918}$$

and equation (914) became a valid real number equation, it is clear that equation (914) has no real number solution for the case of constant or arbitrarily varying angles $\theta_{\mathbb{N}}^{NO}$ and θ_{n}^{NO} because the angles $\theta_{\mathbb{N}p}$ and θ_{np} are small and do not agree with the singularity of γ_{oG} because in this case the effects of the Einstein-Lorentz factor γ_{oG} and its singularity at $v_o = c$ have not been removed by a proper choice of the angles $\theta_{\mathbb{N}}^{NO}$ and θ_{n}^{NO}. Therefore small high-density natural objects should not be expected to undergo superluminal flight. Consider now a large high-density natural object with a natural internal APF, such as the early expanding universe, which is a high-density object compared to the vacuum and which is cutting a large swath through the vacuum. For this case equations (159) and (654) give

$$\sec^{1/3}\theta_{\mathbb{N}p}\cos^{1/3}\theta_{np} = i\gamma\gamma_{oG}^{-1}\xi_{xG}^{NO}\sec^{1/3}\theta_{\mathbb{N}}^{NO}\sec^{1/3}\theta_{n}^{NO} \quad [\ln_{po}, LN_p], [hn_o^{NO}, LN] \tag{919}$$

which also has no possible solutions for constant values or any arbitrary variation of the angles $\theta_{\mathbb{N}}^{NO}$ and θ_{n}^{NO} because the angles $\theta_{\mathbb{N}p}$ and θ_{np} are small and cannot agree with the singularity of γ_{oG} because of the presence of γ_{oG} whose singularity at $v_o = c$ in this case is not compensated by a special choice of the angles $\theta_{\mathbb{N}}^{NO}$ and θ_{n}^{NO}, and also the imaginary index cannot be removed except by a very special choice of the condition in equation (918), and therefore large high-density natural objects also cannot travel at superluminal speeds. Therefore even if a natural object is located within a cosmic APF so that γ has no singularity at $v = c$, the internal APF that induces arbitrary values of the angles $\theta_{\mathbb{N}}^{NO}$ and θ_{n}^{NO} and the arbitrary values of the angles $\theta_k, \theta_x, \theta_\omega$ and θ_t, superluminal motion is prohibited for natural objects by the singularity of the Einstein-Lorentz factor γ_{oG} which is not canceled out at the point $v_o = c$ in equations (914) and (919) and by the appearance of the imaginary index i which also is not canceled from these equations for the range $v_o \geq c$.

5. ULTRAFAST POLYPHASE SPACE VEHICLE DESIGN CONCEPTS. This section gives a brief description of the structure of Maxwell's equations for the electromagnetic APF within a polyphase space vehicle that is required for ultrafast subluminal flight and for superluminal flight and presents a simple calculation of the average internal phase angles θ_{n}^{sv} and $\theta_{\mathbb{N}}^{sv}$ that enter the calculation of the average energy density of the electromagnetic APF that is required to compensate for the effects of the Einstein-Lorentz factor γ_{oL} for ultrafast subluminal flight and the coherent space and time coordinate variation in the electromagnetic APF that is required to compensate for the effects of the Einstein-Lorentz factor γ_{oG} for superluminal flight.

For the i'th component volume of a polyphase space vehicle, Maxwell's equations for zero external APF or for zero self-induced APF can be written as

$$\vec{\nabla} \cdot \vec{B}_i = 0 \tag{920}$$

$$\vec{\nabla} \cdot \vec{D}_i = \rho_{qi} \tag{921}$$

$$\vec{\nabla} \times \vec{E}_i = -\partial \vec{B}_i / \partial t \tag{922}$$

$$\vec{\nabla} \times \vec{H}_i = \vec{j}_i + \partial \vec{D}_i / \partial t \tag{923}$$

For an electromagnetic field that is also an APF, all physical quantities including space and time coordinates are complex numbers in an internal space, and Maxwell's equations are written as[14]

$$\vec{\bar{\nabla}} \cdot \vec{\bar{B}}_i = 0 \tag{924}$$

$$\vec{\bar{\nabla}} \cdot \vec{\bar{D}}_i = \bar{\rho}_{qi} \tag{925}$$

$$\vec{\bar{\nabla}} \times \vec{\bar{E}}_i = -\partial \vec{\bar{B}}_i / \partial \bar{t} \tag{926}$$

$$\vec{\bar{\nabla}} \times \vec{\bar{H}}_i = \vec{\bar{j}}_i + \partial \vec{\bar{D}}_i / \partial \bar{t} \tag{927}$$

where

$$\vec{\vec{B}}_i = (\overline{B}_{\alpha i}) \qquad \vec{\vec{H}}_i = (\overline{H}_{\alpha i}) \qquad \vec{\vec{D}}_i = (\overline{D}_{\alpha i}) \qquad \vec{\vec{E}}_i = (\overline{E}_{\alpha i}) \qquad \vec{\vec{j}}_i = (\overline{j}_{\alpha i}) \tag{928}$$

where $\alpha = x, y, z$ for Cartesian coordinates, and where the complex number physical quantities that appear in Maxwell's equations are written as

$$\overline{B}_{\alpha i} = B_{\alpha i} \exp(j\theta_{B\alpha i}) \qquad \overline{H}_{\alpha i} = H_{\alpha i} \exp(j\theta_{H\alpha i}) \tag{929}$$

$$\overline{D}_{\alpha i} = D_{\alpha i} \exp(j\theta_{D\alpha i}) \qquad \overline{E}_{\alpha i} = E_{\alpha i} \exp(j\theta_{E\alpha i}) \tag{930}$$

$$\overline{\rho}_{qi} = \rho_{qi} \exp(j\theta_{\rho qi}) \qquad \overline{j}_{\alpha i} = j_{\alpha i} \exp(j\theta_{j\alpha i}) \tag{931}$$

This set of equations can be used to describe electromagnetism in the presence of an external APF or a self-inducing APF form of an electromagnetic field.[14]

For one-dimensional electromagnetic waves in the x-direction of the matter of a component of a polyphase space vehicle with zero conductivity for simplicity, Maxwell's equations become

$$(\overline{v}_{pi}^{sv})^2 \partial^2 \overline{E}_{\xi i}^{sv} / \partial \overline{x}^2 - \partial^2 \overline{E}_{\xi i}^{sv} / \partial \overline{t}^2 = 0 \tag{932}$$

$$(\overline{v}_{pi}^{sv})^2 \partial^2 \overline{H}_{\xi i}^{sv} / \partial \overline{x}^2 - \partial^2 \overline{H}_{\xi i}^{sv} / \partial \overline{t}^2 = 0 \tag{933}$$

where $\xi = y, z$ and where

$$\overline{E}_{\xi i}^{sv} = E_{\xi i}^{sv} \exp(j\theta_{E\xi i}^{sv}) \qquad \overline{H}_{\xi i}^{sv} = H_{\xi i}^{sv} \exp(j\theta_{H\xi i}^{sv}) \tag{934}$$

and the complex number phase velocity is given by

$$\overline{v}_{pi}^{sv} = v_{pi}^{sv} \exp(j\theta_{vpi}^{sv}) = \overline{\omega}_i / \overline{k}_i = (\overline{\mu}_i^{sv} \overline{\varepsilon}_i^{sv})^{-1/2} \tag{935}$$

and where $\overline{\omega}_i$ = complex number angular frequency, $\overline{k}_i$ = complex number wave number, $\overline{\mu}_i^{sv}$ = complex number magnetic permeability and $\overline{\varepsilon}_i^{sv}$ = complex number electric permittivity which are represented as

$$\overline{\omega}_i^{sv} = \omega_i^{sv} \exp(j\theta_{\omega i}^{sv}) \qquad \overline{k}_i^{sv} = k_i^{sv} \exp(j\theta_{ki}^{sv}) \tag{936}$$

$$\overline{\mu}_i^{sv} = \mu_i^{sv} \exp(j\theta_{\mu i}^{sv}) \qquad \overline{\varepsilon}_i^{sv} = \varepsilon_i^{sv} \exp(j\theta_{\varepsilon i}^{sv}) \tag{937}$$

and therefore for a self-inducing electromagnetic wave APF or for an electromagnetic wave in an external APF, the components of the complex number phase velocity are

$$v_{pi}^{sv} = \omega_i^{sv} / k_i^{sv} = (\mu_i^{sv} \varepsilon_i^{sv})^{-1/2} \qquad \theta_{vpi}^{sv} = \theta_{\omega i}^{sv} - \theta_{ki}^{sv} = -1/2(\theta_{\mu i}^{sv} + \theta_{\varepsilon i}^{sv}) \tag{938}$$

For electromagnetic waves in a vacuum portion of a polyphase space vehicle that contains an APF the phase velocity is determined by

$$v_{pi}^{sv} = c = \omega_i^{sv} / k_i^{sv} = (\mu_i^{sv} \varepsilon_i^{sv})^{-1/2} \qquad \theta_{vpi}^{sv} = \theta_c = 0 \qquad \theta_{\omega i}^{sv} = \theta_{ki}^{sv} \qquad \theta_{\mu i}^{sv} = -\theta_{\varepsilon i}^{sv} \tag{939}$$

For constant internal phase angles, the wave equations (932) and (933) can be written as two real number wave equations and a phase angle equation for a self-inducing APF electromagnetic wave in matter as follows

$$(v_{pi}^{sv})^2 \partial^2 E_{\xi i}^{sv} / \partial x^2 - \partial^2 E_{\xi i}^{sv} / \partial t^2 = 0 \tag{940}$$

$$(v_{pi}^{sv})^2 \partial^2 H_{\xi i}^{sv} / \partial x^2 - \partial^2 H_{\xi i}^{sv} / \partial t^2 = 0 \tag{941}$$

$$2\theta_{vpi}^{sv} - 2\theta_{xi}^{sv} = -2\theta_{ti}^{sv} \tag{942}$$

where the internal phase angles $\theta_{E\xi i}^{sv}$ and $\theta_{H\xi i}^{sv}$ cancel out to give the phase angle equation (942). Using equation (938) allows the phase angle equation (942) can be rewritten as

$$\theta_{vpi}^{sv} = \theta_{xi}^{sv} - \theta_{ti}^{sv} = \theta_{\omega i}^{sv} - \theta_{ki}^{sv} = -1/2(\theta_{\mu i}^{sv} + \theta_{\varepsilon i}^{sv}) \tag{943}$$

The real number energy density of the electromagnetic waves in the matter of the space vehicle is written as

$$\mathfrak{E}_{emi}^{sv} = 1/2[\mu_i^{sv}(H_{\xi i}^{sv})^2 + \varepsilon_i^{sv}(E_{\xi i}^{sv})^2] \tag{944}$$

and will be used as an index of the strength of the electromagnetic wave APF. For the case of zero APF all internal phase angles are equal to zero. The electromagnetic wave APF generally induces an internal phase angle $\theta_{\mathfrak{N}i}^{sv}$ in the number of molecules or atoms in the material of the polyphase space vehicle and an internal phase angle θ_{ni}^{sv} in the particle number density and this is described by equations (613) through (632).

The self-induced electromagnetic APF in the matter of a portion of a polyphase space vehicle induces a shearing field and a tension field in high-density $[hn_{oi}^{sv}]$ matter, and a shearing field and compression field in $[ln_{oi}^{sv}]$ matter.[9] The static shearing field energy density is written as[9]

$$\mathfrak{E}_{Si}^{sv} = 1/2 G_{oi}^{sv} \tan^2 \theta_{\mathfrak{N}i}^{sv} \qquad \tan\theta_{\mathfrak{N}i}^{sv} = (2\mathfrak{E}_{Si}^{sv} / G_{oi}^{sv})^{1/2} \qquad [ln_{oi}^{sv}],[hn_{oi}^{sv}] \tag{945}$$

The static tension field component energy density of the APF in $[hn_{oi}^{sv}]$ matter is associated with a decrease in particle number density as obtained from equation (630) and (632) as[9]

$$\mathfrak{E}_{Ti}^{sv} = 1/2 K_{oi}^{sv}(\cos\theta_{ni}^{sv} - 1)^2 \qquad \cos\theta_{ni}^{sv} = 1 - (2\mathfrak{E}_{Ti}^{sv} / K_{oi}^{sv})^{1/2} \qquad [hn_{oi}^{sv}] \tag{946}$$

subject to $\mathfrak{E}_{Ti}^{sv} \leq 1/2 K_{oi}^{sv}$. The static compression field component of the energy density of the APF in $[ln_{oi}^{sv}]$ matter is associated with an increase in the particle number density and is obtained from equation (631) as[9]

$$\mathfrak{E}_{Ci}^{sv} = 1/2 K_{oi}^{sv}(\sec\theta_{ni}^{sv} - 1)^2 \qquad \sec\theta_{ni}^{sv} = 1 + (2\mathfrak{E}_{Ci}^{sv} / K_{oi}^{sv})^{1/2} \qquad [ln_{oi}^{sv}] \tag{947}$$

For the case when the static shearing energy is large and the static compression and static tension energies of the matter in a portion of a polyphase space vehicle can be neglected then $\mathfrak{E}_{emi}^{sv} = \mathfrak{E}_{Si}^{sv}$

$$\mathfrak{E}^{sv}_{emi} = 1/2G^{sv}_{oi}\tan^2\theta^{sv}_{Ni} \qquad \tan\theta^{sv}_{Ni} = (2\mathfrak{E}^{sv}_{emi}/G^{sv}_{oi})^{1/2} \qquad [ln^{sv}_{oi}],[hn^{sv}_{oi}] \qquad (948)$$

For the case when the static shearing energy density can be neglected with respect to the static tension energy density for $[hn^{sv}_{oi}]$ matter then $\mathfrak{E}^{sv}_{emi} = \mathfrak{E}^{sv}_{Ti}$ and

$$\mathfrak{E}^{sv}_{emi} = 1/2K^{sv}_{oi}(\cos\theta^{sv}_{ni} - 1)^2 \qquad \cos\theta^{sv}_{ni} = 1 - (2\mathfrak{E}^{sv}_{emi}/K^{sv}_{oi})^{1/2} \qquad [hn^{sv}_{oi}] \qquad (949)$$

and $\mathfrak{E}^{sv}_{emi} \le 1/2K^{sv}_{oi}$, and for the case when the static shearing energy can be neglected with respect to the static compression energy density for $[ln^{sv}_{oi}]$ matter then $\mathfrak{E}^{sv}_{emi} = \mathfrak{E}^{sv}_{Ci}$ and

$$\mathfrak{E}^{sv}_{emi} = 1/2K^{sv}_{oi}(\sec\theta^{sv}_{ni} - 1)^2 \qquad \sec\theta^{sv}_{ni} = 1 + (2\mathfrak{E}^{sv}_{emi}/K^{sv}_{oi})^{1/2} \qquad [ln^{sv}_{oi}] \qquad (950)$$

which is true for static elasticity.

More generally, it has been suggested that the electromagnetic APF energy density $\mathfrak{E}^{sv}_{emi}$ in $[hn^{sv}_{oi}]$ matter of the polyphase space vehicle is distributed between the tension field energy density and the shearing field energy density in the following way[9]

$$\mathfrak{E}^{sv}_{emi} = \mathfrak{E}^{sv}_{Ti} + \mathfrak{E}^{sv}_{Si} \qquad [hn^{sv}_{oi}] \qquad (951)$$

while in $[ln^{sv}_{oi}]$ matter the electromagnetic energy density $\mathfrak{E}^{sv}_{emi}$ is distributed between a static compression field energy density and a static shearing field energy density in the manner

$$\mathfrak{E}^{sv}_{emi} = \mathfrak{E}^{sv}_{Ci} + \mathfrak{E}^{sv}_{Si} \qquad [ln^{sv}_{oi}] \qquad (952)$$

where

$$\mathfrak{E}^{sv}_{Ti} = K^{sv}_{oi}(K^{sv}_{oi} + G^{sv}_{oi})^{-1}\mathfrak{E}^{sv}_{emi} \qquad \mathfrak{E}^{sv}_{Si} = G^{sv}_{oi}(K^{sv}_{oi} + G^{sv}_{oi})^{-1}\mathfrak{E}^{sv}_{emi} \qquad [hn^{sv}_{oi}] \qquad (953)$$

$$\mathfrak{E}^{sv}_{Ci} = K^{sv}_{oi}(K^{sv}_{oi} + G^{sv}_{oi})^{-1}\mathfrak{E}^{sv}_{emi} \qquad \mathfrak{E}^{sv}_{Si} = G^{sv}_{oi}(K^{sv}_{oi} + G^{sv}_{oi})^{-1}\mathfrak{E}^{sv}_{emi} \qquad [ln^{sv}_{oi}] \qquad (954)$$

which combined with equations (945) through (947) gives[9]

$$\cos\theta^{sv}_{ni} = 1 - [2\mathfrak{E}^{sv}_{emi}(K^{sv}_{oi} + G^{sv}_{oi})^{-1}]^{1/2} \qquad 0 \le \mathfrak{E}^{sv}_{emi} \le 1/2(K^{sv}_{oi} + G^{sv}_{oi}) \qquad [hn^{sv}_{oi}] \qquad (955)$$

$$\tan\theta^{sv}_{Ni} = [2\mathfrak{E}^{sv}_{emi}(K^{sv}_{oi} + G^{sv}_{oi})^{-1}]^{1/2} \qquad 0 \le \mathfrak{E}^{sv}_{emi} \le 1/2(K^{sv}_{oi} + G^{sv}_{oi}) \qquad [hn^{sv}_{oi}] \qquad (956)$$

$$\sec\theta^{sv}_{ni} = 1 + [2\mathfrak{E}^{sv}_{emi}(K^{sv}_{oi} + G^{sv}_{oi})^{-1}]^{1/2} \qquad 0 \le \mathfrak{E}^{sv}_{emi} \le \infty \qquad [ln^{sv}_{oi}] \qquad (957)$$

$$\tan\theta^{sv}_{Ni} = [2\mathfrak{E}^{sv}_{emi}(K^{sv}_{oi} + G^{sv}_{oi})^{-1}]^{1/2} \qquad 0 \le \mathfrak{E}^{sv}_{emi} \le \infty \qquad [ln^{sv}_{oi}] \qquad (958)$$

$$\tan\theta^{sv}_{Ni} = 1 - \cos\theta^{sv}_{ni} \qquad \theta^{sv}_{ni} \ge \theta^{sv}_{Ni} \qquad 0 \le \theta^{sv}_{ni} \le \pi/2 \qquad 0 \le \theta^{sv}_{Ni} \le \pi/4 \qquad [hn^{sv}_{oi}] \qquad (959)$$

$$\tan\theta^{sv}_{Ni} = \sec\theta^{sv}_{ni} - 1 \qquad \theta^{sv}_{ni} \ge \theta^{sv}_{Ni} \qquad 0 \le \theta^{sv}_{ni} \le \pi/2 \qquad 0 \le \theta^{sv}_{Ni} \le \pi/2 \qquad [ln^{sv}_{oi}] \qquad (960)$$

The case of a dynamic APF field in which the average kinetic energy is equal to the average potential energy associated with elastic deformation can be described by making the replacement $\mathfrak{E}^{sv}_{mi} \to 1/2\mathfrak{E}^{sv}_{mi}$ in equations (948) through (958), and it is clear that equations (959) and (960) remain unchanged.[9] The polyphase space vehicle is composed of sections made of solids, liquids and gases all of which can be described as $[hn^{sv}_{oi}, SN_i]$ structures, so that the angles $\theta^{sv}_{\mathfrak{N}}$ and θ^{sv}_{n} and φ^{sv}_{sxG} which are required for superluminal flight,are taken to be averages over the entire volume of the space vehicle and over each solid, liquid and gas structure as follows

$$\theta^{sv}_{\mathfrak{N}} = (V^{sv})^{-1}\sum_i \theta^{sv}_{\mathfrak{N}i} V^{sv}_i \qquad \theta^{sv}_{n} = (V^{sv})^{-1}\sum_i \theta^{sv}_{ni} V^{sv}_i \qquad \varphi^{sv}_{sxG} = (V^{sv})^{-1}\sum_i \varphi^{sv}_{sxGi} V^{sv}_i \tag{961}$$

where the total volume V^{sv} of the polyphase space vehicle is given by

$$V^{sv} = \sum_i V^{sv}_i \tag{962}$$

where the volume of the i'th component of the polyphase space vehicle is V^{sv}_i.

As shown in equations (870) and (902), the average phase angle conditions for a polyphase space vehicle, moving in the vacuum, passing through the $v_o = c$ flight barrier are

$$\theta^{sv}_{n} = \theta^{sv}_{\mathfrak{N}} + \left|\theta^{sv}_{V}\right| = \pi/2 \qquad \theta^{sv}_{\mathfrak{N}} = \pi/4 \qquad \theta^{sv}_{V} = -\pi/4 \qquad \varphi^{sv}_{sxG} = \pi/2 \tag{963}$$

and this is compatible with equation (959) that is valid for an $[hn^{sv}_{oi}]$ material in the presence of a static APF with an energy density of $\mathfrak{E}^{sv}_{mi} = 1/2(K^{sv}_{oi} + G^{sv}_{oi})$ or a time dependent APF with an energy density of $\mathfrak{E}^{sv}_{mi} = K^{sv}_{oi} + G^{sv}_{oi}$. Therefore both subluminal and superluminal ultrafast flight is possible for a small high-density polyphase space vehicle. However equations (647), (648), (665), (666), (960) and (963) suggest that an average low-density, $[ln^{sv}_{o}]$, polyphase space vehicle cannot be engineered to attain ultrafast flight. This can be interpreted as saying that systems of particles of matter whose particle number density is less than the particle number density of the local vacuum cannot attain ultrafast speeds. However, a high vacuum contained within a compartment constructed of a high-density, $[hn^{sv}_{oi}]$, material is expected to be carried along to ultrafast speeds by the polyphase space vehicle but will produce drag.

6. CONCLUSION. In this paper Einstein's theory of special relativity and the Lorentz transformations are assumed to be correct and applicable to the case when all space and time coordinates and static and dynamic variables of physics, except the fundamental constants of nature such as $\hbar$, c, G,$\cdots$, are complex numbers in an internal space due to the presence of a cosmic ambient phase field (APF) in the vacuum of space and an artificial APF generated internally in a space vehicle.[5,9,14] A cosmic APF is required to remove the singularity of the zero APF form of the Einstein-Lorentz factor γ_o for motion of a body at the light speed by forming a complex number Einstein-Lorentz factor whose magnitude γ does not have this singularity.[5] This paper applies this generalization to the case of the ultrafast flight of a polyphase space vehicle in whose various components is located a tunable artificial APF in a manner similar to that of a polyphase rocket.[9] Space and time coordinate factors are used used to relate the magnitudes of the complex number space and time coordinates to their corresponding zero APF values. The coordinate factors as seen by an observer in a stationary-

coordinate are related to the coordinate factors as seen by an observer in a moving inertial system by using both the complex number generalization of the Lorentz transformations and the standard real number form of these transformations. The coordinate factors are found to depend on both γ_o and γ so that γ_o appears again, and therefore a polyphase space vehicle with an artificial APF that can be adjusted to counteract the singularity of γ_o at the light speed is also required for ultrafast flight. The cases of subluminal and superluminal ultra fast flight are considered, and it is found that for subluminal ultrafast flight a mechanical, acoustic or electromagnetic APF which can induce critical values of the internal phase angles θ_n^{sv} and $\theta_{\aleph}^{sv}$ within a polyphase space vehicle is sufficient to counteract the singularity of γ_{oL} as the light speed is approached from the subluminal side, while for superluminal flight a mechanical, acoustic or electromagnetic APF which can induce the required critical values of the angles θ_n^{sv}, $\theta_{\aleph}^{sv}$ and φ_{sxG}^{sv}, where always $\varphi_{sxG}^{sv} = \pi/2$ for this case, and critical values of the internal angles of the wave number and angular frequency of the APF wave and the space and time coordinates within a polyphase space vehicle, are required to counteract the singularity of γ_{oG} and the removal of the imaginary index i that appear in the expressions for the apparent space and time coordinates factors in the case of superluminal flight of a space vehicle. In other words the conditions for ultrafast flight are

a. Einstein's theory of special relativity is valid for the case of zero APF.
b. There exists a cosmic APF, however small, in all the vacuum of spacetime, which requires that all physical quantities, including space and time coordinates but with the exception of the fundamental constants of nature, are complex numbers in an internal space.
c. Einsteins idea of the form covariance of the laws of physics is valid for matter and the vacuum in the presence of an APF.
d. A manmade mechanical, acoustic or electromagnetic APF can be generated within the polyphase space vehicle and requires that all physical quantities, including space and time coordinates but with the exception of the fundamental constants of nature, are complex numbers in an internal space.
e. For subluminal and superluminal flight the manmade APF inside the polyphase space vehicle can be tuned to cancel the infinities associated with γ_{oL} and γ_{oG} respectively at the point $v_o = c$.
f. For superluminal flight, a manmade ordinary mechanical, acoustic or electromagnetic field is generated within the polyphase space vehicle that can eliminate the imaginary index i that appears in $\gamma_o = -i\gamma_{oG}$ for $v_o \geq c$.

Therefore ultrafast flight may be possible for manmade objects but it is essentially impossible for natural bodies even if a cosmic APF exists throughout all space.

REFERENCES

1. Hirshfeld, A. and Sinnott, W., Sky Catalogue 2000, Vols. 1&2, Sky Publishing Corp., Cambridge, 1982.

2. Dibon-Smith, R., StarList 2000, John Wiley, New York, 1992.

3. Møller, C., The Theory of Relativity, Oxford University Press, 1955.

4. Aharoni, J., The Special Theory of Relativity, Oxford University Press, 1959.

5. Weiss, R. A., "Maxwell's Equations with Broken Internal Symmetries," article in Gauge Theory of Thermodynamics, K&W Publications, Vicksburg, 1989.

6. Weiss, R. A., Clean Fission, K&W Publications, Vicksburg, 1992.

7. Weiss, R. A., Internal Fission, K&W Publications, Vicksburg, 1995.

8. Weiss, R. A., Vacuum Fission, K&W Publications, Vicksburg, 1998.

9. Weiss, R. A., Polyphase Fission and Polyphase Rockets, K&W Publications, Tucson, 2003.

10. Weiss, R. A., Polyphase Fission Nuclear Power, K&W Publications, Tucson, 2004.

11. Weiss, R.A., "Chemical Reactions of Gases in a Gravitational Field," article in Vacuum Fission, K&W Publications, Vicksburg, 1998 but also reprinted in Polyphase Fission Nuclear Power, K&W Publications, Tucson, 2004.

12. Weiss, R. A., Gauge Theory of Thermodynamics, K&W Publications, Vicksburg, 1989.

13. Rindler, W., Introduction to Special Relativity, Clarendon Press, Oxford, 1991.

14. Weiss, R.A., "Electromagnetic Wave Propagation Suppression," article in Vacuum Fission, K&W Publications, Vicksburg, 1998.

APPENDIX A. SPACE AND TIME COORDINATE FACTORS ASSOCIATED WITH THE LORENTZ-FITZGERALD CONTRACTION AND THE TIME DILATION. For the case of zero APF, equation (144) shows that an interval of length Δx_o as seen by an observer in the stationary-coordinate system who simultaneously measures the end points of the interval is related to the interval of length $\Delta x'_o$ at rest in the moving system as seen from the moving reference frame by the Lorentz-Fitzgerald contraction as follows

$$\Delta x_o = \gamma_o^{-1} \Delta x'_o = \gamma_o^{-1} L_x^{sv} \qquad (\Delta x'_o / \Delta x_o) = \gamma_o \qquad \text{(A1)}$$

where L_x^{sv} = the proper length of a component of the space vehicle. In the presence of a cosmic APF in the spacetime between the two systems of coordinates and an artificial APF generated within the moving space vehicle, the Lorentz-Fitzgerald contraction is written as in equation (182)

$$\Delta\overline{x} = (\overline{\gamma})^{-1} \Delta\overline{x}' = (\overline{\gamma})^{-1} \overline{L}_x^{sv} \qquad (\Delta\overline{x}' / \Delta\overline{x}) = \overline{\gamma} \qquad \text{(A2)}$$

where $\overline{L}_x^{sv}$ = proper complex number length of a component of the space vehicle. The corresponding magnitude equation (A2) is given by

$$\Delta x = \gamma^{-1}\Delta x' = \gamma^{-1}L_x^{sv} \qquad (\Delta x' / \Delta x) = \gamma \qquad \text{(A3)}$$

as shown in equation (182). Then by the definition of the space coordinate factors

$$\Delta x' = C'_{\sigma x}\Delta x'_o = C_{\sigma x}^{sv}\Delta x'_o = C_{\sigma x}^{sv}L_x^{sv} \qquad \text{(A4)}$$

$$\Delta x = C_{\sigma x}\Delta x_o = \gamma^{-1}\Delta x' = \gamma^{-1}C'_{\sigma x}\Delta x'_o = \gamma^{-1}C_{\sigma x}^{sv}L_x^{sv} \qquad \text{(A5)}$$

so that equation (A5) can be rewritten as

$$C_{\sigma x} = \gamma^{-1}C'_{\sigma x}(\Delta x'_o / \Delta x_o) = \gamma_o\gamma^{-1}C'_{\sigma x} = \gamma_o\gamma^{-1}C_{\sigma x}^{sv} \qquad \text{(A6)}$$

which is another way of writing the Lorentz-Fitzgerald contraction.

Consider now the time dilation equation for the case of zero APF as given by equation (145)

$$\Delta t_o = \gamma_o\Delta t'_o = \gamma_o L_t^{sv} \qquad (\Delta t'_o / \Delta t_o) = \gamma_o^{-1} \qquad \text{(A7)}$$

which shows that time runs slower as seen from the stationary system. The complex number time dilation equation for the case when a cosmic APF is present and when the space vehicle contains an artificially generated APF is given as in equation (187) by

$$\Delta\bar{t} = \bar{\gamma}\Delta\bar{t}' = \bar{\gamma}\bar{L}_t^{sv} \qquad (\Delta\bar{t}' / \Delta\bar{t}) = (\bar{\gamma})^{-1} \qquad \text{(A8)}$$

where $\bar{L}_t^{sv}$ is a complex number proper time duration of an event in the space vehicle and as seen by an observer in the space vehicle. The magnitude equation associated with equation (A8) is

$$\Delta t = \gamma\Delta t' = \gamma L_t^{sv} \qquad (\Delta t' / \Delta t) = \gamma^{-1} \qquad \text{(A9)}$$

The definition of the time coordinate factors gives then

$$\Delta t' = C'_t\Delta t'_o = C_t^{sv}\Delta t'_o = C_t^{sv}L_t^{sv} \qquad \text{(A10)}$$

$$\Delta t = C_{\sigma t}\Delta t_o = \gamma\Delta t' = \gamma C'_{\sigma t}\Delta t'_o = \gamma C_{\sigma t}^{sv}L_t^{sv} \qquad \text{(A11)}$$

and solving for $C_{\sigma t}$ in equation (A11) gives

$$C_{\sigma t} = \gamma C'_{\sigma t}(\Delta t'_o / \Delta t_o) = \gamma\gamma_o^{-1}C'_{\sigma t} = \gamma\gamma_o^{-1}C_{\sigma t}^{sv} \qquad \text{(A12)}$$

which is the desired transformation formula for $C_{\sigma t}$.

Equation (A6) is only valid for the conditions describing the Lorentz-Fitzgerald-contraction namely the simultaneity (relative to the stationary system) of the times at which the endpoints of a moving object are measured by an observer in the stationary system. On the other hand, the condition for the validity of the time dilation of equation (A12) is that the time is measured at the same point in the moving coordinate system, so that $\Delta x' = 0$ (the point at the back end of the space vehicle which is attached to the origin of the moving coordinate system). As has been shown in Sections 4 and 5, only the time dilation formula (A12) is of use for application to the ultrafast flight of a polyphase space vehicle.

APPENDIX B. THE SPACE AND TIME COORDINATE FACTORS FOR THE SPECIAL RELATIVISTIC VELOCITY ADDITION FORMULAS FOR SPACETIME THAT IS LOCATED IN AN APF. The complex number velocity addition formula for special relativity theory in spacetime with an APF present is obtained as a simple real number, zero APF, generalization of equations (91) and (92) and are given in equations (194) and (195) as

$$\overline{w} = (\overline{u} + \overline{v})(1 + \overline{uv}c^{-2})^{-1} \qquad \text{(B1)}$$

$$\overline{u} = (\overline{w} - \overline{v})(1 - \overline{wv}c^{-2})^{-1} \qquad \text{(B2)}$$

where $\overline{u}$ = complex number speed of a body as seen by an observer on the moving frame of reference, $\overline{v}$ = complex number relative velocity of the moving and stationary frames of reference, $\overline{w}$ = complex number velocity as seen by an observer in the stationary frame of reference and where c = light speed in the vacuum which is a real number because it is a constant of nature.[5] These complex numbers are represented as[5]

$$\overline{w} = w\exp(j\theta_w) \qquad \overline{u} = u\exp(j\theta_u) \qquad \overline{v} = v\exp(j\theta_v) \qquad \text{(B3)}$$

Then equation (B1) can be written as[5]

$$\overline{w} = w\exp(j\theta_w) = (A + jB)(C + jD)^{-1} \qquad \text{(B4)}$$

$$w = [(A^2 + B^2)(C^2 + D^2)^{-1}]^{1/2} \qquad \theta_w = \theta_N - \theta_D \qquad \text{(B5)}$$

$$\tan\theta_N = B/A \qquad \tan\theta_D = D/C \qquad \text{(B6)}$$

where[5]

$$A = u\cos\theta_u + v\cos\theta_v \qquad \text{(B7)}$$

$$B = u\sin\theta_u + v\sin\theta_v \qquad \text{(B8)}$$

$$C = 1 + uvc^{-2}\cos(\theta_u + \theta_v) \qquad \text{(B9)}$$

$$D = uvc^{-2}\sin(\theta_u + \theta_v) \qquad \text{(B10)}$$

and then equations (B5) and (B6) can be used to determine w and θ_w in terms of the quantities u, v, θ_u and θ_v. Similarly, the following equations are valid for for the inverse relation in equation (B2)

$$\overline{u} = u\exp(j\theta_u) = (A_i + jB_i)(C_i + jD_i)^{-1} \qquad \text{(B11)}$$

$$u = [(A_i^2 + B_i^2)(C_i^2 + D_i^2)^{-1}]^{1/2} \qquad \theta_u = \theta_N^i - \theta_D^i \qquad \text{(B12)}$$

$$\tan\theta_N^i = B_i / A_i \qquad \tan\theta_D^i = D_i / C_i \tag{B13}$$

where[5]

$$A_i = w\cos\theta_w - v\cos\theta_v \tag{B14}$$

$$B_i = w\sin\theta_w - v\sin\theta_v \tag{B15}$$

$$C_i = 1 - wvc^{-2}\cos(\theta_w + \theta_v) \tag{B16}$$

$$D_i = -wvc^{-2}\sin(\theta_w + \theta_v) \tag{B17}$$

and equations (B12) and (B13) can then be used to determine u and θ_u in terms of the quantities w, v, θ_w and θ_v.

The velocity magnitudes u, v and w that appear in equations (B1) through (B17) correspond to the presence of an APF, and can be related to the corresponding zero APF velocities u_o, v_o and w_o by using equations (95), (136) and (95) as follows

$$w = w_o(C_{\sigma x} / C_{\sigma t}) \qquad v = v_o(C_{\sigma x}^v / C_{\sigma t}^v) \qquad u = u_o(C'_{\sigma x} / C'_{\sigma t}) \tag{B18}$$

where $C_{\sigma x}$ and $C_{\sigma t}$ are the space and time coordinate factors associated with the distance between the moving space vehicle and the stationary reference system and are given by equations () through (), $C_{\sigma x}^v$ and $C_{\sigma t}^v$ are the space and time coordinate factors that correspond to the relative velocity between the stationary and moving coordinate systems and where $C'_{\sigma x}$ and $C'_{\sigma t}$ are the space and time coordinate factors that correspond to the motion of the space vehicle relative to the moving reference system. Combining equations (B7) through (B10) with equation (B18) gives

$$A = u_o(C'_{\sigma x} / C'_{\sigma t})\cos\theta_u + v_o(C_{\sigma x}^v / C_{\sigma t}^v)\cos\theta_v \tag{B19}$$

$$B = u_o(C'_{\sigma x} / C'_{\sigma t})\sin\theta_u + v_o(C_{\sigma x}^v / C_{\sigma t}^v)\sin\theta_v \tag{B20}$$

$$C = 1 + u_o v_o c^{-2}(C'_{\sigma x} / C'_{\sigma t})(C_{\sigma x}^v / C_{\sigma t}^v)\cos(\theta_u + \theta_v) \tag{B21}$$

$$D = u_o v_o c^{-2}(C'_{\sigma x} / C'_{\sigma t})(C_{\sigma x}^v / C_{\sigma t}^v)\sin(\theta_u + \theta_v) \tag{B22}$$

Then equations (B4) through (B6) can be used to determine w and θ_w in terms of v_o, u_o, θ_v, and the space and time coordinate factors $C'_{\sigma x}, C'_{\sigma t}, C_{\sigma x}^v$ and $C_{\sigma t}^v$. Similarly, combining equations (B14) through (B17) with equation (B18) gives

$$A_i = w_o(C_{\sigma x} / C_{\sigma t})\cos\theta_w - v_o(C_{\sigma x}^v / C_{\sigma t}^v)\cos\theta_v \tag{B23}$$

$$B_i = w_o(C_{\sigma x} / C_{\sigma t})\sin\theta_w - v_o(C_{\sigma x}^v / C_{\sigma t}^v)\sin\theta_v \tag{B24}$$

$$C_i = 1 - w_o v_o c^{-2}(C_{\sigma x} / C_{\sigma t})(C_{\sigma x}^v / C_{\sigma t}^v)\cos(\theta_w + \theta_v) \tag{B25}$$

$$D_i = -w_o v_o c^{-2}(C_{\sigma x} / C_{\sigma t})(C_{\sigma x}^v / C_{\sigma t}^v)\sin(\theta_w + \theta_v) \tag{B26}$$

and in this way the space and time coordinate factors enter the relativistic theory of the addition of velocities in the presence of an APF.

APPENDIX C. POLYPHASE ROCKETS. A polyphase rocket consists of a chemical rocket that has perhaps six APFs of different intensities and phases applied to six rocket component systems including: (1) the exhaust gas aperture walls, (2) the combustion chamber foramen walls, (3) the walls and atmosphere of the personnel compartment, (4) the solid or liquid fuel and other solid or liquid systems that constitute the main body of a rocket, (5) the exhaust gases in the exhaust tube, and (6) the combustion gases in the combustion chamber.[9] For practical purpose these APFs will be mechanical, acoustic or electromagnetic wave fields. Because the first four systems are high-density solid or liquid systems while the latter two are low-density (relative to the solids) gaseous systems, it has been shown that the application of APFs leads to enhancement of the rockets thrust and an increase in the liftoff parameter of the polyphase rocket. For a conventional rocket, which corresponds to zero APFs, and for the adapted condition of maximum thrust, the liftoff parameter is given by[9]

$$s_L^o = (m_R^o g_E^o)^{-1} \rho_e^o (v_e^o)^2 A_a^o = (m_b^o Q_R^o g_E^o)^{-1} \rho_e^o (v_e^o)^2 A_a^o \qquad \text{(C1)}$$

which is a simple parameter which can be used for practical calculations because rocket liftoff occurs when $s_L^o = 1$, and where m_R^o = conventional total instantaneous mass of a rocket, m_b^o = mass of the solid part of the rocket body and the liquid or solid fuel, $Q_R^o = m_R^o / m_b^o$ = conventional mass factor, g_E^o = conventional acceleration of gravity at the Earth's surface, ρ_e^o = conventional mass density of the exhaust gases in the exhaust tube, v_e^o = conventional velocity of the gases in the exhaust tube and A_a^o = conventional area of the aperture of the rocket exhaust tube. For a polyphase rocket the complex number liftoff parameter for the adapted condition is given by[9]

$$\bar{s}_L = s_L \exp(j\theta_{sL}) = (\bar{m}_R \bar{g}_E)^{-1} \bar{\rho}_e \bar{v}_e^2 \bar{A}_a = (\bar{Q}_R \bar{m}_b \bar{g}_E)^{-1} \bar{\rho}_e \bar{v}_e^2 \bar{A}_a \qquad \text{(C2)}$$

where the complex number mass factor is given by

$$\bar{Q}_R = Q_R \exp(j\theta_Q^R) = \bar{m}_R / \bar{m}_b \qquad Q_R = m_R / m_b \qquad \theta_Q^R = \theta_m^R - \theta_m^b \qquad \text{(C3)}$$

where $\bar{m}_R$ = complex number instantaneous mass of the polyphase rocket, $\bar{m}_b$ = complex number mass of the rocket body including the fuel, $\bar{g}_E$ = complex number acceleration of gravity at the Earth's surface, $\bar{\rho}_e$ = complex number mass density of the exhaust gases in the exhaust tube, $\bar{v}_e$ = complex number exhaust gas velocity and where $\bar{A}_a$ = complex number area of the exhaust area. It has been shown that the magnitude of the complex number liftoff parameter for a polyphase rocket can be written as the following general result[9]

$$s_L = s_L^o S_R \sec\theta_n^b \sec\theta_n^e \qquad s_L \geq s_L^o \qquad \text{(C4)}$$

where s_L^o = liftoff parameter for a conventional rocket as given by equation (C1) and where S_R is a complicated factor that satisfies $S_R \approx 1$, so that approximately

$$s_L = s_L^o \sec\theta_n^b \sec\theta_n^e \qquad s_L \geq s_L^o \qquad \text{(C5)}$$

and therefore liftoff of a polyphase rocket is possible even when the conventional liftoff parameter $s_L^o \leq 1$ provided that the internal phase angles of the particle number densities of the rocket body and exhaust gases θ_n^b and θ_n^e are sufficiently large as to make $s_L \geq 1$.[9]

INDEX

H-I

J-K

L

M

N

O-P